Mechatronics for Safety, Security and Dependability in a New Era

Mechatronics for Safety, Security and Dependability in a New Era

Edited by

Eiji Arai and Tatsuo Arai

ELSEVIER

Amsterdam • Boston • Heidelberg • London • New York • Oxford
Paris • San Diego • San Francisco • Singapore • Sydney • Tokyo

ELSEVIER B.V.
Radarweg 29
P.O. Box 211, 1000 AE Amsterdam
The Netherlands

ELSEVIER Inc.
525 B Street, Suite 1900
San Diego, CA 92101-4495
USA

ELSEVIER Ltd
The Boulevard, Langford Lane
Kidlington, Oxford OX5 1GB
UK

ELSEVIER Ltd
84 Theobald's Road
London WC1X 8RR
UK

© 2007 Elsevier Ltd. All rights reserved.

This work is protected under copyright by Elsevier Ltd., and the following terms and conditions apply to its use:

Photocopying
Single photocopies of single chapters may be made for personal use as allowed by national copyright laws. Permission of the Publisher and payment of a fee is required for all other photocopying, including multiple or systematic copying, copying for advertising or promotional purposes, resale, and all forms of document delivery. Special rates are available for educational institutions that wish to make photocopies for non-profit educational classroom use.

Permissions may be sought directly from Elsevier's Rights Department in Oxford, UK: phone (+44) 1865 843830, fax (+44) 1865 853333, e-mail: permissions@elsevier.com. Requests may also be completed on-line via the Elsevier homepage (http://www.elsevier.com/locate/permissions).

In the USA, users may clear permissions and make payments through the Copyright Clearance Center, Inc., 222 Rosewood Drive, Danvers, MA 01923, USA; phone: (+1) (978) 7508400, fax: (+1) (978) 7504744, and in the UK through the Copyright Licensing Agency Rapid Clearance Service (CLARCS), 90 Tottenham Court Road, London W1P 0LP, UK; phone: (+44) 20 7631 5555; fax: (+44) 20 7631 5500. Other countries may have a local reprographic rights agency for payments.

Derivative Works
Tables of contents may be reproduced for internal circulation, but permission of the Publisher is required for external resale or distribution of such material. Permission of the Publisher is required for all other derivative works, including compilations and translations.

Electronic Storage or Usage
Permission of the Publisher is required to store or use electronically any material contained in this work, including any chapter or part of a chapter.

Except as outlined above, no part of this work may be reproduced, stored in a retrieval system or transmitted in any form or by any means, electronic, mechanical, photocopying, recording or otherwise, without prior written permission of the Publisher. Address permissions requests to: Elsevier's Rights Department, at the fax and e-mail addresses noted above.

Notice
No responsibility is assumed by the Publisher for any injury and/or damage to persons or property as a matter of products liability, negligence or otherwise, or from any use or operation of any methods, products, instructions or ideas contained in the material herein. Because of rapid advances in the medical sciences, in particular, independent verification of diagnoses and drug dosages should be made.

First edition 2007

ISBN-13: 978-0-08044-963-0
ISBN-10: 0-080-44963-8

Printed in Great Britain.

07 08 09 10   10 9 8 7 6 5 4 3 2 1

Working together to grow
libraries in developing countries

www.elsevier.com | www.bookaid.org | www.sabre.org

ELSEVIER   BOOK AID International   Sabre Foundation

TABLE OF CONTENTS

Skill-Assist and Its Related Safety-Oriented Technology 1
Yamada, Y.

From RoboCup to RoboCity CoRE 5
Asada, M.

Welfare

Evaluation of Operability of an Electric Cart Handle without the Bending
Motion of Wrists 9
Inoue, Y., Kurata, J., Uchiyama, H.

Quantification of Dynamic Walking Stability of Elderly by Using Nonlinear
Time-Series Analysis and Simple Accelerometry 13
Ohtaki, Y., Arif, M., Suzuki, A., Fujita, K., Nagatomi, R., Inooka, H.

Development of a Welfare Robot Based on RECS Concept -Task of Setting
a Meal on the Table- 19
Yamaguchi, T., Kawano, H., Takano, M., Aoyagi, S.

Development of a Mobility Aid for the Visually Impaired Using a Haptic
Force Generator 23
Ikeda, T., Matsuda, H., Shiota, Y., Sakamoto, K., Shimizu, Y.

Study of the Design Method of an Ankle-Foot-Orthosis 27
Mine, Y., Takashima, T., Fujimoto, H.

Expected Main Features and Business Model of Healthcare
Partner Robot Based on the Market Analysis 31
Kabe, A.

Evaluation of Human Motor Characteristics in Driving a Wheelchair
with Hand-Rims 37
Ohta, Y., Uchiyama, H., Kurata, J.

Dynamic Modeling for Attendant Propelling Wheelchairs 43
Suzuki, T., Uchiyama, H., Kurata, J.

Development of a Non-Powered Lift for Wheelchair Uses -Mechanism
to Transmit Rotation of Wheels by Many Rollers- 47
Kobayashi, Y., Seki, H., Kamiya, Y., Hikizu, M., Maekawa, M., Chaya, Y.,
Kurahashi, Y.

Guidance of Electric Wheelchair by the Lead Type Operating Device
with Detecting Relative Position to Assistance Dog 53
Uemoto, T., Uchiyama, H., Kurata, J.

Tele-Operation

Development of Master-Slave Robotic System for Laparoscopic Surgery 57
Suzuki, T., Aoki, E., Kobayashi, E., Tsuji, T., Konishi, K., Hashizume, M.,
Sakuma, I.

Workers in Manufacturing Systems

Workplace Tasks Design Support System by Using Computer Mannequin 63
Mitsuyuki, K., Ono, T., Matsumoto, Y., Fukuda, Y., Arai, E.

Simulation and Evaluation of Factory Works Using Musculoskeletal
Human Body Model 67
Sato, T., Arisawa, H.

Development of Measuring Device for Lower Leg Swelling during
Standing Work Tasks 73
Kawano, T., Mizuno, T.

Deformable Object Manipulation

Spreading of Clothes by Robot Arms Using Tracing Method 77
Salleh, K., Seki, H., Kamiya, Y., Hikizu, M.

Indirect Simultaneous Positioning of Deformable Objects without
Physical Parameters or Time-Derivatives 81
Hirai, S.

Planning of Knotting Manipulation 87
Wakamatsu, H., Kato, T., Tsumaya, A., Arai, E., Hirai, S.

Planning

Analyzing and Evaluating Robot Motion Algorithm for Sweeping Task   93
Saito, Y., Ezawa, M., Fukumoto, Y., Ogata, H., Torige, A.

Method for Solving Inverse Kinematics of Redundant Robot under
Restraint by Obstacles   97
Kawamoto, J., Tashiro, K., Takano, M., Aoyagi, S.

CAD/CAM

Expression of Theoretical Design Information and Intention
Transmitting Architecture   103
Takeuchi, K., Tsumaya, A., Wakamatsu, H., Arai, E.

Detection of Uncut Regions in Pocket Machining   109
Seo, M., Kim, H., Onosato, M.

Flexible Process Planning System Considering Design Intentions
and Disturbance in Production Process   113
Han, G., Koike, M., Wakamatsu, H., Tsumaya, A., Shirase, K., Arai, E.

A Study on Calculation Methods of Environmental Burden for NC
Program Diagnosis   119
Narita, H., Norihisa, T., Chen, L., Fujimoto, H., Hasebe, T.

Assembly/Disassembly

Assembly System by Using Prototype of Active Flexible Fixture   125
Yamaguchi, T., Higuchi, M., Nagai, K.

Assembly Sequence Planning Using K-Nearest-Neighbor Rule   129
Murayama, T., Eguchi, T., Oba, F.

Proposal of Ubiquitous Disassembly System for Realizing Reuse
and Recycling in Cooperative Distributed Facilities   133
Tateno, T., Kondoh, S.

Sensors & Actuators

Development of a Micro Tactile Sensor Utilizing Piezoresistors
and Characterization of Its Performance   139
Izutani, J., Maeda, Y., Aoyagi, S.

Development of Sensors Based on the Fixed Stewart Platform   145
Irie, K., Kurata, J., Uchiyama, H.

Microfabrication of a Parylene Suspended Structure and Investigation
of Its Resonant Frequency   149
Yoshikawa, D., Aoyagi, S., Tai, Y.C.

Machining

Direct Prediction of Cutting Error in Finish Endmilling Based
on Sequence-Free Algorithm   153
Kaneko, J., Teramoto, K., Horio, K., Takeuchi, Y.

Development of Curved Hole Machining Method -Size Reduction
of Hole Diameter-   157
Nakajima, T., Ishida, T., Kita, M., Teramoto, K., Takeuchi, Y.

Microchannel Array Creation by Means of Ultraprecision Machining   163
Andou, F., Yamamoto, A., Kawai, T., Ohmori, H., Ishida, T.,
Takeuchi, Y.

Automation of Chamfering by an Industrial Robot (Development
of Positioning System to Cope with Dimensional Error)   169
Tanaka, H., Asakawa, N., Kiyoshige, T., Hirao, M.

Human Behavior Analysis

Interactive Behavioral Design between Autonomous Behavioral
Criteria Learning System and Human   173
An, M., Taura, T.

Human Behavior Based Obstacle Avoidance for Human-Robot
Cooperative Transportation   177
Aiyama, Y., Ishiwatari, Y., Seki, T.

Evaluation Methods for Driving Performance Using a Driving Simulator
under the Condition of Drunk Driving or Talking Driving with a Cell Phone 181
Azuma, Y., Kawano, T., Moriwaki, T.

Computational Model and Algorithm of Human Planning 185
Fujimoto, H., Vladimirov, B., Mochiyama, H.

Humanoids

Safety Design for Small Biped-Walking Home-Entertainment Robot
SDR-4XII 189
Iribe, M., Moridaira, T., Fukushima, T., Kuroki, Y.

Scheduling

A Study on a Real-Time Scheduling of Holonic Manufacturing
System -Coordination Among Holons Based on Multi-Objective
Optimization Problem- 195
Iwamura, K., Seki, Y., Tanimizu, Y., Sugimura, N.

A Study on Integration of Process Planning and Scheduling System
for Holonic Manufacturing System -Scheduler Driven Modification
of Process Plans- 201
Shrestha, R., Takemoto, T., Sugimura, N.

Genetic Algorithm Based Reactive Scheduling in Manufacturing
System -Advanced Crossover Method for Tardiness Minimization
Problems- 207
Sakaguchi, T., Tanimizu, Y., Harada, K., Iwamura, K., Sugimura, N.

A Basic Study on Cost Based Scheduling 213
Sashio, K., Fujii, S., Kaihara, T.

Vision

Search and Pose Recognition of Industrial Components Using
Curvature of Optimized Edge Pixels 219
Goto, K., Saitoh, F.

Vision-Based Navigation of an Outdoor Mobile Robot Using a Rough Map 223
Yun, J., Miura, J., Shirai, Y.

Teaching a Mobile Robot to Take Elevators 229
Iwase, K., Miura, J., Shirai, Y.

Generated Image Feature Based Selective Attention Mechanism
by Visuo-Motor Learning 235
Minato, T., Asada, M.

Precise Micro Robot Bio Cell Manipulation Based on the Microscopic
Image Recognition 241
Misaki, D., Naoto, C., Usuda, T., Fichiwaki, O., Aoyama, H.

New Services & Decision Making in Manufacturing

Service Explorer -A Tool for Service Design- 247
Shimomura, Y., Sakao, T., Hara, T., Arai, T., Tomiyama, T.

A Framework for Service Engineering Based on Hierarchical
Colored Petri Nets 253
Tian, G., Miura, T., Hara, T., Shimomura, Y., Arai, T.

Observables of Opposites Alternatives in Decision Making 257
Yagi, J., Arai, E., Matsumoto, S.

Manufacturing Systems

Enhanced Distributed-Simulation Using ORiN and HLA 261
Inukai, T., Hibino, H., Fukuda, Y.

Object-Oriented Embedded System Development Method for Easy
and Fast Prototyping 265
Vallius, T., Haverinen, J., Röning, J.

Integrated Construction Process Management System 271
Takata, M., Arai, E., Yagi, J.

A Robotized System for Prototype Manufacturing of Castings and Billets 277
Sallinen, M., Sirviö, M.

Towards Human-Profile Based Operations in Advanced Factory Governance Systems: Contemporary Challenges for Socio-Technical Systems Design? Eijnatten, F.M.V, Goossenaerts, J.B.M.	281
Relation Diagram Based Process Optimization of Production Preparation Process for Oversea Factory Sato, S., Inamori, Y., Nakano, M., Suzuki, T., Miyajima, N.	287
Cyber Concurrent Manufacturing Integrated with Process Engineering and 3D-CG Simulation -Product Design, Production System Design, and Workstation System Design as a Case Study on Curtain Wall Construction Work- Tamaki, K.	293

Wireless Communication

Wireless Data Transfer Applied on Hydraulic Servo Karhu, O.I., Virvalo, T., Kivikoski, M.	297
The Challenges on the Development of Mobile Controlled RFID System Soini, M., Sydanheimo, L., Kivikoski, M.	301
Wireless Communication with Bluetooth Hearing Protector Oinonen, M., Myllymäki, P., Ritamäki, M., Kivikoski, M.	305
Development of Local Positioning System Using Bluetooth Hirota, T., Tanaka, S., Iwasaki, T., Hosaka, H., Sasaki, K., Enomoto, M., Ando, H.	309
Analysis of Multiple Object Identification with Passive RFID Penttilä, K., Sydänheimo, L., Kivikoski, M.	313
Modeling Electromagnetic Wave Propagation in Paper Reel for UHF RFID System Development Keskilammi, M., Sydänheimo, L., Kivikoski, M.	317
Effect of Conductive Material in Objects on Identification with Passive RFID Technology: A Case Study of Cigarette Cartons Ukkonen, L., Soini, M., Engels, D., Sydänheimo, L., Kivikoski, M.	323

Control

Current Limiter Complicates the Dynamic Characteristics of Servo Motor 329
Serikitkankul, P., Seki, H., Hikizu, M., Kamiya, Y.

Active Suspension System with High-Speed ON/OFF Valve
(Application of Preview Control with Adaptive Digital Filter) 333
Yamada, H., Muto, T.

Embedded Distributed Sub Control System Based on Hybrid Controller 339
Lammila, M., Virvalo, T., Lehto, E.

Active Noise Cancellation Hearing Protector with Improved Usability 343
Oinonen, M., Raittinen, H., Kivikoski, M.

Suppressing Mechanical Vibrations in a PMLSM Using Feedforward
Compensation and State Estimates 347
Hirvonen, M., Handroos, H.

Characterization, Modeling and Simulation of Magnetorheological
Damper Behavior under Triangular Excitation 353
Cortés-Ramírez, J.A., Villarreal-González, L.S., Martinez-Martínez, M.

Measuring & Monitoring

Soft-Sensor Based Tree Diameter Measuring 359
Hölttä, V.

Study On-Machine Work Piece Measurement on 5-Axis Controlled
Machining Center 365
Nakamura, S., Ihara, Y.

A New Methodology to Evaluate Error Space in CMM by Sequential
Two Points Method 371
He, W.M., Sato, H., Umeda, K., Sone, T., Tani, Y., Sagara, M., Matsuzaki, H.

Pressure Monitoring System of Gland Packing for a Control Valve 377
Wada, M., Naito, M., Hosaka, H., Hirota, T., Okutsu, R., Izumi, K.

Application

Fabrication of a Micro Needle Made of Biodegradable Polymer Material 381
Aoki, T., Izumi, H., Aoyagi, S.

An Effective State-Space Construction Method for Reinforcement
Learning of Multi-Link Mobile Robots 385
Nunobiki, M., Okuda, K., Maeda, S.

Circularly Polarised Rectenna for Enhanced Dual-Band Short-Range
Wireless Power Transmission 389
Heikkinen, J., Kivikoski, M.

Conductive Fibres in Smart Clothing Applications 395
Hännikäinen, J., Järvinen, T., Vuorela, T., Vähäkuopus, K., Vanhala, J.

Design of Low-Clearance Motion Platform for Driving Simulators 401
Mohellebi, H., Espié S., Kheddar, A., Arioui, H., Amouri, A.

High Performance Low Cost Stereo Projector System for Virtual Reality 405
Kosola, H., Palovuori, K.

Analytical and Experimental Modeling of Intra-Body Communication Circuit 409
Terauchi, Y., Hachisuka, K., Sasaki, K., Kishi, Y., Hirota, T., Hosaka, H.,
Fujii, K., Ito, K.

Design of Multi Sensor Units for Searching Inside of Rubble 415
Inoue, K., Yamamoto, M., Takubo, T., Mae, Y., Arai, T.

Mechatronics Design and Development towards a Heavy-Duty
Waterhydraulic Welding/Cutting Robot 421
Wu, H., Handroos, H., Pessi, P.

Qutie-Modular Methods for Building Complex Mechatronic Systems 427
Tikanmäki, A., Vallius, T., Röning, J.

Link Length Control Using Dynamics for Parallel Mechanism
with Adjustable Link Parameters 431
Tanaka, W., Arai, T., Inoue, K., Takubo, T., Mae, Y., Koseki, Y.

A Proposal of the Multimedia Archive System with Watermark Index
for Prevention of Disasters 435
Maehara, F., Tanno, Y.

PREFACE

"Mechatronics" is the technology or engineering field that originated in Japan and has been rapidly developed over the last two to three decades. The industrial robot, which is a typical example of a "mechatronics product", can be found in every field today. Mechatronics products include: intelligent engine and cruise control systems for automobiles; intelligent household electric appliances such as computer controlled sewing machines, washing machines, refrigerators, VTR; auto-focusing cameras; computer peripheral equipment such as printers, magnetic/optical disk drives; information-processing equipment such as digital communication systems, portable telephones; NC machine tools such as the wire-cut electric discharge machine, and NC milling machine; a building control system including air conditioning, automatic doors; semiconductor manufacturing facilities; an automatic ZIP code reading machine; a sorting machine for agricultural or marine products.

The word "mechatronics" was initially defined as integrated engineering in mechanics and electronics. However, the meaning of this word gradually changed to be taken widely as "intellectualization of machines and systems with information technology," these days. Namely, machines and systems, which control their motion autonomously and appropriately based on external information, generate a common technology of mechatronics. According to this definition, the automation of an industrial factory line, such as a machining line, assembly line, or strip mill line, is based on the mechatronics technology. Control systems for train operation and intelligent transportation systems (ITS) for automobiles are also realized by mechatronics technology. It is thus evident that mechatronics has allowed us to achieve high productivity and has enriched our daily life by improving the performance of machines and systems. We are now confronting many issues related to our aging society and sustaining the environment. Mechatronics is expected to play an important role in solving the issues brought by the new era.

New mechatronics will be applied in the support of elderly and handicapped people to lead their lives safely and securely, both in their social activities as well as in their daily lives. Machines and systems should be designed and operated on the basis of analysis and evaluation of human psychology and behavior in addition to considering the coordination of humans and machines. The new mechatronics will interface with human science. The forms and the motions of supporting robots, for example, should be designed after evaluating their psychological effects on users. A monitoring system, which secures peaceful and comfortable lives for citizens, is required to make more detailed observations, however, it should bring no mental and psychological stress to the people being monitored. Safety recovery is also an important application for the new mechatronics, which covers rescue tasks after disasters, destruction of hazardous and abandoned weapons, and restoration of polluted environments.

The 5th International Conference on Machine Automation, held in November 24-26, 2004 at Osaka University in Japan, covered various aspects of mechatronics in a new era, providing new methodologies and tools to design and to build machines and systems for safety, security, and dependability. This book includes selected papers from the conference. The conference was sponsored by Osaka University and the Japanese Council of International Federation for the Promotion of Mechanism and Machine Science. First, we would like to thank the members of the Organizing Committee. Further thanks go to the authors, the members of the Program Committee and the conference participants for their contribution to the success of the conference and this book.
In conclusion, we strongly hope that this book will have a useful shelf life.

The editors,
Eiji Arai and Tatsuo Arai

Skill-Assist and Its Related Safety-Oriented Technology

Yoji YAMADA
Intelligent Systems Institute,
National Institute of Advanced Industrial and Science Technology (AIST),
AIST Tsukuba Central 2, Tsukuba City, Ibaraki, 305-8568, Japan.

Abstract

Skill-Assist is a kind of power assist devices which has been introduced to labor-intensive final assembly lines in automotive industry for helping workers mount such heavy modular parts as instrumental panels on vehicles with high positioning accuracy. The paper briefly describes a current technology which has been already implemented on the controller of the Skill-Assist for putting it to practical use, and a future safeguarding method that has been developed on a laboratory basis. An important issue on putting Skill-Assist devices into practice was resolved by attaining a certain level of safeguarding technology which follows the initial productivity enhancement effort of controlling the devices so as to reflect workers' desired skills in the statistical sense. We can report that no accident or system failure has occurred since commencement of operation. The main part of this paper is focused on description of our laboratory-based study concerning safeguarding scheme against human operational slip. Finally, a comment is made on how to put human-coexistence type service robots into practice from the viewpoint of sceince and technology study.

Index Terms

power assist device, FTA, operation slip, Hidden Markov Model, control policy, service robot

I. Introduction to Skill-Assist

In 1990's, increasing numbers of automobile manufacturers have adopted modular component assembly systems which offer various merits. However, modular components are larger and heavier than individual parts. In view of the fact that an assembly process involving human operators can easily change tasks and flexibly adapt to varying demands, a technology that can reduce the physical workload on human operators during modular component installation is an important element to be incorporated into the assembly process.

Power augmentation in heavy-duty conveyance operation, especially in the horizontal direction, is important because large inertial force is exerted in the phase of positioning a heavy load. It results not only in decreased production efficiency caused by deterioration in an operator's skill; it also engenders a condition among operators that is known as coxalgia. Therefore, we proposed construction of a power assist device called "Skill-Assist" for mounting operations in automobile assembly processes. **Fig. 1** is a picture of Skill-Assist being maneuvered by an operator (a worker) in an actual production line.

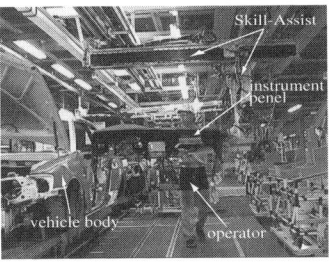

Fig. 1. Overview of a Skill-Assist introduced to a production line

The Skill-Assist varies mechanical impedance depending on phases in a task so that it reflects workers' desired skills in the statistical sense. Concretely, phase-dependent variable impedance control is applied to controlling each of the linear actuator so that a desired mechanical impedance characteristic is reflected and desired operating force is exerted at the handle. Fig. 2(a) shows a sample of operating force pattern in a sequential task of mounting an instrument panel in a practical production line. Each of the phase sequence number ① through ⑥ corresponds to the task motion in Fig. 2(b).

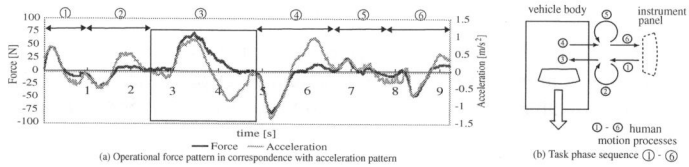

Fig. 2. Operational force pattern in an actual task cycle

II. RISK ASSESSMENT WITH 3-STEP METHODS FOR THE RISK REDUCTION PROCESS OF A SKILL-ASSIST

The guarantee of operator safety while maneuvering a Skill-Assist is a prerequisite. In addition to its benefit as a basic mechanism for improvement in maneuvering performance, Skill-Assist also offers the characteristic of being equipped with an appropriate mechanism for safety improvement. We have examined safety issues associated with such power assist devices as Skill-Assist from various aspects based primarily on FTA (Fault Tree Analysis). We followed the risk assessment protocol and undertook overall risk reduction measures. Fault Tree Analysis (FTA) was conducted initially to identify hazards in the actual operation of installing an instrument panel in a vehicle body. The risk identification results are summarized that there are two distinctive error sources except for participation of third party. Safeguarding implementation of various complimentary protective measures has been incorporated into Skill-Assist based on these results to ensure that Skill-Assist provides a safety level that is acceptable to all parties concerned except for risks attributable to human error, which shall also be addressed when application fields for Skill-Assists are expanded in the future. We will deal with this problem later.

III. DEVELOPMENT OF A METHOD FOR PREVENTING ACCIDENTS CAUSED BY HUMAN ERROR[1]

As stated before, we must consider that methods for preventing accidents resulting from mistaken human actions should also be developed. The necessity of this kind of technology originates in the fact that we can hardly exclude all mechanically hazardous parts around the device as well as its motion: In most cases, the end effector of the device needs to handle objects which may cause human injury even if a device is free from such hazardous parts or tools. However, if hazardous situations in collaborative tasks can be identified clearly in advance, further technologies for securing human safety can be applied so that the device can prevent hazardous accidents caused by human error. Implementing the basic risk estimation protocol, the following situation was categorized as the most hazardous event for our application (Severity=Serious injury, Event Probability=High, Avoidance=Not Likely):

An operator's right hand grabbing a Skill-Assist's handle gets pinched with the center pole of a vehicle body frame when he is operating the Skill-Assist to insert the instrument panel through the frame. We assume an identified hazard point (HP) as illustrated in **Fig. 3**, where trajectories of both reaching and avoiding HP are shown in 2-D space because the z component of the Skill-Assist is fixed during the insertion task phase.

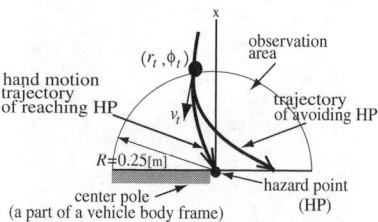

Fig. 3. Hand motion trajectories in the vicinity of the HP

Based on the operator's hand motion trajectories in the observed area, observation data sequences for both trajectory patterns to reach and to avoid HP are expressed with hidden states using HMMs (Hidden Markov Models) for their pattern classification.

The system is made to judge to which model the actual hand motion trajectory belongs from probabilistic evaluation based on two propositions: the proposition D of stating "dangerous operation" and S of "safe operation". "Dangerous operation" leads to a stopping operation (a brake is applied to the Skill-Assist), and "safe operation" to continuing the operation. However, if the output goes out of the pretaught pattern area, the corresponding element of the observation symbol probability distribution becomes 0. Therefore, Dempster-Shafer (abbreviated as DS) theory is applied. We use two distinct operation control policies, Safety-Preservation (SP) policy and Fault-Warning (FW) policy. Operation control is carried out according to a policy corresponding to the observation result of where the hand position lies at a time: Control judgment is made depending on which policy the third proposition X which accepts either S or D as a frame of discernment and where the observation symbol distribution becomes zero under both HP collision and avoidance trajectory models. In the study, this way of observation-space-dependent policy determination is referred to as HMM-OPD.

We performed 10 operation iterations to teach data in alignment with each of the two fixed trajectories of reaching and avoiding HP. We conducted experiments based on the teaching data to verify effectiveness of an accident prevention method by using the operation control with the proposed HMM-OPD. In the first accident prevention experiment, we could successfully prevent all 20 trials of the subject's motion hand movement reaching HP from colliding with it by use of the proposed HMM-OPD method. In the other 20 collision avoidance experiment iterations, it is judged to stop operation totally for 18 times out of 20 repetitions. Such unnecessary halts in operation indicated that productivity might decrease severely in the event of application of HMM-OPD and also imply that either: 1) definition of the FW and SP space was not initially optimized or 2) teaching data concerning safe operation were in short supply which resulted in 4 times of stop operation in FW space. The following section proposes a workability improvement process comprising renewal of both state policies and teaching data as a solution to coping with problems remaining in the accident prevention method proposed so far.

IV. EXTENTION OF HMM-OPD TO WORKABILITY IMPROVEMENT

In the study, the combination of safety and productivity is referred to as workability; optimal workability is defined as the ability to perform a task in the state where safety is secured and there is no unnecessary stoppage of operation by the Skill-Assist. A workability improvement process is carried out in the process of repeating what we call "hazard simulation" and HMM reconstruction is eventually performed.

Generally, it is difficult for an operator to teach initial conditions to optimize an observation space in which a higher priority is strictly determined for either safety or productivity. Therefore, heuristic reconstruction of the observation space is attractive. Next, we performed a hazard simulation process that is characteristic of presenting no danger of real pinch and contributes to minimizing the volume of SP space through demonstrations in off-line operation.

Then, HMM is reconstructed using the constructed space. **Fig. 4** shows results of a collision avoidance experiment after such a workability improvement process in which the operation is carried out along the trajectory pattern toward a HP. In this case, because the volume of FW space is expanded by 12 times of FW·SP space–policy renewal, the FW policy is implemented until the operator's hand reaches $r = 0.09$ m, when another judgment is made to continue the operation. Moreover, it turns out that probability of dangerous operation is equal to 1 at $r = 0.09$ m. Subsequently, a different judgment is made to cease operation; then accident prevention can be performed and eventually $v = 0$ m/s at $r = 0.06$ m. This leads us to infer that the process is a useful method for optimizing the observation space.

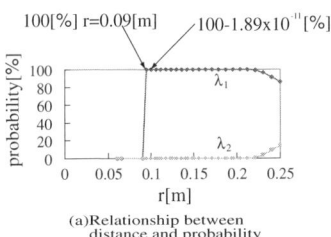
(a) Relationship between distance and probability

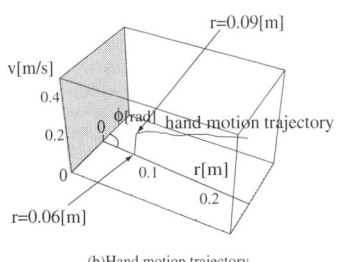
(b) Hand motion trajectory

Fig. 4. Stop operation for collision avoidance after FW·SP space–policy renewal

The second renewal process of teaching data is also implemented successfully, and collision avoidance experiments demonstrated that we could obtain a clear result of 20 continuous successful operation repetitions with no halted operation in either FW or SP space after workability improvement processes.

V. DISCUSSION ON SUCCESSFULLY PUTTING SERVICE ROBOTS TO PRACTICAL USE

Finally, we make a comment on how to put human-coexistence type service robots into practice from the viewpoint of sceince and technology study. **Fig. 5** shows the effect of safety technology of service robots implemented in conformity with international standard and/or quthenticated by certification authority. If exhaustive risk assessment is conducted to yeild safety technology and operation manuals as countermeasures against unacceptable risks, human negligence will not be called into question.

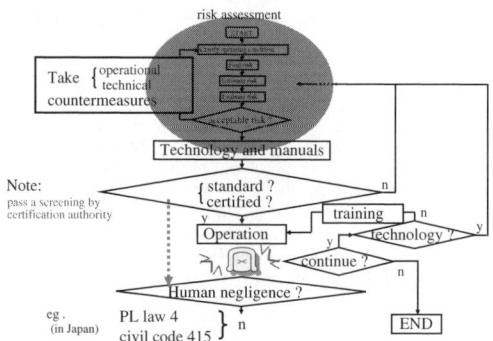

Fig. 5. Toward a good engineering practice with socially acceptable safety technology

In Japan, for example, a right of making a plea in running a risk of development is accepted by Article 4 of Product Liability Law: Referring to the whole knowledge (international standards) of safety technology may allow manufacturers for exemption from being in charge. Moreover, Article 415 of Civil Code may include the cause of an accident as an immunity reason after an exhaustive risk assessment is conducted and 3-step method (securing essential safety, safeguarding countermeasures, and informing users of residual risks) is followed. These articles are considered to encourage manufactures to develop robots as a good engineering practice. From this encouragement, operation of robots will be continued if desired, safer technology will be implemented, and the robots will be accepted by society even after some injury occurs.

VI. SUMMARY

1) Skill-Assist, a human-coexistence power assist devices was introduced to labor-intensive final assembly lines in automotive industry so that it reflected workers' desired skills in the statistical sense.
2) Various safety measures for the Skill-Assist's controller were effectively incorporated into pivotal points of the controller.
3) We studied on a method for preventing hazardous accidents caused by human error as a future technology which allowed the Skill-Assist to detect an operator's mistaken action from a sequence of hand motion trajectory data. The proposed operation control with observation-space-dependent policy determination supported leading of the Skill-Assist to an appropriate judgment depending on the hand motion state.
4) A workability improvement process composed of two renewal sub-processes was also implemented.
5) Finally, a comment is made on how to put human-coexistence type service robots into practice from the viewpoint of sceince and technology study.

ACKNOWLEDGMENT

This work was mainly conducted when the author was with Toyota Technological Institute under financial support of Toyota Motor Company, and he is indebted to the following people: former Prof. Yoji UMETANI, Dr. Tetsuya MORIZONO at the Institute, and Mr. Hitoshi KONOSU, Mr. Toshiharu MITOMA, and his colleagues at Toyota Motor Company Ltd.

REFERENCES

[1] Y. YAMADA, et al., (2004). "Warning: To Err is Human - Working Toward a Dependable Skill-Assist with a Method for Preventing Accidents Caused by Human Error", *IEEE Robotics and Automation Magazine*, **11:2**, 34-45.

FROM RoboCup TO RoboCity CoRE

Minoru Asada

Graduate School of Eng. Osaka University,
Suita, Osaka 565-0871, JAPAN

ABSTRACT

This article presents the brief introduction of Robot World Cup Competition and Conference, in short, RoboCup. The aims and the current activities are introduced. Next, RoboCity CoRE, an inner city RT base, is introduced as a RT experiment field open to public.

KEYWORDS

RoboCup, RoboCupSoccer, RoboCupRescue, RoboCupJunior, RoboCity CoRE, Open Lab., Studio,Safety Verification Field.

INTRODUCTION

RoboCup is an attempt to foster intelligent robotics research by providing a standard problem of which the ultimate goal is to build a team of eleven humanoid robots that can beat the human world cup champion soccer team by 2050. It's obvious that building a robot to play soccer game is an immense challenge; readers might therefore wonder why even bother to propose RoboCup. It is our intention to use RoboCup as a vehicle to promote robotics and AI research, by offering a publicly appealing but formidable challenge [1, 2].

A unique feature of RoboCup is that it is a systematic attempt to promote research using common domain, mainly soccer. Also, it is perhaps the first to explicitly claim that the ultimate goal is to beat human world cup champion team. One of the effective ways to promote engineering research, part from specific application developments, is to set a significant long term goal. When the accomplishment of such a goal has significant social impact, we call this kind of goal a *grand challenge project*. Building a robot to play soccer is not such a project. But its accomplishment would certainly considered as a major achievement in the field of robotics, and numerous technology spin-off can be expected during the course of the project. We call this kind of project a *landmark project*, and RoboCup is definitely a project of this kind.

Since the first RoboCup in 1997 [3], it has grown into an international joint-research project in which about 4000 researchers from 40 nations around world participate (see Figure 1), and it is one the most ambitious landmark projects of the 21st century. RoboCup currently consists of three divisions:

RoboCupSoccer aiming towards the final goal, RoboCupRescue, a serious social application to the rescue activities for any kinds of disasters, and RoboCupJunior, an international education-based initiative designed to introduce young students to robotics.

The RoboCup 2005 competition was the largest scaled one since 1997. About 1800 team members of 340 teams from 31 nations and regions around world participated. Eventually, a total of 182,000 spectators witnessed this marvelous event. To the best of our knowledge, this was the largest robotic event in history. Figure 2 shows a game of the humanoid league, 2 on 2 from RoboCup 2005 Osaka. For more details, please refer to [4, 5, 6, 7, 8, 9, 10, 11, 12, 13, 14, 15].

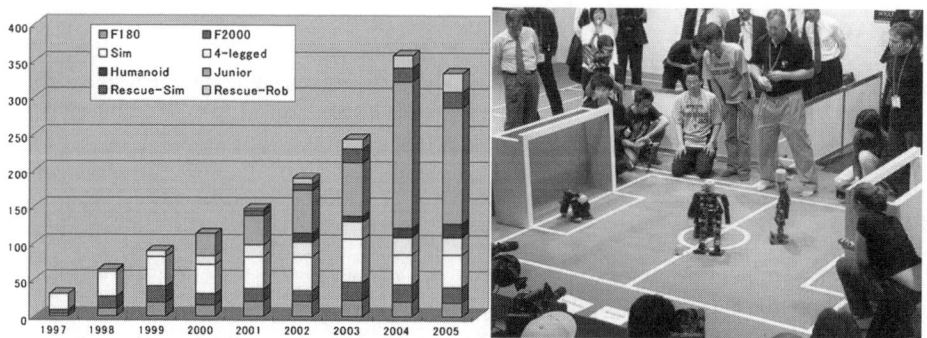

Figure 1: The number of teams (left) and 2 on 2 game of the humanoid in RoboCup 2005

RoboCity CoRE : An inner city RT base

A basic concept of RoboCup are an international joint research, a landmark project: sharing the dream, and open to different disciplines, open to public. Currently, the competition and conference is once a year, and a natural extension of RoboCup concept is to have a permanent place to deploy our activities.

RoboCity CoRE (Center of RT Experiments) is an inner city labs for symbiotic experiments with robots, new partners of our future life. CoRE aims at only one RT base around the world where simultaneous progresses of research, industrialization, and education carry on simultaneously.

Open to public means that researchers, artists, companies, citizens interchange with each other to emerge new ideas, that leads the development of science, technology, and culture. CoRE will be a new cultural symbol of the future high-technological, ecological city.

CONCLUSION

The brief introduction of Robot World Cup Competition and Conference, in short, RoboCup is given. The aims and the current activities are briefly introduced. Next, the idea of RoboCity CoRE, an inner city RT base, is given as a RT experiment field open to public.

Figure 2: An image of RoboCity CoRE

REFERENCES

H. Kitano, M. Asada, Y. Kuniyoshi, I. Noda, E. Osawa, and H. Matsubara. "robocup: A challenge problem of ai". *AI magazine*, 18(1):73–85, 1997.

Minoru Asada, Hiroaki Kitano, Itsuki Noda, and Manuela Veloso. Robocup: Today and tomorrow – what we have learned. *Artificial Intelligence*, 110:193–214, 1999.

Hiroaki Kitano, editor. *RoboCup-97: Robot Soccer World Cup I*. Springer, Lecture Note in Artificial Intelligence 1395, 1998.

I. Noda, S. Suzuki, H. Matsubara, M. Asada, and H. Kitano. Robocup-97 the first robot world cup soccer games and conferences. *AI magazine*, 19(3):49–59, 1998.

Minoru Asada, Manuela M. Veloso, Milind Tambe, Itsuki Noda, , Hiroaki Kitano, and Gerhard K. Kraetzschmar. Overview of robocup-98. *AI magazine*, 21(1):9–19, 2000.

Silvia Coradeschi, Lars Karlsson, Peter Stone, Tucker Balch, Gerhard Kraetzschmar, and Minoru Asada. Overview of robocup-99. *AI magazine*, 21(3):11–18, 2000.

Peter Stone, Minoru Asada, Tucker Balch, Raffaello D'Andrea, Masahiro Fujita, Bernhard Hengst, Gerhard Kraetzschmar, Pedro Lima, Nuno Lau, Henrik Lund, Daniel Polani, Paul Scerri, Satoshi Tadokoro, Thilo Weigel, and Gordon Wyeth. Robocup-2000: The fourth robotic soccer world championships. *AI magazine*, 22(1):11–38, 2001.

Manuela Veloso, Tucker Balch, Peter Stone, Hiroaki Kitano, Fuminori Yamasaki, Ken Endo, Minoru Asada, M. Jamzad, B. S. Sadjad, V. S. Mirrokni, M. Kazemi, H. Chitsaz, A. Heydarnoori, M. T. Hajiaghai, and E. Chiniforooshan. Robocup-2001: The fifth robotic soccer world championships. *AI magazine*, 23(1):55–68, 2002.

Minoru Asada and Hiroaki Kitano, editors. *RoboCup-98: Robot Soccer World Cup II*.

Springer, Lecture Note in Artificial Intelligence 1604, 1999.

Manuela Veloso, Enrico Pagello, and Hiroaki Kitano, editors. *RoboCup-99: Robot Soccer World Cup III*. Springer, Lecture Note in Artificial Intelligence 1856, 2000.

Peter Stone, Tucker Balch, and Gerhard Kraetzschmar, editors. *RoboCup-2000: Robot Soccer World Cup IV*. Springer, Lecture Note in Artificial Intelligence 2019, 2001.

Andreas Birk, Silvia Coradeschi, and Satoshi Tadokoro, editors. *RoboCup 2001: Robot Soccer World Cup V*. Springer, Lecture Note in Artificial Intelligence 2377, 2002.

Gal Kaminka, Pedro U. Lima, and Raul Rojas, editors. *RoboCup 2002: Robot Soccer World Cup VI*. Springer, Lecture Note in Artificial Intelligence 2752, 2003.

Brett Browning Andrea Bonarini, Daniel Polani and Kazuo Yoshida, editors. *RoboCup 2003: Robot Soccer World Cup VII*. Springer, Lecture Note in Artificial Intelligence 3020, 2004.

C. Sammut J. Santos-Victor D. Nardi, M. Riedmiller, editor. *RoboCup 2004: Robot Soccer World Cup VIII*. Springer, Lecture Note in Artificial Intelligence 3276, 2005.

EVALUATION OF OPERABILITY OF AN ELECTRIC CART HANDLE WITHOUT THE BENDING MOTION OF WRISTS

Yuki INOUE, Junichi KURATA and Hironobu UCHIYAMA

Department of Mechanical Systems Engineering Kansai University,
3-3-35, Yamatechou, Suita, Osaka 564-8680, JAPAN

ABSTRACT

Most electric carts employ on oval steering handle mechanism, which is seemed to make operation difficult. When the operator turns the handle, great flexion of wrist joint is observed at both grips, causing distortion of the upper part of the body. As operator moves his upper part to avoid this inconvenience in turning the handle, the center of gravity of the cart and operator is moved to the extreme centrifugal direction. To resolve this problem, we propose a new type of handle, which can be operated with a straight-line motion, without any bending of the wrist joint, a bending torque forced upon wrist joints, a curved shape of upper body and large movement of the center of gravity of cart system. We confirmed that this type of handle was quite comfortable for electric carts by various experimental results.

KEYWORDS

Human behavior, Electric cart, Elderly people, Handicapped people, Four parallel links, Wrist

INTRODUCTION

As motion-support machines for the elderly and physically handicapped, electric carts and wheelchairs are useful. Since electric carts can be easily operated with only a handle and an accelerator lever, they should be very useful. However, most of these carts employ oval steering handle, which requires large flexion of the wrist joint at both grips of the handle. This results large distortion of the upper part of operator's body. And this distortion causes large shift of the center of gravity, and the cart system makes the center of gravity to the outside of the body center axis. In order to resolve this problem, we propose a new type of handle instead of ordinary oval shape, which has a structure of four parallel linkages. In the case of new handle, it can be operated by only forward or backward motion of the operator's arm without any bending or twisting motion of his wrist joint.

Research on the characteristic of upper part of operator's body caused by bending or twisting of the wrist joint has not been reported very much before. The remarkable inquires into arms in progress reported by Mussa-Ivaldi (1985) was the first report about multi-joint stiffness measurements. They disturbed a position of subject's hand from an equilibrium position to the other, and measured the restoring force after new equilibrium had been reached. They gave us the conclusion that the appeared force field was almost equivalent to it of spring. The hand stiffness could be represented graphically by an ellipse shape. The major axis of ellipse was typically oriented in the direction of a straight line that connects the shoulder and middle point of the upper arm. While the stiffness specifies the exerted force due to a position deviation in static equilibrium, the dynamic relation between small force and position variations is explained by the impedance. Some useful results on the impedance of hand have been reported by Dolan (1993) and Tsuji (1995). Both teams used a robot to exert small varying forces on the hand and assumed a second-order mass viscosity stiffness model to fit the force and position data. Flash and Mussa-Ivaldi (1990) showed that the shoulder stiffness varied with the stiffness provided by the biarticular muscles that was used to obtain a polar direction of the stiffness ellipse (towards the shoulder). They were approximately found in their measurements. In general, arm impedance is the resultant of passive dynamics of the arm, intrinsic impedance of activated muscles and reflexive contributions. However, the characteristic of a wrist has not been made clear.

In this report, we described the verification results of operational performance and synthetic view on new type of handle, which was efficient to keep operator's posture stable with simple operation as the forward and backward motion of operator's arm.

STRUCTURE OF LINK TYPE HANDLE

The new type of handle is shown in Figure 1. This handle is composed of four parallel linkage mechanisms. A handle axis is fixed to driven joint, and the handle axis is rotated by forward and backward movement of two riving joints. The rods are used to restrict a motion of driving joint. The grip of a link type handle is always moving in parallel with the body. Bending motion of wrist joint becomes almost unnecessary in all the process of steering operation. Therefore, the stabilized natural seating position posture can always held, and it can operate.

METHODS

Special equipment was developed to examine the performance characteristics of our newly device electric cart maneuvering mechanism. Two torque sensors[20 N · m] and two potentiometers were attached to the steering shaft and one side of the handle grips. The measured values of the steering angle and bending angle of the wrist were analyzed as a function of the turning torque forced to the

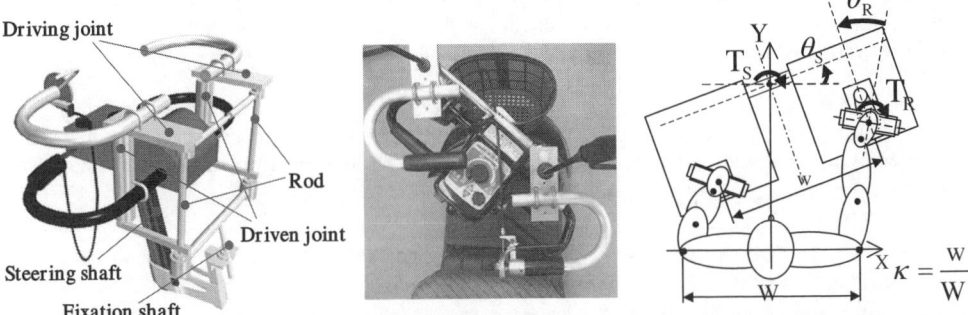

Figure 1: Link type handle

Figure 2: Subject position on measurement device

steering shaft loaded by a spring[1N/mm]. The distance during both grips shows the handle width w, and it is examined about 170, 240, and 430 mm in this experiment. A subject (24 years old, shoulder width 520 mm) gripped the handle of a measurement device tightly. The wrist was fixed to simulate elderly people's function. During each trial, the subject's wrists were kept in a specific posture. The arm posture shaped by shoulders and elbow joints was varied by handling the measurement device, the grips of which were fixed. A subject turns a handle with following the sine wave signal indicated in the oscilloscope may be followed. The rotation angle and torque around a steering axis and the rotation angle and torque in a grip were measured. The rotation angle of steering axis θ_s, torque of steering axis Ts, the rotation angle of grip θ_R, and torque of grip T_R were shown in Figure 2.

RESULTS

When the operator normally turned the conventional oval handle, no restrictions conditions were operated using torsion of an elbow. Therefore, when it was operated without using torsion of an elbow, the amount of compensation was decided by the rotation angle of grip. A subject steered without using torsion of an elbow, and measured rotation angle of grip θ_R.

The measurement results were shown in Figure 3. In case of conventional oval handle operation, the rotation angle of right wrist increased according to the increase in steering angle θ_s. It was very important to reduce the rotation of wrist joint for the elderly people, whose limit angle of bending motion of wrists was smaller than that of young people. The same examination has been done to perform the link type handle. As Figure 3 shows, the staring shaft can be turned with only little bending of a wrist. When κ is 0.83, rotation angle of grip varied in the region from -20 degrees to 0 degrees. When κ is 0.47 and 0.33, it turned out that the rotation angle of a grip decreased extremely. The torque of wrist was measured in case of the conventional oval handle and the link type handle. Experiment results were shown in Figure 4. On conventional oval handle, change of the torque of wrist was few only in the region of steering angles from -20 degrees to 20 degrees, and the torque violently increased out of this region. On link type handle (κ = 0.47, 0.33), the value of torque reduced very much. However, the much wider handle, for example κ = 0.83, resulted little increase of torque to drive handle.

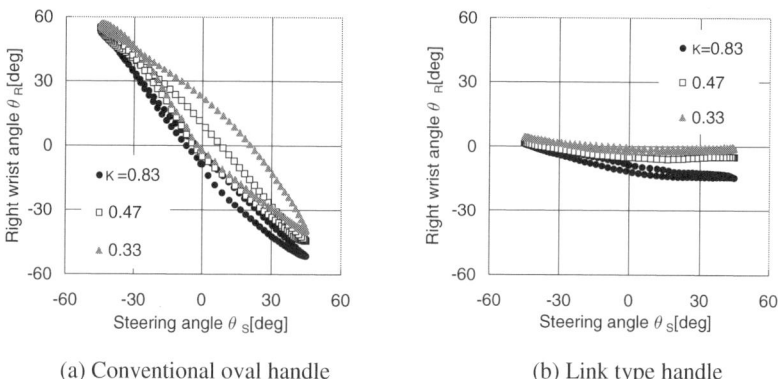

(a) Conventional oval handle (b) Link type handle

Figure 3: Relationship between steering angle and right wrist angle

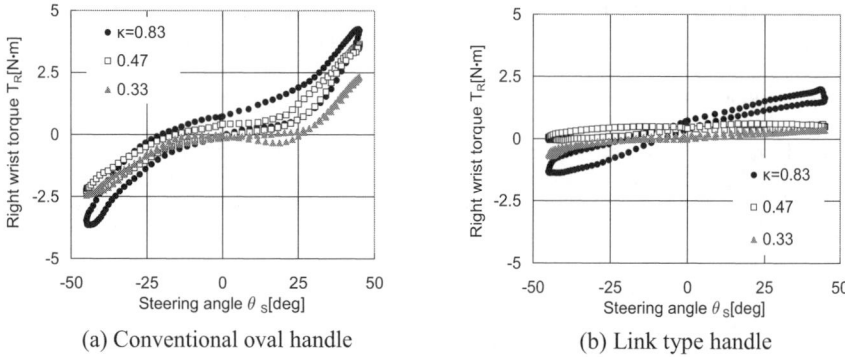

(a) Conventional oval handle (b) Link type handle

Figure 4: Relationship between steering angle and right wrist torque

DISCUSSION

Most electric carts have employed oval steering handle mechanism. On the conventional oval handle, bending motion of wrist was forcused as a major problem. On the other hand, the amount of arm motion was substantially reduced by using link type handle. Therefore, new type handle was efficient not only to decrease the burden, but also to make operator's operation convent. Finally, it was become clear that there was little bending of wrist and no movement of center of gravity of body by the use of link type handle. Therefore, a seating position posture always would be held in the stable state, and comfortable operation could be performed by improvement in operativity.

CONCLUSION

The link type handle mechanism, which was proposed in this report, has not required the rotation of wrist which causes the large torque at the wrist. Since a seating position posture has been held in the natural and stable state, it turns out that run stability could be improved.

ACKNOWLEDGEMENT

Some parts of this report have been helped by the research fund on "Living Assistance Technology for Improvement of Quality of Life" supported by ORDIST (Organization for Research and Development of Innovative Science and Technology) in Kansai University.

REFERENCES

Mussa-Ivaldi FA, Hogan N, Bizzi E (1985). Neural, mechanical and geometric factors subserving arm posture in humans. J Neurosci **1985:5**, 2732-2743

Dolan JM, Friedman MB, Nagurka ML (1993) Dynamic and loaded impedance components in the maintenance of human arm posture. IEEE Trans Sys Man Cybern **1993:23**, 698-709

Tsuji T, Morasso PG, Goto K, Ito K (1995). Human hand impedance characteristics during maintained posture. Biol Cybern **1995:72**, 475-485

Flash T, Mussa-Ivaldi F (1990) Human arm stiffness during the maintenance of posture. Exp Brain Sci **1990:82**, 315-326

QUANTIFICATION OF DYNAMIC WALKING STABILITY OF ELDERLY BY USING NONLINEAR TIME-SERIES ANALYSIS AND SIMPLE ACCELEROMETRY

Y. Ohtaki[1], M. Arif[2], A. Suzuki[3]
K. Fujita[4], R. Nagatomi[5], H. Inooka[1]

[1]New Industry Creation Hatchery Center, Tohoku University,
6-6-04 Aza Aoba, Aramaki, Aoba-ku, Sendai 9808579, JAPAN
[2]Pakistan Institute of Engineering and Applied Science, PAKISTAN
[3]Instruments Technology Research Co. Ltd., Sendai, JAPAN
[4]Center for Preventive Medicine and Salutogenesis, Tohoku Fukushi University, Sendai, JAPAN
[5]Graduate School of Medicine, Tohoku University, Sendai, JAPAN

ABSTRACT

This study presented a technique to assess dynamic walking stability utilizing a nonlinear time-series analysis and a portable instrument. Main objective was to investigate its usefulness in the assessment of elderly walking. The method was consisted of measurement of three-dimensional acceleration of the upper body, and estimation of the Lyapunov exponents, thereby directly quantifying local dynamic stability while walking. Straight level walking of young and elderly subjects was investigated in the experimental study. Effects and efficacies of the interventions for the elderly were demonstrated by the proposed method. The experimental results suggested that the method was useful in revealing degree of improvements on the walking stability.

KEYWORDS

Gait analysis, Walking stability, Portable instrument, Acceleration, Nonlinear time-series analysis, Aging, Medical application.

INTRODUCTION

Falls cause a serious hazard to elderly people. Impaired mobility due to injuries or a fear of falling diminish a person's ability to perform activities of daily living (Maki *et al.* 1991). Hip fracture due to fall accidents amounts to more than 10 % of bed-bound status of the elderly in Japan.

Although falling is a result of complex and multi-factorial problem, lack of postural control is one of the major contributing factors. Aging effects on the sensory feedback have been hypothesized to be a key factor in adjusting posture to maintain their balance against unpredictable external or internal variations of gait. In addition, recent randomized controlled trials that have tested the effectiveness of the intervention for elderly have indicated that exercise training significantly increase their aerobic capacity and muscle strength, which might result in improvement of the postural stability.

Conventionally, clinicians have been assessed personnel walking ability based on performance of static balance tests and measure of simple gait factors (walking speed, cadence, step length, etc.), mostly focusing on quantifying regional amount of body sway, variability of gait factors or joint angles. Those methods provide a practical evaluation, however, the measure of static balance or gait variability itself does not mean that of dynamic stability of walking. Dynamic stability represents a resilient ability to maintain certain continuous cyclic movement by accommodating internal or external perturbations (Hurmuzlu et al. 1994). On the other hand variety of instruments have been used to quantify walking characteristics in a more precise manner, by means of the video-based motion capture system, goniometry, or force plates. However, those methods requires considerable setups, then limited to laboratorial environments. Recently, mechatronics progress made it possible to realize small and low power consumptive accelerometry as a testing tool applicable in the field of medical therapy(Aminian et al. 2002, Ohtaki et al. 2001, 2005). Some advanced algorithms have been also proposed to evaluate gait performances and dynamic walking stability basing on a simple accelerometry(Dingwell et al. 2000, 2001, Buzzi et al. 2003, Arif et al. 2004). Further application of those method to a physical assessment is strongly required to enhance efficiency and effectiveness of interventions. Nevertheless, it is still insufficient to investigate subsequent improvements on walking abilities in terms of the stability of dynamical system.

This study was intended to present a practical method to assess walking stability by using a portable instrument, then to investigate its usefulness in the physical assessment for elderly people. The method employed a measurement of three-dimensional acceleration of the body, and an application of nonlinear time-series analysis which directly assess stability of the dynamical system. Straight level-walking of young and elderly subjects were investigated in the experiment. Moreover, its feasibility in assessing effects and efficacies of the five-month interventions including aerobic exercise training was investigated.

METHODOLOGY

In this study, we focused on local dynamic stability which is defined as a sensitivity of the dynamical system to small perturbations in gait variability which produced by one's locomotor system itself. Lyapunov exponent estimation was applied to evaluate the local dynamic stability of walking. Firstly, state space was reconstructed from the obtained acceleration data after determining appropriate time delay and embedding dimension:

$$y(t) = (x(t), x(t+\tau), \cdots, x(t+(d-1)\tau)). \tag{1}$$

Where, $y(t)$ is the d dimensional state vector, $x(t)$ is the original acceleration data, τ is the time delay, and d is the embedding dimension. A Schematic representation of the reconstruction process was shown in Figure 1. A valid state space must include a sufficient number of coordinates to unequivocally define the state of the attractor trajectories. Time delay τ was determined as a time when autocorrelation coefficient of the data gets lower than the reciprocal value of natural

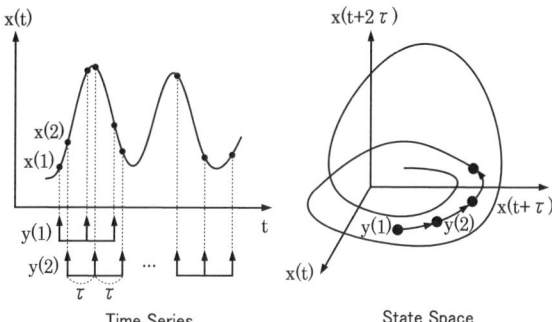

Figure 1: Reconstruction of a attractor trajectory in the state space (in case of the embedding dimension $d = 3$).

log. Embedding dimension d was determined by using the global false nearest neighbors algorithm (Cao 1997). In our case, the embedding dimension was four, to form a valid state space.

Lyapunov exponent quantifies the average exponential rate of divergence of neighboring trajectories in a reconstructed state space. The estimation of the largest Lyapunov exponent performed with the method proposed by Kantz (Kantz 1994). The Lyapunov exponent λ was defined as the following.

$$D(\Delta t) = D(0)e^{\lambda \Delta t}. \qquad (2)$$

The notation $D(\Delta t)$ denotes the displacement between neighboring trajectories after The notation Δt interval. $D(0)$ is the initial distance between neighboring point. Lyapunov exponent λ quantifies the average exponential rate of divergence of neighboring trajectories in a reconstructed state space. A Higher value of Lyapunov exponent indicates a larger divergence of the attractor in the state space, suggesting less stability of the dynamical system. We calculated the exponent from ten steps acceleration data in steady state of the walking trial. Data were analyzed without filtering to avoid complications associated with filtering nonlinear signals.

EXPERIMENT

We developed a portable device consisted of monolithic IC accelerometers (±2 G, ADXL202E; Analog Devices Inc., MA, USA) with 16-bit duty cycle converter, Li-Ionic batteries, micro processor units and CompactFlash card. This equipment is small (100×55×18.5 mm) and lightweight enough to carry without any restriction. The equipment was attached to the center of lower back representing the center of gravity of the body using a back supporter, as shown in Figure 2. Three-dimensional acceleration as lateral, vertical, and anteroposterior direction were measured by the portable equipment with sampling frequency of 100 Hz.

Seven healthy young adults (25.0 ± 1.6 yr.) and fifty-four elderly adults aged (76.7 ± 4.6 yr.) participated in the experiment. All subjects gave signed informed consent. Prior to the experiment, physical conditions and exercise habit were examined by questionnaires. The subjects were instructed to walk at their self-selected speed on a 16 m straight track without any restriction. The beginning and the end of the strait track 3 m were considered as transition phases of the walking. Constant walking phase in middle 10 m of the track was applied to the calculation.

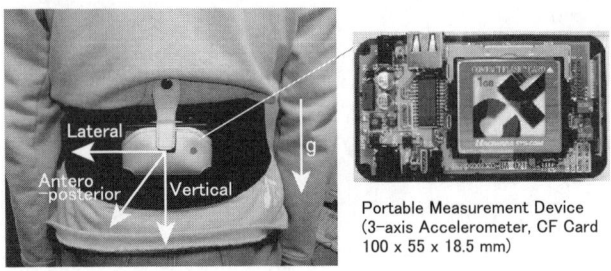

Figure 2: View of a subject wearing the portable instrument by using a back supporter.

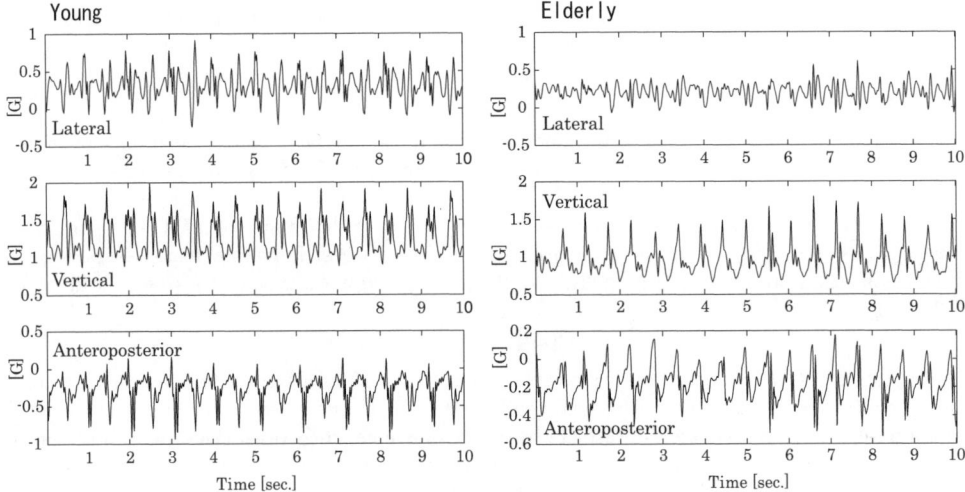

Figure 3: Typical examples of accelerations as measured in healthy young subject and an elderly before the intervention.

Firstly, we investigated the dynamic stability in comparison with the young and the elderly subjects. Secondly, we demonstrated efficacy and effectiveness of the intervention for elderly, quantifying degree of improvements on the walking stability. The intervention program was designed for elderly living in the community through the continuous five-month training conducted by medical doctors and physical therapists. The training program was intended to improve aerobic and physical performance by strengthening the muscular group involved in standing and walking. Subjects attended two-hour classes once a week.

RESULT

Figure 3 shows typical examples of acceleration waveforms as measured in healthy young subject (left) and an elderly subject before the intervention (right). The elderly presented clear fluctuations on the waveforms, resulted in large value of the Lyapunov exponents. Figure 4 illustrates average values of the estimated Lyapunov exponents comparing with the young and the elderly in the direction of lateral, vertical, and anteroposterior respectively. The lower and upperlines of the

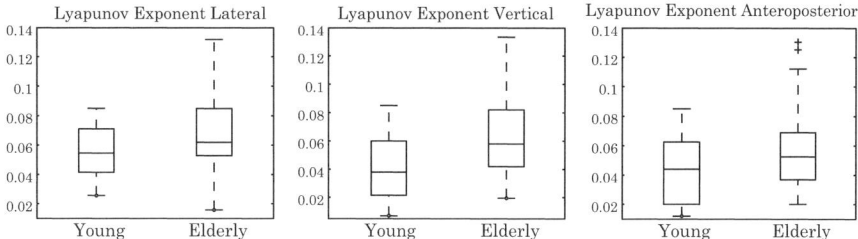

Figure 4: Comparison of Lyapunov exponents between young and all elder subjects. The lower and upperlines of the box are 25th amd 75th percentiles.

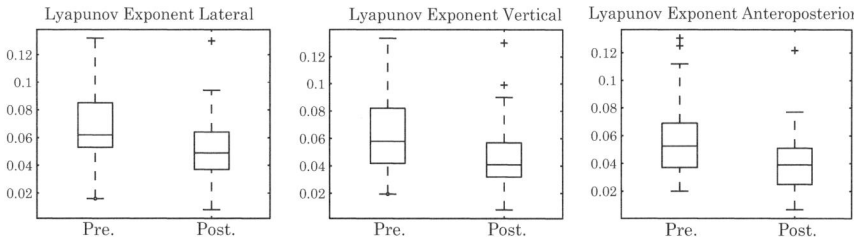

Figure 5: Comparison of Lyapunov exponents between the pre-intervention and the post-intervention subject in the elderly.

box are the 25th and 75th percentiles. The line in the middle of the box is the median. The wiskers shows the extent of the rest of the data. Elderly subjects generally exhibited higher value indicating much instability in all direction, but no statistical significance was observed except in the vertical direction ($p < 0.05$). Figure 5 shows the average value of the estimated Lyapunov exponent comparing with the pre-intervention and the post-intervention in elderly subjects. The post-intervention illustrates significantly smaller value of the exponent in all direction ($p < 0.05$). The result suggested that the method feasibly reveals the effects of the interventions on the improvement of walking stability in elderly.

In the experiment, a short walking distance was chosen to avoid effects of fatigue from elderly persons' walking. It is important to mention that estimation of Lyapunov exponents is sensitive to the data size and the observation time. Therefore, estimation accuracy of Lyapunov exponents was rather low in this study. However, we quantified the exponential rate of divergence of trajectories, which followed trends of Lyapunov exponents. The proposed method was adequate to quantify the nature of the dynamic system while walking. A quantitative measure of the walking stability may provides an essential tool for assessing personnel risk of falls, designing proper treatments, and monitoring progress and efficacy of the intervention.

CONCLUSION

This study presented a technique for assessing dynamic stability of walking using nonlinear time-series analysis with a portable instrument. This method is easily applicable and reliable in the clinical field and daily situations. The experimental results suggested that the proposed method quantify degree of improvements in walking stability, which contributes to ascertain the effectiveness of exercise intervention for elderly. Further application of the present technique may help predicting personal risk of falls.

ACKNOWLEDGMENT

We are grateful to Dr.I.Tsuji at Graduate School of Medicine, Tohoku University, Sendail Silver Center, and Miyagi Physical Therapist Association for their cooperation in our study. This research is grant aided by Japanese Ministry of Education, Culture, Sports, Science and Technology.

References

[1] Aminian K., Najafi B., Bula C., Leyvraz P. F., Robert P. H. (2002). Spatio-temporal Parameters of Gait Measured by An Ambulatory System Using Miniature Gyroscopes. *Journal of Biomechanics* **35:5**, 689-699.

[2] Arif, M., Ohtaki, Y., Nagatomi, R., and Inooka, H. (2004). Estimation of the Effect of Cadence on Gait Stability in Young and Elderly People using Approximate Entropy Technique, *Measurement Science Review* **4**, 29-40.

[3] Buzzi, U. H., Stergiou, N., Kurz, M. J., Hageman, P. A., and Heidel, J. (2003). Nonlinear dynamics indicates aging affects variability during gait. *Clinical Biomechanics* **18**, 435-443.

[4] Cao, L. (1997). Practical Method for Determining The Minimum Embedding Dimension of A Scalar Time Series. *Physica* **D:110**, 43-50.

[5] Dingwell J. B., Cusumano J. P., Sternad D., Cavanagh P. R. (2000). Slower Speeds in Patients with Diabetic Neuropathy Lead to Improved Local Dynamic Stability of Continuous Overground Walking. *Journal of Biomechanics* **33**, 1269-1277.

[6] Dingwell J. B., D. and Cavanagh P. R. (2001) Increased variability of continuous overground walking in neuropathic patients is only indirectly related to sensory loss. *Gait and Posture* **14**, 1-10.

[7] Hurmuzlu Y., Basdogan C. (1994). On the Measurement of Dynamic Stability of Human Locomotion. *Journal of Biomechanical Engineering* **116**, 30-36.

[8] Kantz, H. (1994). A robust method to estimate the Lyapunov exponent of a time series. *Physics Letters* **A:185**, 77-87.

[9] Maki B. E., Holliday P. J., Topper A. K. (1991). Fear of Falling and Postural Performance in The Elderly. *Journal of gerontology: Medical sciences* **46**, 123-131.

[10] Ohtaki Y., Sagawa K., Inooka H. (2001). A Method for Gait Analysis in A Daily Living Environment Using Body-mounted Instruments. *JSME International Journal C* **44:4**, 1125 - 1132.

[11] Ohtaki Y., Susumago M., Suzuki A., Sagawa K., Inooka H. (2005) Automatic Classification of Ambulatory Movements and Evaluation of Energy Consumptions Utilizing Accelerometers and a Barometer. *Journal of Microsystem Technologies* **11:8-10**, 1034-1040.

DEVELOPMENT OF A WELFARE ROBOT BASED ON RECS CONCEPT
- TASK OF SETTING A MEAL ON THE TABLE -

T. Yamaguchi, H. Kawano, M. Takano and S. Aoyagi

Systems Mangement Engineering, Kansai University
3-3-35, Yamate-cho, Suita, Osaka 564-8680, Japan

ABSTRACT

In this paper, a home robot is developed based on RECS (Robot Environment Compromise System) concept. This concept aims to share the technical difficulties with robot and environment. In other words, RECS modifies the environment in order that the robot task would be possible or easy. Among the various domestic tasks, setting a meal on the table is focused. This task comprises preparing a meal and loading tableware with meals to a tray at a kitchen, transporting this tray to the table, putting the tableware on the table, and pouring bottled drinks to cups on demand.

KEYWORDS

RECS concept, Setting a meal, Robot, Image processing, Barcode mark

INTRODUCTION

An aging society is progressing at high speed in advanced countries, especially in Japan. In the middle in the twenty-first century, it is predicted that two or three adult persons should support one aged person, and the care workers are apparently in short supply. As one of the countermeasure to this, technical assistance by a robot is much expected. The present level of the robot technology is not sufficient for performing multiple complicated tasks autonomously. Considering this, authors have already proposed RECS (Robot Environment Compromise System) concept[1] as shown in **Fig. 1**. This concept aims to share the technical difficulties with robot and environment. In other words, RECS modifies the environment in order that the robot task would be possible or easy. This modification should be minimized to the extent that the human beings who live and work in the environment do not feel uncomfortable and are not obstructed.

SYSTEM OF WELFARE ROBOT

A welfare and home robot based on RECS concept is developed as shown in **Fig. 2**. This robot comprises a mobile platform and a 7 DOF articulated manipulator arm on it. This platform is equipped

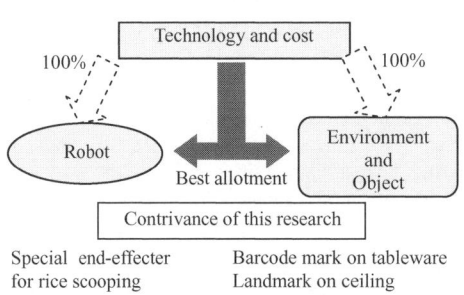

Figure 1: RECS concept

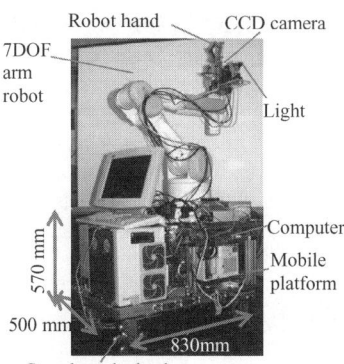

Figure 2: Welfare and home robot

with four star-shaped wheels, which assists this robot to step over a slight bump or level differences in the environment[1]. A CCD camera with a light is attached to the tip of the robot arm. All driving mechanisms of the platform and the arm are controlled by a computer on the robot. Localization of the robot in the indoor environment[2] and recognition of the tableware by image processing[3] are also carried out by this computer.

PREPARING MEAL BY ROBOT

In order to realize the task of filling rice into a bowl, a special rice scoop for a robot is developed as shown in **Fig. 3**. Full length of this tool is about 310 mm and width of it is about 100 mm. The head of it can be linearly expanded and contracted within the range of about 50 mm by utilizing a spring inside it. A robot can scoop up cooked rice in easy operation owing to this flexible mechanism, namely the robot needs only to put this special scoop into the rice and move it horizontally. Irrespective of residual amount of the rice in the rice cooker, the robot is allowed to take down this scoop to the same height always as shown in **Fig. 4**. Furthermore, the rice can be carried without falling, since it is contained in a bag shaped cloth. After the scoop is positioned over a bowl, the robot only needs to turn the scoop upside down for filling the rice into the bowl. The robot can realize the fine round shape of the rice only by pushing a cup shown in **Fig. 5** into the rice.

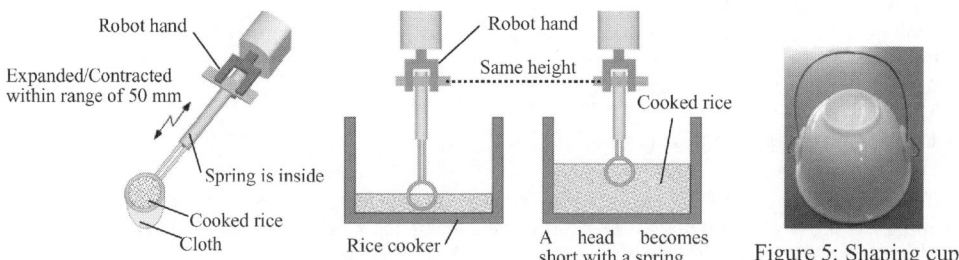

Figure 3: Rice scoop for robot   Figure 4: How to use a rice scoop for robot   Figure 5: Shaping cup

SETTING TABLEWARE BY ROBOT

Based on RECS concept, four barcode marks are given to the edge of the tableware as shown in **Fig. 6**. Since this mark is made of retro-reflective material, it is conspicuous only when illuminated by a light.

A picture taken by a CCD camera and binarized with an appropriate threshold level when illuminated is shown in **Fig. 7**. Only marks are conspicuous and the image processing becomes drastically easy compared with processing the raw image data. A robot can recognize the type of the tableware by reading the barcode information. Also the robot can calculate the position of the tableware by utilizing image data of four marks. By using four marks, the reliability of reading barcode is also improved.

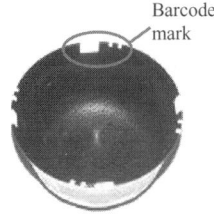

Figure 6: Tableware of this research

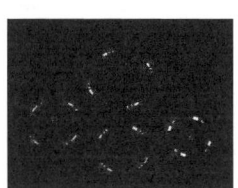

Figure 7: Binary image

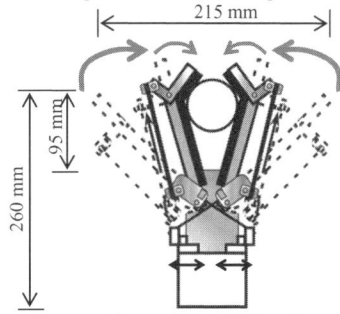

Figure 8: Robot hand

A finger of robot hand is not able to put into inside of the tableware. And, the size of tableware is various. To solve these problems, a robot hand is developed as shown in **Fig. 8**. This hand is equipped with a link mechanism, and the tip of a finger is bent automatically so as to be fitted to the object outer shape. Owing to this mechanism, the robot hand can hold an object firmly with four contact points.

EXPERIMENT

The task of preparing rice by using the developed rice scoop for a robot is carried out. The situation of this task is shown in **Fig. 9**. The contents of the task are following (1)-(5):
(1) The rice scoop is put into the rice in a rice cooker.
(2) The rice scoop is moved horizontally and the rice is scooped up.
(3) The rice is transported and positioned over the bowl.
(4) The rice scoop is turned upside down and the rice is filled into the bowl.
(5) The rice is roundly shaped by the shaping cup.
Cooked rice is fairly filled into a bowl as shown in Fig. (6). From this result, it is proved that this method is effective for preparing the rice.

The task of setting tableware is carried out by the robot as shown in **Fig. 10**. The contents of the task are following (1)-(6):
(1) At a kitchen, the robot takes a picture of the tableware and recognizes it.
(2) The robot grasps the tableware by the developed robot hand with link mechanism.
(3) The robot loads the tableware to a tray.
(4) The robot transports the tray to the table. Landmarks on ceiling are employed for localizing the robot.
(5) The robot sets the tableware on the table.
(6) The robot pours bottled drinks to a cup.
Seeing the series of performed tasks shown in **Fig. 10**, it seems possible in future that the task of setting a meal on the table is realized by applying RECS concept.

CONCLUSION

As a part of development of a welfare robot based on RECS concept, the task of setting a meal on the

Figure 9: Situation of the task with a rice scoop for a robot

Figure 10: Situation of the setting tableware by a robot

table is focused in this study. The summary is as follows:
1) A special rice scoop for a robot is developed, and the method of preparing rice by using this tool is proposed.
2) The method of setting tableware on the table is proposed. An image processing method of the tableware using barcode marks is developed. A special grasping robot hand is also developed.
3) Experiments of both preparing a meal and setting tableware are carried out. The results show the good possibility of applying RECS concept to these tasks.

ACKNOWLEDGEMENT

This work was partially supported by JSPS (Japan Society for the Promotion of Science).KAKENHI (16310103), MEXT (Ministry of Education, Culture, Sports, Science and Technology).KAKENHI (17656090), the Kansai University Special Research Fund, 2004 and 2005.

REFERENCES

1) Takano M., Yoshimi T., Sasaki K. and Seki H. (1996). The development of the inside movement robot system based on the RECS concept. *Journal of Japan Society for Precision Engineering*, **62:6**, 1334-1338.
2) Aoyagi S., Kiguchi Y., Tsunemine K. and Takano M. (2001). Position and orientation measurement of a mobile robot by image recognition of simple barcode landmarks and compensation of inclinations. *The Journal of The Institute of Electrical Engineers of Japan*, **C121:2**, 375-384.
3) Aoyagi S., Kinomoto K., Ieuji S. and Takano M. (2000). The recognition and handling of the tableware by the robot based on the RECS concept. *The Journal of The Institute of Electrical Engineers of Japan*, **C120:5**, 615-624.

DEVELOPMENT OF A MOBILITY AID FOR THE VISUALLY IMPAIRED USING A HAPTIC FORCE GENERATOR

T. Ikeda [1,2], H. Matsuda [1], Y. Shiota [1], K. Sakamoto [2] and Y. Shimizu [2]

[1]Department of Rehabilitation Engineering, Polytechnic University
[2]Systems Engineering, University of Electro-Communications

ABSTRACT

This paper describes the development of a "virtual cane," a kind of walking aid for the visually impaired, using a haptic force generator consisting of a gyroscope and its controller. It is shown theoretically that the haptic force generator can change the superficial moment of inertia of a rod. Good agreement between the analytical and experimental results indicates that the haptic force generator provides one of the essential functions of the virtual cane.

KEYWORDS

virtual cane, visually impaired, haptic illusion, gyroscope, haptic force generator

INTRODUCTION

The white cane is currently used by the visually impaired as a mobility aid. Visually impaired individuals often face difficulty in using the white cane in places such as supermarkets, where products or other people often present barriers to free movement. Several mobility aids have been developed in order to overcome this problem(Funakubo and Hatsuyama, 1995). Such devices detect obstacles by means of infrared rays or ultrasonic waves and indicate the presence of obstacles by vibration. Although these devices are useful, thorough training is required to become skilled in the use of such devices. If a device reproduces the inner force sense to the same degree as would be perceived using a white cane, then the user can visibly perceive the existence of obstacles without contacting them. We herein refer to such a device as a "virtual cane."

The main component of the virtual cane is a haptic force generator. This paper focuses on a haptic generator using a gyroscope. It is shown theoretically that the haptic force generator can change the superficial moment of inertia of a rod. This means that the rod length perceived by grasping and wielding a rod can be changed, because the perceived length of a rod is proportional to its moment of inertia(Turvey, 1996).

SUPERFICIAL MOMENT OF INERTIA

One can perceive the length of a rod by simply grasping one end of the rod and gently wielding it, without even seeing the rod. This kind of touch is referred to as "dynamic touch"(Turvey, 1996). It has been reported that the perceived length of a rod by dynamic touch increases exponentially with the moment of inertia of the rod(Chan, 1994). These results imply that perceived length can be changed freely by varying the moment of inertia of the rod. In the following, we demonstrate that it is possible to vary the moment of inertia of a rod superficially, without actually varying its shape or mass distribution.

Consider a rod that is grasped by one end and wielded around the wrist joint, which is assumed as the origin. If the external torque $\mathcal{T}_{gX}$ is applied to the rod in addition to the torque applied by the wrist, $\mathcal{T}_X$, as shown in Figure 1(a), then the equation of motion of the rod is given as

$$\mathcal{T}_X - mgh\cos\phi + \mathcal{T}_{gX} = I_0 \dot{\Omega}_X. \tag{1}$$

where m is the mass of the rod, g is the acceleration due to gravity, h is the center of the mass of the rod, I_0 is the moment of inertia of the rod, and ϕ and $\Omega_X(=\dot{\phi})$ are the rotation angle and angular velocity, respectively, of the rod rotating around the wrist joint.

Next, the external torque $\mathcal{T}_{gX}$ is generated so as to be proportional to the angular acceleration $\dot{\Omega}_X$ as

$$\mathcal{T}_{gX} = -I_s \dot{\Omega}_X \tag{2}$$

where I_s is an arbitrary constant. Equation 1 reduces to

$$\mathcal{T}_X - mgh\cos\phi = (I_0 + I_s)\dot{\Omega}_X. \tag{3}$$

Comparing Eqn. 3 and Eqn. 1, the moment of inertia changes from I_0 to $I_0 + I_s$ by applying the external torque, $\mathcal{T}_{gX}$, according to Eqn. 2. We refer to the moment of inertia, I_s, as the superficial moment of inertia. Thus, the external torque $\mathcal{T}_{gX}$, satisfying Eqn. 2, can be generated.

It is desirable to design the virtual cane as a non-installing and non-equipping device in order to make it possible to use the virtual cane in various locations. This requirements of the design can be fulfilled using a gyroscope. Thus, external torque $\mathcal{T}_{gX}$ is generated by a haptic force generator consisting of a gyroscope.

Figure 1(b) shows the haptic force generator represented as a simplified model consisting of a rod and a rotor. The rotor is attached firmly to the rod, but can be tilted in the plane perpendicular to the rod axis. The external torque, $\mathcal{T}_{gX}$, generated by the spinning rotor can be derived using the coordinate systems having a common origin and the Euler angles (ϕ, θ, ψ) shown in the figure. The XYZ-coordinate system is the global coordinate system, in which the external torque $\mathcal{T}_{gX}$ is represented as the component along the X axis. It is assumed without any loss of generality that the device is wielded in the YZ-plane of the global coordinate system. The x'y'z'-coordinate system is obtained by a rotation of the XYZ-coordinate system about the X-axis through an angle ϕ. Consequently, the y'-axis of the x'y'z'-coordinate system coincides with the central axis of the rod. The x''y''z''-coordinate system, in which the z''-axis coincides with the rotor axle,

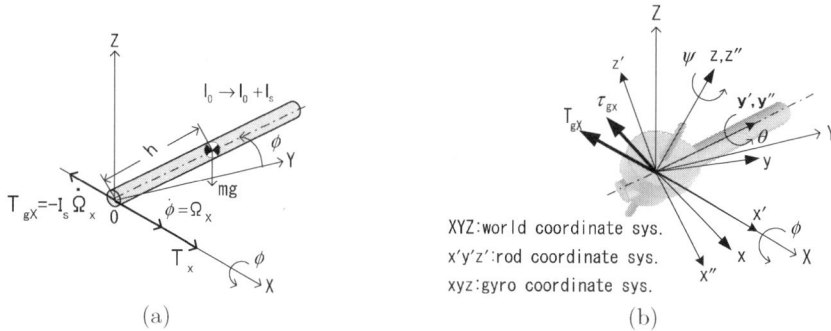

Figure 1: Rod rotated by applying torque. (a) External torque $\mathcal{T}_{gX}$, in addition to $\mathcal{T}_X$; (b) External torque $\mathcal{T}_{gX}$ generated by the gyroscope

is obtained by a rotation of the x'y'z'-coordinate system about the y'-axis through an angle θ. Similarly, the xyz-coordinate system is obtained by a rotation of the x''y''z''-coordinate system about the z''-axis through an angle ψ. The derivatives of the Euler angles with respect to time, $\dot{\phi}$, $\dot{\theta}$ and $\dot{\psi}$, represent angular velocities about the X-, y'- and z''-axes respectively.

Since the rotational axis of the rotor coincides with the z-axis of the xyz-coordinate system, the angular momentum of the spinning rotor has a nonzero component only in the z-axis. The angular momentum $\boldsymbol{L}_g$ is given by

$$\boldsymbol{L}_g = (0, 0, I_g \dot{\omega}_g)^T \tag{4}$$

where I_g is the moment of inertia of the rotor and $\dot{\omega}_g$ is the angular acceleration of the rotor.

Tilting the spinning rotor generates a gyro-moment $\boldsymbol{\tau}_g$, as expressed by the following equation:

$$\boldsymbol{\tau}_g = \boldsymbol{L}_g \times \boldsymbol{\omega}. \tag{5}$$

where $\boldsymbol{\omega}$ is the angular velocity represented in the xyz-coordinate system.

Substituting Eqn. 4 into Eqn. 5 and then transforming coordinate systems, the gyro-moment in the XYZ-coordinate system, $\boldsymbol{\mathcal{T}}_g$, is obtained as follows:

$$\begin{pmatrix} \mathcal{T}_{gX} \\ \mathcal{T}_{gY} \\ \mathcal{T}_{gZ} \end{pmatrix} = I_g \omega_g \begin{pmatrix} -\dot{\theta} \cos\theta \\ \dot{\phi} \cos\phi \cos\theta + \dot{\theta} \sin\phi \sin\theta \\ \dot{\phi} \sin\phi \cos\theta + \dot{\theta} \cos\phi \sin\theta \end{pmatrix}. \tag{6}$$

The first component of the gyro-moment $\mathcal{T}_{gX}$ in Eqn. 6 gives the external torque satisfying Eqn. 2. From Eqn. 2 and Eqn. 6, the following relation is obtained:

$$I_s \dot{\Omega}_X = I_g \omega_g \frac{d}{dt} \sin\theta. \tag{7}$$

Equation 7 provides one method of controlling the tilt angle θ of the spinning rotor in order to generate the external torque $\mathcal{T}_{gX}$ satisfying Eqn. 2. For this method, a desired superficial moment of inertia, I_s, is given as follows:

$$\theta = \sin^{-1}\left(\frac{I_s}{I_g \omega_g} \int \dot{\Omega}_X \, dt\right), \tag{8}$$

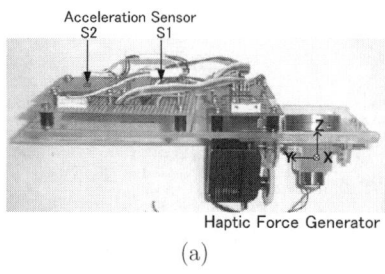 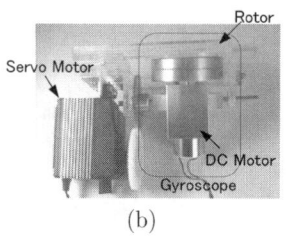

Figure 2: Experimental device for haptic illusion (size: 80 × 225 × 88 mm, weight: 0.386 kg). (a) Overall view; (b) Haptic force generator

This equation indicates that the method based on Eqn. 8 requires measurement of the angular acceleration $\dot{\Omega}_X (= \ddot{\phi})$ of the rod.

DEVICE FOR HAPTIC ILLUSION

Figure 2 shows an experimental device for haptic illusion of rod length. The device consists of a haptic force generator consisting of a gyroscope, as shown in Figure 2(b), and two three-axial acceleration sensors of the piezo-resistance type (H560, Hitachi Metals, Ltd.) to detect the angular acceleration $\dot{\Omega}_X$. The haptic force generator consists of a gyroscope and a precession controller. The DC motor rotates the rotor of the gyroscope at high speed. The precession controller consists of a servo motor and a pulse width modulation circuit to control the servo motor. The precession of the gyroscope is applied by the servo motor. By grasping the part of the haptic force generator and wielding it around the X-axis, the gyro-moment is produced around the X-axis.

An experiment was performed in order to distinguish differences in superficial moment of inertia. The subjects were able to easily discriminate changes in the superficial moment of inertia produced by the device. However, subjects were not able to perceive changes in the superficial moment of inertia as a change in the haptic illusion of rod length. A great deal of time is required in order to correlate a change in the superficial moment of inertia and a change in the haptic illusion of rod length because the perceived length is affected by the complicated shape of the device.

CONCLUSIONS

This paper described the development of a haptic force generator using a gyroscope as a non-installing and non-equipping device. It was shown theoretically that the haptic force generator was able to superficially vary the moment of inertia of a rod. A device for haptic illusion of rod length was developed based on this theory. Good agreement between the analytical and experimental results indicates that the haptic force generator provides one of the essential functions of the virtual cane.

REFERENCES

Funakubo H. and Hatsuyama Y. (1995). *Assistive Technology*, Sangyo-Tosho.
Turvey M.T. (1996). Dynamic touch. *American Psychologist* **51:11**, 1134–1152.
Chan T.C. (1994). Haptic perception of partial-rod lengths with the rod held stationary or wielded. *Perception and Psychophysics* **55:5**, 551–561.

STUDY OF THE DESIGN METHOD OF AN ANKLE-FOOT-ORTHOSIS

Yasuhiro Mine[1], Takamichi Takashima[2], Hiroshi Fujimoto[3]

[1] National Vocational Rehabilitation Center
(The Graduate School of Human Sciences, Waseda University)
4-2 Namiki, Tokorozawa-City, Saitama, 359-0042 Japan
[2] National Rehabilitation Center for Persons with Disabilities
4-2 Namiki, Tokorozawa-City, Saitama, 359-0042 Japan
[3] The Faculty School of Human Sciences, Waseda University
2-579-15 Mikajima Tokorozawa-City, Saitama 359-1192 Japan

ABSTRACT

Ankle-Foot Orthoses (AFO) are orthotics for patients having pathological gait represented by the stroke. In this study three-dimensional digital data from a physical object was extracted by using 3D-digitizer. If we have a 3D digitized orthotic model, the shape can be modified easily and Finite Element Analysis (FEA) and Stereo-Lithography (Rapid-Prototyping System:RPS) can be easily done. Continuing the research, we then focused on patients suffering from hemiplegics stroke in order to develop a method of fabricating the AFO, which fit patients even more. The major purpose of this study is to produce an orthotic design method for fitting each patient by measuring spastic muscle action and by deducing parameters for an orthosis from the data. We will have to consider how the outcomes of measurements should be processed as parameters in the design of 3DCAD.

KEYWORDS

Spasticity, Hemiplegic gait, Rehabilitation, Orthosis, 3Ddesign, Measurement

INTRODUCTION

Ankle-Foot Orthoses (AFO) are orthotics for the patients who have the pathological gait represented by the stroke. They were using for the rehabilitation approach, and they support the pathological gait. The conventional method of fabricating the plastic AFO as follows: a plaster cast of the leg is made, then a thermoplastic resin is molded, with the basic form of the orthosis shaped by trimming, and finally the fixtures are attached. The orthosis is adjusted by means of having the patient wear the orthosis. In this study, three-dimensional digital data from a physical object was extracted using a 3D-digitizer. The process of digital scanning is made possible when some control lines are drawn on the plaster model, and then the model is scanned along these control lines. If we have a 3D digitized orthotic model, the shape can be modified easily and FEA and RPS can be easily done.

For example, The 3D-data of orthosis for the FEA provide a surface model. The element type used is the Parabolic Triangle. Boundary conditions of load and restraint are determined at the heel-off point in the gait cycle. Because of the size limit in the table of the stereo-lithography (Rapid Prototyping System), we made models of the AFOs that were 50% in size. It took 12 hours to produce 2 models. This time, although the resin like the property of polypropylene is being used for the orthosis of RPS, it is necessary to develop a synthetic resin whose toughness and tenacity are fit for orthosis. The shapes of AFO have been conventionally modified in such a way as to provide rigidity to the orthoses through the inclusion of several cords under the plastic seat for stress relaxation. Three-dimensional digitalized data, in contrast, can modify the shapes by using the Sweep Function. Permeability to air, which has been sought after, can be obtained by this new method. hemiplegic stroke tend to show involuntary movements and muscle spasms called "Clonus" in response to external stimulation. The attachment of AFO is known to alleviate those spasms, but the process has not been elucidated fully yet. When an orthosis is prescribed, the Brunnstrom method is applied in evaluating the grades of paralysis. But the grade of paralysis is not a factor in the manufacture of the orthosis. And it is hard to say that it is an orthosis suitable for the patient. The major purpose of this study is to be reflected in the design method of the orthostic fitting each patient and to be quantified by measuring spastic muscle action.

Methods

In order to measure the gait characteristics of spastic muscle action with hemiplegic stroke, the Spastic Measurement Orthosis (SMO) was made. The measuring method was as follows: a load cell was attached to the back of an orthosis, the ankle-joint was kept at 90 degrees by walking, and the force weighing on the circumference of the ankle was measured. The experiment was performed by the combination of 3-D motion capture (VICON512 OXFORD METRIX) and force platforms (AMTI) and the SMO. This experiment system is shown in Fig.1. The load-cell can measure the force of ankle plantarflexion and ankle dorsiflexion in the circumference of an ankle during gait. The signal from the force platforms of the VICON was time-synchronized with the load data taken in the PC. Kinematical data were obtained from a set of a reflective marker sampling at 60 Hz placed on the ankle joint of the orthosis. The subjects were patients with stroke with spastic paralysis as well as healthy persons. Each subject was instructed to walk on the force platform.

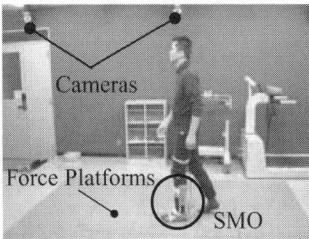

Figure1: The experiment system

Results

The situation of the load-cell attached to the back of the orthosis during walking is as follow. If the ankle joint is plantarflexion, the load-cell measured the compressive loads (the negative loads). If the ankle joint is dorsiflexion, the load-cell measured the tensile loads (the positive loads). In the experiment, a normal gait and a spastic gait imitated by the healthy person were performed. The spastic gait is to be imitated the paralysis of ankle plantarflexion due to strokes. Fig.2 shows the measurement results for the normal gait, and for the spastic gait. The value of the vertical axis shows the compressive load and the tensile load in the gait cycle. This graph shows two-gait cycles of the stance phase and the swing phase. The peak of the load applied to the circumference of the ankle joint in the stance phase in normal gait was 28kgf of ankle Dorsiflexion; on the other hand, the peak of the load in the spastic gait was 8kgf of ankle dorsiflexion. The load by ankle plantarflexion was about -3kgf of the position where the swing phase and stance phase change; on the other hand, the spastic gait was about −31kgf in swing phase. Fig.3 shows the moment of the circumference of the ankle joint when using the orthosis for this experiment and a motion capture system. In the motion capture system, the moment is calculated from the ankle joint position of a reflective marker and the floor reaction force. Using the orthosis in the experiment, the moment is calculated from the ankle joint position of the orthosis and the force measured by the load-cell. Since there is no floor reaction-force data during the swing phase in the force platform, the moment of this phase is not calculated. The normal gait shows that the foot enters the stance phase from the heel strike; on the spastic gait, the foot enters from a foot-flat by a drop foot tendency. The difference between the normal gait and spastic gait is clearly shown in a comparison.

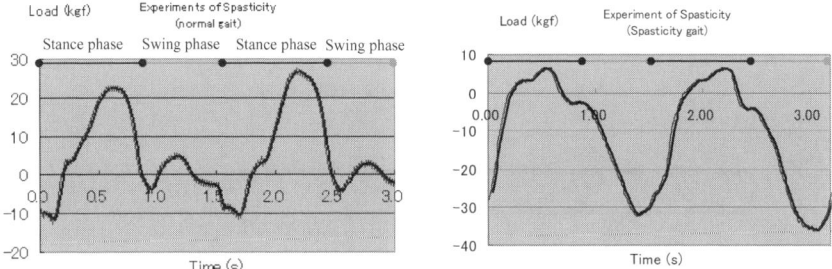

Figure2: The load of SMO

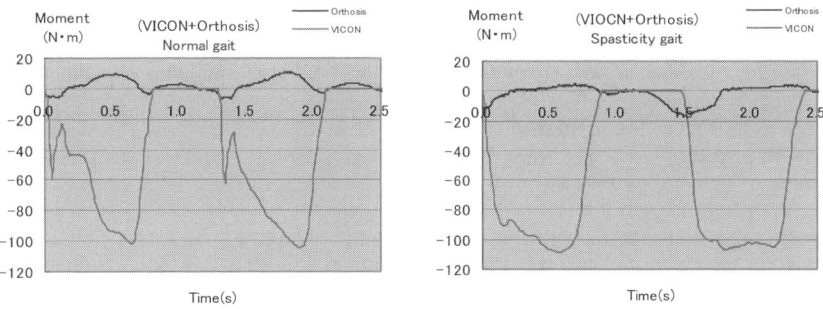

Figure3: The result of VICON and SMO

Discussion

Fig.4 shows a comparison of the force in the circumference of the ankle during gait and the gait cycle. The relationship between each event in the gait and the force in the circumference of an ankle is as follows:
1: Acceleration begins as soon as the foot leaves the ground, and at this time, the force of kicking the ground causes the ankle plantarflexion.
2: Midswing occurs when the foot passes directly beneath the body, and in order to shake out the foot, the force causes the ankle dorsiflexion.
3: Deceleration stabilizes the foot in preparation for the next heel strike, and the force causes the ankle plantarflexion.

4: The heel strike corresponds to the foot contacting the ground, and the force causes the ankle plantarflexion rapidly.
5: Since the force at the time of midstance is in the neutral position of the load-cell, it is zero.
6: Heel-off occurs as the heel loses contact with ground, and the force is the maximum of the ankle plantarflexion.
Although this is a healthy person's condition, the patient's drop-foot tendency increases the force of ankle plantarflexion. The difficulty in walking occurs due to this drop-foot, which occasionally causes a fall during gait.

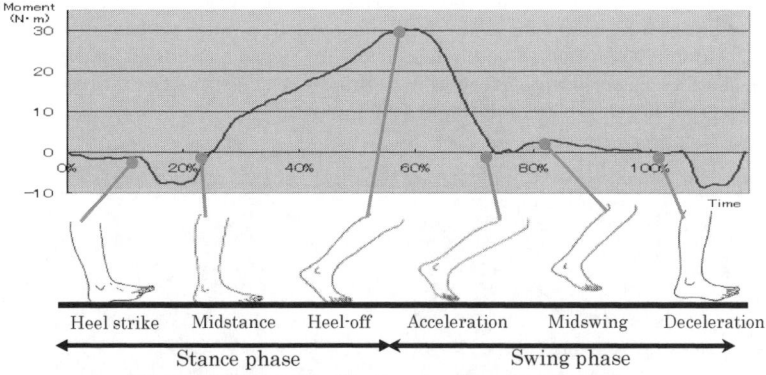

Fig. 4: Comparison with the gait (Healthy person)

The relation between the moment and the visco-elasticity of the muscle of the circumference of the ankle is shown. It is thought that the moment in the circumference of an ankle joint is generated by tension in the muscles of the foot. Since it measured keeping the ankle joint at 90 degrees in this experiment, inertia was very minute and can be omitted. The visco-elasticity can be found using the formula for the moment of the circumference of this leg joint and this experiment. If this visco-elasticity value is used for the parameter of the visco-elasticity of an orthosis, it is thought that orthodontics of the foot can be performed appropriately.

Conclusions

The purpose of this study was to experimentally determine the feature point of ankle during gait. We will have to consider how the outcomes of measurements should be processed as parameters in the design of 3DCAD. When we measure the spasticity in the gaits, it is important to take special care to measure the tendency to counter the force of drop foot. This study is an initial stage in 3D_designing system of AFOs.

References

1. A.D.Pandyan. C.I.M. Price.H. Rodgers. M.P.Baranes. G.R.Johnson. (2001).Biomechanical examination of a commonly used measure of spasticity. Clinical Biomechanics: 2001.16.859-865
2. Christopher. L.Vaughan, Brian L. Davis, Jeremy C.O'Connor. (1992). Dynamics of Human Gait. Human Kinetics Publishers.
3. Sergio T. Fonseca, Kenneth G. Holt, Elliot Saltzman, Linda Fetters. (2001). a dynamical model of locomotion in spastic hemiplegic cerebral palsy: influence of walking speed. Clinical Biomechanics.16. 793-805
4. Y.Ebara, T.Tubota, T.Tuchiya, T.Nosaka, S.Yamamoto. (2002).Biomechanics of an AFO. MDP.
5. Y.Ebara, S.Yamamoto. (1997). Analysis of Gait by joint moments. Clinical walk analysis study group.MDP.

EXPECTED MAIN FEATURES AND BUSINESS MODEL OF HEALTHCARE PARTNER ROBOT BASED ON THE MARKET ANALYSIS

Akiyoshi Kabe

Department of Health Science and Social Welfare, School of Human Sciences, Waseda University
2-579-15 Mikajima, Tokorozawa, Saitama, 359-1192 JAPAN

ABSTRACT

The new role of "Mechatronics" in the 21st century, is to develop the robots in new category, supporting people directly. The concept of Healthcare Partner Robot is derived from the market analysis in the working group to make business plan, and several contacts to potential users and related officers of cooperating municipalities/institutes in a few countries. The configuration of Healthcare Partner Robot system, and expected main functions, target range of potential users, an acceptable cost, an example of business model with related development strategies are explained. The further evaluation of this concept will be done continuously.

KEYWORDS

Communication system, Human behavior analysis, Human robot collaboration, Human-machine interface, Mobile robot, Service robot, Supporting robot

1. INTRODUCTION

The new role of "Mechatronics" in the 21st century is the one of the main technology to support people directly by intelligent machines and systems with sensor, information technology, network, and human sciences. Many prototypes of robots for medical and welfare applications, have been developed. But much less number of those robots are introduced as final product to the market due to technical problem [1]. If we focus not only on the technical issues but also on the marketing issues, it should be possible to find an appropriate application as final product for these prototypes of robots, as we found many applications for industrial robots in the 20th century. We propose to focus on the specific application area for these robots based on both technical and marketing analysis, and to integrate as "Healthcare Partner Robot System", by specific strategies of product development, contributing to the healthcare of elderly people. This paper provides the basic analysis of the healthcare service market, and proposes the expected main features and an example of business model of healthcare partner robot according to the market demand.

2. MARKET ANALYSIS

2.1 Healthcare, Homecare market in Japan

The percentage of the population of Japanese society over 65 years old, will be 25% range in 2014 (about 17% in 2000), and reach to about 35% in 2050. The population of the working-age group, 15 – 64 years old, is expected as 62% in 2014 (about 68% in 2000), and about 54% in 2050 [2]. This shows that 1 elderly person is currently supported by 4 working-age persons in 2000, but will be supported only by 2.5 persons in 2014 , and by 1.4 persons in 2050.

When we focus on the actual number of elderly people, who really needs care from other persons in meal/ toilet/ bath, more than 70 min per day, ranked as "Demand Level 3 or more (Heavy Users)", are 1,254 thousands in 2003, increasing 100 thousands per year from 2000 as shown in Table 1 [3]. If we assume that it increase by same number for 10 years, it will increase by 1 Million people as heavy user. As we have a look on "Demand Level 2 or less (Light Users)" people, who needs care partly for about 30 min up to 60 min per day, are 2,191 thousands in 2003, increasing 310 thousands per year, and will increase by 3.1 Million people after 10 years.

TABLE 1
THE TREND OF CERTIFIED NUMBER OF NEED-CARE PERSONS
[thousand persons]

YEAR	<2000>	<2001>	<2002>	<2003>
SUM	2182	2582	3029	3444
Need-care Level 0	291	320	398	499
Level 1 (light)	551	709	891	1056
Level 2 (medium)	394	490	571	636
Level 3 (heavy)	317	358	394	426
Level 4(very he.)	339	365	394	419
Level 5 (super)	290	341	381	409

Consequently, there will be demand for caring power from 1 Million "Heavy Users", and 3.1 Million "Light Users". The demand from "Heavy Users" are described as follow, such as
 - Needs care partly/ fully in meal, toilet, bath/ clothes or for all movement
We see many prototypes of equipments for these applications, but see much less number of final products in the market, for example in the product guidebook of International Home Care & Rehabilitation Exhibitions 2003 [4]. The development cost to realize features to support these applications, can be higher than the cost for "Light Users"-related applications. On the other hand,

most of welfare equipments are categorized as "Consumer Products", like TV, furniture, others.

2.2 Focus on the demand from "Light Users"

The other target users, "Light Users", can take care of themselves mostly, and tends to do by themselves as far as possible. Reviewing the tasks of Service Dogs, which are already working to support these people with wheelchair, the examples of simple tasks are as follows; [5]

1)Pick up something ordered by voice command   2)Pass to user or bring to trash box
3)Pull sliding door by pulling towel   4)Touch on some switches of wall

The basic part of Service Dog's features, can be Typical features of the robot for "Light Users", working for daily care, because of this market acceptance.

2.3 Healthcare Partner Robot for "Light Users"

As reviewed in the tasks of Service Dog, the new category of robot is proposed with the following key points to support "Light Users", as shown in Figure 1.

a)Simple features close to consumer products
b)Easy to use only by simple commands to reduce high cost/ intelligent modules
c)Use inexpensive, high reliability Hardware proven in the Factory Automation market.
d)Prepare Healthcare menus software on the platform of Human-Machine Interface and network.

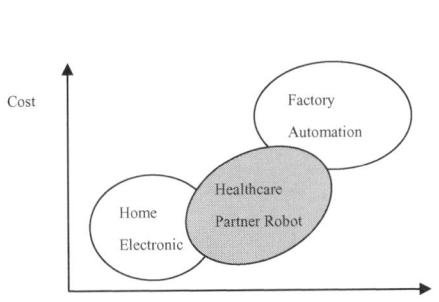

Figure 1. Positioning of Healthcare Partner Robot   Figure 2. An Example Partner Robot

2.4 A quick feedback to Healthcare Partner Robot concept from potential users

We made demonstration of an example Partner Robot, shown in Figure 2 developed by PFU, as candidate healthcare function and made interview about robot features mainly to working-age and elderly people of Tokorozawa city in March 2004. This robot is "Mobile Agent Robot", controlled by handy phone or PC remotely, and moves around according to instructions, generates pre-recorded voice, call to pre-registered phone number to indicate these events. After showing this robot, the feedback from public citizens, 39 in total, is compiled in Table 2 and Table3.

TABLE 2
EXPECTED APPLICATION FIELD FOR HOME USE

	1st	2nd	3rd
Expected Application Field	General Assistance at Home (40%)	Care, Welfare (34%)	Security (19%)

TABLE 3
DO NOT LIKE THE ROBOTS SUCH AS ;

	1st	2nd	3rd
General Features	Too Emotional (53%)	Touching body at care in toilet, bath (33%)	Difficult to use (3%)
External Shape	Machine without cover (15%)	Animal type (15%)	Human type (13%)

3. HEALTHCARE PARTNER ROBOT SYSTEM

As mentioned before, the main role of "Healthcare Partner Robot" is to cover the major part of " A BIT OF CARE" requests mainly from "Light Users", while absence of helpers, or in the night. Figure 3. shows the general view of "Healthcare Partner Robot".

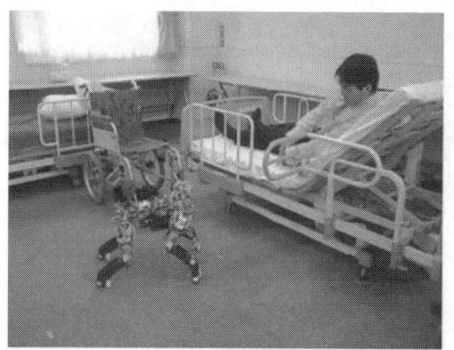 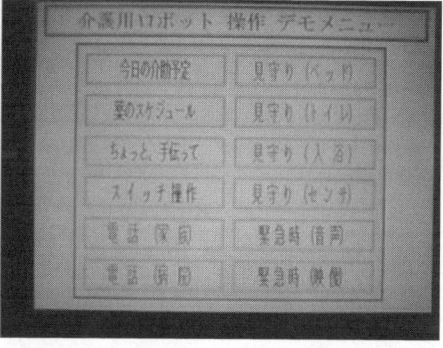

Figure 3. General View of "Healthcare Partner Robot"   Figure 4. An Example of Healthcare Menu

3.1 Main Components of System

(1) A touchpanel is used as Human Machine Interface, selecting menu on the touch screen to execute the menu by robot and related equipments, as shown in Figure 4.
(2) A voice recognition software is used to recognize the user made command from the pre-registered users, and execute the related macro-operation.
(3) Robot hardware usually has several sensors, such as vision, infra-red, mechanical switch for security, for interaction to external event[6].
(4) The bio-metric sensors, are located either close to the person to be cared or attached on the robot hardware.

3.2 Expected main Functions

(1) Monitoring of health condition, human behavior and supporting the related communications
   - According to the trigger from bio-metric sensor, or processing result of vision sensor for human behavior, the robot inform to the caring persons at remote side by calling to handy phone.
   - The robot helps the communication between elderly people and caring persons by phone.
(2) "Pick and Pass"
   - According to the request from the touch panel, the robot provides the view on the touch panel, from CCD. The user touches the target to be picked up, control the robot by inching motion.
   - The user controls the pick up motion of the robot, until it is done successfully.
(3) "Push or Touch a switch", and other variations
   - The variation of other functions as combination of going to the requested location, and performing the related motions, are executed in same way.

4. A BUSINESS MODEL OF HEALTHCARE PARTNER ROBOTS, AND THE RELATED DEVELOPMENT STRATEGIES

4.1 Value Analysis of Healthcare support

The caring helpers get the salary under the condition that they take a rest in the night for example, without exceptional cases. The value which Healthcare Partner Robot provides, is the opportunities and the free time to take a rest in the night for those who are working for continuous care, like family members, helpers in the various care homes.

As the varieties of " A BIT OF CARE" requests from "Light Users" are covered by Healthcare Partner Robot, the elderly people can take care by themselves, while the helpers are NOT available. This is the value as complement to the existing care system, both for caring, and elderly people.

4.2 Acceptable Cost comparing with other available equipments, and similar case; Service Dog

There are NO equivalent equipment to compare the exact same value. But there are similar equipments as physical effects. The typical cost of these are,
   1)Lift for transfer among bed/wheelchair etc. in home   0.7 Million yen
   2)Bed with electrical moving function   0.4 Million yen
The similar case, service dog, cost about 3 Million yen for training, and has capability to execute about 10 customized functions. If we calculate by each function, it cost about 0.3 Million yen. Therefore, the typical acceptable cost for Healthcare Partner Robot is around from 0.3 to 1.0 Million yen for a few major functions.

4.3 An example of Business Model and related development strategies

(1) Rental Sales to several channels for Hospitals, Care Facilities, and Personal Use, through technical support agents. But the real customer claims come in directly.
(2) Healthcare Menu Software works on the same platform of touch panel through the network, which

communication protocol of application layer, is standardized. This make it possible to exchange the healthcare menu software, and related robot motion programs, as mutual assets. Finally, it connects the market of each country, by exchanging application software menu, result in the cost reduction of robot hardware and software development cost.

5. CONCLUSION

The Healthcare Partner Robot concept, with the features listed above and the cost, is a candidate to contribute to elderly people care activities, as described. The further evaluation by potential users, will be done in Japan, and Europe, especially at Healthcare facilities for elderly people, and the municipality of cooperating cities. Some of the expected main functions are already available. But we continue to integrate the remaining part as bio-metric sensor and the touch panel.

REFERENCES

1. Waseda Univ. Humanoid Project, The Humanoid Robot, Nikkan-Industrial Newspaper, 1999, pp.28-32
2. National Institute of Population and Social Security Research, *Population Projections for Japan: 2001 – 2050*, Tokyo, 2002, p.3
3. Care and Insurance WG of Social Insurance Committee, *The Actual Status of Care Insurance System: The material 1 of 2nd meeting,* Tokyo, 2003, pp. 1-3
(http://www.mhlw.go.jp/shingi/2003/07/s0707-4b1.html),
4. The association of public relations for health and welfare, *the product guidebook of welfare equipments in the 30th Inte'l Home Care & Rehabilitation Exhibition 2003*, Tokyo, 2003, pp. 1-396
5. Y. Kimura, Mainichi Newspaper Team of Hanshin Branch, *The Service Dog: Sincere*, Tokyo, 2003, pp. 99 – 101
6. T.Kase, S.Nishiyama, T. Kubota, *Development of a Light Utility Robot Concept Model, "Maple"*, 22nd Annual Conf. of the Robotics Society of Japan, 3D23, 2004

EVALUATION OF HUMAN MOTOR CHARACTERISTICS IN DRIVING A WHEELCHAIR WITH HAND-RIMS

Yuki OHTA, Hironobu UCHIYAMA and Junichi KURATA

Department of Mechanical System Engineering, Kansai University
3-3-35, Yamatechou, Suita, Osaka 564-8680, JAPAN

ABSTRACT

A self-propelled type of wheelchair with hand-rims has a good drive assembly that is human operator and offers the convenience of mobility. The wheelchair propulsion of this type, however, causes fatigue to the operator. As a solution of this problem, wheelchairs, which are equipped with a power assist control system, are needed. In this research, the model based on the quantitative human-wheelchair system was proposed for establishing its design manual. The major factor to be considered was the influence of propelling force to the velocity of wheelchair caused by their inertia. The subject participated in a follow-up test in order to adjust the velocity of the wheelchair to the reference velocity. The motor characteristics of operators in driving a wheelchair was described in detailed with results of evaluation by using the special designed test machine.

KEYWORDS

Human-machine system, Power assist, Human behavior, Wheelchair propulsion

INTRODUCTION

A self-propelled wheelchair is one of useful assistive device in the user's life. Moving with it gives many people, who have difficulty to walk, not only the convenience but also the motivation to improve their own quality of life (QOL) and activity of daily living (ADL), and expands their field of activities as motivation increases (Bengt (1993)). Long travel and drive, however, have potential for fatigue because of its propelling operation form. Propelling operation with hand-rims requires especially longtime and cycling motion of upper-limb and puts too heavy physical burden on their muscle, reported by Rodgers et al (2000). For these reasons, recently a self-propelled wheelchair with an electric power-assist function is still getting more and more popular to maintain their own QOL and reduce fatigue.

An electric power-assist function for the wheelchair has to work to their needs of propelling motion and reduce their physical fatigue. The upper-limb fatigue have been researched. Fatigue is caused from the propelling operation that the operator generates the propelling force, reported by Cooper (1998) and Rodgers et al (2003), and the operation with the motion such as hold and release should be continuously repeated during propulsion, reported by Veeger et al (1998). Moreover, the propelling motion in different situation have been researched by Hildebrant et al (1970), who reported the force required to propel over the slope is larger than the propelling-force on the flat road. As referred to example above, the wheelchair propulsion has individual autonomic motion that the operator adjusts each velocity and force. So, the power-assist function is necessary to analyze the quantitative human characteristics of the human-wheelchair system. However, since many wheelchair researches are still not summarized much quantitatively, they do not give much good design of this function.

In this research, we propose the system based on the quantitative human-wheelchair system model for establishing design manual of the power-assist function. The first factor to be considered should be the influence of propelling force to velocity by inertia, we consider.

METHODS

Model

The basic concept model of the human-wheelchair system is shown in Figure 1. In this system, human on a wheelchair propels it and be considered a source of power. The operator consider senses a propulsion velocity of wheelchair (ω_d [rad/s]) at the thought of such virtual input as goal or reference velocity (ω_s [rad/s]), gives a propelling torque (τ_d [Nm]) independently adjusted as a follow-up to it to hand-rims, and gets the propulsion velocity again.

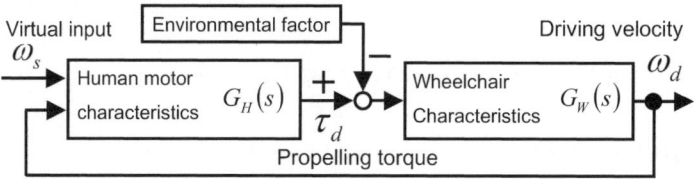

Figure 1: A block diagram description of the human-wheelchair system

Instrumentation and Subject

In this research, we made equipment to evaluate human characteristics of the human-wheelchair system (shown in Figure 2). The equipment is composed of double roller structure. The roller is 160 mm in diameter, and two rollers are attached in one axis, and the width between 2 axes is 320mm. One of the axes is connected to flywheels, and another axis is connected to the brake mechanism, through belt-pulley mechanism. The flywheels generate the inertia as the equivalent inertia weight of rotation by adjustment of two drums (a drum is 330 mm in diameter and has a mass of 80 kg). The brake mechanisms generate the resistance such as friction resistance and inclination resistance on the real road using the adjustment of band brake. The wheelchair is fixed in a regular position and the rear wheels rotate on rollers of two axes. The wheelchair witch weight 14.6 kg has rear wheels of 22 inches in diameter. The air pressure of real wheels is 0.29 MPa, and the width between the two wheels is 535 mm (Kawamura-cycle co., Hyogo, Japan).

The fundamental measurement elements are the propelling torque (τ [Nm]), the propulsion velocity (ω [rad/s], 1[rad/s]≒1[km/h] in this regard), and human physical strain that is estimated by heart rate variation during the wheelchair propulsion. The hand-rim, which is connected to a right rear wheel of wheelchair, is instrumented with torque transducer (Sohgoh Keiso co., Kanagawa, Japan) for propelling torque measurement. A rotary encoder (Ono Sokki co., Kanagawa, Japan) is attached in one roller axis, and measures propulsion velocity. An oscilloscope puts in front of wheelchair to provide visual feedback of propulsion velocity and reference velocity as control objectives. A wireless heart rate monitor (Polar Erectro oy, Finland) records heart rate (HR [bmp]) during wheelchair propulsion.

We asked one subject to try a propelling test on grounds that subject of this research might need heavy work in some case on the test. The subject was a healthy 22 years old woman, weighting of 56 [kg], with a resting heart rate (RHR [bpm]) of 64 [bpm].

Test Protocol

In order to measure the propelling torque to the propelling velocity and evaluate human fatigue during wheelchair propulsion under different inertia, a subject was asked to propel the test wheelchair on the equipment, and in the follow-up test. The follow-up test is a test tracking and fixing the propulsion velocity at the reference velocity, for using displayed the voltage signal value of both velocities on the oscilloscope. A subject propelled the wheelchair for 6 minutes at a set constant velocity. Subjects were at rest while seated for about 15 minuets before testing. Six minutes was enough time to attain a steady heart rate. The data gave the propelling torque and the propulsion velocity, by determined at the respective velocity (from 0 to 4.0 km/h and every 5 km/h), and changing inertia applied by flywheels adjustment, that is a value on rear wheel axis is equated with the total mass of an operator and the test wheelchair, for example, we used inertia values shown in Table 1.

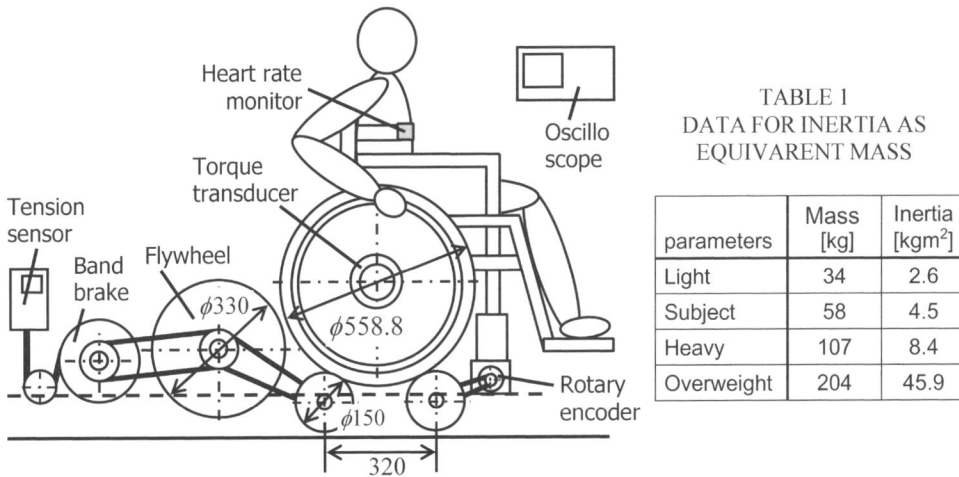

TABLE 1
DATA FOR INERTIA AS EQUIVARENT MASS

parameters	Mass [kg]	Inertia [kgm^2]
Light	34	2.6
Subject	58	4.5
Heavy	107	8.4
Overweight	204	45.9

Figure 2: Schematic diagram of the wheelchair test system showing the method used to apply inertia to the rollers

Statical Analysis and Evaluation of propelling motion

Human operator gets steady motion in case of the propulsion on steady environment although the propelling torque is intermittently transferred to the propelling wheel and the propelling motion is not stable. In this steady motion, the each response of the torque and the velocity until 5 to 30 seconds after starting to propel with initial velocity = 0 [rad/s] is steady averaged value of one propelling cycle that is hold and release with using hand-rims. In this paper, we determined that the average of the torque and the velocity after heart rate is stable, however, the time to attain a steady heart rate is later.

We evaluated the changes of each element in the steady state with the following two values. The self-propelled wheelchair propulsion is a cyclic movement to which a power balance can be applied (Van der Woude et al (1998)) and Power output (P [W]) during wheelchair propulsion is equaled to energy consumed of the body (Cooper (1998)). In this paper, we define the meaning of power ratio as the ratio of propelling momentum to driving one at one cycle and power output for a given inertia value is derived from each cyclic data average that is the measured propelling torque and the rear wheel angular velocity:

$$P = \tau_d \times \omega_d \quad (\because \omega \cong \omega_s) \tag{1}$$

Relative exercise intensity was generally used to estimate physical strain for individualized program by HR response and expressed as a percent of the heart rate reserve (I [%]), determined by equation 2 (Janssen et al (1994)).

$$I = (THR - RHR)/(MHR - RHR) \times 100\% \tag{2}$$

The HR is measured during the test. HR takes steady response by the middle of the test. In this research, The HR for about 1 minute before finishing the test is defined as THR [bpm], the HR at rest is defined as RHR [bpm] when the subject gets up and MHR [bpm] is maximum HR, determined by

$$MHR = 220 - \text{Age of a subject} \tag{3}$$

RESULTS

The propulsion velocity fluctuations to the propelling torque of one cycle in the steady state at $\omega_s = 3.0$ [rad/s] is depicted in figure 3 as an example of the results in the follow-up test. The amplitude of propelling velocity (ω_a [rad/s] shown in figure 3), which derived peak velocity to averaged propelling velocity of one cycle from experimental data, peak propelling torque (τ_m [Nm] shown in figure 3), and frequency of the propelling cycle ($1/t_d$ [cycle/s]) to each reference angular velocity differed from the each inertia and this relation is shown in figure 4. Relation to the propelling velocity average of one cycle (ω_d [rad/s]) and the propelling torque average (τ_d [Nm]) under different inertia is shown in figure

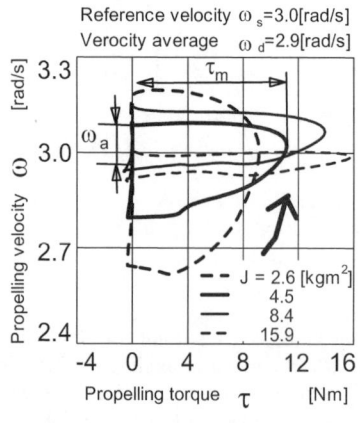

Figure 3: Relation to velocity and torque in the steady state

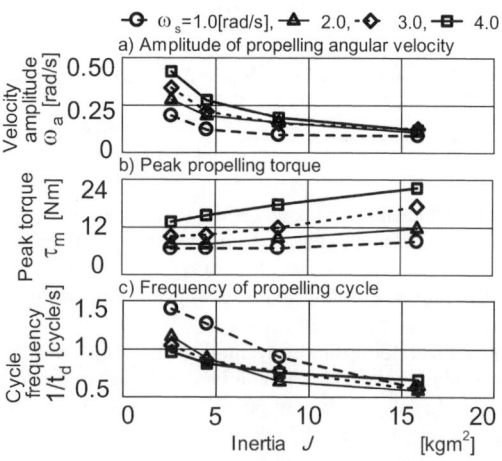

Figure 4: Propelling elements under different inertia in the steady state

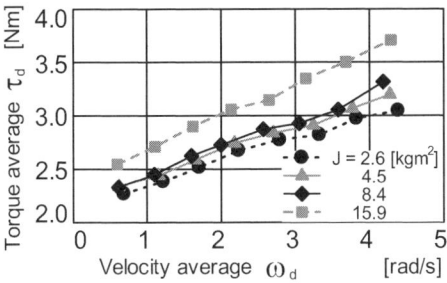
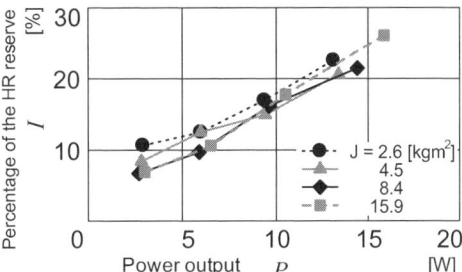

Figure 5: Torque average to each averaged velocity in the steady state

Figure 6: HR reserve to power output for evaluation of wheelchair propulsion in the steady state

5. For evaluation, figure 6 shows relation derived power output (P [W]) and heart rate reserve (I [%]) in the test of figure 5.

DISCUSSION

The wheelchair equipped with the power assist control system is needed development or an improvement to a design based on the human-wheelchair model, not on a trial and error. Generally, this system of the wheelchair with hand-rims, which transmit human power as source of the propelling force, should contain not only assist power but also adjustment element corresponding to change of the resistances that cause fatigue. For one of resistances that should be considered, we should consider about inertia, which is resistance to change of movement and which changes with total mass of the human-wheelchair system. For realization of this system, many relations that occur between human and the wheelchair should be studied and evaluated quantitatively.

First, we made equipment for this study and we confirmed usefulness for the evaluation of some characteristics at straight propulsion. Next, we obtained the steady state velocity-torque characteristic. Finally, we acquired and verified the some relation between inertia power and human elements in the human-wheelchair system. These contents and considerations are described to following about each.

As described by the result (Figure 3, 4), the propelling peak torque tended to vary inversely as the propelling velocity amplitude that the change of inertia affects. Because large inertia resistance is equivalent to increase the driving resistance, speed variation and propelling frequency, which make operator feel like bad driving quality and too much operation, decrease. The propelling peak torque of each cycle, however, increases for the effect of inertia. We have then deduced that heart rate reserve increase in proportion to inertia increase from the results. On the other hand, averaged torque and velocity of one cycle were not affected by inertia (figure 5). The case of overweight ($J=15.9[kgm^2]$), however, revealed an increased torque and we have considered that the reason was influence of resistance caused by inertia on the equipment. In the steady state, theoretically of relationship between averaged torque and velocity predict that these values were not influenced by inertia and we made sure it. From these results, we have considered that heart rate reserve to power output varies with inertia. The experimental result had little change to inertia (shown in figure 6) and did not agree to the theoretical estimates. This result showed that the variation of heart rate has a tendency similar to the power. Moreover, we found that an element, which varied heart rate by propulsion under different inertia, dose not only depend on the peak torque but also depend on the frequency of cycling motion and each element is adjusted by human self-directive ability of that kind.

At this experiment assumed on the simulated flat straight road, we tested the influence of inertia as one of the resistances. As the result, although we confirmed human-wheelchair system has an adjustment element, we hardly found what constitute the adjustment element to inertia. The wheelchair with power-assist function considering the inertia was and reported by Uchiyama et al (1997) and this system has inertia as virtual mass only release time of hand-rim operation. However this development was on a trial and error and we consider that design guideline of the function of its kind treats quantitatively is better. For this reason, our future research is that human adjustment element based on model decides quantitatively the relationship with propelling motion and heart rate reserve.

CONCLUSION

We made an equipment of the propelling test for the self-propelled wheelchair that enable human to realize changes occurred by the propelling situation. We found experimentally the human basic characteristic that is the influence of propulsion and the adjustment of motion by the inertia in the follow-up test that was used the wheelchair with hand-rims.

ACKNOWLEDGMENTS

Some parts of this report have been helped by the research fund on "Living Assistance Technology for Improvement of Quality of Life" supported by ORDIST (Organization for Research and Development of Innovative Science and Technology) in Kansai University. Authors would like to thank staff of the practical factory of Kansai University for their technical support. And we also thank many colleagues for their support during our work.

REFERENCES

Bengt Engstrom. (1993). *ERGONOMICS Wheelchairs and Positioning*, Posturalis, Sweden (First Japanese language edition by Miwa-Shoten Ltd., Tokyo.)
Rodgers MM, R.E. Keyser, E.R. Gardner and P.J. Russell. (2000). Influence of trunk flexion on biomechanics of wheelchair propulsion. *J Rehabil Res Dev* **37:3**, 283-295.
Cooper RA. (1998). *Wheelchair Selection and Configuration*, Demos Medical Publishing (First Japanese language edition 2000 by Igaku-Shoin Ltd., Tokyo.)
Rodgers M.M., K.J. McQuade, E.K. Rasch, R.E. Keyser and M.A. Finley. (2003). Upper-limb fatigue-related joint power shifts in experienced wheelchair users and non wheelchair users. *J Rehabil Res Dev* **40:1**, 27-38.
Veeger DH and Woude LH. (1998). Wrist motion in handrim wheelchair propulsion. *J Rehabil Res Dev* **35:3**, 305-313.
Hildebrant G, E.D. Voigt, D. Bahn, B. Berendes and J. Kroger. (1970). Energy costs of propelling wheelchair at various speeds - cardiac response and effect on steering accuracy. *Arch Phys Med Rehabil* **51**, 131-136.
Van der Woude L.H.V., K.M.M. Hendrich, H.E.J. Veeger, G.J. Van Ingen Schenau, R.H. Rozendal, G. De Groot and A.P. Hollander. (1998). Manual wheelchair propulsion: effects of power output on physiology and technique. *Med. Sci. Sports Excerc.* **20:1**, 70-78.
Janssen T.W.J., C.A.J.M. van Oers, H.E.J. Veeger, A.P. Hollander, L.H.V. van der Woude and R.H. rozendal. (1994). Relationship between physical strain during standardized ADL tasks and physical capacity in men with spinal cord injuries. *Paraplegia,* **32**, 844-859.
Uchiyama A and Fukui T. (1997). Control system design of power assist wheelchair. *JSME,* **97:34**, 204-208. (Written by Japanese language)

DYNAMIC MODELING FOR ATTENDANT PROPELLING WHEELCHAIRS

Tatsuto Suzuki [1], Hironobu Uchiyama [2] and Junichi Kurata [2]

[1] Department of Mechanical Engineering, Maizuru National College of Technology,
Maizuru, Kyoto 625-8511, JAPAN
[2] Department of Mechanical Engineering, Kansai University
Suita, Osaka 564-8680, JAPAN

ABSTRACT

The purpose of this study is to establish the model of attendant propelling tasks in order to design for safe and low load attendant's propelled wheelchair with assist system. We propose the model of attendants with assumption that attendant and wheelchair is basic motor-load system. The validation of the proposed model identified by experiments shows well corresponding to real attendant's propelling behaviour. From the results, the proposed model is used effectively for safe and low load design of assisted wheelchairs.

KEYWORDS

Attendant propelling wheelchair, Assist system, Attendant's model, Safe and low load design

INTRODUCTION

Attendant's propelled manual wheelchairs provide the opportunities of social activities for those who cannot propel wheelchairs themselves. But it's hard for attendants to propel wheelchairs with the people to move place to place with keeping safe against around traffic. So attendants strongly desire the wheelchair easy to propel and stop. Cremers (1989) developed the attendant's propelled wheelchairs with assist system for reducing the attendant's loads and now some type of them are commercialized. The recent wheelchairs with assist system have motorized wheels driven by the controller which detects propelling force of grips and continuously determines amount of assisting force exerted by the motorized wheels. Recently, attendants want high assist force because our society becoming aged and low birth rate brings deficit of young attendants, so attendants become aged gradually. Rising assist force causes rapid and unstable wheelchair's movement, because small changes of propelling force provide large torque of auxiliary wheels by the controller. In addition, it's easy to expect that various uneven surfaces, such as thresholds, openings, gaps, cracks and holes on road dynamically change the propelling force of wheelchairs. Therefore, it's difficult for attendants to regulate the movement of the wheelchairs. Our final purpose of this study develops the controller for high assisted and very safe wheelchairs. To achieve it, we need attendant's model to develop the controller. In this paper, we propose, identify and validate the model with experiments

MODELING FOR ATTENDANT PROPELLING

There are some previous studies to investigate the propelling behaviour: Resnick (1995) studied the maximal and sub-maximal condition of propelling carts. Al-Eisawi (1999) studied the steady load of propelling manual carts on some road surfaces, but attendant's models have been not proposed until now. Figure 1 shows the model that we propose. We assume that an attendant and a wheelchair are basic motor-load system. The model of the attendant has pushing force F – walking speed Vh characteristic, like the torque-rpm characteristic of motors. The model of the wheelchair has driving resistance r(Vc): Vc is wheelchair speed. The attendant's model has also other three dynamic elements, pushing motion dynamics, following wheelchair dynamics and reducing force against relative distance. The pushing motion dynamics describes time response of exerting force by human muscles and it is assumed a 2nd-order mechanical system. The following wheelchair dynamics describes attendant's behaviour for following wheelchair, which is assumed a tracking control system of walking speed against wheelchair speed. The controller of this element is assumed PID controller, and human body element is assumed a 1st-order lag system with time constant Tp. The reducing force against relative distance describes a phase lead compensator against relative distance ΔL, because human usually uses feedforward control. The wheelchair's model has a centre body mass m with driving resistance r(Vc).

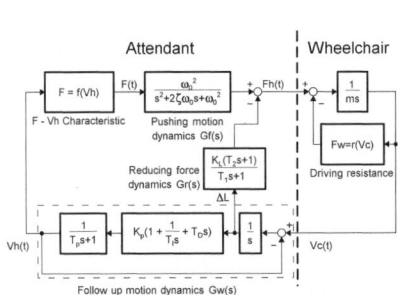

Figure 1   Model of attendant - wheelchair system

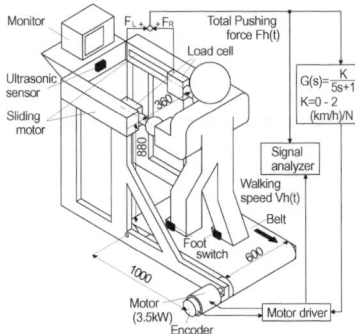

Figure 2   Pushing motion analyzer with estimating function of suitable manipulation

EXPERIMENT FOR IDENTIFICATION

We produce an experimental system showing Figure 2 to identify the model parameters. This treadmill has grips with load cells for detecting propelling force, and sums both grip forces to output total force signal. The grips are fixed on slider motors at the same positions of wheelchairs. The wide belt of the treadmill is motorized and the motor is so strong that subjects cannot disturb it. We identify the model with only one subject, because we focus on the bilateral relationship between four elements in the model. The subject is 22years healthy male having no functional disorders. First, to identify the F-Vh characteristic, we add a feedback element to simulate the load of wheelchairs. We assume the load L in proportional to wheelchair's speed V, so it shows L=(1/K)V, here K is a coefficient and shows the strength of load. We obtain F-Vh characteristic with several different K and 1st order lag system to stabilize the subject's propelling. Second, to identify the pushing motion dynamics, we examine pushing force response. The grips move forward 1km/h when the subject pushes over a threshold level to simulate starting wheelchairs. Third, to identify the following motion dynamics, we examine the step responses of body movement against the step forward movement of the sliders. The subject's movement is detected by ultrasonic sensor fixed in front of experimental system. Meanwhile, we record the reducing of pushing force to identify the reducing pushing force dynamics.

RESULTS

Identification of Model Elements

Figure 3 shows the result of F-Vh characteristic with K=0 - 2N/(km/h). White circle markers show measured propelling points against K. At low load(small K), the subject walks fast, 3km/h but pushes weakly, about 12.5N. With increasing load, larger K, the walking speed decreased and the pushing force increased gradually. A dotted line shows the estimated F-Vh characteristic, F=86-23Vh. The black circle makers show mechanical propelling power calculated from the F-Vh characteristic. The max power of the subject was 30W at 2km/h. Figure 4 shows the result of pushing force responses. The vertical axis of Figure 4 is normalized by each max value. All responses had rapid increase and after that fall off immediately, because the subject dropped pushing force after the grip forward movement. We assumed these responses as step response and estimated the parameters, damping factor ζ=0.8506 and natural frequency ω_n =6.603. Figure 5(a) shows the result of following response. The vertical axis of Figure 5(a) is normalized by each final value. A dotted line shows the step input of grip's step forward movement. The subject began to follow to the grip movement lately, and then the subject's body stopped with overshooting, because of body mass. We estimated the parameters of the following motion element. Thick line shows the estimated response, which has Tp=0.5063, Kp=2.4987, Ti=2.6606 and Td=0.2140. Figure 5(b) shows the result of the reducing force against the relative distance. This result was recorded with Figure 5(a) simultaneously. A thin and dotted line shows the step input of the grip movement. A thick and dotted line shows the relative horizontal distance between the grips and the position of the subject's body. The late response of the body movement was found in the short period at starting. Thin lines show falling pushing force for the increasing of the relative distance. The pushing force starts to fall at same time of increasing the relative distance and then rises oppositely. Then, the pushing force almost returned to initial force. We estimated the parameters, T1=0.01, T2=0.3672 and KL=-0.0957.

Validation of the model

Figure 6 shows the validating result of the model in a period from starting to driving steadily. We compare between the model and experments under the same wheelchair's conditions that the mass is 100kg and the driving resistance identified by experiments on flat linoleum is $r(Vc) = 10.2\exp(-1.84Vc) + 1.38Vc + 8.74$ The subject exerted large force until the wheelchair speed reached about 3km/h. Then attendant drove it at about constant speed. Two leg motion of walking provide some periodic changes only on the pushing force. But there is no periodic change on the wheelchair speed, so that the wheelchair mass was very large. The simulation result in the upper graph of the Figure 6 almost corresponded to the experimental result despite with some differences. The lower graph of the Figure 6 shows calculated result of relative speed and distance between the attendant and the wheelchair. The relative speed and distance increased with starting wheelchair. After that, the attendants began to follow the wheelchair, so the relative speed shows minus value and relative distance began to decrease. Finally, Both the relative speed and distance was adjusted to zero gradually.

DISCUSSION

We found that F-Vh characteristic showing the Figure 3 has performance curve like other motor's one.

At low load, the attendant eases to pushing, so keeps the walking speed fast. With increasing load, the pushing becomes harder and the large pushing force needs long time period of foot's touching on the ground, so the walking speed becomes slower. The mechanical power of the attendant is so small that the assisted system is needed for most of attendants. The pushing force responses showing in the Figure 4 slow, because attendants push carefully against unknown loads. The following responses against grip movement in the Figure 5(a) and (b) provide that attendants cannot keep its relative distance and propelling force. Attendants delay to response against the grip movement and adjust its position slowly despite the force have already adjusted. We expected from these results that human couldn't reproduce its position and forces exactly and settle them within certain range. The phenomenon of the falling force was well found in the fast slider speed condition, because responding against fast object was more difficult. Despite of the facts, well corresponding between the model and the experiment was found in the Figure 6. Neglecting dynamics, such as sudden dropping strength dynamics, causes some differences. This time experiment carried out only one direction, such as increasing force, moving forward. It is probably need to investigate the experiments of the opposite directions, because human does not always have only one linear characteristic. Lately, the proposed model describes attendant's behavior, mainly the pushing force and the relative distance very well. We will estimate and assess the load and the safe of attendants with the proposed model.

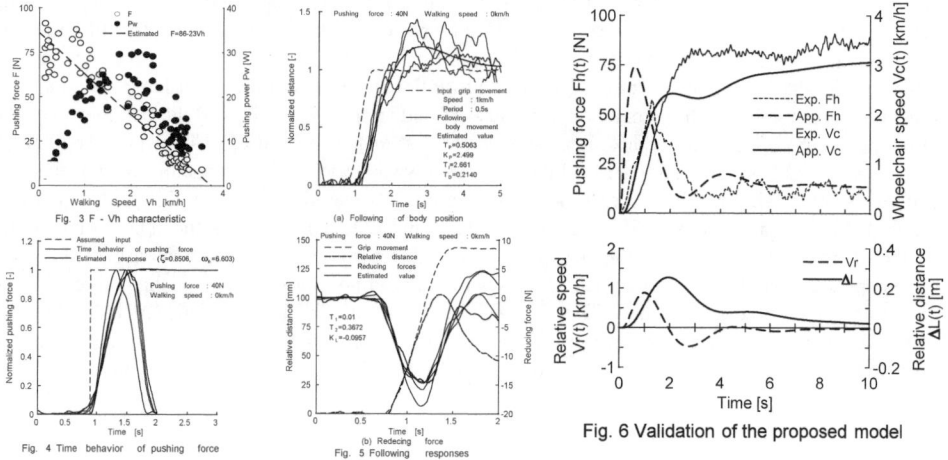

Fig. 3 F - Vh characteristic
Fig. 4 Time behavior of pushing force
Fig. 5 Following responses
Fig. 6 Validation of the proposed model

CONCLUSIONS

We proposed the model to expect the attendant's behavior for the safe and low load design of the assisted wheelchair with high assist. The validation of the proposed model shows well corresponding against the experiment. The model can describes attendant's behavior on various conditions. Therefore, the model is useful for the controller design of assisted wheelchairs.

REFERENCES

Al-Eisawi, K. W., Kerk, C. J., Congelton, J. J., Amendola, A. A., Jenkins, O. C., Gaines, W. (1999), Factors affecting minimum push and pull forces of manual carts, *Applied Ergonomics* **30**, 235-245

Cremers, G. B. (1989), Hybrid-powered wheelchair : a combination of arm force and electrical power for propelling a wheelchair, *Journal of Engineering and Technology* **13**, 142-148

Resnick, M. L., Chaffin, D. B. (1995), An ergonomic evaluation of handle height and load in maximal and submaximal cart pushing, *Applied Ergonomics* **26**, 173-178

DEVELOPMENT OF A NON-POWERED LIFT FOR WHEELCHAIR USERS
- MECHANISM TO TRANSMIT ROTATION OF WHEELS BY MANY ROLLERS -

Y. Kobayashi [1], H. Seki [1], Y. Kamiya [1], M. Hikizu [1], M. Maekawa [2],
Y. Chaya [3] and Y. Kurahashi [3]

[1] Department of Mechanical Systems Engineering, Kanazawa University,
Kakuma, Kanazawa, 920-1192, Japan
[2] Industrial Research Institute of Ishikawa,
2-1 Kuratsuki, Kanazawa, 920-8203, Japan
[3] Fujiseisakusho Co., Ltd.,
Ha 195 Akai, Nomi, 920-0101, Japan

ABSTRACT

Wheelchair users need lifts to climb up / down steps at entrances with small spaces. Lifts driven by motors or hydraulic equipments are large and expensive. They also need switches to start / stop actuators. The aim of our study is to develop a compact and non-powered lift for wheelchair users. We have already made a lift driven by wheels of a wheelchair on it, however, it has some problems. Because wheelchair direction was fixed, a user must enter the lift backward in case of ascent. Complicated mechanism must be equipped so that small front casters can pass through the lift stage and large rear wheels can drive the lift. Therefore, a new non-powered lift using many rollers is proposed to improve these problems.

KEYWORDS

Support system, Power assist, Lift, Wheelchair, Mechanism, Welfare tools

1. INTRODUCTION

In Japan, private houses usually have doorsteps at entrances. It is difficult for wheelchair users to climb up / down such steps without attendants. If the height of a step is less than 150 mm, manual wheelchair users can go it over by lifting front casters [1]. However, it requires a user's skill. Generally, the height of doorsteps at the entrances are from 200 mm to 500 mm. One solution is to place a slope, but it needs so much place for a wheelchair user to climb up easily (The slope should be less than 10 degrees) [2]. Another solution is to use the lift which moves vertically as shown in Figure 1. Since most lifts are driven by electrical motors or hydraulic actuators, it makes the lifts large, heavy and expensive. It asks users or attendants for switching operation to start / stop the lifts. Entrances

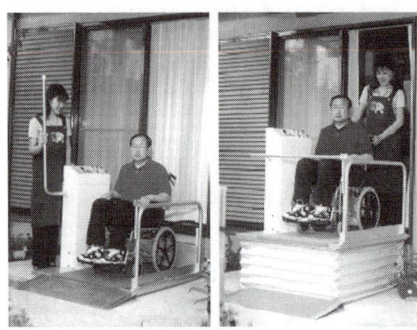

Lift produced by Fujiseisakusho Co., Ltd
Figure 1: Powered lift for a wheelchair

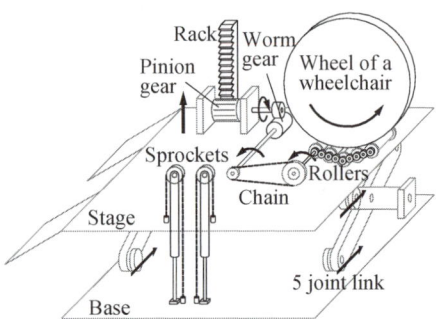

Figure 2: Mechanism of the non-powered lift driven by wheels of a wheelchair

should be reconstructed to place the lifts.

We have developed a non-powered, lightweight and compact lift which doesn't require any operation by attendants [3]. We have already made a lift driven by wheels of a wheelchair on it, however, it had some problems. Because wheelchair direction was fixed, a user must enter the lift backward when ascending. Complicated mechanism must be equipped so that small front casters can pass through the lift stage and large rear wheels can drive the lift. Therefore, a new non-powered lift using many rollers is proposed to improve these problems.

2. MECHANISM OF THE NON-POWERED LIFT

The new mechanism of the proposed lift is shown in Figure 2. After a wheelchair goes into the lift stage till the rear wheels are located on the rollers, the rear wheels can rotate the rollers by friction without moving the body of the wheelchair. This rotation is transmitted to a rack / pinion gear via a worm gear and it makes the lift stage up or down. The worm gear has a role to prevent the stage from falling down if the wheels slip on rollers or the user stops to rotate the rear wheels. The stage is kept horizontally by a link mechanism. This lift works automatically when a wheelchair goes into the stage, and a wheelchair can goes out from the stage by rotating the rear wheels on the rollers locked automatically when the lift movement is completed. The lifting height can be adjusted to the step by limiting the movable length of the rack / pinion gear.

Five rollers are placed in an arc for one wheel. One reason is to distribute the load from rear wheels and make the deformation of their wheels small. The deformation of the wheels prevents their rotation. Another reason is to prevent the rear wheels from running over rollers and to enable the small front casters to pass on them. If we consider only driving rollers, the minimum number of rollers are two for one rear wheel. But small front casters fall between rollers. If plates are placed between rollers, rear wheels can't contact with rollers. Because directions of a wheelchair are reverse between the ascent and the descent as shown in Figure 3, four sets of rollers are arranged lengthwise and crosswise for a wheelchair to go forward into the lift stage when both ascending and descending. Then, all rollers are connected by gears and shafts and they have flanges for wheels not to slip sideways. Since both rear wheels and front casters are on rollers, the front casters are rotated by the rear wheels via the rollers.

Proposed lift has many advantages. This lift doesn't require any switching operations by users because the lift is driven by rotating wheels of wheelchairs by him/herself. Since the lift doesn't have heavy actuators, it is compact and lightweight. So lift can be carried comparatively easily and it is also suitable for temporary or rental use. The lift can be used for both manual and powered wheelchairs. Since the lift doesn't have any electrical parts, it has water-resistance and easy maintenance. It can

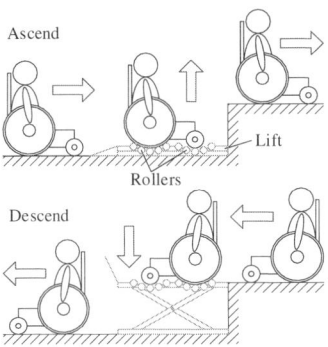
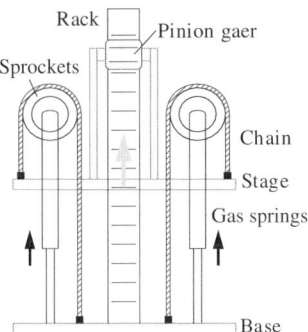

Figure 3: Motion of the lift when ascending / descending

Figure 4: Mechanism to decrease driving torque

work under outdoor, power failure and some disaster. A problem is that the lift can't ascend / descend without a wheelchairs. One wheelchair user can't use this lift after another. It must be used personally.

3. MECHANISM TO DECREASE DRIVING TORQUE

Driving torque to ascend the lift is larger than that to descend it. This isn't efficient because mechanical parameters of driving parts should be determined under the condition of the maximum torque. In order to decrease the difference of driving torque between ascent and descent, assist mechanism with gas springs are attached. In comparison with coil spring, gas spring has a characteristic that the reaction force doesn't change so much while extending. Though it is good to cancel out the constant load, its length is over twice as long as its stroke. By applying the principle of a moving pulley, the assist mechanism with long stroke can be realized as shown in Figure 4. It consists of short gas springs, chains and sprockets. It can double the stroke of gas springs, however, the reaction force of gas springs should be two times as large as that without this mechanism. Then, it uses double the number of the gas springs. When there is no wheelchair on the stage, the worm gear holds the stage against the gas spring force.

4. MECHANICAL ANALYSIS

The ascending (/ descending) speed and the driving torque are analyzed. The ascending height of the stage y and ascending speed $\dot{y}$ become

$$y = \frac{D_p \cdot D \cdot i}{2 \cdot d} \omega \quad , \quad \dot{y} = \frac{D_p \cdot D \cdot i}{2 \cdot d} \omega = \frac{D_p \cdot i}{d} v \tag{1}$$

where θ is the rotation angle of rear wheels of a wheelchair, D is the diameter of rear wheels, d is the diameter of rollers, i is the total ratio of the worm gear and sprockets, D_p is the diameter of the pinion gear, ω is the angular velocity of the rear wheels, and v is the running velocity of the wheelchair as shown in Figure 5. When the rear wheels are rotated at a constant velocity, the lift stage ascends uniformly. If rear wheels are rotated at the velocity of 0.3 revolution per second (1.88 rad/s), which assumes a manual wheelchair for example, the ascending speed becomes 10 mm/s in the case of D = 570 mm (22 inches), D_p = 28 mm, d = 30 mm, i = 1/50. Assume that the velocity of a powered wheelchair is v = 6 km/h = 1.67 m/s, the ascending speed becomes 31 mm/s.

The driving torque of the rear wheels T is expressed by

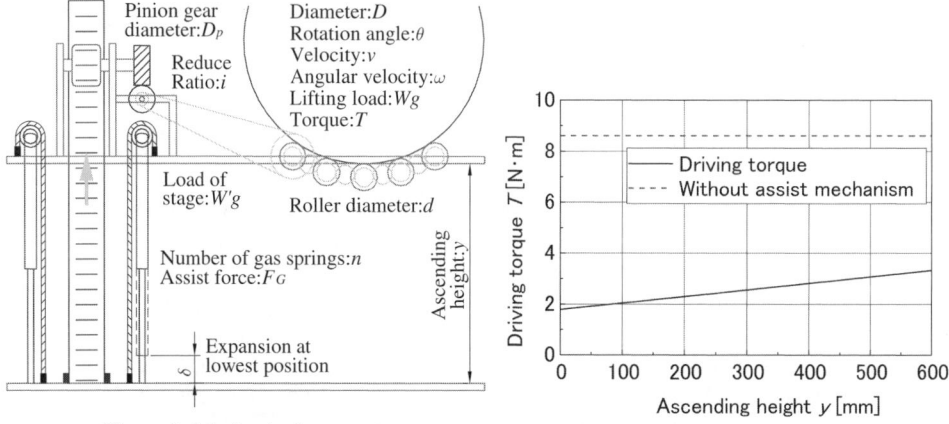

Figure 5: Mechanical parameters   Figure 6: Variation of driving torque

$$T = \frac{D_p \cdot D \cdot i}{2 \cdot d}(Wg + W'g - F_G) \quad , \quad F_G = \left[F_{max} - \frac{F_{max} - F_{min}}{S}\left(\frac{y}{2} + \delta\right)\right]\frac{n}{2} \qquad (2)$$

where g is the acceleration of gravity, Wg is the load (human + wheelchair), $W'g$ is the load of the stage, F_G is the force by assist mechanism, F_{max} and F_{min} are the maximum and minimum reaction force of a gas spring respectively, S is the stroke of a gas spring, δ is the initial stroke of the gas springs (The stage is at the lowest position) and n is the number of gas springs. In the case of $W = 90$ kg, $W' = 75$ kg, $F_{max} = 654$ N, $F_{min} = 490$ N, $S = 340$ mm, $\delta = 27$ mm, $n = 4$, the driving torque is shown in Figure 6. The driving torque is 3.3 Nm at maximum when the ascending height is 570 mm. If the lift has no assist mechanism ($F_G = 0$), it becomes $T = 7.8$ Nm. This shows that the assist mechanism is very effective.

5. CONDITION TO DRIVE ROLLERS

When the rear wheels drives rollers, they slip or run over the rollers if the transmitted torque is too large. If these are happened, rotation can't transmitted from the rear wheels to the rollers. The maximum transmitted torque changes according to the size and placement of the rollers. These parameters are represented by the contact angle α, which is the angle between wheels and rollers. The relationship around the wheels and rollers is shown in Figure 7, where R is the wheel radius, β is the ratio of the load at the rear wheels.

The moment around the roller I must be negative for the rear wheels not to run over the rollers. Because the rear wheels contact with only the roller I the instant that they run over the rollers, only F_1 is considered. Therefore, this condition is expressed by

$$T < \beta WgR \sin \alpha \qquad (3)$$

The condition for wheels not to slip on rollers is that the driving torque doesn't exceed the friction force between the wheels and rollers, i.e.

$$T < \mu \beta WgR \qquad (4)$$

where μ is the friction coefficient, and the friction force at rollers I ~ V is approximated to $\mu \beta Wg$ all

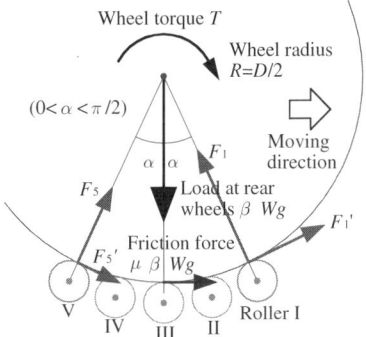
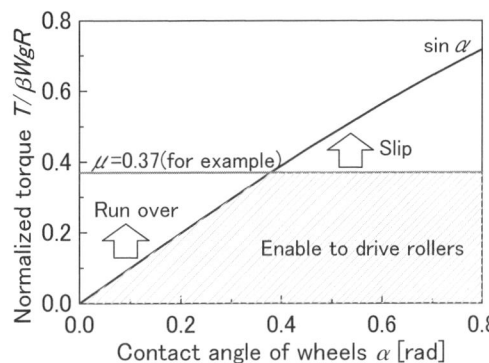

Figure 7: Statics between rollers and wheels   Figure 8: Relationship between drive, slip and run over

together. These conditions are shown in Figure 8. If α is small, the rear wheels run over the rollers before slip on them. If α is large, slip occurs earlier than running over rollers. In the case of $W = 90$ kg, $\alpha = 0.33$ rad, $\beta = 0.66$ and $\mu = 0.37$, the torque to cause running over is 53 Nm, and that to cause slip is 61 Nm. Since the driving torque to lift up $W = 90$ kg is 3.3 Nm, the wheels doesn't run over the rollers nor slip on them while the stage is ascending. But these analysis are under quasi-static conditions, so dynamic effects, as wheels are rotated roughly for example, make it easier to slip or run over the rollers. Therefore, margins should be considered in design.

6. EXPERIMENTS

A non-powered lift has been made on trial as shown in Figure 9. The specification is shown in Table 1. The wheelchair with rear wheel's diameter of 570 mm takes 18 revolutions of wheels to ascend the height of 600 mm. If a user rotates the rear wheels 0.3 revolution per second, the ascending speed of the lift stage is 10 mm/s, and the stage ascends the height of 600 mm in 1 minute. The developed lift was succeeded to lifting a wheelchair with a user and continuous motion of a wheelchair from going into the stage to going out of it was executed smoothly as shown in Figure 10. The developed lift was tested by both manual wheelchairs and powered wheelchairs.

We measured the driving torque of the rear wheels while the lift stage is ascending. The torque measured sensors were made and they were attached between the wheels and hand rims as shown in Figure 11. When a user acts the forces at the hand rims to rotate wheels, sensors of thin cylinders are distorted and they are measured by strain gages. The measurements were done by handicapped persons who use manual wheelchairs usually. We measured the forces while a user goes into the stage, ascends / descends, and goes out of it. The driving force at a hand rim is 0.7 kgf in calculation, however, the measured forces are about 8 kgf and 6 kgf while ascending and descending respectively. This seems to be caused by the loss by the transmission and the deformation of wheels, the resistance by front casters, which are rotated by the rear wheels via the rollers, and dynamic effects by the motion that a user rotate the wheels discontinuously. However, measured force when running on flat floor is about 5 kgf, so the driving force is as same as or little larger than that.

7. CONCLUSION

Non-powered lift driven by wheels of a wheelchair has been proposed for wheelchair users. The front casters can pass smoothly through the rollers by placing 4 sets of rollers. And it enables a user go into / out of the lift stage in the forward direction when both ascending and descending. Since the

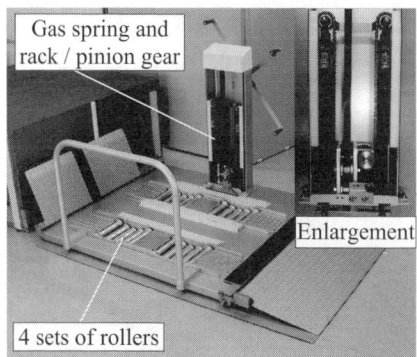

Figure 9: Developed lift driven by wheels

Lifting weight (Human + wheelchair)	90 kg (Normal) 150 kg (Maximum)
Lowest height of stage	50 mm
Maximum height of stage	620 mm
Size of stage	1000 × 1000 mm
Lift weight	110 kg
Assist force by gas springs	100 kgf (Approx.)
Driving torque of a wheel	3.3 Nm (Maximum)
Roller diameter	30 mm
Reduce ratio	1/50
Pinion gear diameter	28 mm
Contact angle	0.33 rad
Wheel diameter of standard wheelchair	570 mm

TABLE 1
SPECIFICATION OF THE DEVELOPED LIFT

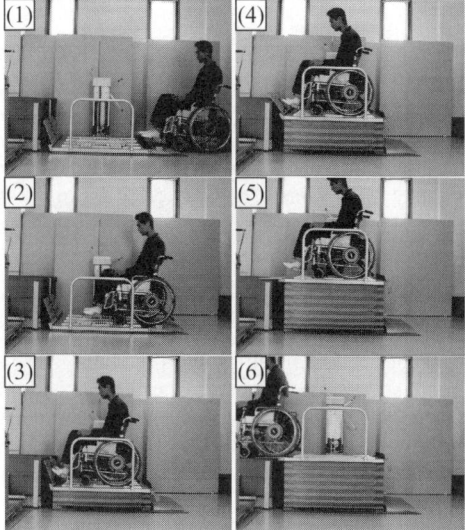

Figure 10: Motion of the non-powered lift for a powered wheelchair to climb up

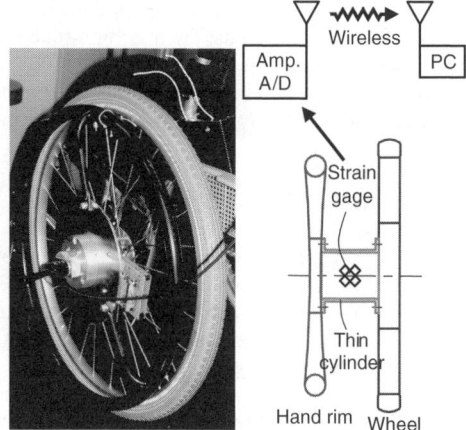

Figure 11: Sensor to measure hand rim torque

mechanism to decrease driving torque has been also proposed, the lift can be ascended by the force as almost same as the force for a wheelchair to run on flat floor.

REFERENCES

1. Bengt Engström (1993). *ERGONOMICS wheelchairs and Positioning*, Posturalis
2. Selwyn Goldsmith (1967). *Designing for the disabled*, Royal Institute of British Architects
3. H. Seki, Y. Kobayashi, Y. Kamiya, M. Hikizu and M. Maekawa (2002) . DEVELOPMENT OF A NON-POWERED LIFT FOR WHEELCHAIR USERS. *Proc. of 4th Int. Conf. on Machine Automation*, 275-282

GUIDANCE OF ELECTRIC WHEELCHAIR BY THE LEAD TYPE OPERATING DEVICE WITH DETECTING RELATIVE POSITION TO ASSISTANCE DOG

T. Uemoto, H. Uchiyama and J. Kurata

Department of Mechanical Systems Engineering, Kansai University
3-3-35, Yamatechou, Suita, Osaka 564-8680, Japan

ABSTRACT

A guidance control method to let an electric wheelchair follow an assistance dog is proposed. In this method, electric wheelchair employs the guidance unit composed of a lead, a winder and two potentiometers. The lead connected to the winder is reeled out or in as the relative position between the assistance dog and the guidance unit is changing. The length and the direction angle of the lead are detected by two potentiometers. Both translational and rotational signals used to control the electric wheelchair are generated by these two detected information. In this report, we described an opinion about an adjustment of the control system by some results of simulated experiment.

KEYWORDS

Electric wheelchair, Assistance dog, Guidance control, Human friendly machine, Optimum control

INTRODUCTION

The number of electric wheelchair's user is growing in these years. Some users of electric wheelchair hope to choose the most suitable input device for themselves from various types. Control stick, the typical input device for electric wheelchair, is designed for common user. Therefore, it's probably not true that this device is fitted well to each user. We proposed the push button type of input device and the bi-state operating controller as one example for diversification of input device, Maeda (2002) and Uemoto (2003). After the Law Concerning Assistance Dogs for the disabled was executed in October 2003 by Japanese government, the expectation for activity of an assistance dog has been swelled. Now, we are focusing our attention on assistance dogs and their owners. Assistance dog performs the request of picking up of a thing, assistance of attachment and detachment clothes and change of posture, standing up and the support in the case of a walk, opening and/or closing of a door, operation of a switch, the rescue in case of emergency and so on. Additionally, assistance dog sometimes leads a wheelchair. However, this work forces the head and back of assistance dog a great corporal burden, for example Coppinger (1995). In this report, we propose the device for an assistance dog guiding an electric wheelchair in order to make this burden mitigate, and also as one proposal for the diversification of input device, Maeda (2003). We confirmed fundamental mobility of electric wheelchair by simulated experiments, and described the result and knowledge.

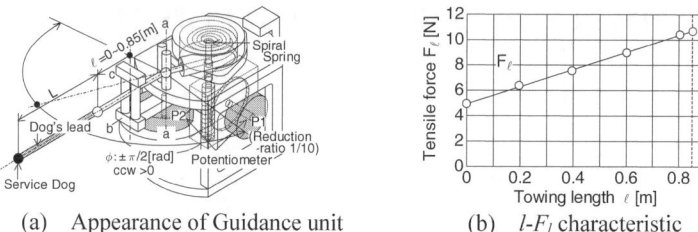

(a) Appearance of Guidance unit   (b) l-F_l characteristic

Figure 1: Guidance unit and l-F_l characteristics

GENERAL DESCRIPTION ABOUT ELECTRIC WHEELCHAIR AND GUIDANCE UNIT

Figure 1 shows the Guidance unit that can detect the distance and direction between this and assistance dog. This unit is fixed on the tip of left armrest of an electric wheelchair. The lead is connected with the harness worn to assistance dog in the place $L+\ell$ [m] away from a-a axis. Here, L is the radius of the area for support work and 1[m] this time. While assistance dog stays in this area, electric wheelchair can't drive. When the assistance dog is out of this area, the lead is reeled out or in as the relative position between the assistance dog in walk is changing. The maximum length ℓ_m of the lead is 0.85[m], because of the safety. The rotation angle of winding drum is converted into voltage by the potentiometer $P1$ through the worm gear, the slowdown ratios of which is 1/10. By the same way, the direction angle φ is detected by the potentiometer $P2$ through the lever b. The measured value φ ranges from $-\pi/2$ to $\pi/2$. We can adjust the range of tension F_ℓ, that is generated in reeling out the lead, by the selection of spiral spring. At present, the tension F_ℓ ranges from 5 to 11[N] shown in figure 1(b). The specification of electric wheelchair we used is as follows; the diameter of front wheel is 150[mm], it of rear wheel 560[mm], the width 0.54[m], the full length 1[m], two DC motors that drive right and left rear wheel independently and the maximum velocity 4.5[km/h] in forward driving.

BLOCK DIAGRAM OF GUIDANCE CONTROL SYSTEM

Figure 2 shows the block diagram of Guidance control system that consist of Guidance unit, driving system, and a fundamental gain adjustment system. The measured value ℓ is the integral of relative straight velocity between the guidance unit and assistance dog and also the measured value φ is the integral of relative angular velocity between them. Characteristics of driving system could be approximated to first-order lag element, and the time constant T_W is 0.22[sec], when total weight of electric wheelchair and passenger is 100[kg]. The control system consists of translational control

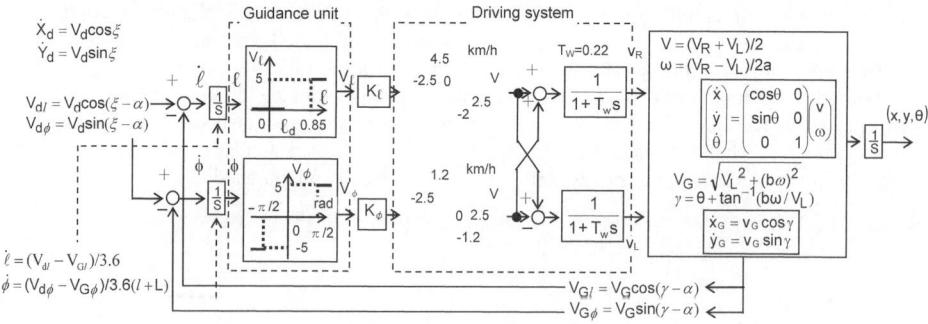

Figure 2: Block diagram of Guidance control system

element (upper part of the block diagram) and rotational control element (lower part). Translational signal generates translational velocity V, and rotational signal generates difference of velocity between left wheel and right wheel.

EVALUATION ABOUT FOLLOWING CHARACTERISTICS IN STRAIGHT DRIVING

As the first step of optimum adjustment, we determined the translational gain K_ℓ that affects the following characteristics. Under the condition when an assistance dog began to walk straight with V_d=2.5[km/h], some time responses of moving speed of electric wheelchair were calculated as in figure 3(a). In the case of K_ℓ=0.1, the maximum velocity of wheelchair was less than the velocity of assistance dog. It resulted that the electric wheelchair could not follow the assistance dog. In the case of K_ℓ=0.277, the length of lead has not been extended to limit and the velocity response was very smooth. Also, electric wheelchair could move on following to assistance dog in K_ℓ=0.8. However, the velocity of wheelchair exceeded dog's speed, and passenger's riding comfort might become worse by speed adjustment. In figure 3(b), some typical values of time response are shown as a function of the translational gain K_ℓ. An extent of an oblique line in figure 3(b) shows the area where the electric wheelchair could not follow the dog. The lager K_ℓ, the earlier the wheelchair's speed would be in a steady. However, since the amount of overshoot P_m was increased, the fluctuation of velocity became larger. In the case of K_ℓ=0.277, the P_m was kept small and an electric wheelchair could follow an assistance dog even with faster speed, for example V_d=3[km/h].

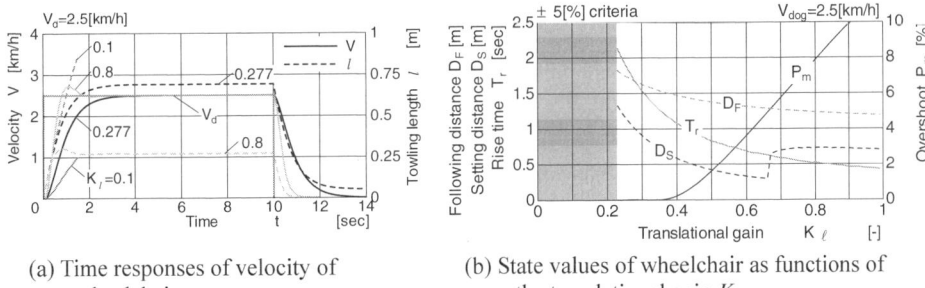

(a) Time responses of velocity of wheelchair

(b) State values of wheelchair as functions of the translational gain K_ℓ

Figure 3: Simulation results to fix the translational gain K_ℓ

EVALUATION OF FOLLOWING CHARACTERISTICS IN ROTATIONAL DRIVING

In order to determine the rotational gain K_ϕ, we considered one situation. An assistance dog walks straight, makes one rotation on keeping constant turning radius after that, and return to walk straight again. This situation can be deal with one part of turning corner. We used one evaluation value P described in equation (1).

$$P = \int_0^{2\pi} \frac{\left|R_W(\beta)^2 - \rho_d^2\right|}{\rho_d^2} d\beta \qquad (1)$$

Here, the value β was rotational angle to center of clearance circle O_d drawn by assistance dog, the ρ_d turning radius of assistance dog and the R_W distance between the center point of assistance dog's clearance circle O_d and the center point of electric wheelchair, shown in figure 4(a). Some simulation results were obtained, when the dog's speed was V_d=2.5[km/h] and turning radius of assistance dog was ρ_d=3[m]. In the case of small K_ϕ, the electric wheelchair turns right or left with large turning radius compared with the assistance dog's one, because the generated rotational signal was not enough to make velocity difference large. Especially, when K_ϕ was too small, the length of lead exceeded the limit and became impossible to follow an assistance dog. On the other hand, in the case of large K_ϕ, the turning radius of electric wheelchair was smaller than that of assistance dog. Therefore,

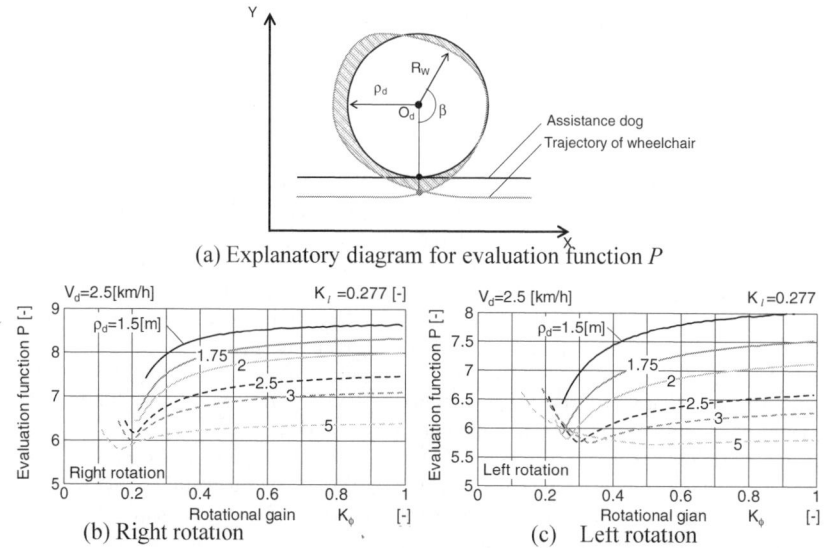

Figure 4 Simulation results to fix the rotational gain K_ϕ

electric wheelchair would turn inside an assistance dog. Both trajectories of them were overlapped at $K_\phi=0.33$ in left rotation and $K_\phi=0.25$ in right rotation. These condition on K_ϕ results the fluctuation of evaluation function shown in figure 4(b) and 4(c). The smaller evaluation function P implies the better condition of rotational gain K_ϕ. From the minimum points of these results, the rotational gain $K_\phi=0.25$ in right rotation and $K_\phi=0.25\sim0.3$ in left rotation might be determined. We evaluated driving trajectories in practical route many times and determined the optimum rotational gain $K_\phi=0.3$.

DISCUSSION AND CONCLUSIONS

We proposed the device and control system for an assistance dog guiding an electric wheelchair. And we can expect the mitigation of assistance dog's corporal burden by using this device and system. In addition, matching of Guidance unit and conventional driving system of electric wheelchair was copleted with only a fundamental gain adjustment system. In present situation of the development of electric wheelchair, major company develops all elements of electric wheelchair, body of wheelchair, driving system, control system, and input device. Moreover, if input device is changed, they tend to redesign most part of electric wheelchair. But, if consider about only input device and its compliance for conventional driving system of electric wheelchair, we can reduce the cost of development. Ultimate of this concept is standardization of connecting system between input device and driving system of electric wheelchair. And we can develop suitable input device for each user.

References

Maeda M and so on (2002), Evaluation of Traveling performance and Optimal Adjustment of an Electric wheelchair employing Bi-state Operation, *Mechanical Engineering Congress Japan*, **2002:7**, 167-168

Maeda M and so on (2003), Performance estimation of the Electric Wheelchair on Guidance control with Service dog, *Mechanical Engineering Congress Japan*, **2003: 5**, 147-148

Uemoto T and so on (2003), Driving characteristics of Electric Wheelchair by the Binary controller detecting motion of head, *Mechanical Engineering Congress Japan*, **2003:5**, 149-150

Coppinger R(1995), Dog Studies Program and Lemelson, *Center for Assistive Technology Development*, Hampshire College, 3-11

DEVELOPMENT OF MASTER-SLAVE ROBOTIC SYSTEM FOR LAPAROSCOPIC SURGERY

T. Suzuki[1], E. Aoki[1], E. Kobayashi[1], T. Tsuji[1], K. Konishi[2], M. Hashizume[3], and I. Sakuma[1]

[1] Institute of Environmental Studies, Graduate School of Frontier Sciences, The University of Tokyo
7-3-1, Hongo, Bunkyo-ku, Tokyo, 113-8656, Japan
[2] Department of Disaster and Emergency Medicine, [3] Department of Innovative Medical Technology,
Graduate School of Medical Sciences, Kyushu University,
3-1-1, Maidashi, Higashi-ku, Fukuoka, 812-8582, Japan

ABSTRACT

Laparoscopic surgery is widely performed as a less traumatic minimally invasive surgery. It, however, requires experiences and skills for surgeons. For realizing high quality and preciseness of surgical operation, we developed a new compact slave robot in a master-slave system. It consisted of manipulator positioning arm, forceps manipulator, and bending forceps. We integrated them into a slave robot with seven DOFs. *In vivo* experiment was conducted to evaluate the basic motion and the feasibility.

KEY WORDS

Minimally invasive surgery, laparoscopic surgery, computer assisted surgery, surgical robot, medical robot, master-slave system, RCM mechanism, pivot motion, robot forceps, and bending forceps

IINTRODUCTION

Laparoscopic surgery is widely performed as a means of minimally invasive surgery. In this method, surgeons cut 3-4 holes on the abdominal wall, and entire operations are conducted inside the abdominal cavity through the incision holes using rigid thin scope (laparoscope) and long-handled surgical tools such as forceps, scalpel (Figure 1). Compared with the conventional laparotomy requiring large incision on the abdomen, laparoscopic surgery has benefits for patients because of its small invasion; reduction of postoperative pain and hospital stay time. This patient-friendly technique, however, is rather difficult and cannot be applied to all cases, mainly because the limited degrees of freedom (DOF) of forceps eliminate the dexterity of surgeons. Forceps have only four DOFs (two-DOF pivot motion for orientation of forceps, and two DOFs for insertion and rotation of forceps). Procedure is operated symmetrically around the incision hole, so that surgeon gets confused (Figure 1). Responding to these issues, master-slave surgery-assisting robotic manipulators with maneuverable robotic arms and laparoscope

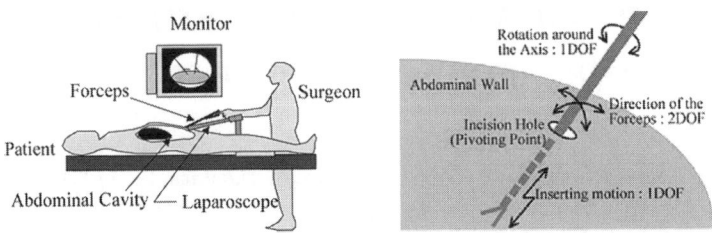

Figure 1: (left) laparoscopic surgery (right) limited DOFs of forceps

holder, such as da Vinci® surgical system, have been developed and clinically applied. These robots enhanced the dexterity and ability of surgeons beyond the limit of human hand, and enabled precise operation that could not be realized using conventional forceps (Hashizume M., et al. (2004)). They, however, have some problems, such as; large size for conventional operating room, occupation of the space above the abdomen by robotic manipulators, and collision with manipulators or surgeons. Thus, the purpose of this study was to develop a compact surgical-assisting robot with enough working space. It should function as a slave robot near the patient body (Kobayashi Y., et al. (2002)). We developed a new robotic system with three forceps that corresponded to both hands of surgeon and one hand of assistant, and evaluated the feasibility in *in-vitro* and *in-vivo* situations.

METHOD

A new robot system consisted of three modules; manipulator-positioning arm, forceps manipulator, and robotized forceps with a two-DOF bending joint and a grasper. We assumed that target organ was mainly liver and that the half weight of the liver, 6[N], should be manipulated, and we made it as the required specification for this manipulator.

Manipulator-positioning arm

There are two kinds of surgical robotic manipulators in the point of mechanical setting up; one is "suspending" arm (ex. da Vinci®), and the other is "bedside" arm (ex. ZEUS®(Marescaux J., et al. (2002))). Bedside arm has advantages in its small size. It is, however, clear that setting-up procedure is complicated because of less flexibility in alignment of each arm. Although suspending arm is large in size, flexible and intuitive positioning is possible. Thus, we adopted suspending arm for our surgical manipulator. We used a commercialized surgical microscope arm (CYGNUS, Mitaka Kohki, Japan) for a platform of manipulator-positioning arm to reduce development time (Figure 2). It had six DOFs using parallel linkage mechanism for the position and spherical joint for posture. Each joint could be driven passively and had disk brake with pneumatic-releasing mechanism. It had an advantage in the point that it kept braking and never release in case of electric power down. We used this arm for the rough positioning of the whole manipulator. This system aimed to operate three forceps manipulators equivalent to the surgeon's both hands and an assistant's hand. Thus, at the distal end of microscope arm, three 6-DOF arms were mounted for precise positioning of each forceps manipulator. It consisted of two parts: selective compliance assembly robot arm (SCARA) with passive 2-DOF horizontal positioning and active positioning using linear actuator, and passive spherical joint with pneumatic-releasing breaking. The pneumatic-releasing braking mechanisms could be released only by pushing two buttons at the same time, so that the brake would never be released by accident. This kind of redundant switch system is necessary to enhance the safety. The working space of each arm was R300[mm]*H200[mm]. Six-DOF arms enabled the intuitive arbitrary positioning. The position and orientation was measured using optical tracking system (Polaris®, Northern Digital Inc. Canada). No angle sensor such as rotary encoder was mounted at the joint of manipulator-positioning arm.

Forceps manipulator

Forceps manipulator had four DOFs equivalent to the conventional laparoscopic surgery (Figure 1). It adopted Remote Center of Motion (RCM) mechanism so that forceps could rotate around the pivot point (the incision hole) without any mechanical joint at the center of rotation. In this study, we used "two linear actuator mechanism" as an RCM mechanism using a couple of ball screws and servomotors. When the feeding ratio of two actuators was constant, the lines running through the tips of both linear actuators crossed at a certain point, and it was the pivot point (Figure 3)(Kim D., et al. (2002)). In Figure 3(center), "Rotation1" was realized by rotating the whole manipulator using a servomotor, and "Rotaion2" was realized using "two linear actuator mechanism". The center of pivot motion was located at the intersection of each rotational axis. The insertion of the forceps along the shaft was realized by "double-stage mechanism". It consisted of a couple of linear stage. One was mounted on the other, and it expanded like a ladder truck (Figure 3). This mechanism realized double-long traveling distance of the linear stage, and the size of manipulator was miniaturized. In this manipulator, the

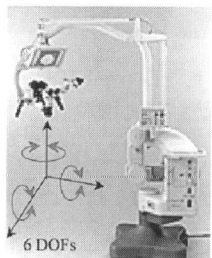 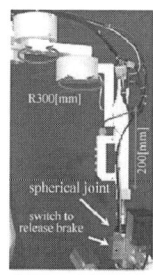 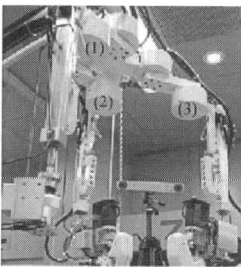

Figure 2: (left) microscope arm, (center) six-DOF precise positioning arm, (right) three arms for each manipulator

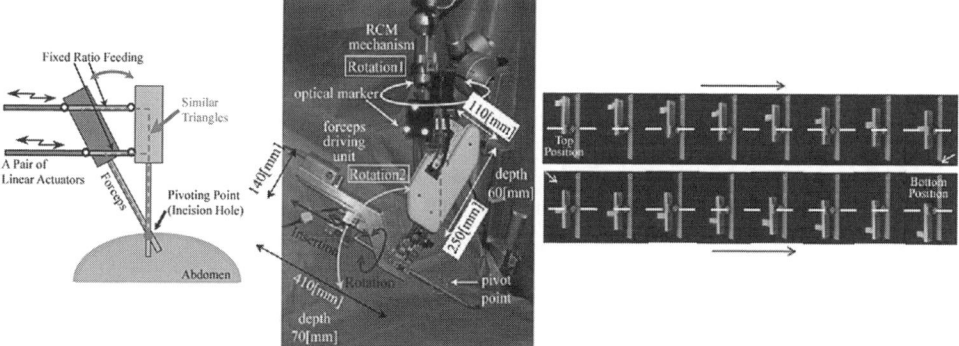

Figure 3: (left) RCM mechanism, (center) prototype, (right) double-stage mechanism

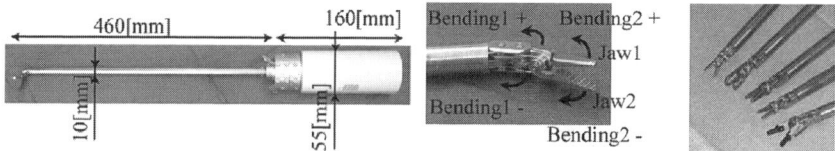

Figure 4: (left) prototype, (center) bending joint and grasper, (right) various kinds of forceps

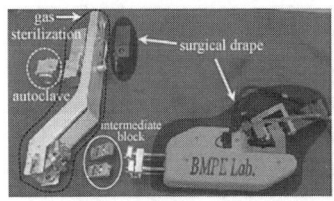 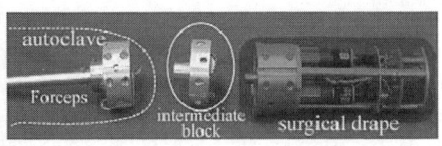

Figure 5: separation for sterilization; (left) forceps manipulator, (right) bending forceps

inserting distance of 300[mm] was realized using the 150[mm]-long linear stage. Rotational motion of the forceps around the shaft was realized by transmitting the rotation of the motor with gear. The size of RCM part was W250*H110*D60[mm^3], and forceps driving unit was W140*H410*D70[mm^3]. The whole weight was 4.5[kg].

Robotized forceps with a two-DOF bending joint and a grasper

We adopted minimal six DOFs to follow the motion of surgical procedure by surgeon. Robotized forceps had two DOFs for bending motion and one DOF for the grasping motion of the jaw (Figure 4), resulting seven DOFs by integrating four-DOF forceps manipulator. Bending and grasping motions were actuated with a tendon mechanism, which had advantages in its mechanical simplicity and compactness and wide working range. Contrary, elongation of the wire might lead to the decrease of tension and the poor controllability. Tension control mechanism should be implemented to maintain the tension, however, it will cause the increase of size and complexity. We used plastic wire made of polyarylate with low elastic property (approximately 0.8%)(Gravity Jigging, Fujino line Corp., Japan). Each bending motion was independent from each other mechanically, and the path length of the wire was constant despite of the bending angle of the other joint (Nishizawa K., et al. (2004)). We manufactured various kinds of instruments; needle holder, grasper, and soft tissue grasper (Figure 4). Bending forceps consisted of "forceps part" and "motor driving unit". The forceps was 10[mm] in diameter, 460[mm] in length. The motor driving unit was 55[mm] in diameter and 160[mm] in length. The whole weight was 0.7[kg].

Separation mechanism for clinical application

All surgical tools used in the operating room must be sterilized by steam (120[deg C], 2[atm]). Here, electric motors and circuit boxes could not be sterilized because they did not have water- and heat-resistant property. Thus, we adopted "three parts method"(Hefti J.L., et al. (1998)); forceps manipulator and bending forceps were separated into modules; autoclave-compatible part and non-sterilized part. Non-sterilized part was wrapped with surgical drapes (sterilized clothes) except mechanical interface. At the mechanical interface, a sterilized intermediate block is attached. Hereby, non-sterilized parts are not exposed, and sterilized module is attached to the intermediate block. With this method, sterilized module can be connected to non-sterilized part via intermediate block without contamination. Forceps manipulator was separated into the RCM motion module (non-sterilization) and forceps-driving module (sterilization compatible). They were connected using intermediate block. In the case of bending forceps, it could be separated into forceps module (autoclave compatible) and motor driving unit (non-sterilization) (Figure 5).

EXEPERIMENTAL RESULTS

Forceps manipulator

Basic performance evaluation of the forceps manipulator; working range, force, torque, and speed, was conducted (TABLE 1, Figure 6). The speed was measured under two conditions: one was conducted with the acceleration time of 25[msec], the other was 1000[msec]. This was because the maximum speed of a stepper motor depended on the acceleration time. We aimed to set the system control frequency to be 10[Hz] (100[msec/cycle]), and we assumed that the acceleration time was at most quarter of cycle, 25[msec]. The speed of 1000[msec] acceleration was measured as a case of enough long acceleration time. The working space was sector form whose radius was 340[mm] and whose vertex angel was 180[deg] in horizontal plane, and vertical depth was 360[mm] (Figure 6).

Bending forceps

We also measured the working range, positioning accuracy, backlash, maximum speed, and torque of bending forceps (TABLE 2, Figure 6). Bending angle for "Bending1" and "Bending2" was 304[deg] and 201[deg] respectively. The jaw for grasping opened up to 201[deg]. Low positioning accuracy was caused because of not only wire elongation and tension decrease, but also loose knot during the measurement. Output torque of "Bending1", "Jaw1", and "Jaw2" were equivalent to the force; 1.9, 3.0, and 3.7[N] respectively. We found abrasion and break of the wire, showing that friction occurred between the wire and path or pulley and that the transmission efficiency was reduced by the friction.

In-vivo experiments

We conducted *in-vivo* experiments on a swine to evaluate the system in simulated clinical environment. Setting-up time was less than 30 minutes and we thought it clinically feasible. The slave robot was controlled using a master manipulator by a surgeon (Mitsuishi M., et al. (2003)). Each DOF had enough working range, however, we found some problems. One problem was that; we inserted the surgical tool through trocar, an outer tube with air sealing sleeve to avoid gas leakage. This air sealing fitted the shaft of forceps tightly, so that inserting motion was obstructed. We also had a problem about the direction of trocar. The initial direction of trocar was not necessarily directed toward the target. In such a case, friction force between inner wall of trocar and outer surface of forceps disturbed the motion of forceps. This issue will be solved by integrating trocar into the forceps manipulator.

DISCUSSIONS AND CONCLUSIONS

We developed a new compact robotic system as a slave robot in a master-slave system. It consisted of three modules; manipulator positioning arm, forceps manipulator, and robotized forceps with a two-DOF bending joint and a grasper. Manipulator positioning arm realized intuitive easy setting up by the combination of rough and precise positioning. Forceps manipulator realized four-DOF motion of the forceps around the incision hole with wide working space (Figure 6). As for a RCM mechanism, "two liner actuator mechanism" realized mechanical

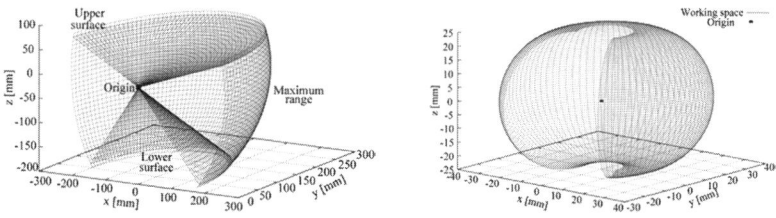

Figure 6: Working space; (left) forceps manipulator, (right) bending forceps

fixation of the pivoting point at the trocar port. For the inserting motion of the forceps, we adopted "double–stage mechanism" inspired by ladder truck. It realized double-long working distance comparing with the length of linear stage. Wire-driven bending forceps with two DOFs had wider working space than human hand, thus motion space was enough to follow surgical procedures. Tendon-driven bending forceps had two independent joints that realized easy control and stable motion. We also implemented separation mechanism for sterilization considering the clinical application. The separation mechanism showed another benefit in the bending forceps; we could easily exchange the various instruments (Figure 4) depending on the surgical scenario by just connecting instruments to the motor driving unit. Slave manipulator was built by integrating those modules. As the results of evaluations and in-vivo experiments, our new robot was feasible as a slave robot in the master-slave surgical robot system for laparoscopic surgery. As future works, we will also integrate an electric cautery forceps and a laser coagulator as end effectors on this robotic system. They will work as novel therapeutic devices different from conventional surgical robot just following the surgeon's hand motion.

This study was supported by the Research for the Future Program JSPSRFTF99I00904.

TABLE 1
EVALUATION RESULTS OF FORCEPS MANIPUTLATOR

		working range	max. speed (25/1000[msec])		torque or force
RCM Mechanism	Rotation1	171.27 ± 0.05[deg]	216[deg/sec]	230[deg/sec]	3.6[Nm]
	Rotation2	62.52 ± 0.12[deg]	91[deg/sec]	93[deg/sec]	3.6[Nm]
Forceps	Insertion	287.30 ± 0.06[mm]	120[mm/sec]	160[mm/sec]	5.5[N]
	Rotation	356.55 ± 0.13[deg]	7.5[round/sec]	17.5[round/sec]	0.48[Nm]

TABLE 2
EVALUATION RESULTS OF BENDING FORCEPS

	working range[deg]	accuracy[deg]	backlash[deg]	max.speed[deg/sec]	Torque[mNm]
Bending1(+)	+147.7	-3.7 ± 0.7	14.6 ± 0.1	161	16.0
Bending1(-)	-156.2	-3.6 ± 0.9	14.6 ± 0.1	161	16.0
Bending2(Jaw1)	+96.3	-5.8 ± 0.5	16.4 ± 0.4	120	31.5
Bending2(Jaw2)	-104.3	-4.3 ± 0.8	20.1 ± 0.3	120	40.7

REFERENCES

Hashizume M., et al. (2004). Robotic surgery and cancer: the present state, problems and future vision, *Japanese Journal of Clinical Oncology*, **34:5**, 227-237

Hefti J.L., et al. (1998) Robotic three-dimensional positioning of a stimulation electrode in the brain, *Journal of Computer Aided Surgery*, **3:1**, 1-10.

Kim D., et al. (2002). A new, compact MR-compatible surgical manipulator for minimally invasive liver surgery, *5th Medical image computing and computer-assisted intervention – MICCAI2002, LNCS2489 Springer*, 164-169.

Kobayashi Y., et al. (2002). Small occupancy robotic mechanisms for endoscopic surgery, *5th Medical image computing and computer-assisted intervention – MICCAI2002, LNCS2489 Springer*, 75-82.

Marescaux J., et al. (2002). Transcontinental robot-assisted remote telesurgery: feasibility and potential applications, *Annals of Surgery*, **235:4**, 487-492

Mitsuishi M., et al. (2003) Development of a remote minimally-invasive surgical system with operational environment transmission capability, *IEEE International Conference on Robotics and Automation*, 2663-2670.

Nishizawa K., et al. (2004). Development of interference-free wire-driven joint mechanism for surgical manipulator systems, *Journal of Robotics and Mechatronics*, **16:2**, 116-121.

WORKPLACE TASKS DESIGN SUPPORT SYSTEM BY USING COMPUTER MANNEQUIN

Keiji Mitsuyuki[1], Toshihide Ono[1], Yusaku Matsumoto[2],
Yoshiro Fukuda[3], Eiji Arai[4]

[1] Production Engineering Department, DENSO CORPORATION,
Aichi, 448-8661, JAPAN
[2] INFORMATION SERVICES INTERNATIONAL-DENTSU, LTD.
Tokyo, 33620-5399, JAPAN
[3] Department of Computer Science, Graduate School of Engineering, Hosei University,
Tokyo, JAPAN
[4] Department of Manufacturing Science, Graduate School of Engineering, Osaka University,
Osaka 565-0871, JAPAN

ABSTRACT

On designing manual operation processes, the 3-D computer simulation model of the workplace with a computer mannequin can help a process planner consider Kaizen ideas to improve workplace tasks. For using the 3-D computer simulation with the computer mannequin in practice, this paper proposes a new modeling system, which enables users to modify the workplace tasks easily and quickly. The system can record the time, the motion code based on MTM, and the posture of the computer mannequin integrally step by step while teaching those data to the computer mannequin. The recorded data regenerate the simulation semi-automatically, and the users remodel the manual operation process concentrating on only differences between the initial model and the improved model.

KEYWORDS

Human behavior analysis, Manual operation process, Design methodology, Computer mannequin,

1. INTRODUCTION

In the field of manual operation processes, the continuous improvement activity (Kaizen) is implemented after starting production to reduce worker's operation time. But the Kaizen after starting production is too late to face the ongoing global competition. Therefore, on designing them, it is important for a process planner to modify and optimize the workplace tasks as a pre-emptive Kaizen.

A 3-D computer simulation model of the workplace with a computer mannequin can present the worker's behavior, estimate the operation time based on MTM(Method Time Measurement), Maynard(1984), and assume the worker's posture at the virtual place. The process planner can analyze the operation time and the worker's posture through the simulation, consider Kaizen ideas, and also implement such ideas on the virtual manual operation process to confirm their effectiveness. However, in the conventional 3-D computer simulation with the computer mannequin, e.g. Jack(UGS PLM Solutions), DELIMA Ergo(Dassault Systems), it requires to teach the computer mannequin his tasks from scratch in order to remodel the workplace tasks according to the Kaizen ideas. This modeling takes a lot of time for trial and error. Therefore, using the conventional 3-D computer simulation with the computer mannequin is tough for the process planner from a practical standpoint. In order to reduce the remodeling time, this paper proposes a new modeling system, which enables users to modify the workplace tasks easily and quickly. To confirm the efficiency of the proposed system, a test case is tried out.

2. REQIREMENT FOR WORKPLACE TASK DESIGN SUPPORT SYSTEM

In order to plan efficient manual operation processes in the process planning phase, the process planner makes the Kaizen activity in the virtual manual operation processes on the computer shown in Figure 2. On the 3D graphics, the manual operation process is presented, and waste of time on hand, waste in transportation, waste of movement, overburden of working posture are found by observing the animation on the computer or evaluating the calculated time for each task. To resolve the wastes, Kaizen ideas are applied to the virtual manual operation process, and the effects of the ideas are evaluated. Until the effect reaches the target of cycle time and the appropriate worker's burden, Kaizen activity is repeated on the computer. Then, the actual workplace is constructed based on the improved virtual workplace. The expected system is the workplace task design support system that enables process planners to execute such virtual Kaizen activities easily and rapidly.

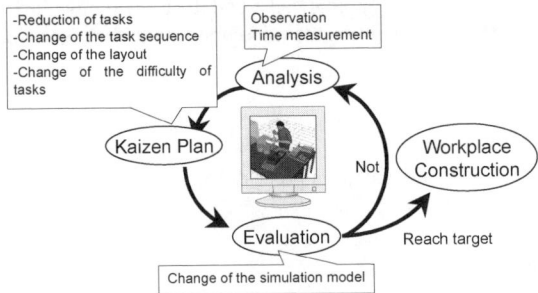

Figure 1: Kaizen activity on computer

3. WORKPLACE TASKS DESIGN SUPPORT SYSTEM

In order to reduce input effort, this study defines the data for reuse to generate simulation models, and proposes a method to reuse this data.

3.1 Basic model structure

A virtual manual operation process consists of a computer mannequin and a workplace. The model structure is as follows.
1) Computer mannequin
The posture of computer mannequin **M** is defined as $P = g(\mathbf{M}, \mathbf{S}, \mathbf{H_r}, \mathbf{H_l})$. $\mathbf{S} = (x_s, y_s, z_s)$: Stand point,

H_r: Position and angle of a right hand, H_l: Position and angle of a left hand, g: Function to generate a posture based on inverse kinematics.

2) 3-D object for workplace

The layout of 3-D object for workplace is defined as $\mathbf{L} = (L_{o1}, L_{o2}, \ldots, L_{on})$. $\mathbf{L}_{Oj} = (x_{Oj}, y_{Oj}, z_{Oj}, \theta_{xOj}, \theta_{yOj}, \theta_{zOj})$: Layout of 3-D object Oj, n: Number of 3-D objects.

3.2 Generation of status of manual operation process

The Scene, when the computer mannequin does a task, is called as a task scene. The status of manual operation process of a task scene k is defined as equation 1.

$$\mathbf{W}_k = \mathbf{W}_0 + \sum_{i=1,k}(\Delta \mathbf{L}_i, \Delta \mathbf{P}_i, \Delta t_i) \tag{1}$$

$\Delta \mathbf{L}_i$: Change of layout from task scene i-1to i

$\Delta \mathbf{P}_i$: Change of posture of computer mannequin $\mathbf{M}$ from task scene i-1to i

Δt_i : Time from task scene i-1to i calculated from T_k (Motion code for task description of a task scene k) and D_{Tk} (Motion code for task difficulty of a task scene k).

In order to make a simulation model, task scenes $\mathbf{W}_k$ are generated through modeling interface to teach the posture of the computer mannequin and to build, place, and move objects. The generated task scenes $\mathbf{W}_k$ are recorded. The recorded task scenes from $\mathbf{W}_0$ to $\mathbf{W}_{end}$ are divided by *tunit* (e.g. 1/30 second), which is a unit of time slice on animation, in order to generate ($\Delta \mathbf{L}_t$, $\Delta \mathbf{P}_t$, *tunit*) as an animation file. The upper part of figure 2 shows the conventional system structure.

3.3 Proposed method to reuse status-generating procedure

In addition to $\mathbf{W}_k$, the layout of 3-D objects, $\mathbf{L}_k$, the posture, $\mathbf{P}_k$, motion code for task description, T_k, motion code for task difficulty, D_{Tk}, target object of task T_k, O_{Tk}, and, positions and angles of both hands, $\mathbf{H}_{rk}$ and $\mathbf{H}_{lk}$, are recorded all together as the status-generating procedure record data, $\mathbf{W}_k^*$.

$$\mathbf{W}_k^* = (\mathbf{W}_k, \mathbf{L}_k, \mathbf{P}_k, T_k, \mathbf{S}_k, D_{Tk}, O_{Tk}, \mathbf{H}_{rk} \text{ or } \mathbf{H}_{lk}) \tag{2}$$

Figure 2 shows the developed system structure. In order to reuse the previously recorded $\mathbf{W}_k^*$ to regenerate $\mathbf{W}_k^*$', it is necessary to choose reusable data of $\mathbf{W}_k^*$. Figure 3 shows the algorithm for judging reuse of $\mathbf{W}_k^*$. The recorded data are fully reused in the case a) and e), and partly reused in the case c) and d). Therefore, the remodeling effort can be reduced with this algorithm.

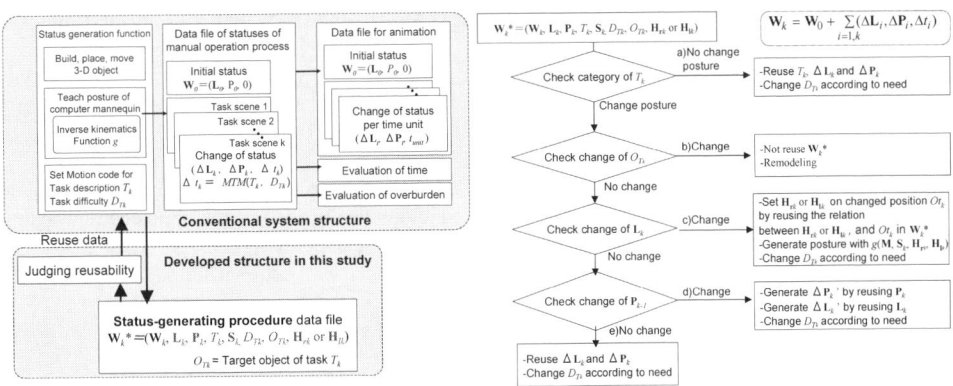

Figure 2: Developed system structure   Figure 3: Algorithm for judging reuse

3.4 Implementation

This proposed workplace tasks design support system is developed based on Jack in this study.

4. CASE STUDY

4.1 Test case

Figure 4 shows a test case. A worker sets a main part on a pallet, and then assembles a part A on it with a screw B, and a part C. After assembling, the worker flicks the switch in order to move the pallet to the next process. The initial plan of that process is shown in the left picture in figure 4. The improved plan has five differences including the positions of the box for main parts, the box for screws B, the box for parts C, the screwdrivers, and the switch. Also the sequence of assembling a part A is changed to after a part C. The right picture in figure 4 shows this improved plan.

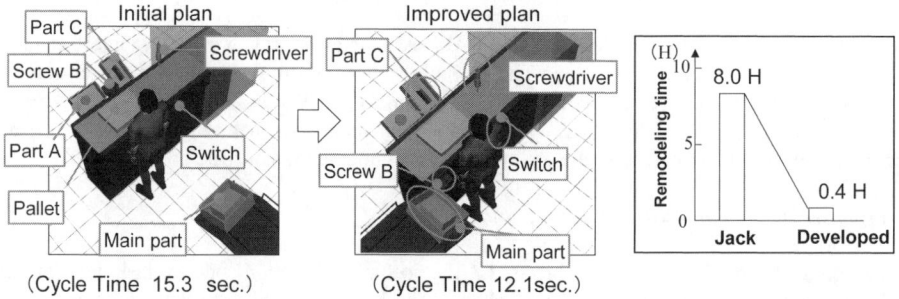

Figure 4: Test result

4.2 Test result

By building simulation models of the initial plan and the improved plan, it becomes clear that the improved plan shortens worker's operation time by 3.2 seconds. Through this case, time taken for remodeling the initial plan into the improved plan is compared between the cases using the original Jack and the developed system. The result is that the original Jack took 8.0 hours, and the developed system took 0.4 hours. The developed system can remodel the simulation plan more than 10 times as fast as the original Jack.

5. CONCLUSION

This study shows that remodeling time, which consists of the time to teach 3-D computer mannequins and the time to move 3-D objects, can be reduced dramatically by reusing the recorded status-generating procedure. The proposed workplace tasks design support system provides easier and quicker remodeling for planning manual operation processes than conventional simulation systems.

6. REFERENCES

Maynard H. B., et al. (1948), Methods-time measurement, New York, McGraw-Hill Book Co., USA.

SIMULATION AND EVALUATION OF FACTORY WORKS USING MUSCULOSKELETAL HUMAN BODY MODEL

Takako Sato, Hiroshi Arisawa

Graduate School of Environment and Information Sciences, Yokohama National University,
79-7, Tokiwadai, Hodogayaku, Yokohama, 240-8501, JAPAN

ABSTRACT

Optimum design of work motions is one of the most important issues to construct human-machine co-existing systems. Traditional Ergonomics tried to evaluate pain/fatigue on an observation base. However, generic evaluation method does not exist and Ergonomics has not discussed the mechanism of pain/fatigue. So we proposed individual musculoskeletal human model and proposed Info-Ergonomics concept. This paper overviews Info-Ergonomics concept and focuses on modeling and description of musculoskeletal human bodies.

KEYWORDS

Human Body Modeling, Musculoskeletal Model, Human Motion Simulation, Ergonomics, Human Body Database

1. INTRODUCTION

Ergonomics is one of the fields which pursues the physiological and physical comfortability of various human works, such as factory workers, sports players, and rehabilitation patients. Among them, optimum design of work motions is one of the most important issues in human-machine co-existing systems. Traditional Ergonomics have been facing this problems from the viewpoint of measuring characteristics (i.e. human body shape, weight of each body segment, range of motion of joints) of human bodies. Then they evaluate pain/fatigue on an observation (questionnaire) base. However this approach has many problems. First, generic method to formalize human bodies and to describe problems do not exist. Specific model and methods have been developed in case by case. Second, Ergonomics are just observing correlation between human posture/motion and pain/fatigue, but have not discussed the mechanism of pain/fatigue.

On the other hand, if we can construct precise musculoskeletal human body in the individual level, we can simulate bone muscle action by captured posture/motion and evaluate pain/fatigue in a series of works.

So we proposed individual musculoskeletal human model "BBHM" (Bone Based Human Model) and proposed Info-Ergonomics concept, which means "Information model based Ergonomics".

This paper overviews Info-Ergonomics concept firstly, then will be focusing on modeling and description of musculoskeletal human bodies.

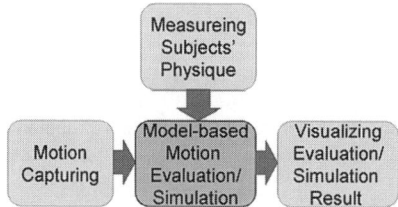

Figure 1: Concept of Info-Ergonomics Simulation System

2. INFO-ERGONOMICS SIMULATION

As mentioned above, Info-Ergonomics is a concept which can simulate pain/fatigue on computer-based human mockup model. The basic idea of Info-Ergonomics established as a wide-range application of Real World Database (RWDB) system [1]. RWDB is integration of 4 component of technologies, Video Capturing, Model based Analysis, Database Processing and Computer Vision. Similarly in the Info-Ergonomics Simulation, Capturing human body motion, Human model creation, Model based analysis/simulation and Visualization are fundamental technologies. This concept is summarized in Figure 1. The function and target of each module is as follows.
- Measuring subjects' physique
  In order to evaluate/simulate human body motions precisely, customized human body model must be required. When measuring subjects' physique, not only body size but the body characteristics (range of motion, muscle strength, and so on) should be included.
- Motion capturing
  Camera-based motion capture system is adequate because it can detect human posture at each time point without disturbance.
- Model-based Motion evaluation/simulation
  Using the customized human model mentioned above, load/fatigue estimation must be done in musculoskeletal level for each time-point-posture of a work motion, provided by motion capturing system.
- Visualizing Evaluation/Simulation result
  In order to help intuitive understanding of simulation/evaluation result, some 3D CG systems which can display all bones and muscles with textures in real-time way are required. Also, coloring bone/muscle segments depending on pain/fatigue level is highly recommended.

When we realize total system, we must develop each device, software, and design data format in detail. Especially the data format which bridges functions and functions has an important role. As a result, total system has been designed as a data flow map as shown in Figure 2.
Detail functions of major boxes will be discussed in later sections.

3. PRECISE HUMAN MODEL AND THE DESCRIPTION METHOD

In order to achieve precise evaluation/simulation reflecting individuality(body size, flexibility, physical condition and so on),   creating precise human-mockup is the most important issue. However, as human body has a very complicated structure, it is impossible to implement all factors of human bodies such as bones' shape, positions to connect bone and muscle, maximum muscle force, and so on. Therefore selecting essential parameters to execute human simulation and the measurement methods of those parameters should be   considered keenly. Also, another important issue is the description method of individual human body, i.e. model description methodology.
From now on, we will be focusing on core technologies and data format to descirbe measurement

results.

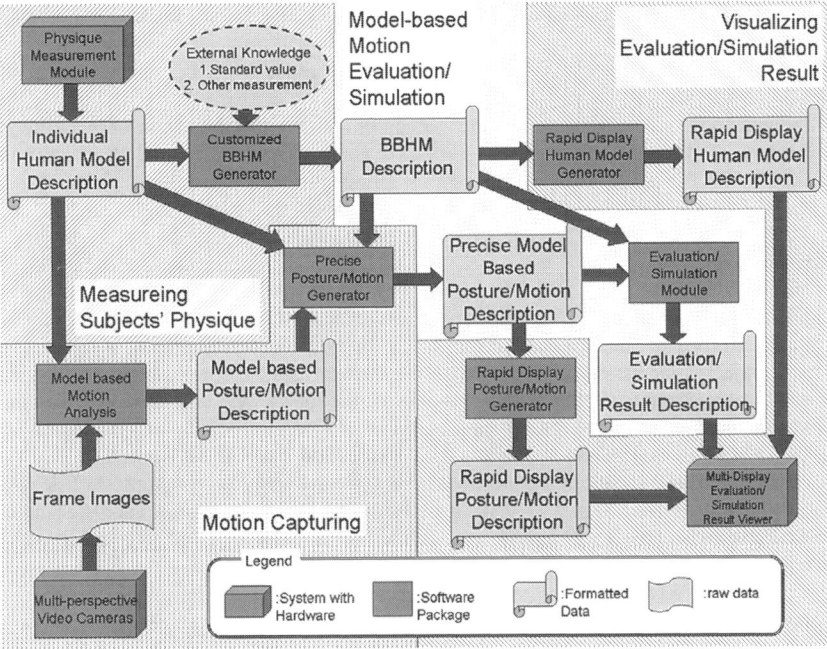

Figure 2: Data Flow Map of Info-Ergonomics Simulation System.

3.1. Measuring Subjects' Physique

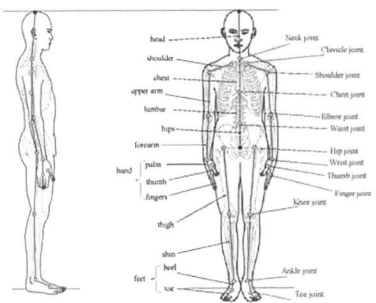 

Figure 3: subject's mockup for motion capture   Figure 4: motion capture results

In order to evaluate/simulate individual posture/motion taking individuality (age, statue, life style, flexibility, and so on) into consideration, we need to measure subjects' individuality at first.
Among many measurement items showing idividuality, two items below are definitely related.
- Segments' length, width, center of joint
  These parameters affect the geometric position of human body model.
- Flexibility, muscle strength
  Even if the physique is the same, an athletic ability can be extremely different depending on these parameters.

Ideally, it is necessary to measure the sizes, shapes and positions of all a subject's bones or muscles in

order to imitate precise human body model with subject's individually. However, it is not possible without CT/MRI because bones and muscles can't be observed from outside. It is too expensive to use CT/MRI except for medical treatments. On the other hand, our strategy suggests to detect only small number of parameters.
So we tried to reduce parameters to small number of observable ones such as length of upper arms, shoulder width and so on.
We divide human body into a number of segments according to major joints, then we measure the size of body segments. As for range of motion of joints, we adopt measuring method standardized by the Japanese Orthopedic Association and the Japanese Association of Rehabilitation Medicine. About center of joints, we use estimation methods suggested by medical statistics[2][3][4].

3.2. Motion Capturing

In order to detect human motions (or posture sequences) precisely, existing motion capturing systems force us to put a number of "markers" on the surface of body. However, of course, it restricts flexible movement of subjects in a variety of situations. On the other hand, motion capture without markers cannot guarantee the accuracy. This is the reason we introduced "Model based Posture Analysis" method. In the method we use subjects' physique.
So we developed motion capture method with a few markers and image processing.
In order to enable us to capture motion with few markers, we use subjects' physique. At first, we make subjects' mockup with markers. Then we make the mockup to do various postures and we compare not only markers' positions but also the mockup's outline and the captured image's region of subject. Figure 3 shows subjects' mockup.
Result of motion capture with subjects' mockup and 6 markers (right and left of hip, both wrist, both ankle) is shown in Figure 4. By the grace of marker, we are able to capture subject's motion with high accuracy.

3.3. Motion evaluation/simulation

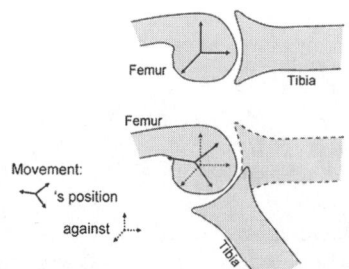 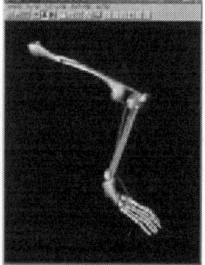

Figure 5: Tibial sliding against femur   Figure 6: Musculoskeletal human model (Right leg)

Detected motions from capturing devices in the section 3.2 are very brief ones. They are "solid model" level description shown in Figure 3, which has only 17 joints and 43 degrees of freedom.
On the other hand the musculoskeletal model used for precise evaluation/simulation must be much more complicated. A human body has approximately 200 bones and 600 muscles in total. After excluding mutually fixed bones such as cranium, number of actual bones is 45.
Furthermore, another type of complexity problem exists. The model shown in Figure 3 is solid-link type. However, in the real joints, a center of rotation is not always a fixed point. For instance, in knee joint, tibia rotates with sliding against femur. (Figure 5) As a result we must develop converting algorithm from motion captured data to the real bone-based motions.
Joint motions in the model shown in Figure 3 can be regarded as a perspective motion of human body.

Several estimating methodologies of real movement of bones using such perspective motions have been proposed[5][6]. Using one of those method, we developed the algorithm of conversion. The posture data format for one BBHM joint shown in Table 1.

Also, data about BBHM bones and muscles are described as shown in Table 2, 3. From those data the spatial positions of bones and muscles of each point of time can be calculated in geometrical model Additionally, using Hill-type model[7], one of the well-known model of muscle, we can estimate maximum muscle force of each timepoint.

4. EXPERIMENTAL RESULTS AND DISCUSSION

As discussed above, we can estimate muscle length minute by minute from physique and motion data. Then, in order to estimate maximum muscle force of each time point, we use Hill-type model which models Rheological parameters of muscle. By using Hill-type model, we are able to know relation among muscle length, muscle contraction speed, muscle diameter, and maximum muscle force.

As a simple example estimating muscle force, we estimated muscle forces of right legs when the subject was walking. There are about 34 muscles, but here, we extract 3 muscles which are mainly used when subject walks. Figure 6 shows those muscles and bones of right leg. Estimation results are shown in Figure 7. We are able to know that muscles are repeatedly contract synchronizing with walking.

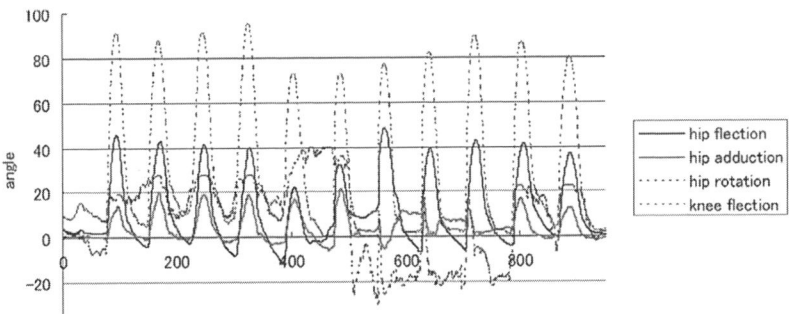

Figure 7: Estimation Result

5. CONCLUSION

In this paper, we proposed musculoskeletal human body model for motion evaluation/simulation. We also surveyed principles of motion capturing and physique measuring. Another Important issue, description method of physique and posture are discussed with partially completed examples.

REFERENCES

[1] H. Arisawa, T. Tomii, H. Yui and H. Ishikawa. (1995). Data model and architecture of multimedia database for engineering applications. *IEICE Trans. Inf. & Syst.* **E78-D:11**, 1362-1368.

[2] Clinical Gait Analysis Forum of Japan. (1992). *DIFF Data Interface File Format (DIFF) User's Manual*, Clinical Gait Analysis Forum of Japan.

[3] Davis, R. B., Ounpuu, S., Tyburski, D. and Gage, J. R. (1991) A gait analysis data collection and reduction technique. *Human Movement Sciences* **10:5**, 575-587.

[4] Vaughan, C. L., Davis, B. L., and O'Conners, J. C. (1992). *Dynamics of Human Gait*, Human Kinetics Publisher.

[5] I. Kapandji (1974). *Physiology of the Joints: The Trunck and the Vertebral Column (Trunk & Vertebral Column)*, Churchill Livingstone.
[6] Furukawa, D., Mori K., Suenaga, Y. (2002) Human Spine Posture Estimation from 2D Frontal and Lateral Views Using 3D Physically Precise Spine Model. *Proceedings of MVA2002*. 224-227.
[7] Delp S. L., Loan P.(1995) A graphics-based software system to develop and analyze models of musculoskeletal structures, *Comp. Biol. Med.* **25:1**, 21-34.

TABLE 1
BBHM BASED POSTURE DESCRIPTION FORMAT

data name	format	sample data	semantics
TIMEPOINT	float	1.333	time point
SEGID	four figures int	0003	bone ID number
MOVEMENT	float x 3	0.02, 0.01, 0	movement (Figure 5)
ROTATION	float x 4	0.866, 0.354, 0.25, 0.25	quaternion in local coordinate

TABLE 2
BBHM DESCRIPTION FORMAT (BONE)

data name	format	sample data	semantics
SEGID	four figures int	0003	bone ID number
SEGNAM	string	right shinbone	bone name
VRTNUM	int	1082	vertex number
VRTn	float x 3	0.1282 0.343 1.849	local coordinate of n-th vertex($0 < n \leq$ VRTNUM)
SRFNUM	int	1254	surface number
SRFLOOPr	float x (X+1)	487 486 489 -1	surface loop($0 < n \leq$ SRFNUM, X>2)(CCW)
LENGTH	float	8.3 4.8 48.2	bone length (sagittal-horizonal axis, frontal-horizonal axis, vertical axis)
NXTSEGNUM	int	2	next bones' number
NXTSEGIDm	four figures int	4	bone ID of next m-th bone($0 < m \leq$ NEXTSEGNUM)
NXTSEGCRDm	float x 3	-63.3 -39.6 -23.43	local coordinate of connect point of next m-th bone ($0 < m \leq$ NEXTSEGNUM)
MSLVRTNUM	int	4	muscle attached vertex number
MSLVRTCRDp	float x 3	-59.2 -64.9 -19.23	local coordinate of p-th muscle attached vertex ($0 < p \leq$ NEXTSEGNUM)

TABLE 3
BBHM DESCRIPTION FORMAT (MUSCLE)

data name	format	sample data	semantics
SEGID	four figures int	1002	muscle ID number
SEGNAM	string	right muscle of psoas major	muscle name
VRTNUM	int	4	number of attachment(fixed end, mobile end, via points)
BONEIDq	four figures int	0001	attaching bone ID($0 < q \leq$ VRTNUM)
MSRVRTIDq	int	3	local coordinate ID(=p) of BONEIDq
MAXFORCE	float	10.52	maximum force of natural length
LENGTH	float	28.3	muscle length of fundamental standing position

DEVELOPMENT OF MEASURING DEVICE FOR LOWER LEG SWELLING DURING STANDING WORK TASKS

T.Kawano and T.Mizuno

Department of Industrial and Systems Engineering, Setsunan University,
Neyagawa, Osaka 572-8508, JAPAN

ABSTRACT

In this study a measuring device called as SWELL (Sensoring Wire for Edema of Lower Leg) has been developed in order to measure the lower leg swelling of the worker during the standing work tasks. The device consists of a flexible wire, a coil spring, and a flat spring with a strain gauge. The wire is wound around the lower leg with proper tension and the strain of the flat spring produces when the leg swelling develops. The characteristics of the device, such as linearity, hysteresis, drift, and thermal behavior, are demonstrated. Experiments were carried out to measure the swelling of the lower leg for thirteen subjects. The results showed that SWELL was continuously able to measure the leg swelling defined as the increase of the circumference length of the leg, the magnitudes of the swelling increased almost linearly, and the measured average magnitudes were 3.0mm after standing for 30 minutes. In addition, a strong correlation was shown between the swelling magnitude estimated by digital camera method and the one measured with SWELL.

KEYWORDS

Lower Leg Swelling, Measuring Device, Strain Gauge, Standing Work, Human Directed Manufacturing System

1. INTRODUCTION

The assembly lines and cell production system consist predominantly of workers' standing work tasks. However, standing works often cause leg swelling and fatigue. Furthermore long-term leg swelling increases the risk of developing pathological reactions such as varicose vein, thrombosis and pulmonary embolism[1]. In order to decrease such fatigue and risks the workplace so as to reduce the leg swelling should be designed in the standing work tasks. However, the adequate measuring device for the leg swelling has not been around in actual workplaces. In order to evaluate the swelling such factors are considered as lower-limb skin temperature[2], venous pressure[3], lymph flow[4], tissue fluid pressure [5], ultrasound and venography[6], and bioelectrical impedance[7]. However, almost previous studies have just used a tape measure to measure the magnitude of the swelling itself. The measurements with the tape measure vary widely depending on the examiners. The measurement of

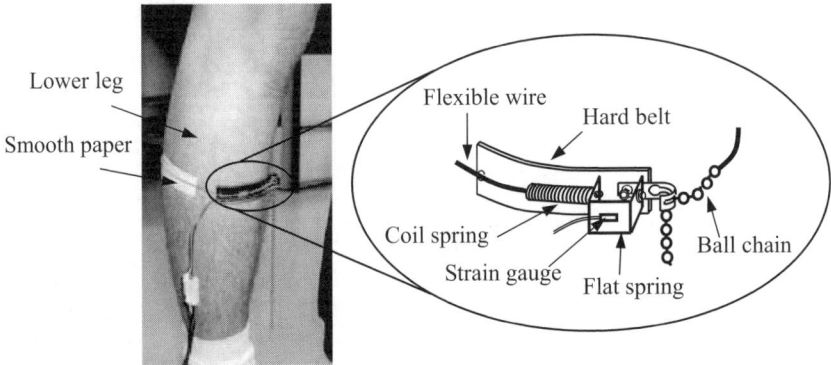

Figure 1: Measuring device SWELL for lower leg swelling

the volume of the lower leg by means of water displacement volumetry (WDV) has also been reported[8]. This method is attended with practical difficulties particularly in the workshop. The purpose of this study is to develop a measuring device utilized for the worker's lower leg swelling in the workplaces and to perform the characteristic tests of the device. Finally experiments were carried out to apply the device to standing work tasks for thirteen subjects.

2. MEASURING DEVICE

Figure 1 shows a measuring device of the leg swelling developed in this study. It is named as SWELL (Sensoring Wire for Edema of Lower Leg). The device mainly consists of a flexible wire, a coil spring, a flat spring, and a strain gauge(T44-FA2-350-11-VS015C). The wire and the flat spring are made of stainless steel. The both ends of the flat spring($10 \times 20 \times 0.3t$) are bent at 90 degrees. The coil spring and the wire($220 \times 1.0 \phi$) are connected to the one end of the flat spring. The other end of the flat spring is fastened to the hard belt. When the wire is wound around the lower leg with proper tension, the flat spring produces strain as the leg swelling develops. The magnitude of the leg swelling is estimated by measuring the strain of the flat spring. Here the leg swelling is defined as follows: The magnitude of the leg swelling is expansion of the maximum circumference length of the lower leg. The unit is "mm". The ball chain is used to fix the wire to the other end of the flat spring adjusting the length of the wire. The wire refrains from digging into the skin of the leg due to the smooth paper belt in the figure. And the smooth paper makes the wire move smoothly when the circumference length increases. A linear relationship was found between the strain and the expansion of the coil spring after repetitions of the measurements. The spring constant of the coil spring is 11.16gf/mm. It is considered that the value is too small to inhibit the swelling of the leg.

3. CHARACTERISTIC TESTS

Calibrations of SWELL were performed. Nonlinearity(1.51%) and hysteresis(1.50%) within 10mm expansion were both sufficiently small. The proportionality constant was 0.13V/mm. When SWELL was wound around the leg, the value of the output voltage vs. circumference of the leg became smaller (0.11V/mm) due to the friction between the wire and the smooth paper.

Strain gauge has a potential for a drift caused by the passage of time and temperature rises. First, the drift in the course of time was examined. While the circumference length of a concrete cylinder(ϕ 100

×200) was continuously measured using SWELL for 30 minutes, the drift of the output voltage did not occur. Next, the circumference of the same concrete cylinder heated successively with an electric heater was measured again using SWELL for 30 minutes. The output of SWELL reduced from 0 to –0.2mm as the temperature of the concrete surface rose gradually from 28 to 36°C. It is supposed that the amount of change in the output of SWELL is sufficiently small because the surface temperature of the lower leg is around 32°C.

4. SWELLING DURING PROLONGED STANDING

Experiments were carried out to measure the swelling of the lower leg for ten male and three female subjects, aged 21 to 24. The subjects were asked to keep standing without moving their legs and with peg operations by their hands for 30 minutes. For control experiments, they were also asked to keep sitting without doing anything for 30 minutes on the other day. Figure 2 shows typical changes in the leg swelling for 30 minutes. The magnitudes of the swelling increase almost linearly. Thus SWELL can continuously measure the leg swelling. The measured average magnitudes for thirteen subjects were 3.0mm after standing for 30 minutes. In order to verify the result of the measurement with SWELL, non-contact measurement using a digital camera was conducted[9]. The pictures were taken from the right-hand side of the subjects. Two pictures taken before and after being swollen were superimposed by manual operations using image-processing software in order to measure the increase

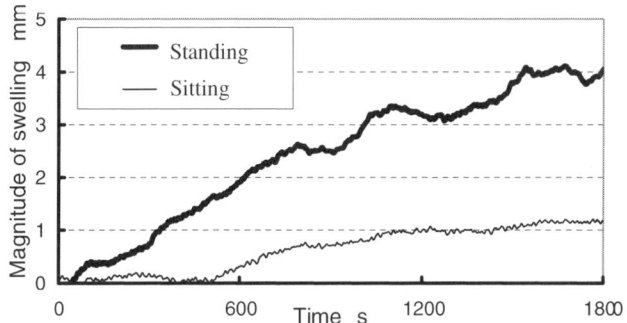

Figure 2: Typical changes in the leg swelling (21 years old, male)

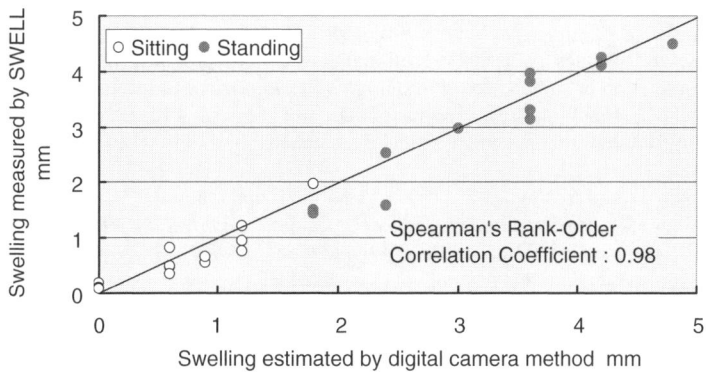

Figure 3: Correlation between the swelling estimated by digital camera method and the one measured by SWELL (Maximum value after 30 min)

of the leg diameter. On the assumption that the leg is circular in cross section and the swelling develops uniformly, the magnitude of the swelling defined in this study is roughly estimated by the calculation that the increase of diameter is multiplied by 3. Figure 3 shows the correlation between the magnitude of the swelling estimated by the digital camera method and the one measured by SWELL. There is a strong correlation between them.

5. CONCLUSION

A measuring device (SWELL) for lower leg swelling of the human was developed and the accuracy of the device was checked. Experiments were carried out to measure the swelling of the lower leg in the standing work tasks. The results are summarized as follows:

(1) SWELL consists of a flexible wire, a coil spring, and a flat spring with a strain gauge. The strain that the flat spring produces is measured as the leg swelling develops.
(2) SWELL has the following characteristics: A linear relationship exists between the voltage outputs of the strain amplifier and the expansion within 10mm of the coil spring. The strain gauge has no drift in the course of time. As the temperature of the object to be measured rises, the output of SWELL does not increase but decreases slightly. The amount of decreasing is sufficiently small around the temperature of the lower leg.
(3) SWELL can continuously measure the leg swelling defined as the increase of the circumference length of the leg. The magnitudes of the swelling increase linearly, and the measured average magnitudes are 3.0mm after standing for 30 minutes.

6. REFERENCES

[1] Winkel J. (1987). On the significance of physical activity in sedentary work. Knave B. and Widebäck, P.G. eds. *Work with display units 86*. Elsevier Science Publishers, North-Holland, 229-236.
[2] Noddeland H., Winkel J. (1988). Effects of leg activity and ambient barometric pressure on foot swelling and lower-limb skin temperature during 8h of sitting. *European Journal of Applied Physiology*, **57**, 409-414.
[3] Pollack A.A. and Wood E.H. (1949). Venous pressure in the saphenous vein at the ankle in man during exercise and changes in posture. *Journal of Applied Physiology* **1**, 649-662.
[4] Olszewski W., Engeset A., Jaeger P.M., Sokolowski J. and Theodorsen L. (1977). Flow and composition of leg lymph in normal men during venous stasis, muscular activity and local hyperthermia. *Acta Physiologica Scandinavica*, **99**, 149-155.
[5] Kirkebo A., and Wisnes A. (1982). Regional tissue fluid pressure in rat calf muscle during sustained contraction or stretch. *Acta Physiologica Scandinavica*, **114**, 551-556.
[6] Johanning J.M., Franklin D.P., Thomas D.D. and Elmore J.R. (2002). D-dimer and calf circumference in the evaluation of outpatient deep venous thrombosis, *Journal of Vascular Surgery*, **36:5**, 877-880.
[7] Seo A., Kondo Y. and Yoshinaga F. (1997). A portable apparatus for monitoring leg swelling by bioelectrical impedance measurement, *Journal of Occupational Health*, **39**, 150-151.
[8] Deursen D.L., Deursen L.L.J.M., Snijders C.J. and Goossens R.H.M. (2000). Effect of continuous rotary seat pan movements on physiological oedema of the lower extremities during prolonged sitting, *International Journal of Industrial Ergonomics*, **26:5**, 521-526.
[9] Kawano T. and Tada T. (2003). Characteristics of Swelling Around Lower Leg During Standing Work Tasks, *Proc. of XVth Triennial Congress of the IEA*, **2**, 395-398.

SPREADING OF CLOTHES BY ROBOT ARMS USING TRACING METHOD

Khairul Salleh, Hiroaki Seki, Yoshitsugu Kamiya and Masatoshi Hikizu

Department of Mechanical Systems Engineering, Kanazawa University,
Kakuma-machi, Kanazawa, Ishikawa, 920-1192 JAPAN

ABSTRACT

This paper proposes a method of clothes spreading using two robot arms with sensors equipped grippers and a fixed CCD camera. This research is focused on getting the robot arms to find and hold two side-by-side corners of a rectangular towel, one by using a simple image processing method and the other by using a unique tracing method. The tracing method in our context is a method of tracing the towel's edge by a robot arm based on the feedbacks from sensors and also images from the CCD camera. By using this method, we have succeeded in spreading a rectangular shaped towel.

KEYWORDS

Edge tracing, deformable object, edge of clothes, corner, image processing, home service robot

INTRODUCTION

Unlike factory robots that do routine work and mostly handle rigid objects, home service robots must be able to cope with changes to the surroundings and be able to manipulate soft objects. Unfortunately, the development of robots that can accomplish complicated housework that requires complicated manipulation is difficult and is still in the early stages although urgently required. We focused on this problem and chose clothes spreading and folding as an example. In daily life, spreading clothes is an action conducted usually before hanging one's clothes to dry or folding it. The reason why clothes are hard to manipulate is because they are deformable objects. There are several theoretical researches concerning clothes manipulation but very few have been proven practically [1][2]. We decided then to do a practical research using our own suggested method. The basic idea of clothes spreading is to hold the corner(s) of the clothes in order to spread or fold it [1]. The problem is how to find the corner(s). A lot of past researches failed because of this. Although the usage of CCD camera plays a big role in finding a corner [2][3], the corner(s) of clothes will not necessarily be visible to the camera. This will result in the robot not holding the true corner of the clothes thus resulting in improper handling of the clothes. By using tracing method to find the second corner, we will prove that the second corner found is the one next to the corner initially founded and not the one across it.

TOWEL SPREADING SYSTEM

Figure 1 shows a scene in towel spreading. In this research, two robot arms, a Js2 from Kawasaki Heavy Industries with 6 D.O.F and an RCH-40 from Yamaha with 5 D.O.F, both equipped with grippers that were designed to manipulate clothes are used. The grippers have force sensors to detect the force being applied to the fabric and also 4 infrared sensors each in order to detect whether or not the fabric is inside the gripper. Figure 2 shows the details of the gripper. The grippers were designed so that they can stand up to 400gf of force and at the same time being sensitive to force change. The gap between the two fingers of a gripper (written as gap of a gripper later on) is controlled using encoder of a motor shown. A CCD camera is also used to detect the corner(s) of the towel and is fix positioned in front the two robot arms as shown in Figure1. The images taken by the CCD camera are in 8-bit gray scale and 640x480 dots in size. An image processing board TRV-CPW5 is also equipped.

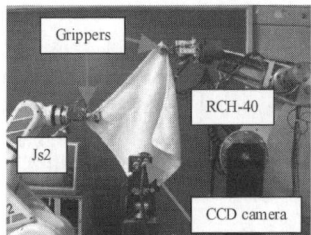
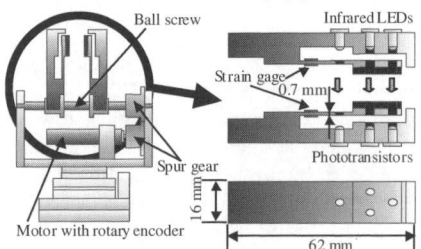

Figure 1: Clothes spreading system   Figure 2: Gripper designed for clothes spreading

FINDING & GRASPING OF THE FIRST CORNER

In this chapter, we will explain to you our proposed method on how to detect and hold the first corner of a rectangular towel using simple image processing method. After the towel is picked up randomly and being held in the air, the lowest point to the ground is usually one of the four corners of the towel. In order to locate this corner, CCD camera is used. This process is shown in Figure 3. First, difference image photometry method is used to erase the background and leave only the outline of the towel by taking 2 pictures of the towel, the latter one taken after the robot has slightly moved downwards (about 5mm). The two images are then subtracted and the resulting image (difference image) will be produced. The image is then binarized and dilation and erosion operations are applied to reduce noise. The resulting image is then labeled and the lowest point within the labeled data with more than a certain amount of pixels (small areas are considered noise) is recognized as a corner of the towel.

1. 2 images of towel are taken at slightly different positions
2. Difference image of the images
3. Binary image, after dilation and erosion operations
4. Labeling and the lowest point is determined
6. Corner found

Figure 3: Image processing in locating first corner

After the corner has been located, the gripper holding the towel (Gripper A) will rotate at the wrist and search for the best angle for the other gripper (Gripper B)) to grip and hold the corner successfully. Since the robot arms and the CCD camera are positioned at right angles to each other, the more of the area of the towel above the founded corner (lets say G) seen by the CCD camera, the more parallel the corner is to the gripper B, meaning the easier gripper B can grasp the corner. This is shown in Figure 4. After calculating the coordinate of the corner using simple triangulation method, gripper B will then attempt to grasp the corner and judge whether it has successfully grasped it or not by using the infrared sensors. In case of failure, the robot can be programmed to repeat the process from any point.

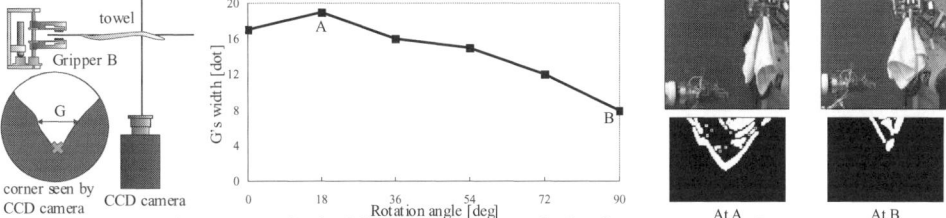

Figure 4: Method of finding the best angle for first corner grasping

FINDING & GRASPING OF THE SECOND CORNER USING TRACING METHOD

After gripper B has successfully gripped the first corner, gripper A will then release the towel and trace the edge of the towel starting from beneath the first corner and stops after reaching the second corner. The flow of this tracing process is shown in Figure 5. First, a possible corner of the towel is determined within a range just below gripper A using the CCD camera as shown in Figure 6. The range is set at 100x80 dots, the topmost right point set at 10 dots below the tip of gripper A on screen. The distance between gripper A and the possible corner which is the point furthest left is then calculated. Gripper A will then start tracing the edge of the towel by smartly using the feedbacks from the infrared sensors as shown in Figure 6. Force control of 20gf is applied to gripper A during this process. The reason why gripper A will eventually find a corner next to the one being grasped is due to its movement pattern. As you can see in Figure 6, the ideal feedback pattern is A. Feedback pattern B is received when the gripper is gripping too deep. When this occurs, the gripper will move away from the towel until the pattern returns to A. It is hard though to keep the pattern at A because there is a chance that the towel will slip away from the gripper. Once the towel starts slipping away from the gripper, it is hard for the robot to prevent the towel from slipping entirely. We thought about this problem and suggested that during feedback pattern A, gripper A should move slightly to the right. By doing this, the feedback pattern will eventually turn to B and then the gripper will move away from the towel as mentioned earlier. By using this pattern, gripper A will be tracing along the edge of the towel and ensures it to reach the corner next to the one being grasped, not the one across it.

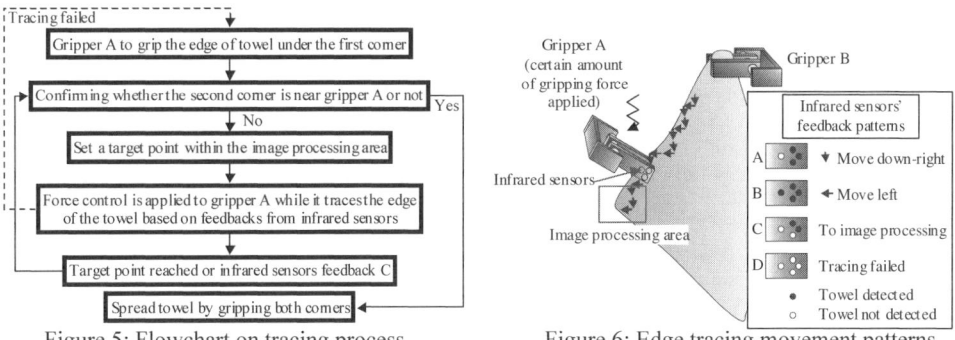

Figure 5: Flowchart on tracing process   Figure 6: Edge tracing movement patterns

EXPERIMENTS

Experiments were conducted 20 times for each process during towel spreading using a 32cm x 32 cm size towel. Experiment results have shown 90% success rate for the first edge to be correctly detected. The percentage for the edge to be successfully gripped is 65%. The main failure reason is the failure to find the best gripping angle for gripper B. Experiment results for edge tracing has shown 70% success rate. The main reason of failure in tracing is the failure to determine the possible corner correctly. This is due to the noise within the image(s) taken. Readjusting the threshold parameter can solve the problem. Figure 7 shows an example of gripper A's movements (in this case Js2) during edge tracing. This proves that our proposed method for edge tracing and corner confirmation have been successful.

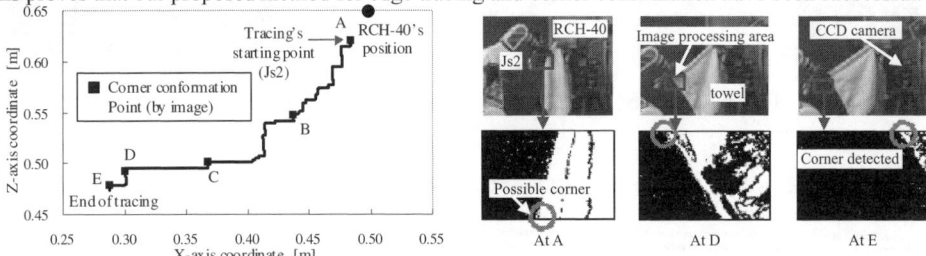

Figure 7: Example on gripper movement pattern and corner confirmation during tracing

We have also conducted experiments for the whole process. The success rate of the whole process without any failure stands at 50% and it took approximately 4 minutes to spread the towel. Since failure can be detected and all the processes can be repeated over and over again, the success rate can eventually reach 90% and over. Figure 8 shows the scene for the whole process.

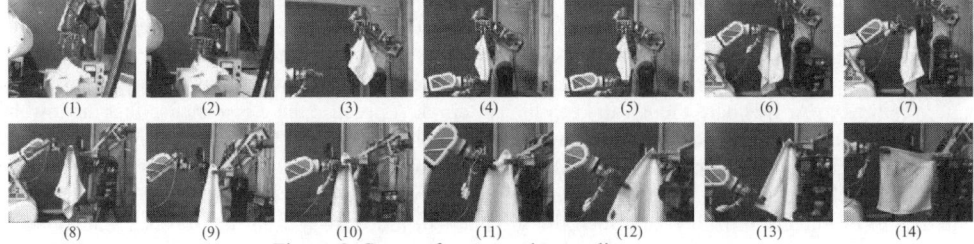

Figure 8: Scenes from towel spreading process

CONCLUSIONS

A method of towel spreading using a unique edge tracing method using a combination of infrared sensors feedbacks and CCD camera images is presented. Manipulating deformable objects such as clothes are indeed difficult and challenging, but with the right ideas, we have shown that clothes manipulation can be successfully accomplished to a certain rate.

REFERENCES

[1] Hamajima K. and Kakikura M. (1998). Planning Strategy for Task of Unfolding Clothes. Proc. of the 16th Annual Conf. of the Robotics Society of Japan, 389-390.
[2] Ono E., Kita N. and Sakane S. (1997). Unfolding a Folded Using Information of Outline with Vision and Touch Sensor. Journal of the Robotics Society of Japan, **15:2**, 113-121.
[3] Smith P.W., Nandhakumar N. and Tamadorai A.K. (1996). Vision Based Manipulation of Non Rigid Objects. Proc. of IEEE Int. Conf. Automat, **4**, 3191-3196.

INDIRECT SIMULTANEOUS POSITIONING OF DEFORMABLE OBJECTS WITHOUT PHYSICAL PARAMETERS OR TIME-DERIVATIVES

Shinichi Hirai
Dept. Robotics, Ritsumeikan Univ.
Kusatsu, Shiga 525-8577, Japan

ABSTRACT

This paper describes the control of indirect simultaneous positioning of a viscoelastic 2D object without any physical parameter of the manipulated object. Applying continua modeling of isotropic viscoelastic deformation, I first show that the positioning can be performed successfully by a simple integral control among positioned and manipulated displacements without physical parameters. Then, I show that a redundant system, where the number of manipulated displacements exceeds the number of positioned displacements, performs the positioning successfully.

KEYWORDS

deformation, positioning, feedback control, continua, manipulation

1 INTRODUCTION

Many manipulative operations that deal with deformable objects result in a positioning of multiple points on a deformable object [Taylor et al. 1990, Henrich and Wörn 2000]. In this positioning, multiple points on a deformable object should be guided to their desired locations simultaneously. Moreover, it is often impossible to manipulate the positioned points directly. For example, one operation called *linking* is popular in garment manufacturing. In linking of fabrics, loops at the end of a fabric must be matched to loops of another fabric so that the two fabrics can be sewed seamlessly. These points cannot be manipulated directly since a sewing needle is guided along the matched loops. Mating of a flexible part in electric industry also results in the positioning of mated points on the object. These points cannot be manipulated directly since the points in a mating part contact with a mated part. Consequently, we find that a positioning of multiple points on a deformable object is one of fundamental operations in the manipulation of deformable objects. Since the positioned points cannot be manipulated directly, the guidance of positioned points must be performed by controlling some points except the positioned points, as illustrated in

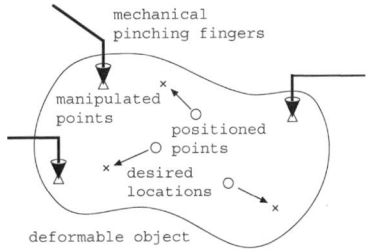

Figure 1: Indirect simultaneous positioning of deformable object

Figure 1. This operation is referred to as *indirect simultaneous positioning*, which is abbreviated as ISP.

An iterative control law based on a roughly estimated physical model of an extensible object has been proposed [Wada et al. 1998]. It has experimentally shown that the positioning can be performed successfully despite of the discrepancy of physical parameters between an actual object and its model. Simple PID-control has been successfully applied to the ISP [Wada et al. 2001]. The former requires roughly estimated physical parameters of a manipulated object and the latter requires time-derivatives of sensor signals, which may cause instability of the ISP process. In this paper, I will apply continua modeling of a viscoelastic object to the indirect simultaneous positioning and will show that a simple integral control based on a distance-based mapping among positioned and manipulated points performs the positioning successfully without any physical parameter of the object.

2 INDIRECT SIMULTANEOUS POSITIONING

Let us describe a deformable object by a set of triangles or tetrahedra. Then, the object deformation can be represented by a set of nodal points. Assume that positioned points and manipulated points are involved in the nodal points. Let $\boldsymbol{u}_i = [u_{i,x}, u_{i,y}]^\mathrm{T}$ be the displacement vector of nodal point P_i. Some displacements of nodal points should be guided to their desired values in an ISP. These displacements are referred to as *positioned displacements*. This guidance should be performed by controlling some displacements except positioned displacements. These displacements are referred to as *manipulated displacements*. Displacements except positioned displacements or manipulated displacements are referred to as *non-positioned non-manipulated displacements*. Consequently, we can classify a set of displacements into three subsets; 1) manipulated displacements, 2) positioned displacements, and 3) non-positioned non-manipulated displacements. For example, three points marked as circles should be guided to their desired locations marked as crosses in a positioning illustrated in Figure 2-(a). This guidance is performed by controlling three points marks as triangles. Thus, a set of positioned displacements is given by $u_{5,x}$, $u_{5,y}$, $u_{6,x}$, $u_{6,y}$, $u_{10,x}$, and $u_{10,y}$ while a set of manipulated displacements is given by $u_{3,x}$, $u_{3,y}$, $u_{4,x}$, $u_{4,y}$, $u_{14,x}$, and $u_{14,y}$. The desired values of positioned displacements can be computed from the initial coordinates and the desired coordinates of positioned points. In a positioning illustrated in Figure 2-(b), three points marked as circles should be aligned on a target line perpendicular to the x-axis. Note that we must guide the x-coordinate of the three points to the x-intercept of the line, while we do not have to control the y-coordinate of the three points. Thus, a set of positioned displacements in this example is given by $u_{5,x}$, $u_{6,x}$, and $u_{10,x}$. Displacements $u_{5,y}$, $u_{6,y}$, and $u_{10,y}$ are involved in non-positioned non-manipulated displacements. The desired values of positioned displacements can be computed from the initial x-coordinate of positioned points and the x-intercept of the target line.

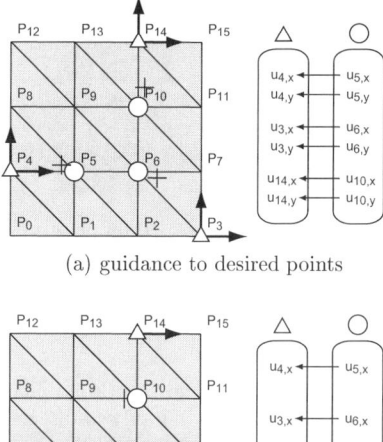

(a) guidance to desired points

(b) guidance to desired lines

Figure 2: Manipulated, positioned, and non-positioned non-manipulated displacements

3 CONTROL LAW

Assume that a vision system can measure the current values of positioned displacements. This implies that the current value of positioned displacements can be measured through a vision system. Moreover, recall that mechanical fingers pinch an extensible object and no slip between the fingers and the object occurs. Namely, the current value of manipulated displacements can be computed from the motion of mechanical pinching fingers.

Let us define a mapping from a set of positioned displacements to a set of manipulated displacements. Let u_i be a positioned displacement and u_i^* be its goal displacement. Determine a manipulated displacement u_j corresponding to each positioned displacement u_i. Then, let us apply the following simple control law:

$$u_j = -K_I \int_0^t (u_i - u_i^*) \, \mathrm{d}t, \tag{1}$$

where K_I denotes integral gain. Recall that a vision system can measure positioned displacement u_i. This equation computes the commanded value of manipulated displacement u_j.

The above equation provides a continuous control law. Let us derive a discrete control law. Assume that positioned displacement u_i can be measured at time interval T. Let u_i^k and u_j^k be the positioned displacement and the manipulated displacement at the k-th time interval $[kT, (k+1)T]$. Then, the above continuous control law turns into a discrete control law as follows:

$$u_j^{k+1} = u_j^k - K_I(u_i^k - r_i^*). \tag{2}$$

Namely, the commanded value of manipulated displacement u_j^{k+1} at the next time interval is computed from the current value of manipulated displacement u_j^k and the current error of positioned displacement $u_i^k - u_i^*$. Note that the these control laws include no physical parameters of a positioned object. This implies that no identification of physical parameters is needed.

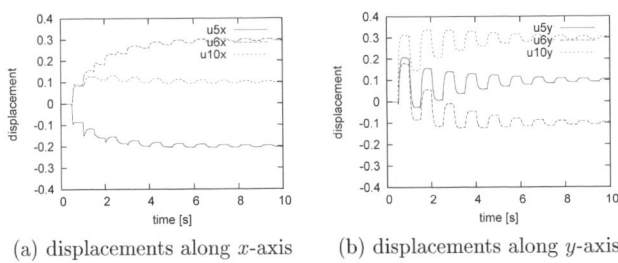

(a) displacements along x-axis (b) displacements along y-axis

Figure 3: Motion of positioned points

4 SIMULATION

Viscoelastic deformation has been extensively studied in solid mechanics and finite element analysis. Let us briefly describe the dynamic modeling of two-dimensional viscoelastic deformation. Note that the deformation modeling is not for the control law of an ISP but for the simulation of an ISP process.

Let $\boldsymbol{\sigma}$ be a pseudo stress vector and $\boldsymbol{\varepsilon}$ be a pseudo strain vector. Stress-strain relationship of 2D isotropic viscoelastic deformation is formulated as $\boldsymbol{\sigma} = (\lambda I_\lambda + \mu I_\mu)\boldsymbol{\varepsilon}$, where $\lambda = \lambda^{\mathrm{ela}} + \lambda^{\mathrm{vis}}\mathrm{d}/\mathrm{d}t$ and $\mu = \mu^{\mathrm{ela}} + \mu^{\mathrm{vis}}\mathrm{d}/\mathrm{d}t$. Elasticity of the object is specified by two elastic moduli λ^{ela} and μ^{ela} while its viscosity is specified by two viscous moduli λ^{vis} and μ^{vis}. Matrices I_λ and I_μ are matrix representations of isotropic tensors, which are given as follows in 2D deformation:

$$I_\lambda = \begin{bmatrix} 1 & 1 & 0 \\ 1 & 1 & 0 \\ 0 & 0 & 0 \end{bmatrix}, \quad I_\mu = \begin{bmatrix} 2 & 0 & 0 \\ 0 & 2 & 0 \\ 0 & 0 & 1 \end{bmatrix}.$$

The stress-strain relationship can be converted into a relationship between a set of forces applied to nodal points and a set of displacements of the points. Let $\boldsymbol{u}_\mathrm{N}$ be a set of displacements of nodal points. Let J_λ and J_μ are connection matrices, which can be geometrically determined by object coordinate components of nodal points. Replacing I_λ by J_λ, I_μ by J_μ, and $\boldsymbol{\varepsilon}$ by $\boldsymbol{u}_\mathrm{N}$ in the stress-strain relationship of a viscoelastic object yields a set of viscoelastic forces applied to nodal points as $(\lambda J_\lambda + \mu J_\mu)\boldsymbol{u}_\mathrm{N}$. Introducing $\boldsymbol{v}_\mathrm{N} = \dot{\boldsymbol{u}}_\mathrm{N}$, a set of viscoelastic forces is given by $K\boldsymbol{u}_\mathrm{N} + B\boldsymbol{v}_\mathrm{N}$, where $K = \lambda^{\mathrm{ela}}J_\lambda + \mu^{\mathrm{ela}}J_\mu$ and $B = \lambda^{\mathrm{vis}}J_\lambda + \mu^{\mathrm{vis}}J_\mu$.

Let M be an inertia matrix and $\boldsymbol{f}$ be a set of external forces applied to nodal points. Let us describe a set of geometric constraints imposed on the nodal points by $A^\mathrm{T}\boldsymbol{u}_\mathrm{N} = \boldsymbol{b}$. The number of columns of matrix A is equal to the number of geometric constraints. Let $\boldsymbol{\lambda}$ be a set of constraint forces corresponding to the geometric constraints. A set of dynamic equations of nodal points is then given by

$$M\dot{\boldsymbol{v}}_\mathrm{N} = -K\boldsymbol{u}_\mathrm{N} - B\boldsymbol{v}_\mathrm{N} + \boldsymbol{f} + A\boldsymbol{\lambda}.$$

Applying the constraint stabilization method [Baumgarte 1972] to the constraints specified by angular velocity ω, system dynamic equations are described as follows:

$$\begin{bmatrix} I & & \\ & M & -A \\ & -A^\mathrm{T} & \end{bmatrix} \begin{bmatrix} \dot{\boldsymbol{u}}_\mathrm{N} \\ \dot{\boldsymbol{v}}_\mathrm{N} \\ \boldsymbol{\lambda} \end{bmatrix} = \begin{bmatrix} \boldsymbol{v}_\mathrm{N} \\ -K\boldsymbol{u}_\mathrm{N} - B\boldsymbol{v}_\mathrm{N} + \boldsymbol{f} \\ 2\omega A^\mathrm{T}\boldsymbol{v}_\mathrm{N} + \omega^2 A^\mathrm{T}(\boldsymbol{u}_\mathrm{N} - \boldsymbol{b}) \end{bmatrix}. \quad (3)$$

Note that the above linear equation is solvable since the matrix is regular, implying that we can sketch $\boldsymbol{u}_\mathrm{N}$ and $\boldsymbol{v}_\mathrm{N}$ using numerical solver such as the Euler method or the Runge-Kutta method.

Let us simulate an indirect simultaneous positioning by taking a simple example illustrated in Figure 2. Two-dimensional deformation of a viscoelastic object is described by nodal points P_0 through P_{15}. Let us guide three points P_5, P_6, and P_{10} to their desired location by controlling

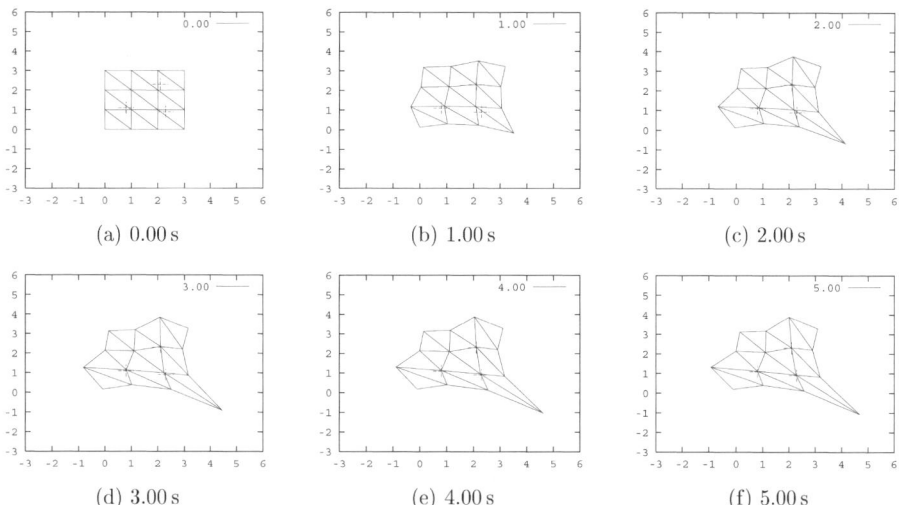

Figure 4: Process of indirect simultaneous positioning to desired points

three points P_3, P_4, and P_{14}. Positioned displacements are $u_{5,x}$, $u_{5,y}$, $u_{6,x}$, $u_{6,y}$, $u_{10,x}$, and $u_{10,y}$ and manipulated displacements are $u_{3,x}$, $u_{3,y}$, $u_{4,x}$, $u_{4,y}$, $u_{14,x}$, and $u_{14,y}$, as illustrated in Figure 2-(a). Let us introduce a distance-based mapping from the positioned displacements to the manipulated displacements. Control law is then formulated as follows:

$$\bm{u}_3 = -K_I \int_0^t (\bm{u}_6 - \bm{u}_6^*) \, dt, \quad \bm{u}_4 = -K_I \int_0^t (\bm{u}_5 - \bm{u}_5^*) \, dt, \quad \bm{u}_{14} = -K_I \int_0^t (\bm{u}_{10} - \bm{u}_{10}^*) \, dt.$$

The corresponding discrete control law is given by

$$u_{3,x}^{k+1} = u_{3,x}^k - K_I(u_{6,x}^k - u_{6,x}^*), \quad u_{4,x}^{k+1} = u_{4,x}^k - K_I(u_{5,x}^k - u_{5,x}^*), \quad u_{14,x}^{k+1} = u_{14,x}^k - K_I(u_{10,x}^k - u_{10,x}^*),$$
$$u_{3,y}^{k+1} = u_{3,y}^k - K_I(u_{6,y}^k - u_{6,y}^*), \quad u_{4,y}^{k+1} = u_{4,y}^k - K_I(u_{5,y}^k - u_{5,y}^*), \quad u_{14,y}^{k+1} = u_{14,y}^k - K_I(u_{10,y}^k - u_{10,y}^*).$$

Elastic and viscous moduli are $\lambda^{\text{ela}} = 7.0$, $\lambda^{\text{vis}} = 4.0$, $\mu^{\text{ela}} = 5.0$, and $\mu^{\text{vis}} = 2.0$. Density is given by $\rho = 0.2$. Positioned displacements are measured at time interval $T = 0.5$. Let desired values of the positioned displacements be $\bm{u}_5^* = [-0.20, 0.10]^{\text{T}}$, $\bm{u}_6^* = [0.30, -0.10]^{\text{T}}$, and $\bm{u}_{10}^* = [0.10, 0.30]^{\text{T}}$. Motion of the positioned displacements is plotted in Figure 3. Gain is given by $K_I = 1.7$. Vibration comes from the viscoelastic nature of the object. Despite of the vibration, the positioned displacements converge to their desired values, as shown in the figure. Deformed shapes during the positioning process are described in Figure 4. Crosses in the figures denote the desired values of the positioned displacements. As shown in the figure, the positioned displacements converge to their desired values.

Let us guide the x-coordinates of P_5, P_6, and P_{10} to their desired values by controlling the x-coordinates of P_3, P_4, and P_{14}. Positioned displacements are $u_{5,x}$, $u_{6,x}$, and $u_{10,x}$ and manipulated displacements are $u_{3,x}$, $u_{4,x}$, and $u_{14,x}$, as illustrated in Figure 2-(b). The discrete control law is then given by

$$u_{3,x}^{k+1} = u_{3,x}^k - K_I(u_{6,x}^k - u_{6,x}^*), \quad u_{4,x}^{k+1} = u_{4,x}^k - K_I(u_{5,x}^k - u_{5,x}^*), \quad u_{14,x}^{k+1} = u_{14,x}^k - K_I(u_{10,x}^k - u_{10,x}^*).$$

Let desired values of the positioned displacements be $u_{5,x}^* = 0.20$, $u_{6,x}^* = -0.20$, and $u_{10,x}^* = -0.20$. Deformed shapes during the positioning process are described in Figure 5. Dotted lines in the figures denote the desired values of the positioned displacements. As shown in the figure, the positioned displacements converge to their desired values.

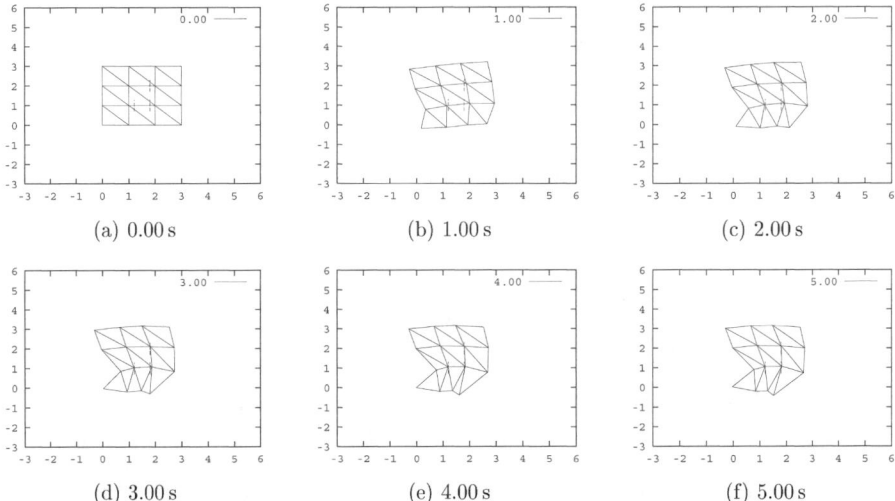

Figure 5: Process of indirect simultaneous positioning to desired lines

5 CONCLUDING REMARKS

I have applied continua modeling of a viscoelastic object to the indirect simultaneous positioning and have simulated the positioning process. I have shown that a simple I-control based on a distance-based mapping among positioned and manipulated displacements performs the positioning successfully without any physical parameter. Note that an iterative control law proposed in this paper requires no time-derivatives of sensor signals and works well with time interval around 0.5, which is larger than the sampling time in video frame rate. In addition, I have shown that a redundant system, where the number of manipulated displacements exceeds the number of positioned displacements, performs the positioning successfully.

I am going to investigate the stability of the positioning process based on the continua modeling and its finite element approximation of viscoelastic deformation. Experimental verification is also a future issue. I will investigate the positioning of an inelastic object including hysteresis and rheological deformation.

References

Baumgarte, J. (1972). *Stabilization of Constraints and Integrals of Motion in Dynamical Systems.* Computer Methods in Applied Mechanics and Engineering **1**, 1–16.

Henrich, D. and Wörn, H. eds. (2000). *Robot Manipulation of Deformable Objects.* Springer–Verlag.

Taylor, P. M. et al. (1990). *Sensory Robotics for the Handling of Limp Materials.* Springer–Verlag.

Wada, T., Hirai, S., and Kawamura, S. (1998). *Indirect Simultaneous Positioning Operations of Extensionally Deformable Objects.* Proc. IEEE/RSJ Int. Conf. on Intelligent Robots and Systems, 1333–1338.

Wada, T., Hirai, S., Kawamura, S., and Kamiji, N. (2001). *Robust Manipulation of Deformable Objects By a Simple PID Feedback.* Proc. IEEE Int. Conf. on Robotics and Automation, 85–90.

PLANNING OF KNOTTING MANIPULATION

Hidefumi Wakamatsu [1], Tsunenori Kato [1], Akira Tsumaya [1], Eiji Arai [1], and Shinichi Hirai [2]

[1] Dept. of Manufacturing Science, Graduate School of Eng., Osaka University
2-1 Yamadaoka, Suita, Osaka 565-0871, Japan
[2] Dept. of Robotics, Ritsumeikan University
1-1-1 Noji Higashi, Kusatsu, Shiga 525-8577, Japan

ABSTRACT

A planning method for knotting/unknotting manipulation of deformable linear objects is proposed. Firstly, topological states of a linear object are represented as finite sequence of crossings and two attributes of each crossing. Secondly, transitions among the topological states are defined. Then, we can generate possible sequences of crossing state transitions, that is, possible manipulation processes from the initial state to the objective state. Thirdly, a method for determination of grasping points and their moving direction is proposed to realize derived manipulation processes. Finally, criteria for evaluation of manipulation plans are introduced to reduce the candidates of manipulation plans.

KEYWORDS

Linear Objects, Manipulation, Knotting, Unknotting, Planning

INTRODUCTION

Deformable linear objects such as tubes, cords, wires, and threads are used widely; not only for data transmission or for object transportation but also for fixing or packing of objects including themselves. Such manipulative tasks include knotting. On the other hand, self-entwining of linear objects should be avoided during their manipulative processes. Therefore, it is important for linear object manipulation to analyze knotting or entwining. There are some studies on rope tying (Inoue 1984, Hopcroft 1991, Matsuno 2001, Morita 2003). In these studies, knotting manipulation of a linear object could be realized by a mechanical system, but how to knot is given. To make a bowknot, for example, we manipulate a linear object dexterously by using several fingers of both hands for bending, twisting, holding, and/or binding. However, how to make a bowknot of us is not unique because it depends on our physical makeup and experience. If knotting/unknotting process of a linear object can be modeled, it is useful for design of knotting/unknotting system with mechanism unlike human arms/hands and for planning suitable for such system. Therefore, in this paper, we propose a planning method for knotting/unknotting of deformable linear objects.

REPRESENTATION OF KNOTTING/UNKNOTTING PROCESS

In this section, we propose a representation method of knotting/unknotting of a deformable linear object. First, let us project the 3D shape of a linear object on a plane. The projected 2D curve may cross with itself. Crossings in the projected curve can specify the crossing state. Next, let us number crossings along the projected curve from one endpoint to the other. One endpoint of the projected curve is defined as the left endpoint E_l, and the other is defined as the right endpoint E_r in this paper. In addition, whether each crossing is involved in the upper part or in the lower part is specified. Symbol C_i^u describes the i-th crossing point is involved in the upper part while C_i^l denotes it is in the lower part. Furthermore, we categorize the crossings into two: left-handed helical crossing C_i^- and right-handed helical crossing C_i^+. The sequence of symbols at individual crossing points determines the crossing states of a linear object. Let us describe a segment between C_i and C_j as $_i^p L_j^q$ where p and q indicate whether the segment is an upper part (then $p,q=u$) or a lower part (then $p,q=l$) at crossing C_i and C_j. Terminal segments adjoining the left and the right endpoints are described as L_i^p and $_j^q L$, respectively. Consequently, we can represent the crossing states of a knotted linear object by a sequence of crossing point symbols.

Knotting/unknotting process of a linear object corresponds to changing the number of its crossings. In order to change the crossing state of a linear object, an operation must be performed on the object. Therefore, a state transition corresponds to an operation that changes the number of crossings or permutes their sequence. In this paper, four basic operations are prepared as shown in Fig.1. Operation I, II, and IV increase or decrease the number of crossings. Let us divide operation I into two: crossing operation CO_I and uncrossing operation UO_I. Crossing operation CO_I increases the number of crossings while uncrossing operation UO_I decreases the number. Crossing operation CO_{II} and CO_{IV} and uncrossing operation UO_{II} and UO_{IV} are defined as well. Operation III does not change the number of crossings but permutes their sequence. Operation III is referred to as an arranging operation AO_{III}. Then, a manipulation process can be represented as transitions of crossing states. It corresponds to iteration of crossing, uncrossing, or arranging operations.

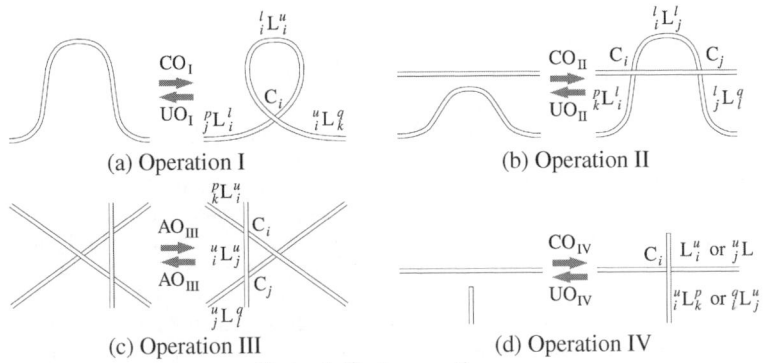

Figure 1: Basic operations

MOTION PLANNING IN KNOTTING/UNKNOTTING MANIPULATION

In order to accomplish one of possible processes, we have to grasp, move, and release the object during the processes. Whether the crossing state of the object changes as expected or not depends on grasping points and their moving direction. Let us define a qualitative manipulation plan as a sequence of crossing state transitions including grasping points and their moving direction to realize each state transition. In this section, we explain a procedure to determine adequate grasping points and their moving direction for one state transition. In addition, for detailed planning and actual execution of manipulation, we have to narrow down qualitative manipulation plans. Therefore, we introduce criteria to evaluate qualitative manipulation plans.

Definition of Actions

Uncrossing operations delete a crossing by moving its upper part or lower part. Let us define a segment between deleted crossing points or between a deleted crossing point and an endpoint as a target segment. Then, we assume that a target segment or its adjacent segments in each crossed state should be grasped in order to realize each uncrossing operation. Furthermore, we define the approaching direction of a manipulator with respect to the projection plane: from the front side or the back side. Next, let us consider moving direction of a grasping point to realize each operation. Generally, a rigid body in the 3D space has 3 DOF in translation and 3 DOF in rotation. Note that the translation along the projection normal does not change the crossing state of a linear object. Omitting this translation, we apply 2 DOF in translation along the projection plane and 3 DOF in rotation into the knotting/unknotting of a linear object. Then, we can select a set of grasping points and their corresponding DOF to perform individual basic operations. In this paper, this set is referred to as an *action*. Fig.2 shows possible actions to realize uncrossing operations. A circle with dot, a circle with cross, and a open circle represent a point to be grasped from the front side, the back side, and whichever side, respectively. We can also derive actions for crossing and arranging operations. Thus, actions, that is, adequate sets of grasping points and their corresponding DOF to realize each operation can be determined. Consequently, possible qualitative manipulation plans, that is, sequences of crossing state transitions and actions for each state transition, can be generated by a computer system when the initial and the objective crossing state of a linear object are given.

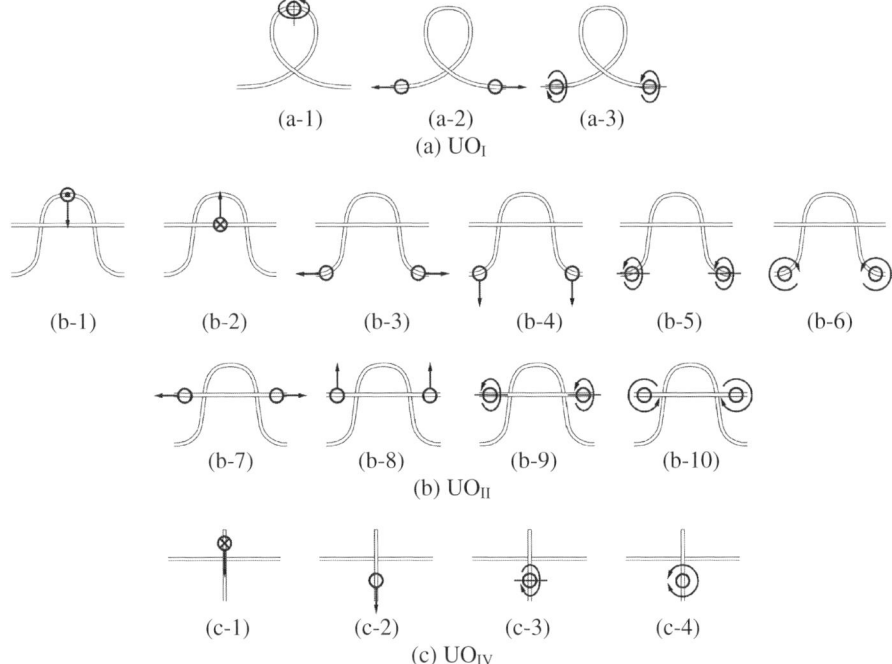

Figure 2: Actions for uncrossing operations

Evaluation of Manipulation Plans

We introduce criteria to evaluate generated qualitative manipulation plans. First, let N_t be the number of state transitions through one sequence. In this paper, we prefer a sequence including fewer intermediate states, that is, fewer state transitions because it takes much time to finish the required manipulation when a selected sequence includes many state transitions. Note that a knotting/unknotting process corresponds to increase/decrease of crossings of a linear object. Recall that operation II generates/deletes two crossings while operation I and IV generates/deletes one crossing. Then, we find that a sequence including the more operations II consists of the fewer intermediate states. Next, let N_c be the changing times of grasping points through one sequence. When a grasping point never change during manipulation, position and direction of a linear object at the grasping point corresponds to those of fingertips of a manipulator obviously. So, estimation of the object shape is not needed once the manipulator grasps the object. However, if a grasping point changes during manipulation, position and direction of a segment to be grasped in the next operation must be estimated in the detailed planning. Furthermore, it takes much time to change a grasping point. Therefore, a sequence in which grasping points are not changed frequently is preferable. By considering these criteria, we can reduce the number of candidates of qualitative manipulation plans.

EXAMPLE OF POSSIBLE PROCESS GENERATION

In this section, we show an example of possible unknotting process generation by a computer system. Fig.3 shows a required manipulation. It corresponds to untying a slip knot. The initial state in Fig.3(a) is represented as E_l-C^{u-}_1-C^{l-}_2-C^{l+}_3-C^{u+}_4-C^{u-}_5-C^{l-}_1-C^{u-}_2-C^{l-}_5-C^{l+}_4-C^{u+}_3-E_r and the objective state in Fig.3(b) is represented as E_l-E_r. Assuming that only uncrossing operations can be used, that is, without AO_{III}, 14 crossing states and 39 state transitions are derived as shown in Fig.4. Thus, possible knotting/unknotting processes of a linear object can be generated automatically when the initial and the objective states are given.

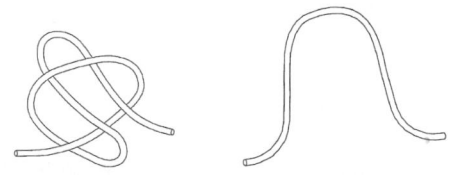

(a) Initial state   (b) Objective state
Figure 3: Required manipulation – untying slip knot –

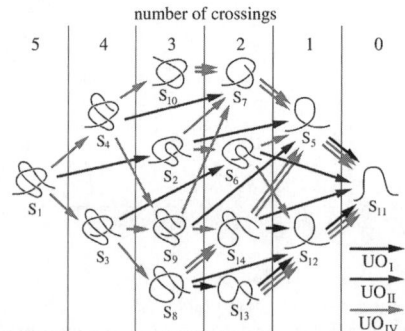

Figure 4: Generated possible unknotting processes

The number of crossings in the initial state is five, and that in the objective state is zero. We can reduce the number of crossings from five to zero by applying two operations UO_{II} and one operation UO_I or UO_{IV} at least. Their possible sequences are described as follows:

$$OQ_1 : UO_{II} \rightarrow UO_{II} \rightarrow UO_I/UO_{IV}, \quad (1)$$
$$OQ_2 : UO_{II} \rightarrow UO_I/UO_{IV} \rightarrow UO_{II}, \quad (2)$$
$$OQ_3 : UO_I/UO_{IV} \rightarrow UO_{II} \rightarrow UO_{II}. \quad (3)$$

Then, we check whether the required process can be realized by applying these three uncrossing operations to the object in the above orders or not. In Fig.4, the following sequences of state transitions SQ_1, SQ_2, and SQ_3 correspond to operation sequences OQ_1, OQ_2, and OQ_3, respectively.

$$SQ_1 : S_1 \rightarrow S_2 \rightarrow S_5 \rightarrow S_{11}, \quad (4)$$
$$SQ_2 : S_1 \rightarrow S_2 \rightarrow S_6 \rightarrow S_{11}, \quad (5)$$
$$SQ_3 : S_1 \rightarrow S_3 \rightarrow S_6 \rightarrow S_{11}. \quad (6)$$

If the required process can not be realized with two operations UO_{II} and one operation UO_I or UO_{IV}, we check that with one operation UO_{II} and three operations UO_I and/or UO_{IV}. In general, we check repeatedly whether a knot with n crossings can be unknotted by applying $_{x+y}C_y$ combination of x operations UO_{II} and y operations UO_I and/or UO_{IV} with decreasing x and increasing y so that they satisfy $2x+y=n$ until a sequence of operations to unknot it is found. Thus, we can efficiently derive manipulation processes including fewer state transitions, that is, processes with lower N_t without generating the whole graph including all possible processes.

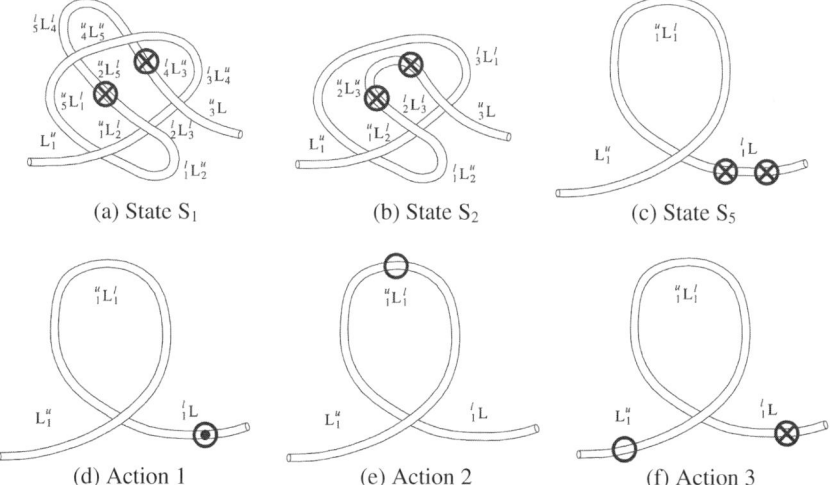

Figure 5: Consideration of changing times of grasping points

Next, we select adequate actions so that a manipulation process has fewer changing times N_c of grasping points. Let us consider sequence SQ_1. For the first transition from state S_1 to state S_2, assume that segments $^u_2L^l_5$ and $^l_4L^u_3$ are grasped from the front side as shown in Fig.5(a) and moved to perform operation UO_{II}. Then, grasped segments become equivalent to segment $^u_2L^u_3$ in state S_2 as shown in Fig.5(b). State S_2 can be changed into state S_5 by moving segment $^u_2L^u_3$. After that, it is found that segment ^l_1L in state S_5 is grasped from the front side by two manipulators as shown in Fig.5(c). There are three actions to change the state into S_{11}. Action 1 is to regrasp segment ^l_1L from the back side for operation UO_{IV} as shown in Fig.5(d). Action 2 is to release segment ^l_1L and to grasp segment $^u_1L^l_1$ as shown in Fig.5(e) for operation UO_I or for operation UO_{IV}. Action 3 is to grasp segment L^u_1 keeping segment ^l_1L grasped for operation UO_I as shown in Fig.5(f). Anyway, we have to change grasping points for the last transition from state S_5 to state S_{11}. Consequently, in the above plans to perform sequence SQ_1, $N_c=1$ and it is minimum. We can also derive the minimum N_c for sequence SQ_2 and sequence SQ_3. The former is $N_c=2$ and the latter is $N_c=1$. This implies that sequence SQ_2 should be eliminated from adequate manipulation plans. Thus, we can narrow down candidates of manipulation plans by considering N_t and N_c. After that, quantitative analysis (Wakamatsu 2004) should be performed in order to check whether a selected manipulation can be realized practically or not considering physical properties of a linear object such as rigidity. Thus, we conclude that our proposed method is useful for planning of knotting/unknotting manipulation of deformable linear objects.

CONCLUSIONS

A planning method for knotting/unknotting manipulation of deformable linear objects was proposed. First, knotting/unknotting processes of a linear object were represented as a sequence of finite crossing state transitions. Next, grasping points and their moving direction to perform each state transition were defined. Then, possible qualitative manipulation plans can be generated by a computer system when the initial state and the objective state of a linear object are given. Finally, criteria for evaluation of generated manipulation plans were introduced. By considering them, we can narrow down candidates of manipulation plans.

REFERENCES

Hopcroft J.E., Kearney J.K., and Krafft D.B. (1991). A Case Study of Flexible Object Manipulation. *Int. J. of Robotics Research,* **10:1**, 41--50.

Inoue H. and Inaba M. (1984). Hand-eye Coordination in Rope Handling, *Robotics Research: The First International Symposium*, MIT Press, 163--174.

Matsuno T., Fukuda T., and Arai F. (2001). Flexible Rope Manipulation by Dual Manipulator System Using Vision Sensor, *Proc. of International Conference on Advanced Intelligent Mechatronics*, 677--682.

Morita T., Takamatsu J., Ogawara K., Kimura H., and Ikeuchi K. (2003). Knot Planning from Observation, *Proc. of IEEE Int. Conf. Robotics and Automation*, 3887--3892.

Wakamatsu H., and Hirai S. (2004). Static Modeling of Linear Object Deformation based on Differential Geometry, *Int. J. of Robotics Research,* **23:3**, 293--311.

ANALYZING AND EVALUATING ROBOT MOTION ALGORITHM FOR SWEEPING TASK

Yuki Saito, Masanori Ezawa, Yusuke Fukumoto, Hiroyuki Ogata, and Akira Torige

Department of Mechanical Engineering, Seikei University
3-3-1 Kichijojikitamachi, Musashino-shi, Tokyo, Japan

ABSTRACT

Recently, sweeping robots, such as cleaning robots for home use, have been actively developed. Most of these robots generate their motion by reactive algorithm with easy sensors mounted. However, with these algorithms, the sweeping task would be done uneven to the sweepable areas. So there are still rooms for developing a more efficient algorithm. To develop an efficient algorithm, evaluation method is in need. However, the methodology to evaluate the efficiency of state informed algorithm has not yet established. Cleaning robots do not always clean the same environment, so the difficulty of this research is that consideration of various kinds of environments is in need. We developed sweeping robot simulator to analyze the existing sweeping algorithms, and reveal their inefficient features to improve the existing algorithms. The algorithms that commercial robot cleaners install were analyzed, and those efficiencies were estimated.

KEYWORDS

sweeping task, motion planning, robot cleaner, planning-based method, reactive method

I. INTRODUCTION

Generating motion is important for the robots over doing some kind of task. Especially, generating motion to cover certain area utilizing the size of the robot is used to realize robotic tasks such as sweeping floors [1].
The method of generating sweeping task motion can be roughly classified into two groups. One is called *planning-based method* and the other is called *reactive method*. Planning-based method plans the sweeping task motion from the information of the environment given beforehand to carry it out. This method can consider optimization of the amount of the task of robots. Reactive method generates task motion on the fly based on information that was gained by sensors. It is often installed on the robots that are on sale or under development. Base of those robots' motion is generated by the combination of moving forward and reflecting in random as it collides. Here, we will name this motion, *air-hockey motion*.
Sweeping robot use can be classified into two; for domestic and for affair. To consider about the latter, because owner uses it as a substitute for an employee, high quality task performance will be required. So the robot will be equipped with lots of sensors and functions, though it will cost. The environment could be regarded as static, and also, the algorithm to sweep vast areas is in need. From this point of view, planning-based approach is suitable to it, because planning-based approach often assumes that the environment is static, and can generate effective motion to sweep a large area.
The specification on demand of robot cleaners between for domestic and for affair completely differs. To consider about the sweeping robot for domestic, the operation is asked to be simple, thus

operations such as inputting the map of the environment should be avoided. And to hold down the cost, quantity of sensors installed should be minimized. Coping the trouble within the case of human moving the robot to the place somewhere else, also known as *kidnapped robot problem* [2] will also be needed. In these perspectives, it is likely to adopt reactive method rather than planning-based.

While commercial robot cleaners for domestic have various devices to sweeping algorithms, there is still room for improvement in the reactive motion algorithm. Frequency of sweeping, for example, tends to left uneven.

The main target of this research is to promote the efficiency of reactive motion algorithm installed in robot cleaner by establishing the way to evaluate the motion algorithm.

II. INITIAL CONDITION SET UP

This paper discusses mainly about the aspect of sweeping motion. Aspects dependent on hardware such as vacuum performance or speed of sweeping robot are not the subjects to argue here.
To clarify the problems, we assume the conditions of sweeping task written below.
- Sweeping certain place would be done in a passage.
- The robot can cover the whole sweeping area.
- To compare the algorithms, specifications of the robot are unified.

Assuming that sweeping task of a place should complete in one passage, high performance would be gained by minimizing overlap, and spreading the areas swept in early step. Therefore, the algorithm can be evaluated by inspecting the *sweeping rate*, completed swept area divided by the area of sweep-able.

To evaluate the sweeping rate, factors that give influence should be controlled. The considerable factors of the sweeping robots are shape of the room and robot, and motion of the robot.

We searched robot contests to refer the way to evaluate sweeping algorithm and searched realty to know the sizes and shapes of the rooms of typical type in Japan [3] [4]. Living rooms of 12-mat room (about $18.56m^2$) are chosen. 6 typical rooms are selected. Those rooms, expressed in Fig. 1, were set referring show room. The effect of furniture give to sweeping motion differs.

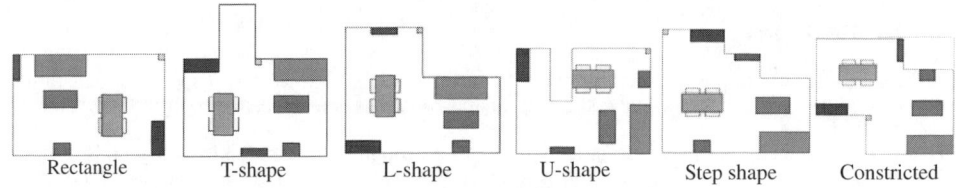

Rectangle   T-shape   L-shape   U-shape   Step shape   Constricted

Fig. 1 Room settings.

Shape and motion of the sweeping robot are determined by referring to commercial robot cleaners. Assuming the case of realizing air-hockey motion and wall following motion, the robot needs to be equipped with bump sensor and hug sensor. And two motions, moving forward and turning, are enough to prepare for it. The specification of the robot is set as Fig. 2.

Shape	circle
Diameter (m)	0.35
Velocity (m/s)	0.32
Truning speed (deg/s)	90
Installed sensors	bump sensor
	hug sensor

Fig. 2 Specification of the robot.

III. SWEEPING TASK SIMULATOR

Simulator was made to evaluate the sweeping algorithm under the condition described in Section II to make it possible to estimate the different algorithm under common condition easily.

Swept area of floor space at a certain period can be obtained by calculating the square measure of passage area at the moment. However, the floor space does not always equal to the area that the robot can sweep. So, calculating the square measure of the room that the robot can sweep is in need to search the achievement of sweeping task. In order to know it, first, the configuration space [5] is constructed by checking collision of the robot in every position. Then the square measure of the workspace associated to simply connected area including the initial configuration is calculated to obtain the amount of sweep-able area.

IV. ANALYSIS OF AIR-HOCKEY MOTION

First, the basic air-hockey motion algorithm was installed in the simulator, and the transitions of sweeping rate for each room are observed. Data were taken 100 times for each room and averages of those were calculated. Results of those rooms are shown in Fig. 3. The plots indicate the percentage of the sweeping rate at certain time in the same room, starting at the same place.

According to the plots, 90% of sweeping was generally achieved in around 20 min., and 95% was done in about 40 min. in 12 mat rooms. As the transition curve can be approximated into logarithm curve, the performance of the task can be quantitatively estimated using logarithm of the rate of un-swept area δ;

$$\delta = -\log(1 - \text{sweeping rate}) \approx at + b \quad (1)$$

The slope a signifies the quickness of the sweeping task, and the intercept b shows the initial boost of the sweeping rate. Each value obtained from simulation is shown as Table 1.

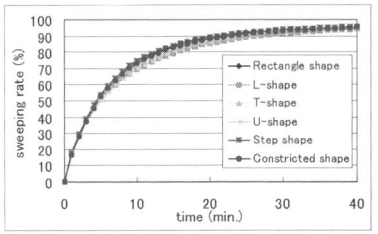

Table 1 Values of slopes and intercepts.

	Rectangle	L-shape	T-shape
slope	0.033	0.029	0.032
intercept	0.211	0.189	0.187

	U-shape	Step shape	Constricted
slope	0.033	0.033	0.032
intercept	0.158	0.222	0.210

Fig. 3 Sweeping each room in air-hockey motion

V. EFFECT OF CHANGING INITIAL POSITION

The change in the initial position of the robot may affect on the sweeping rate. Here, by starting from different initial position shown in Fig. 4 (a), influence to the changes of sweeping rate is researched.
To make the condition as different as possible, initial position was set at the center of the room distant from the obstacles, under the table where the obstacles concentrate, and at the corner of the room near the obstacle. Sweeping task was executed 100 times using air-hockey motion from each initial position to see the changes of sweeping rate. Fig. 4 (b) shows the sweeping rate average of each initial position. From Fig. 4 (b), amplitude is settled in 9% at each moment, and considerable change in sweeping rate by changing the initial position of the robot couldn't be seen from the result.

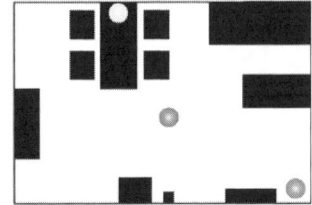 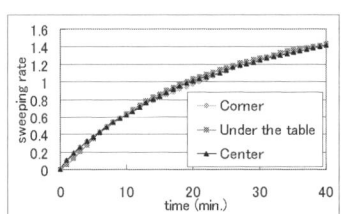

(a) Initial positions used in the simulation. (b) Sweeping point starting from different initial position.
Fig. 4 Initial positions used in the simulation.

VI. ANALYSIS OF COMERCIAL ROBOT ALGORITHM

In this section, effects of the motions when combining motions installed on commercial robot cleaners with air-hockey motion are analyzed. Therefore the motions that were mounted on the robot cleaners were imitated on the simulator. The motion algorithms mounted on the simulator are as follows.
Algorithm I; robot goes straight to the wall, and make a round trip along the walls of the room. Then switch to air-hockey motion.
Algorithm II; first, take air-hockey motion for 60 sec. Continue the motion till it collides to the wall. Then switch to the wall following motion for 30 sec. After that, it parts from the wall in random angle, and switches to the air-hockey motion again. Those two algorithms are repeated regularly.

The result of the simulation is expressed in Fig. 5, and the values of the slope and intercept of the least-squares lines are shown in Table 2.

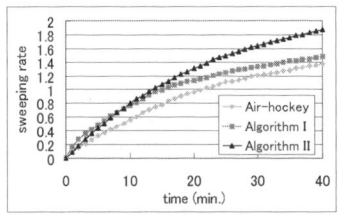

Fig. 5 Result of different algorithm.

Table 2 Slopes and intercepts of the least-squares lines.

	Air-hockey	Algorithm I	Algorithm II
slope a	0.0399	0.0318	0.0459
intercept b	0.1219	0.3757	0.2875

From Table 2, though intercept of the algorithm I is larger than the others, its slope takes almost the same value to the air-hockey algorithm. This shows that algorithm I gained high sweeping rate at the initial stage, however, the capability of sweeping task will not change to air-hockey motion algorithm after. On the other hand, slope of the algorithm II is larger than the others. This result indicates that the capability of sweeping task of algorithm II is much better than the others.
To search this background, sweeping motion was analyzed in detail.

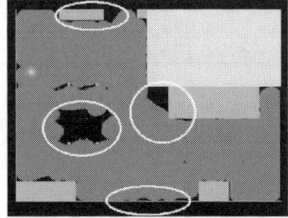 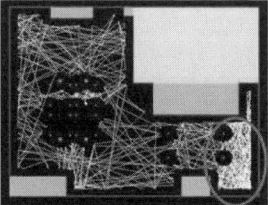

(a) Completed sweeping area.   (b) Trajectory of (a).
Fig. 6 Completed sweeping area and its trajectory.

Fig. 6 (a) shows the area of completed sweeping task of air-hockey motion of 40 min. The place of gradation in circle is the position of the robot ended up with. Fig. 6 (b) shows the trajectory of (a). With air-hockey motion, places around the border of the rooms tend to left un-swept from Fig. 6 (a). Sweeping along the wall at the beginning solves this problem. In addition, it seems that there is an effect to raise the sweeping rate by avoiding the overlaps.
On the other hand, Fig. 6 (b) testifies to presence of unevenness in the task. The robot cannot slip out from the places in the case when the robot enter the place where the entrance is small, and that causes the robot move around the same area. It is considerable that probability of the robot slipping out from the space written above would rise if it switches to wall following motion regularly. And as a result, the efficiency of sweeping task would rise.

VII. CONCLUSION

In this paper, we analyzed and evaluated the sweeping algorithm of reactive method on the purpose of developing efficient sweeping algorithm. A simulator was made to compare and to evaluate the algorithms in the same condition. The effects of the algorithm of commercial robot were estimated by comparing the algorithm of basic air-hockey motion and the algorithm of the commercial robot.

REFERENCES

[1] J. Ota, D. Kurabayashi, T. Arai (2001). *Introduction to Intelligent Robots*, Corona, JP.
[2] S.P. Engelson et al. (1992), *Error correction in mobile robot map learning*, ICRA, pp. 2555-2560.
[5] T. Lozano-Pérez (1983), *Spatial planning: a configuration space approach*, IEEE Trans. Comp. 32, pp. 108-120.
[3] http://www.botlanta.org/
[4] http://www.servicerobots.org/cleaningrobotscontest/index.php

METHOD FOR SOLVING INVERSE KINEMATICS OF REDUNDANT ROBOT UNDER RESTRAINT BY OBSTACLES

J. Kawamoto, K. Tashiro, M. Takano and S. Aoyagi

Systems Mangement Engineering, Kansai University
3-3-35, Yamate-cho, Suita, Osaka 564-8680, Japan

ABSTRACT

For the purpose of collision avoidance, a new efficient method for solving inverse kinematics of a redundant robot (all joints are rotational) is proposed. This method is divided into three procedures, which are position synthesis by 3 joints, orientation synthesis by 3 wrist joints, and collision avoidance by other remaining joints. Each of these three procedures can be calculated analytically. After series of three procedures, the positioning error is occurred, since they are not independent from each other. Therefore, the calculation is carried out iteratively until sufficient convergence is obtained. A simulator based on this method is developed. A robot of 14 DOF can successfully pass through two cylindrical holes in two thick walls, while realizing high positioning and orientating accuracy of its end effector.

KEYWORDS

Redundant robot, Path planning, Simulator, Collision avoidance

INTRODUCTION

For the purpose of collision avoidance, a redundant robot with many rotational joints is necessary. The transition of configuration of this redundant robot, which is called simply as "path" in this paper, should be planned so as that it can avoid obstacles and realize desired goal position and orientation of its end effector. Several researches are carried out to investigate how to decide a path of a redundant robot. In these researches, among many possible paths, one solution path is selected generally by minimizing some evaluation function [1, 2]. Here, it is focused how to use redundant DOF (Degree of freedom) effectively. On the other hand, this research assumes the situation that the robot configuration is firmly restrained by obstacles, and all DOF are used for realizing goal pose (position and orientation) of the end effector and avoiding these obstacles, namely there are no excess DOF. Under this condition, the path planning is not important issue, instead, the method how to decide the joint angles which satisfy the goal pose of the end effector and the restrained (specified) configuration for avoiding the obstacles. Namely, it is necessary to solve inverse kinematics (synthesis) of the robot. In this research, an effective calculation method to solve this synthesis problem under the restraint by obstacles is newly proposed.

Usually Newton method using Jacobian matrix has been used for solving this problem. However, this method needs large number of iterations and takes long computing time until sufficient convergence of solution is obtained. This computing time becomes longer as the total joint number, namely redundancy, becomes larger. Also, the solution cannot be obtained unless the initial value of iterating calculation is taken as an appropriate value in the vicinity of the true value. Moreover, it gives only one solution depending on the initial value, while there may be many solutions such as right hand configuration and left hand configuration, etc.

Considering these circumstances, a new efficient method is proposed, which solves the inverse kinematics by utilizing analytical solution partially [3, 4]. In this paper, a simulator of robot movement is developed based on this method. It is shown by this simulator that a 14 DOF robot can successfully pass thorough two cylindrical holes in two thick walls, and realize a final given pose precisely.

ROBOT MODEL

In the case that obstacles exist, DOF number of [6 + restrained DOF number by obstacles] is totally required for avoiding obstacles and realizing the pose. For example, when a robot arms avoid a cylindrical hole in a thick wall, 4 DOF is restrained as shown in **Fig. 1**. Considering this, when the robot avoids two cylindrical holes on two thick walls, 14 DOF is required. Namely $4 \times 2 = 8$ DOF is necessary for passing through two cylindrical holes, and 6 DOF is necessary for realizing the objective pose, and totally 8+6=14 DOF is required. **Figure 2** shows an example robot structure with 14 rotational joints, of which joint composition is RPP'PP'PP'PP'PP' RPR, where R, P, P' mean rotational joint, pivot joint, pivot joint perpendicularly intersect P joint, respectively.

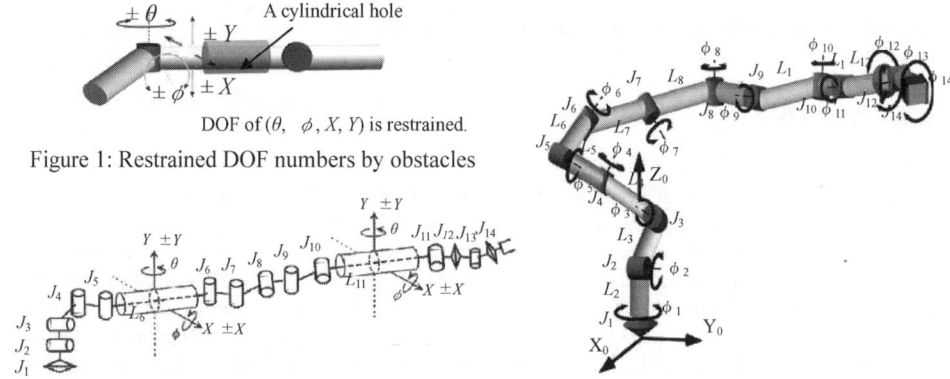

Figure 1: Restrained DOF numbers by obstacles

Figure 3: Necessary joint numbers in front of each wall

Figure 2: Robot model

METHOD FOR SOLVING INVERSE KINEMATICS

Overview

A new efficient method is proposed, which solves the inverse kinematics by utilizing analytical solution partially. It is possible to synthesize the end effector's position analytically by using 3 joints among n joints, where n is number of DOF (degree of freedom). Similarly it is possible to synthesize analytically the end effector's orientation by using wrist 3 joints. Also, it is possible to synthesize the configuration analytically, which avoids collision such as passage through wall gaps, holes, etc., by using other remaining joints. In this new method, these three analytical syntheses are carried out iteratively until sufficient convergence is obtained.

For example, the case that 14 DOF robot (already shown in **Fig. 2**) passes through two holes is assumed. It is possible to synthesize the end effector's position analytically by using 3 joints (Φ_3) among 14 joints, and it is possible to synthesize analytically the end effector's orientation by using wrist 3 joints (Φ_4). Also, it is possible to synthesize the configuration analytically, which avoids two cylindrical holes by using other remaining joints (Φ_1, Φ_2), where $\Phi_1, \Phi_2, \Phi_3, \Phi_4$ are as follows:

$$\Phi_1 = (\phi_1, \phi_2, \phi_4, \phi_5)^T \tag{1}$$

$$\Phi_2 = (\phi_6, \phi_7, \phi_9, \phi_{10})^T \tag{2}$$

$$\Phi_3 = (\phi_3, \phi_8, \phi_{11})^T \tag{3}$$

$$\Phi_4 = (\phi_{12}, \phi_{13}, \phi_{14})^T \tag{4}$$

Let the restraint condition for passage through the first hole be described as H_1, which is concretely the cluster of four equations regulating the position of J_5 axis and L_6 link (see **Fig. 3**). Also, let the restraint condition for passage through the second hole be described as H_2, which is the cluster of four equations regulating the position of J_{10} axis and L_{11} link (also see **Fig. 3**). And, let the objective position of the end effector (workpiece) be $r_w^{(0)}$, the objective orientation be $E_w^{(0)}$. They are functions of $\phi_1, \phi_2, \cdots, \phi_{14}$ and let they be expressed as follows:

$$H_1 = f_1(\Phi_1 \; ; \; \Phi_2, \Phi_3, \Phi_4) \tag{5}$$

$$H_2 = f_2(\Phi_2 \; ; \; \Phi_1, \Phi_3, \Phi_4) \tag{6}$$

$$r_w^{(0)} = f_3(\Phi_3 \; ; \; \Phi_1, \Phi_2, \Phi_4) \tag{7}$$

$$E_w^{(0)} = f_4(\Phi_4 \; ; \; \Phi_1, \Phi_2, \Phi_3) \tag{8}$$

In these expressions, $f_1(\Phi_1 \; ; \; \Phi_2, \Phi_3, \Phi_4)$ means Φ_1 is dominant for realizing the objective result compared with Φ_2, Φ_3, Φ_4. These equations can be analytically solved by fixing non-dominant variables (Φ_2, Φ_3, Φ_4 in this case) to appropriate constant values. Let these solutions be expressed as follows:

$$\Phi_1 = f_1^{-1}(H_1 \; ; \; \Phi_2, \Phi_3, \Phi_4) \quad (\Phi_2, \Phi_3, \Phi_4 = \text{const.}) \tag{9}$$

$$\Phi_2 = f_2^{-1}(H_2 \; ; \; \Phi_1, \Phi_3, \Phi_4) \quad (\Phi_1, \Phi_3, \Phi_4 = \text{const.}) \tag{10}$$

$$\Phi_3 = f_3^{-1}(r_w^{(0)} \; ; \; \Phi_1, \Phi_2, \Phi_4) \quad (\Phi_1, \Phi_2, \Phi_4 = \text{const.}) \tag{11}$$

$$\Phi_4 = f_4^{-1}(E_w^{(0)} \; ; \; \Phi_1, \Phi_2, \Phi_3) \quad (\Phi_1, \Phi_2, \Phi_3 = \text{const.}) \tag{12}$$

In this new method, these four analytical syntheses are carried out iteratively until sufficient convergence is obtained. If the convergence is not enough, the combination of jonts for positioning end effector and avoiding collision (combination of joints for orientating end effector is fixed to the wrist 3 joints) is changed and the iterative analytical synthesis is carried out again.

The features of this method are as follows: 1) iteration number is very small and computing time is reduced to about one tenth of that computed by Newton method, 2) the initial value of iterative calculation is defined analytically (it is surely in the vicinity of true value), 3) it gives all solutions while Newton method gives only one solution, so one can choose the most adequate solution for his task among them.

Setting of Middle Configurations

The Path (the transition of configuration of the robot) is generated from goal position and orientation. Wrist 3 joints of $\phi_{12}, \phi_{13}, \phi_{14}$ are used only for realizing desired goal orientation. Five joints are necessary before a wall to pull away a link through a hole, since four DOF shown in **Fig. 1** and one

DOF along the hole axis must be specified. Until the work piece (end effector) is pulled away thorough the second hole (see **Fig. 4(d)**), J_1–J_5 exist before the first wall. Therefore, five joints of J_6–J_{10} exist before the second wall before J_{11} is pulled away through this hole. After J_{11} is pulled away through this hole, six joints of J_6–J_{11} exist before the wall. Therefore, it is necessary to select five joints among the six to pull away link L_{12}. As the configuration on which the combination of joint angles for realizing the passage through the hole can be changed and must be selected, ten middle configurations are set on the way from a given initial configuration (see **Fig. 4 (l)**) to the final configuration (see **Fig. 4 (a)**). They are concretely as follows:

(1) *The 10th middle configuration*: All joints after J_{10} are positioned on the axis of the second hole, and J_{10} is before the wall and J_{11} is after the wall (See **Fig. 4 (b)**).
(2) *The 9th middle configuration*: All joints after J_{11} are positioned on the axis of the second hole, and J_{11} is before the wall and J_{12} is after the wall (See **Fig. 4 (c)**).
(3) *The 8th middle configuration*: All joints after J_{11} are positioned on the axis of the second hole, and the work piece is before the wall (See **Fig. 4 (d)**).
(4) *The 7th middle configuration*: Joints of J_6 and J_7 are positioned on the axis of the first hole before and after the wall, respectively. Joints after J_8 are arbitrarily set as far as the collision against the second wall does not occur (See **Fig. 4 (e)**).
(5) *The 6th middle configuration*: Joints of J_7 and J_8 are positioned on the axis of the first hole before and after the wall, respectively (see **Fig. 4 (f)**).
(6) *The 5th middle configuration*: Joints of J_8 and J_9 are positioned on the axis of the first hole before and after the wall, respectively (see **Fig. 4 (g)**).
(7) *The 4th middle configuration*: Joints of J_9 and J_{10} are positioned on the axis of the first hole before and after the wall, respectively (see **Fig. 4 (h)**).
(8) *The 3rd middle configuration*: Joints of J_{10} and J_{11} are positioned on the axis of the first hole before and after the wall, respectively (see **Fig. 4 (i)**).
(9) *The 2nd middle configuration*: All joints after J_{11} are positioned on the axis of the first hole, and J_{11} is before the wall (see **Fig. 4 (j)**).
(10) *The 1st middle configuration*: All joint after J_{11} are posited on the axis of the first hole, and the work piece is before the wall (see **Fig. 4 (k)**).

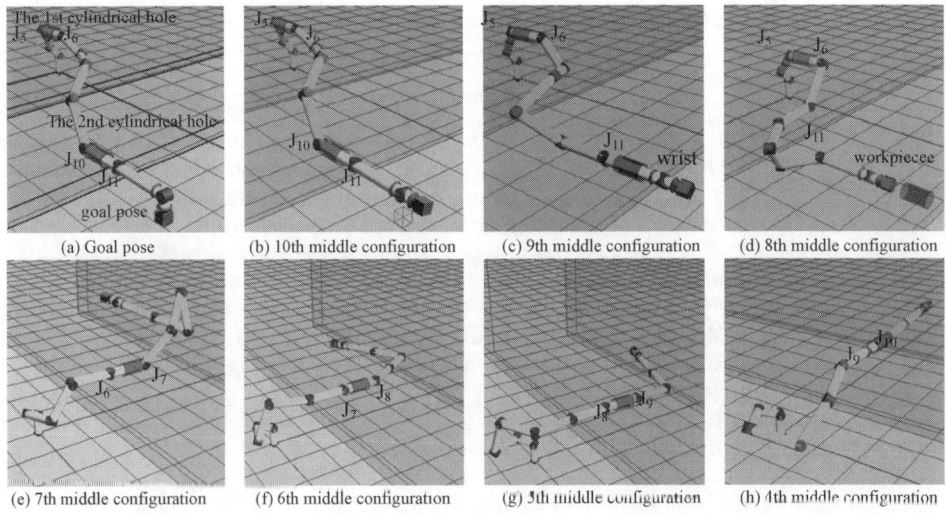

(a) Goal pose (b) 10th middle configuration (c) 9th middle configuration (d) 8th middle configuration
(e) 7th middle configuration (f) 6th middle configuration (g) 5th middle configuration (h) 4th middle configuration

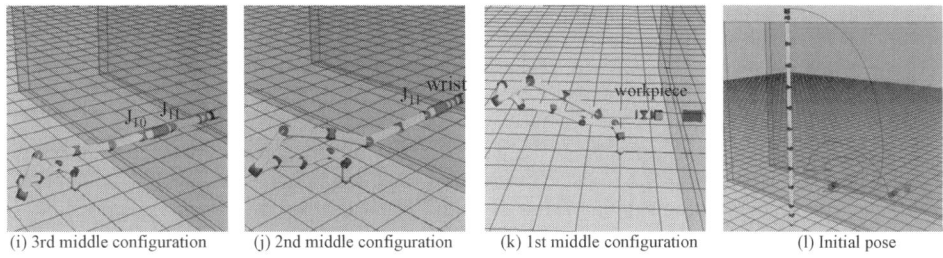

(i) 3rd middle configuration (j) 2nd middle configuration (k) 1st middle configuration (l) Initial pose

Figure 4: A result of simulation (14 DOF robot passes through two cylindrical holes in two thick walls)

Combination of Joint Angles used for Passage thorough Hole

For example, how to decide the combination of joint angles for realizing the passage through the hole from the 9th middle configuration to the 8th one is explained here. In this transition of configuration, J_{11} is positioned before the wall on the axis of the second hole, and the work piece is pulled out through the hole. On the 9th middle configuration, 6 joints of J_6 - J_{11} are before the second wall, and J_{11} is must be used for keeping the axis direction of link L_{12} same as that of the second hole as shown in **Fig. 5**. Considering this, there are 5 joints of $J_6 - J_{10}$ possible to be used for solving this problem. Also, considering that 5 joints (including J_{11}) are necessary before the wall (see the previous paragraph), 4 (=5-1) joints are necessary besides J_{11}. Eventually, there are possible 5 $(= {_5}C_4)$ combinations, namely 4 joints can be selected among 5 joints of $J_6 - J_{10}$. The combinations are shown in **Talbe 1**. In this table, KA joint is positioned on the axis of the hole by K1, K2, K3 joints. Orientation of the axis of KA joint is decided by K4 joint angle. The link (KA+1) is positioned on the axis of the hole by KA joint angle. R1 is excess joint and is fixed to an appropriate value.

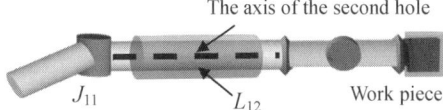

Figure 5: An enlarged figure of the 9th middle configuration

TABLE 1
Combination of joints for realizing path from 9th to 8th middle configuration

	K1	K2	K3	K4	KA	R1
①	6	7	8	10	11	9
②	6	7	9	10	11	8
③	6	8	9	10	11	7
④	6	7	8	9	11	10
⑤	7	8	9	10	11	6

Solving Inverse Kinematics and Realizing Path

For realizing the path from the 9th middle configuration to the 8th middle configuration, first, the joint combination of ① in **Table 1** is employed. The desired configuration is changed gradually by moving the link L_{12} step by step with each 1/50 distance of the hole length. At each desired configuration, the inverse kinematics is solved. If it is happened that the solution can not be converged because of the ill setting (which means there are no solutions for the inverse kinematics under this condition), then the combination of ② in **Table 1** is employed and the inverse kinematics is solved. Like this, the joint combination is changed in turn from ① to ⑤ until the solution is obtained. By carrying out the processes mentioned above, if the work piece is successfully pulled out of the second hole from the 9th middle configuration to 8th one, then the passage through the first hole is focused, which is namely realization of the path from the 8th middle configuration to the 7th one. In this passage, J_5 is positioned before the first wall on the axis of the hole. In this case, there is only one joint combination ($J_1 - J_5$ are necessary and sufficient). The inverse kinematics is solved on each configuration, which is on the path and divided by 1/50 length of the hole length, until J_6 is pulled out of the first hole.

As the same way as mentioned above, by realizing the path between every adjacent middle configuration, the total path from the goal configuration to the initial one is obtained, in which the robot successfully passes through two holes. The path from the initial configuration to the goal one is easily obtained by tracking the obtained path reversely.

Based on the proposed method, a computer simulator is developed for solving the inverse kinematics of redundant robot under restraint by obstacles. Windows XP is adopted as an operating system and Visual C++ is adopted as a programming language. The CPU is Pentium 4 (1.70GHz) and memory is 256 MB. The total computing time for the solution in the case of this paper is about 3 minutes. The simulation results can be graphically shown in a computer display. An example of the simulation result is already shown in **Fig. 4**.

CONCLUSION

For the purpose of collision avoidance, a new efficient method for solving inverse kinematics of a redundant robot is proposed. The summary is as follows:
1) This method is divided into three procedures, which are position synthesis by 3 joints, orientation synthesis by 3 wrist joints, and collision avoidance by other remaining joints. Each of these three procedures can be calculated analytically. After series of three procedures, the positioning error is occurred, since they are not independent from each other. Therefore, the calculation is carried out iteratively until sufficient convergence is obtained.
2) As the configuration on which the combination of joint angles for realizing the passage through the hole can be changed and must be selected, middle configurations are set on the way from a given initial configuration to the final configuration. The method for selecting the combination of joint angles is also proposed.
3) A simulator based on this method is developed. A robot of 14 DOF can successfully pass through two cylindrical holes in two thick walls, while realizing high positioning and orientating accuracy of its end effector.

ACKNOWLEDGEMENT

This work was partially supported by JSPS (Japan Society for the Promotion of Science).KAKENHI (16310103), MEXT (Ministry of Education, Culture, Sports, Science and Technology).KAKENHI (17656090), the Kansai University Special Research Fund, 2004 and 2005.

REFERENCES

[1] Nakamura Y., Hanafusa H. and Yoshikawa T. (1987). Task-Priority Based Redundancy Control of Robot Manipulators. *The International Journal of Robotics Research* **6:2**, 3-15.
[2] Hirukawa H. and Kitamura S. (1987). A collision Avoidance Method for Robot Manipulators based on Safety First Algorithm and the Potential Function. *Journal of the Robotics Society of Japan* **5:3**, 3-11.
[3] Takano M. (1958). A New Effective Solution for Inverse Kinematics Problem (Synthesis) of a Robot with Any Type of Configuration. *J. Fac. Eng., Univ. Tokyo (B)* **38:2**, 107-135.
[4] Kawamoto J., Aoyagi S. and Takano M. (2003). Path Planning of Collision Avoidance for Redundant Robot -Passage through Hole-. *Proceedings of the 21th Annual Conference of the Robotics Society of Japan*, CD ROM no. 2113.

EXPRESSION OF THEORETICAL DESIGN INFORMATION AND INTENTION TRANSMITTING ARCHITECTURE

Kazuhiro Takeuchi [1] and Akira Tsumaya, Hidefumi Wakamatsu, Eiji Arai [2]

[1] Fujitsu Software Technologies Ltd.
Shizuoka, 422-8572, Japan
[2] Department of Manufacturing Science, Graduate School of Engineering, Osaka University
Osaka, 565-0871, Japan

ABSTRACT

In recent years, 3D-CAD systems have been rapidly improving. However, the principal improvements are focused on shape modeling and operability, so that it is still remained to be solved how to handle the design information and intention generated within the design process.

Our objective is to smoothly and broadly support the design process more by proposing an architecture to handle the design information and intention from the upstream design stage to the downstream process. Especially, it is important to analyze the design information which should be handled in the upstream design stage and to transmit this information accurately up to the detailed design stage where the geometries are decided. In this paper, the important information in the upstream design stage and the framework to handle these kinds of information is discussed.

KEYWORDS

Geometric Modeling, Design Information, Design Intention, Design Methodology, Human Cooperative, Computer aided Design

1. BACKGROUND AND OBJECTIVES

3D-CAD systems have been rapidly improving. However, the principal improvements have been focused on geometric modeling and developing user-friendly operational improvements, while neglecting improvements to the treatment of design information and the intention generated in the design process. It is important for 3D CAD systems to become a means for delivering widely used design information to support the designer to the detailed design with design information and intention which was considered in the conceptual design phase [e.g. 1].

Many studies for conceptual design were performed that focused on modeling and it's intention in the conceptual design stage [2][3][4][5][6]. The research for synthesis of each functional design was discussed in [7] and researches of treatment of qualitative information are discussed in [8][9].

Our objective is to propose an architecture to accurately transmit the design information and intention from the upstream to the detailed design stage. For this purpose, we propose the principal architecture by introducing an integrated model with geometrical and intentional information in [10][11]. In this paper, we discuss about important design information at the upstream design stage. This information is important for design requirements but is not detailed yet. Moreover, expression of this design information by the proposed architecture is discussed. In particular, the space where an object does not exist, spatial representation and an application of this architecture including the behavior of the system is discussed. As a result, accurately transmitting the design information and the intention considered at the upstream to detailed design stage becomes possible.

2. SUBSTANCE

To achieve our objective, it is necessary to be able to handle the design information and intention as well as transmit this information to the downstream design phase accurately. In many designs, in the beginning, the outline of the entire product is decided and the design process gradually becomes more detailed. First, we explain the outline and features of a principal architecture. Secondary, important design information and intention at the upstream design stage is considered. Especially, at the design upstream stage the expression of shape, arrangement and functionality are vague. However, this information is a principal requirement for the product and the most important information for designing a final product.

3. POINTS OF PRINCIPAL ARCHITECTURE

The points of principal architecture are concisely described.
- An accurate transmitting framework for design information and intention attaching to geometric elements. This is the mechanism to perceive what was changed and how to change. Where, an edge, face, solid, etc. are objects, and the deletion, division, merging, etc. are the types of change.
- Single design information attaching to a single object and the relational design information attached between objects.
- Enables setting the behavior definition for each design information..
- Behavior definition can evaluate the types of change, mass property and special vector of an object.
- Behavior definition, the transmitting method of the design information and the reaction of systems that will reject an operation or signal alarm output, etc. can be defined.

This proposed principal architecture enables to transmit the design information and intention accurately and enable to define the system reaction for each design information. To handle the design information and intention, the system has a new component; that is the Design Information Processing Component. An outline of each subcomponent is described in the followings.

The flow of processing when the element is changed is shown below.
Step 1; Edit Sensor finds the kind of design change and the target
Step 2; Definition Interpreter interprets the content of the behavior definition that is related with the target and the design information.

   Step 2.1; Definition Interpreter interprets the behavior definitions.

Step 2.2; According to the behavior definition, the system decides the system behavior that includes action for designer and maintenance of the design information etc

4. UPSTREAM DESIGN STAGE REQUIREMENTS FOR PRINCIPAL ARCHITECTURE

During the upstream design stage, the main purpose is to achieve the functional requirements. Shapes, positions, etc. are very simple or vague. However, this information is very important to achieve the main requirements and should be observed in the subsequent design stages. Therefore, to support the design process flow it is important to handle simple or vague information and to transmit this information to the downstream process. Moreover, the case that a simple geometric element expresses some function, that will become a more detailed model or a space function. Thus, handling this space is one of the important items to support during the design process.

Geometrical simplicity consideration

At the upstream design stage, geometric elements express a sub-assembly or part, even if the geometric element is very simple like a line or plane. For example, when a line shows an axis in the upstream design stage, it is necessary to be able to set the design information to a line, surface roughness, material type, weight limitation, etc.. Thus, the mechanism should have the capability to set the design information to targets regardless of geometrical type, where geometrical type means edge, face or solid. The principal architecture fulfills this functionality.

However, it is important to consider is the case of geometric type change; that is not only the case of change of the element itself, but also the case of geometric type change, it is necessary to transmit the design information and intention to the final shape from the simple initial shape. This is a requirement for the framework, transmitting the design information defined in an initial element to a newly generated element.

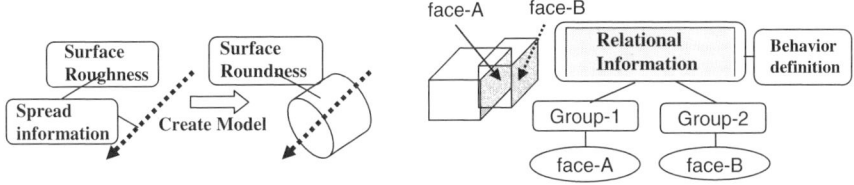

(1) Spread Information   (2) Relational Design Information
Figure 1: Image of spread information and relational design information

To consider the methods of transmitting information, we classify the design information as follows.
1) Model design information
    a) Single design information (EX: weight limitation, volume limitation etc.)
    b) Relational design information (EX: boundary information etc.)
2) Element design information included in the model
    a) Single design information
        Information should spread to newly generated elements by using the initial element. For example, surface roughness defined to the initial axis element should be migrated to the newly generated face when a rotated solid is generated by specifying the initial axis. In this case, there are two patterns; one is spreading to all generated faces unconditionally, or to specify the generated face to spread. Fig. 1-(1) shows an example.

b) Relational design information
For the case of geometric type change, the system should handle the capability to maintain the members of groups. Where, relational design information consists of two groups in Fig. 1-(2). If parallelism is defined between two initial lines, the system should add the axis of the rotated object as a group member when the rotated object is generated.

Consideration of fuzziness concerning positioning

We consider the two types of fuzzy positioning. One is to define rough position; this is a case to possible to define the space in which it can exist. The other is to define relative position. Naturally, there is a case to define both. In the proposed architecture, this is able to be defined as the relational design information between a target model and space. The relative positioning between targets, it is possible to define the big or small conditions as Fig2-(2). Fig. 2-(1) shows patterns of relative conditions. To define several conditions for each coordinate, it is able to define the relative condition between targets. Where, MinX means the minimum x-coordinate extent and MaxX means the maximum x-coordinate extent.

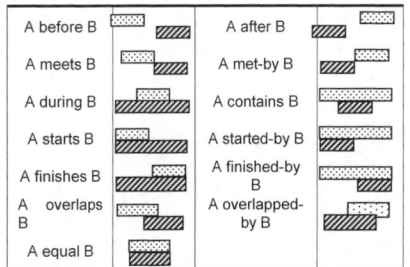

```
< behavior definition> <name>Relative positioning </name>
  < characteristic value of element editing method>
    < group characteristic value> <group no>1</group no>
      < characteristic value>MaxX</ characteristic value>
    </ group characteristic value>
    <comparison ope ><![CDATA[=<]]></comparison ope>>
    < group characteristic value> <group no>2</group no>
      < characteristic value>MaxX</ characteristic value>
    </ group characteristic value>
  </ characteristic value of element editing method>
```

Figure 2: Patterns of relative position for interval and example of x-coordinate behavior definition

Consideration for expression of function

In this section, it is discussed about two functional representations.
1) Behavior
Under certain situations, it is thought about the function as behavior. For example, a motor which generates a rotary motion, the influence of the rotary motion has on the models is not considered. This idea thinks an importance of potential influence. Thus, it is able to handle this design information as a single design information in the proposed architecture.
2) Action
This idea is that the function is some action for the targets. Therefore, it is possible to express by using a verb and object. Then, it is able to handle this design information as relational design information. Thus, the propose architecture can express the function as a behavior or an action.

Consideration for expression of space

Existence space where object can exist is a typical example of space. The space can be greatly classified into two types. One is the space which relates directly to the arrangement of an object, existence space or the space according to movement of object, etc. The other type is pure space, which itself has some design meaning, midair or a cavity in a target, a closed space surrounded by several object and the space which shows flows etc
1) Space which relates directly to object with substance (Territory of geostationary and movement)
2) Space which is defined by surrounding it with several objects (The existence space of a fluid or

gas)
This is a pure space and is defined as a space including a specified point.

Thus, both spaces are defined as a geometrical data. Therefore it is possible handle the space as a target for attaching design information and the intention. The expression of the space which relates directly to an object with substance is possible to treat the relational design information between the target object, space and pure space is possible to treat the single design information as a point. Fig.3-(1) shows the space which shows tracks of object and Fig. 3-(2) shows a case of personal computer and shows the space of air flow for cooling and Fig. 4 shows a example of pure space.

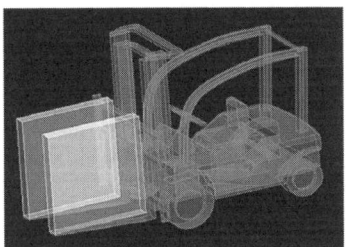 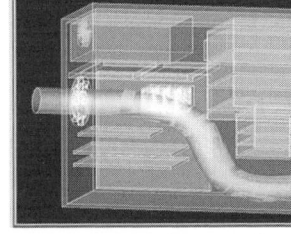

(1) Tracking space   (2) The space of air flow for cooling
Figure 3: Example of the space

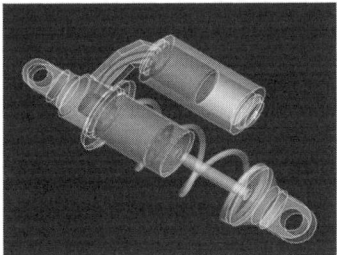

Figure 4: Example of closed space

Moreover, to handle the air flow and a closed space accentually, it is necessary for the mechanism to evaluate the space conditions, opening, closing or penetrating. For example, Fig. 4 shows a suspension part and the space in which oil is filled. The capability to check the open or closed state of this space is very important. It explains the judgment of the opening and closing space, as follows. For simplicity, all of the parts are solid models.

Proposition: Determination the open or closed state of space
  Judgment
    First, we show several definitions
    P : Point included in space to be judged , Bi (i=1,2,,,,n): Parts which compose the suspension
    H : The minimum hexahedron including the all parts
    He : The hexahedron which expands +e (>0) for each coordinate. BD(He) : Boundary set of He
      Then, if we take the differences of all parts from He, in general it becomes several solids.

$$\bigcup_{j=1}^{m} Sj = He - \bigcup_{i=1}^{n} Bi$$

    So, point P is included in Sk for some k. At that time, we can judge the state of space including point P as follows.
      If $b \in Sk$ for some $b \in BD(He)$

Then the specified space is opened, else the specified space is closed
End of judgment.

5. SUMMARIES AND CONCLUSION

In this paper, we proposed the important items at the upstream design stage and shows the expressions based on the principal architecture and its extension. Thus, proposed architecture is extensible and can transmit the design information and intention from the upstream to the downstream design stage. In the upstream design stage, shape and positioning are very simple or vague. To handle this information, we introduced the migratory information and proposed the expression of relative positioning and functions. To handle this information and to transmit this information to the downstream design stage is very effective to achieve the main design intention.

Moreover, it is proposed the treatment of spaces, especially the classification of the space and the judgment of the space state. In the actual design process, it is very important to transmit design information and intention from the upstream design stage to the detailed design stage. This is very important and effective not only the efficiency (reduction of design error or redo), but also for achieving the product concept and the main customer requirements.

The proposed architecture is extensible and accurate to transmit the design information. This architecture is one of the effective approaches to support the design process with the design information and intention.

REFERENCES

[1] Yoshikawa.H and Tomiyama.T (1989,1990):,Intelligent CAD, Asakura-syoten, Tokyo Japan
[2] Pahl.G and Beitz.W(1988), Engineering Design Systematic Approach, Springer-Verlag, Berlin
[3] Arai.E, Okada.K, and Iwata.K(1991), Intention Modeling System of Product Designers in Conceptual Design Phase, Manufacturing Systems, Vol.20, No.4, pp.325-333
[4] Umeda.Y, Ishii.M, Yoshioka.M, Shimomura.Y, and Tomiyama.T(1996), Supporting Conceptual Design Based on the Function- Behavior- State Modeler, Artificial Intelligence for Engineering Design, Analysis, and Manufacturing, Vol.10, No.4, pp.275-288
[5] Stone.R.B, Wood.K.L(2000), Development of a Functional Basis for Design, Journal of Mechanical Design, and Vol.122, pp359-370
[6] Arai.E, Akasaka.H, Wakamatsu.H, and Shirase.K(2000), Description Model of Designers' Intention in CAD System and Application for Redesign Process, JSME Int. J. Series C, Vol.43, No.1, pp.177-182
[7] Chakrabarti.A (ed.)(2000), Engineering Design Synthesis - Understanding, Approaches, and Tools, Springer-Verlag, London
[8] Liu.J, Arai.E and Igoshi.M(1995), Qualitative Kinematic Simulation for Verification of Function of Mechanical products, Trans JSME(C), Vol61, No585, pp.2159-2166, Japanese
[9] Liu.J, Amnuay.S, Arai.E and Igoshi.M(1996), Qualitative Solid Modelling : 1st Report, Qualitative Solid Models and Their Organization, Trans JAME(C)), Vol62, No599, pp.2897-2904, Japanese
[10] Takeuchi.K, Tsumaya.A, Wakamatsu.H, Shirase.Kand Arai.E(2003), Expression and Integrated Model for Transmission of Design Information and Intention, Proc. 6th Japan-France Congress on Mechatronic, pp83-88
[11] Takeuchi.K, Tsumaya.A, Wakamatsu.H and Arai.E(2004), Extensibility for Integrated Model of Geometrical and Intetional Information, JUSFA 2004, JL013

DETECTION OF UNCUT REGIONS IN POCKET MACHINING

Manseung Seo[1], Haeryung Kim[1] and Masahiko Onosato[2]

[1] Department of Robot System Engineering, College of Engineering, Tongmyong University,
535 Yongdang-dong, Nam-gu, Busan 608-711, Korea
[2] Graduate School of Information Science and Technology, Hokkaido University,
Kita-14, Nishi-9, Kita-ku, Sapporo, Hokkaido 060-0814, Japan

ABSTRACT

Upon realization of the fact that uncut regions exist if there is an intersection between a previous tool envelope and a current tool envelope, this study is initiated. As a key concept, the Tool envelope Loop Entity (TLE) is devised to treat every trajectory made by the tool radius as an ordinary offset loop. The TLE concept enables the offset curve generation method to be extended further as a distinctive method in which uncut region detection is done through an identical way of offsetting. To ensure the method works, a prototype system is implemented and evaluated with the tool path generation obviating uncut regions. The result verifies that the proposed method fulfils technological requirements for uncut free pocketing.

KEYWORDS

Pocket, Offset, Offset Loops, Uncut Region, Clean up Curve, Tool Path.

INTRODUCTION

It is not easy to find an efficient method for tool path generation free from uncut regions. In the literature, to solve uncut problems, Held et al. (1994) employed a specific adjustment on successive offset distance through the Voronoi diagram approach and Park & Choi (2001) took local care on tool trajectories through the pair-wise intersection approach. Recently, for offset curve generation, Seo et al. (2004) proposed the Offset-loop Dissection Method (ODM) based on the Offset Loop Entity (OLE) concept, which enables the method to be implemented easily into the system at any condition, regardless of the number of offsets, the number of intersections, and even the number of islands. Recognizing the robustness and flexibility of the ODM and realizing the fact that uncut regions exist if there is an intersection between a previous tool envelope and a current tool envelope, we extend the ODM to uncut region detection. For the adoption of the ODM, we define the Tool envelope Loop

Entity (TLE), i.e., the trajectory made by the tool radius, as a key concept corresponding to the OLE to treat every tool envelope as an ordinary offset loop. The uncut region detection method, namely the extended ODM is proposed. The conspicuous feature of the devised method is that uncut regions are detected in an identical way of offsetting and the clean up curves are treated as ordinary offset loops. Through this study, the problem of obviating uncut regions is resolved.

GENERATION OF OFFSET CURVE FOR POCKETING

To focus the present study on the detection of uncut regions, offset curve generation for pocketing without or with islands is briefly discussed through an illustrated example shown in Fig.1. The boundary of the pocket is defined as the Contour curve Entity (CE) and the sequential linkage of the CEs is defined as the Contour Loop Entity (CLE) as shown in Fig.1(a), by assuming that a CLE is constructed only with lines and circular arcs. Imagining that a circle with a radius that equals the offset distance is rolling on the CE, the trajectory of the center of the circle is defined as the Offset curve Entity (OE), and the sequential linkage of OEs is defined as the inborn OLE as shown in Fig.1(b). In pocket machining, there is a strong possibility that the inborn OLE is formed into an open loop having local and global self-intersections that result in undesirable cuts. The local OLE reconstruction is performed inserting additive OEs or by dissecting intersections in two adjacent OEs to create one crude OLE and to discard four open OLEs as shown in Fig.1(c). However, the crude OLE is intersected globally by itself at three points as shown in Fig.1(d). Detecting an intersection and applying a dissection on the crude OLE, the OLE is decomposed into one simple OLE and one crude OLE. By the second dissection, the OLE is decomposed into one simple OLE and one crude OLE. By the third dissection, the OLE is decomposed into two simple OLEs. Finally, all OLEs become simple OLEs as shown in Fig.1(e). The simple OLE obtained by the global OLE reconstruction may still not be appropriate as an offset curve for machining. The characteristics of OLE, i.e., closeness and orientation, need to be examined to confirm the validity of OLE for continuity and proper direction of the tool path. Fixing the orientation of a CLE to be counterclockwise, two OLEs are selected as valid OLEs, since they are completely closed and counterclockwise. Then, the valid OLEs in Fig.1(f) are kept to play the role of an offset curve for pocketing and the role of CLEs in the next offsetting turn.

One of the salient features of the ODM is the applicability. The offset curve generation method for one OLE works as the method for multiple OLEs. To ensure the merits, the ODM is applied to the generation of an offset curve for a pocket with islands, by shifting the object of intersection detection, dissection, and validation, from one OLE to multiple OLEs. Using an illustrated example of offset curve generation for a pocket with an island, the ODM is evaluated. Figure 1(g) shows the CLEs from one pocket and one island in dotted line, and two simple pocket OLEs and one simple island OLE in solid lines. At an intersection, a pocket OLE and an island OLE are dissected, and reconnected into one combined OLE conserving orientations and vice versa. Then, applying a dissection one more time at the other intersection and reconnecting again, one combined OLE is decomposed into two combined OLEs as shown in Fig.1(h). Performing OLE validation with the rule that the characteristic of the pocket OLE is transferred to the combined OLE when a pocket OLE and an island OLE are combined into an OLE, two valid OLEs are kept to play the role of offset curves for pocketing and the role of CLEs in the next offsetting turn as shown in Fig.1(i). Thus, the ODM works for a pocket with islands.

DETECTION OF UNCUT REGIONS

Uncut regions appear mainly on two occasions. The first is due to the improper selection of tool diameter for pocket boundary. There is no way to avoid this kind of uncut, unless the other tool is selected. The second is due to the complexity of pocket geometry under the offset distance properly fixed for tool diameter and high speed milling. It is avoidable, and is still worthwhile to develop a better way of obviation. Upon realization of the fact that uncut regions exist if there is an intersection between a previous tool envelope and a current tool envelope, the ODM is extended to the uncut region detection and clean up curve generation based on the TLE concept, which enables the ODM to be easily applied to uncut region detection. The method, namely the extended ODM, is proposed by shifting the object of ODM from OLEs to TLEs.

To verify the extended ODM, the entire process of uncut region detection and clean up curve generation is evaluated through an illustrated example shown in Fig.2. Figure 2(a) shows the previous $[(n-1)^{th}]$ tool path, the current $[(n)^{th}]$ tool path, the inward trajectory made by the previous tool path (previous TLE), and the outward trajectory made by the current tool path (current TLE). By taking a glance at Fig.2(a), we easily notice that the uncut region exists if there is an intersection between previous TLE and current TLE. Moreover, by imaging that the previous tool path to be like a pocket CLE and the current tool path to be like an island CLE, the previous TLE may be considered as a pocket OLE and current TLE may be considered as an island OLE, and then, we could see that those exactly match as shown in Fig.2(b). Therefore, we just need to carry out the ODM to detect the uncut regions upon OLE/TLE concepts. After the previous/current TLEs construction, the TLE reconstruction is processed as we did in the offset curve generation of the pocket with one island in Fig.1. Then, non-intersecting simple TLEs are obtained as shown in Fig.2(c). Performing TLE validation with the rule that the characteristic of the previous TLE is transferred to the combined TLE when a previous TLE and a current TLE are composed into a TLE, four simple TLEs with clockwise orientation are discarded. Finally, four valid TLEs corresponding to the boundaries of uncut regions are kept to play the role of the clean up curve. The clean up curves are then appended to current valid OLEs taking the shortest line segment for the construction of an uncut free tool path, as shown in Fig.2(d). Here, we may conclude that the extended ODM is flexible and robust enough to generate offset curves for uncut free pocket machining with islands.

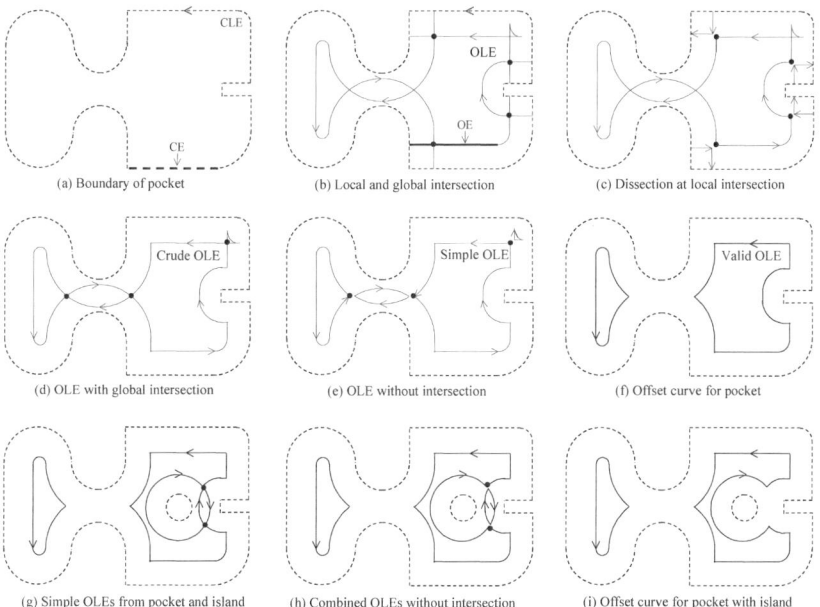

Figure 1: Offset curve generation procedures for a pocket with an island

RESULTS AND DISCUSSION

In order to verify the salient features of the extended ODM, a prototype system is implemented using C language and Open GL graphic library. The screen image of an uncut free tool path obtained from the implemented system is shown in Fig.3. The uncut regions are detected and then attached to the offset contours. The result of the implemented system verifies that the devised method is robust enough to generate uncut free tool paths.

CONCLUSIONS

In this study, we proposed the extended ODM for uncut free tool path generation. The OLE/TLE concept enables the ODM to possess robustness and flexibility. The distinctiveness comes from the facts: 1) The entire procedure is systematically integrated using the OLE/TLE, 2) Every procedure deals only with the OLE/TLE, and 3) Each procedure is designed based on the OLE/TLE. Thus, through this study the problem obviating uncut regions is resolved and the high speed milling becomes feasible.

REFERENCES

Held M., Lukacs G. and Andor L. (1994) Pocket machining base on contour-parallel tool paths generation by means of proximity maps, *Computer Aided Design*, **26:3**, 189-203.

Park S. and Choi, B. (2001). Uncut free pocketing tool-paths generation using pair-wise offset algorithm, *Computer Aided Design*, *33:10*, 739-746.

Seo M., Kim H. and Onosato M. (2005) Systematic approach to contour-parallel tool path generation of 2.5-D pocket with islands, *Computer-Aided Design and Applications*, **2:1**, 213-222.

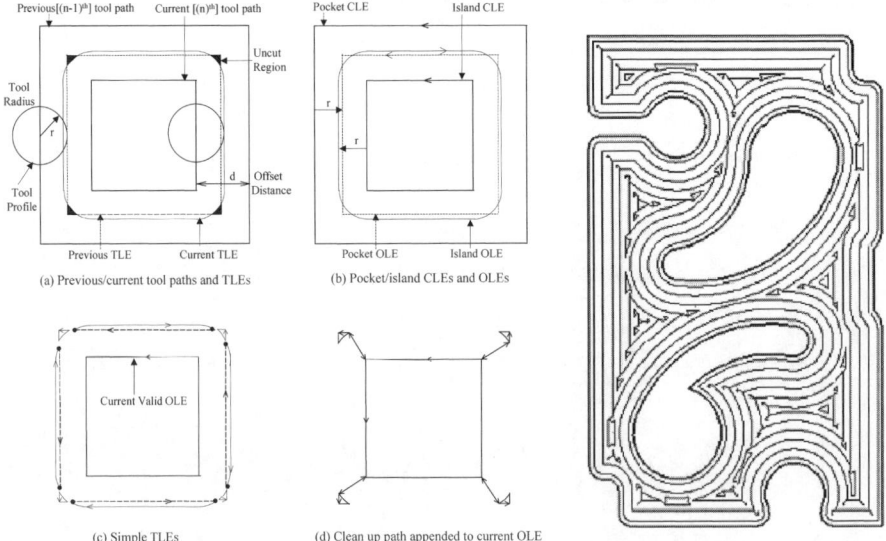

Figure 2: Uncut region detection procedures Figure 3: Uncut free tool path

FLEXIBLE PROCESS PLANNING SYSTEM CONSIDERING DESIGN INTENTIONS AND DISTURBANCE IN PRODUCTION PROCESS

G. Han[1] M. Koike[2] H. Wakamatsu[1] A. Tsumaya[1] E. Arai[1] and K. Shirase[3]

[1] Department of Manufacturing Science, Graduate School of Eng., Osaka University
2-1 Yamadaoka, Suita, Osaka, 565-0871, Japan
[2] Department of Systems Design, College of Industrial Technology
1-27-1 Nishikoya, Amagasaki, Hyogo, 661-0047, Japan
[3] Department of Mechanical Engineering, Faculty of Eng. Kobe University
1-1 Rokkodai, Nada, Kobe, Hyogo, 657-0013, Japan

ABSTRACT

Improvement of machining process planning is an effective way to reduce manufacturing time and cost, and to achieve the desirable functions which are described by designers. This paper proposes a machining process planning system which can flexibly perform process planning, considering design intentions and dealing with disturbances in the manufacturing process by choosing the optimum plans from multiple candidates. The core of the mechanism consists of (1)Extraction of Total Removal Volume(TRV), (2)Decomposition of the TRV into Minimum Convex Polyhedrons (MCP) (3)Recomposition of MCPs into feasible manufacturing features sets(MF set), (4)Recognition of manufacturing feature(MF), (5)Determination of machining sequences by considering various constraints, and (6)Comparison of each candidate containing a certain MF set and machining sequence to obtain the most optimum plan. All the functions are realized and implemented on DLL format compiled in Visual C++ and SolidWorks API.

KEYWORDS

Computer Aided Process Planning, Manufacturing Feature, Machining Sequencing

1. INTRODUCTION

Process planning plays a key role in modern manufacturing. And it provides the functions which translate designers' intentions and finished parts' specifications into technologically feasible plans describing how to manufacture a functional part efficiently and precisely. The task of automatically generating a process plan from a solid model representation of a part is normally subdivided into several activities such as: selection of the machining operations and so on. A process plan should primarily consist of a Manufacturing feature (MF) set which describes the most suitable removal volume set and a machining sequence which are considered optimum for the design intentions and the current manufacturing conditions. Most of current manufacturing systems perform fixed process planning which often leads to provide "fixed plans" for production. Those plans are only applicable in the situation where no errors and disturbances are found during the manufacturing process and no alterations are made to facilities in workshop [1]. Moreover, in some cases, because manufacturing features interpretations are predefined in a fixed way, only small number of plans can be generated as candidates. In addition, those outputted process plans are usually proven not the most efficient and precise for manufacturing. Because a great deal of useful embedded information in the part model is ignored, the determined sequences often fail to satisfy the desirable functions. As a result, the flexibility of process planning becomes an essential and effective way to create more candidates for resolving this problem. To realize the flexibility, our proposed system generates more functionally and technically satisfactory candidates. Finally, the most optimum process plan will be chosen from the candidates by comparing machining time of each plan.

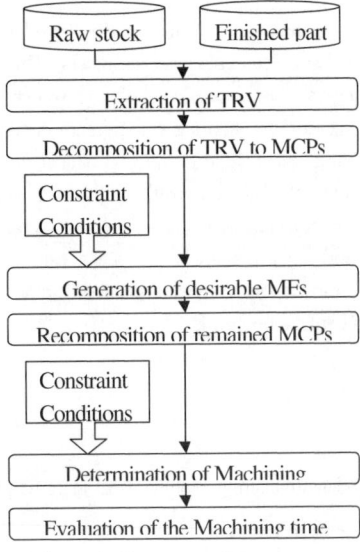

Figure 1: Core parts of the system

2. SYSTEM ARCHITECTURE

This system provides functions of generating one or more candidates of MF set to suit variable machining circumstances, sequencing the MFs and determining the best process plan which can realize the designed part, respecting the desired quality at high efficiency. The overall goal of this flexible process planning system is obtained through the following main steps shown in Fig.1.. The manufacturing feature recognition is executed based on judging the number of the open faces of the feature, by retrieving and modifying the familiar cases from database, case-based reasoning decides machining conditions including tools, cutting conditions, tool path and so on for individual features [2].

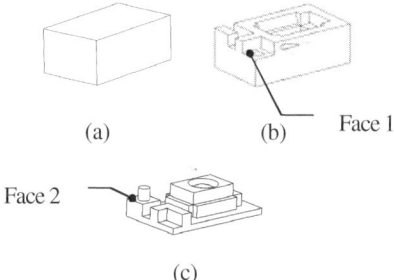

Figure2: Extraction of TRV (a) raw stock (b) resigned product (c) extracted TRV

3 FEATURE INTERPRETATIONS

This system offers multiple feature interpretations, which are represented in the form of MF set through the following steps:

3.1 Extraction of TRV

Process planning starts with the extraction of the removal area which is mainly composed by the planes and cylindrical surfaces in this system. The removal area is computed through difference between the raw stock and finished part. The volume generated in this subtraction process is named Total Removal Volume (TRV). Some parts with complex shapes usually offer TRV composed of more than one removal volume, these volumes are defined as SRV (Sub Removal Volumes) which will be handled respectively. Fig.2 shows an example of the extraction of TRV composed of four SRVs, and one of the faces (Face 1) in the part model and its corresponding face (Face 2) in TRV share the same attributed information.

3.2 Decomposition of TRV into MCPs

For generating enough sets of machinable MFs to cope with diversified facility circumstances and disturbances found in workplace, each SRV will be decomposed into Minimum Convex Polyhedrons (MCP) which can be recomposed into multiple sets of manufacturing features in the next steps. In this system, decomposition is performed by the cutting planes that are generated referring to all the planar faces in each SRV. Every planar face which belongs to SRV is extended enough to split SRV (as in Fig.3). Cylindrical faces will not be considered to create cutting faces. Then system randomly selects one cutting face to bisect SRV and if the SRV is intersected with this cutting face, several new volumes which have one or more created faces will be generated. At the same time, some faces which are attributed with constraints information in the SRV are split into several small faces in separate MCPs. The information is to be inherited from parent faces to new-created faces for delivering the demands information about part manufacturing to later steps. Then the procedures above repeats itself by utilizing other cutting faces to cut all cuttable new-born volumes and original SRVs until all the cutting faces are used. The example about decomposition of the former TRV is shown in the Fig.3 (b).

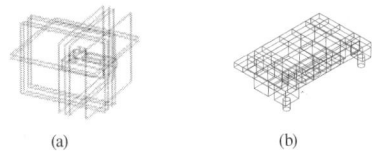

Figure 3: TRV decomposed into Minimum Convex

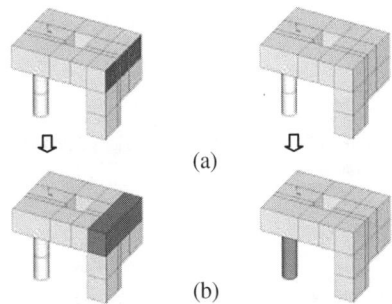

Figure 4: Attributed MCPs and generated MFs

3.3 Generation of the desirable MFs

Manufacturing feature each of which is removed with a single machining operation is a combination of a number of MCPs. Because the tool condition and cutting conditions keep unchanged without tool exchange, machining MCPs attributed with the same demand information as one MF can guarantee the high quality. The MFs(MF set) which can actualize the requirements are generated by recomposing the demand-attributed MCPs. System gathers the MCPs which are demanded by the same description, and combine them into one machinable MF. For example, two cylindrical MCPs with same concentricity and four MCPs sharing the same face which is required by the same surface finish are shown in Fig.4 (a), and the desirable features generated are shown in (b) respectively.

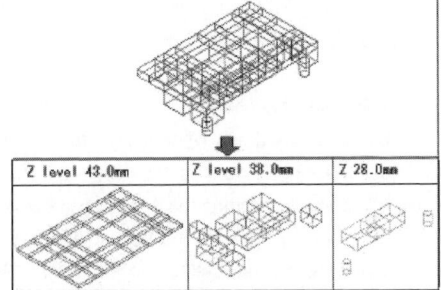

Figure 5: MCPs in different levels

3.4 Recomposition of remained MCPs to MF sets

In this step, the uncombined MCPs without any demand attribution are recomposed to obtain several sets of MFs. Merging these MCPs in different ways leads to different MF sets. MCPs that generated through decomposition are grouped into distinguished levels according to their geometrical position. MCPs whose top Z axis-perpendicular faces share the same Z coordinate value are defined as same level MCPs. An example of remained MCPs, which are classified into 3 levels are illustrated in Fig.5. Because tool properties such as length and strength restrict the sizes of machinable MFs, recomposition is to be executed level by level to avoiding creating MFs which are machinably unavailable in TAD (Tool Approach Direction).

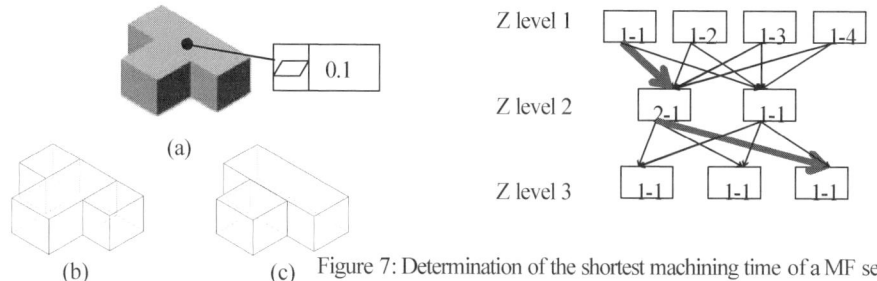

Figure 7: Determination of the shortest machining time of a MF set

(a) SRV with one face demanded by the same constraint of flatness which is valued 0.1
(b) Three MFs ought to be machined continually
(c) Two MFs ought to be machined continually
Figure 6: Determination of machining sequence

4 MACHINING SEQUENCE

One of the important and difficult activities in process planning is the determination of sequence which causes high-quality parts to be produced efficiently. For producing the part here are more than one set of features available to be chosen. Even for one of such sets of MFs, there are many ways to sequence these features for machining. But the utilization of all the possible MF set as removal area descriptions to determine the optimum process plans is rather time-consuming because the huge number of alternatives will overload the system. The constraints in workplace environment and design intentions are considered to eliminate the improper MF sets before they are further used for process planning, Because the majority of current systems focus too much on creating sequences based on part geometry, and fail to utilize other information which describes the designers' intentions, The final sequence plans often dissatisfy the requirement of qualities and functions, or are relatively time-consuming. Based on the constraint rules, which are developed and applied, the constraints obtained from the designer's intentions or the factory environment will be used to resolve this problem. Due to tools' restrictions in length and hardness, machining the MFs that are too large in TAD should be avoided. Therefore in this system sequencing is executed in each level. The solution of one MF set begins with recreating ID numbers to identify remained MFs in one level and sorting all these MFs in this level to generate all possible machining sequences as candidates. The vast number of feasible sequences will become evident through this mean. Without consideration of the constraints in manufacturing, it would be possible for a level composed of N manufacturing features to be processed from one of N factorial sequences. An obvious choice would be to represent a sequence as a string, whose elements are ID of features in a level of this MF set. But in reality this number of the alternatives is reduced by the feasible constraints. Appropriate sequences of each level are extracted from these choices. All the feasible sequences are checked based on geometry constraints, tolerance constraints, and quality constraints. Finally only the satisfactory sequences are picked out for machining time evaluation. Main constraints taken into considerations in this system are: Cylindricity, flatness, dimension tolerance, concentricity, surface finish. The MFs that satisfy the same constraints are to be continually machined. So the strings described by the correctly sorted numbers, whose order represents machining sequence are delivered to the next step. Then the decoding process is applied, translating each code into the string of the features. At last, a number of process plans which comprised of a set of feature interpretation and its machining sequence are provided for optimum plan determination. A simple example about two MF sets desired to be machined continually are shown in Fig.6.

5. OPTIMUM PROCESS PLAN

Because the determination of feature interpretation and sequencing are based on the requirements in qualities and functions, in this system machining time is used as the major criterion in effectiveness evaluation to decide optimal or near-optimal plan. The factors that affect the machining time involve (a) cutting condition generated by case-base reasoning in this system, (b) path length estimated by considering the sizes and machining sequences of the MFs, (c) the effect of surface quality. The machining time consists of cutting time, tools exchanging time and the time cost when tools travel between manufacturing features. The total machining time in a level of a MF set is calculated with the following equation.

$$T_{level} = T_{Feature} + T_{Tool\_exchange} + T_{Travel}$$

Where T(level) is the time cost in the process of machining all the MFs of this level. T(Feature) is the time spent on removing MFs, T (tool_exchange) is time for exchanging tools, and T(travel) stands for the time used in traveling the tools between MFs. Until this step one MF set still possesses more than one appropriate machining sequence each of which cause different machining time. The calculated machining times of every level in one MF set are aligned as Fig.7. The nodes in the figure show the machining time of every sequenced level in every MF set, the two numbers in the node indicate the level number and the machining sequence number respectively, the time which are spent on traveling tools between levels are taken into account as well. The path with the minimum time in the tree means the most efficient machining flow of this MF set. Compared with other MF sets, the corresponding process plan with the shortest machining time is decided as optimum plan for manufacturing this part.

6. CONCLUSION

By taking into account the designer's intentions and making use of the functional and technical constraints, the system proposed in this paper can provide the most optimum process plan for manufacturing the designed part.

REFERENCES

[1] Nagafune N., Kato Y., and Matsumoto T.(1998). Flexible Process Planning based on Flexible Machining Features. *JSME journal* **75**, 127-128.

[2] Shirase K., Nagano T., Wakamatsu H., and Arai E.(2000). Automatic Selection of Cutting Conditions Based on Case-Based Reasoning. *Proceedings of 2000 International Conference on Advanced Manufacturing Systems and Manufacturing Automation*, 524-528

A STUDY ON CALCULATION METHODS OF ENVIRONMENTAL BURDEN FOR NC PROGRAM DIAGNOSIS

H. Narita[1], T. Norihisa[2], L. Y. Chen[1], H. Fujimoto[1] and T. Hasebe[2]

[1]Graduate School of Engineering, Nagoya Institute of Technology,
Gokiso-cho, Showa-ku, Nagoya, Aichi, 466-8555, Japan
[2]OKUMA Corporation,
5-25-1, Shimokoguchi, Oguchi-cho, Niwa-gun, Aichi, 480-0193, Japan

ABSTRACT

Some activities for environmental protection have been tried to reduce environmental burdens in a lot of fields. Manufacturing field is also required to reduce them. Hence, prediction system of environmental burden for machining operation is proposed based on LCA (Life Cycle Assessment) policy. This system can calculate environmental burden (equivalent CO_2 emission) due to the electric consumption of a machine tool, the cutting tools status, the coolant quantity, the lubricant oil quantity and the metal chips, and provide the information of the accurate environmental burden of the machining process by considering some activities related to the machine tool operations. In this paper, the development status of prediction system is described. As a case study, two NC programs that manufacture simple shape are also evaluated to show the feasibility of it.

KEYWORDS

Environmental burden, Life Cycle Assessment, Production cost, Machine tool operation, Virtual machining, NC program diagnosis

INTRODUCTION

Manufacturing technologies pursuing the sustainable development are required due to the evident environmental impacts like global warming, ozone layer depletion and acidification, so manufacturing system has to be reassessed from the view point of environmental protection. Hence an accurate evaluation system of environmental impacts for manufacturing is required. But it is difficult to evaluate environmental impacts because we can not recognize them. In this research, a prediction system of the environmental burden for a machining operation is proposed based on LCA (Life Cycle Assessment) (SETAC, 1993) policy for future manufacturing system. This kind of system will enable engineers to decide the machining strategies, to generate the production scheduling and to evaluate the new manufacturing technologies with considering the environmental impact. In this paper, a conceptual architecture and a system design of the environmental burden calculation system are introduced first. Then, calculation algorithm of the environmental burden due to the machine tool operation is proposed and the feasibility of it is shown through a case study. Furthermore, using the cost data, NC programs are evaluated from the view points of the global warming and the production costs, and low environmental burden and low cost machining operations are discussed

SYSTEM OVERVIEW

Figure 1 shows an overview of the proposed evaluation system of environmental burden for machining operation. A work piece information, some cutting tools information and an NC program are input to the analysis model, the activities related to the machine tool operation and the machining process are estimated. Then, the electric consumption of a machine tool, the cutting tool status (tool wear), the coolant quantity, the lubricant quantity, the metal chip quantity and other factors are evaluated. Here, other factors correspond to the electric consumption of light, the air conditioning and so on. Using these estimated information and the emission intensities data and the resource data, the environmental burden is calculated, when a product is manufactured. The emission intensities data means the parameters required for the calculation of environmental burden. These emission intensities are prepared according to impact category such as the global warming, the ozone layer depilation and so on. The resource data also means the machine tool specification data, cutting tool parameter, etc. for the estimation of machining process. This system can calculate the environmental burden in various cutting conditions, because the machining process is evaluated properly. This is a novel feature of the system.

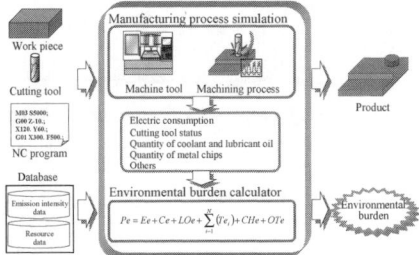

Figure1: Processing flow of the prediction system developed in this research

CALCULATION ALOGORITHM

The total environmental burden is calculated by equation (1). The calculation algorithm of environmental burden is the following.

$$Pe = Ee + Ce + LOe + \sum_{i=1}^{N}(Te_i) + CHe + OTe \qquad (1)$$

Pe: EB of machining operation [kg-GAS]
Ce: EB of coolant [kg-GAS]
Te: EB of cutting tool [kg-GAS]
OTe: EB of other factors [kg- GAS]
Ee: EB of machine tool component [kg-GAS]
LOe: EB of lubricant oil [kg-GAS]
CHe: EB of metal chip [kg- GAS]
N: Number of tool used in an NC program

EB: Environmental burden

Electric consumption of machine tool (Ee)

The environmental burden due to the machine tool electric consumption is expressed by equation (2). In equation (2), the electric consumption of the servo motors and the spindle motor is varied dynamically according to the machining process, so the electric consumption of these motors are calculated with considering the table weight, the friction coefficients of the slide way, the ball screw lead, the transmissibility of the ball screw, the axial friction torque, the cutting force and the cutting torque. Here, these are also predicted by cutting process model (Narita, et. al, 2002). This cutting process model concept can be applied to square end milling operation, ball end milling operation, turning operation and so on. Using these models, various cutting processes can be evaluated.

$$Ee = k \times (SME + SPE + SCE + CME + CPE + TCE1 + TCE2 + ATCE + MGE + VAE) \qquad (2)$$

k: CO_2 emission intensity of electricity [kg-GAS/kWh]
SME: EC of servo motors [kWh]
SPE: EC of spindle motor [kWh]
SCE: EC of cooling system of spindle [kWh]
CME: EC of compressor [kWh]
CPE: EC of coolant pump [kWh]
$TCE1$: EC of lift up chip conveyor [kWh]
$TCE2$: EC of chip conveyor in machine tool [kWh]
$ATCE$: EC of ATC [kWh]
MGE: EC of tool magazine motor [kWh]
VAE: Vampire energy of machine tool [kWh]
EC: Electric consumption

Coolant (Ce)

There are two types cutting fluid, so two equations are proposed for Ce evaluation. First, the water-miscible cutting fluid is explained. The coolant is generally used to enhance the machining performance, and circulated in a machine tool by a coolant pump until the coolant is updated. During the period, some coolants are eliminated due to the adhesion to the metal chips, so the coolant is supplied for the compensation. The dilution fluid (water) is also reduced due to the vapor. So, the equation (3) is adapted to calculate the environmental burden. Second, the water-insoluble cutting fluid is explained. In this case, the discharge rate is an important factor. Hence, the equation (4) is applied.

$$Ce = \frac{CUT}{CL} \times \{(CPe + CDe) \times (CC + AC) + WAe \times (WAQ + AWAQ)\} \qquad (3)$$

CUT: Coolant usage time in an NC program [s]
CL: Mean interval of coolant update [s]
CPe: EB of cutting fluid production [kg-GAS/L]
CDe: EB of cutting fluid disposal [kg-GAS/L]
CC: Initial coolant quantity [L]
AC: Additional supplement quantity of coolant [L]
WAe: EB of water distribution [kg-GAS/L]
WAQ: Initial quantity of water [L]
$AWAQ$: Additional supplement quantity of water [L]

$$Ce = \frac{CUT \times CS}{3600 \times 1000} \times (CPe + CDe) \qquad (4)$$

CS : Discharge rate of cutting fluid [cc/h]

Lubricant oil (LOe)

Lubricant oil is mainly used for spindle and slide way, so two equations are introduced. The minute amounts of oil are supplied to the spindle part in an interval time. For the lubricant of the slide way, the certain amount of the oil is also supplied by pump in an interval time. So, the following equations are adapted to calculate the environmental burden due to lubricant oil. These equations can be adapted oil-air lubricant and the grease lubricant.

$$LOe = Se + Le \qquad (5)$$

$$Se = \frac{SRT}{SI} \times SV \times (SPe + SDe) \tag{6}$$

$$Le = \frac{LUT}{LI} \times LV \times (LPe + LDe) \tag{7}$$

Se: EB of spindle lubricant oil [kg-GAS]
Le: EB of slide way lubricant oil [kg- GAS]
SRT: Spindle runtime in an NC program [s]
SV: Discharge rate of spindle lubricant oil [L]
SI: Mean interval between discharges [s]
SPe: EB of spindle lubricant oil production [kg-GAS/L]
SDe: EB of spindle lubricant oil disposal [kg-GAS/L]
LUT: Slide way runtime in an NC program [s]
LI: Mean interval between supplies [s]
LV: Lubricant oil quantity supplied to slide way [L]
LPe: EB of slide way lubricant oil production [kg-GAS/L]
LDe: EB of slide way lubricant oil disposal [kg-GAS/L]

Cutting tool (Te)

Cutting tools are managed from the view point of tool life. So, the tool life is compared with the machining time to calculate the environmental burden in one machining. Also, the cutting tools, especially for solid end mill, are made a recovery by re-grinding, so these points are considered to construct environmental burden equation.

$$Te = \frac{MT}{TL \times (RCN+1)} \times ((TPe + TDe) \times TW + RCN \times RCe) \tag{8}$$

MT: Machining time [s]
TL: Tool life [s]
TPe: EB of cutting tool production [kg- GAS /kg]
TDe: EB of cutting tool disposal [kg- GAS /kg]
TW: Tool weight [kg]
RN: Total number of recovery
RCe: EB of tool recovery [kg- GAS]

Metal chip (CHe)

Metal chips are recycled to material by an electric heating furnace. This materialization process has to be considered. This kind of equation is supposed to consider material kind, but an electrical intensity of this kind of electric heating furnace is represent by kWh/t, so the equation constructed in this research is calculated from the total metal chip weight.

$$CHe = (WPV - PV) \times MD \times WDe \tag{9}$$

WPV: Work piece volume [cm^3]
PV: Product volume [cm^3]
MD: Material density of work piece [kg/cm^3]
WDe: EB of metal chip processing [kg-GAS/kg]

CASE STUDY

In order to show the feasibility of developed system, a case study is introduced. Then, the impact category is set to global warming to calculate the environmental burden. In this research, CO_2, CH_4 and N_2O are evaluated based on Japanese data, which are decided from environmental report, technical report, home page and industrial table (Tokyo Waterworks, 2002, Tokyo Electric Power Company, 2002, Nansai, 2002, Mizukami, 2002). Here, CH_4 and N_2O emission is converted to equivalent CO_2 emission using the characterization factors and total CO_2 emission is evaluated. Here, the global warming potential (GWP) of 100 years (IPCC, 1995) is used for the characterization factors. The other emission matters related to global warming are ignored, because there are no emissions about the machining operations. In this case study, machine tool is MB-46VA (OKUMA Corp.), cutting tool is carbide square end mill with 2 flutes and 30 deg. helical angle and workpiece is medium carbon steel (S50C). The simple product shown in Figure 2 is evaluated.

The dry machining, the MQL machining and the Wet machining are evaluated in this case study. Here, the life of cutting tool is assumed to be extended to 2 times of original one. The analyzed results are shown in Figure 3. The equivalent CO_2 emission of wet machining is largest and one of dry machining is smallest in this comparison. Using this system, this kind of comparison can be carried out easily from NC program. Here, the detailed discussion is tried based on the analyzed results. The portion of electric consumption is highest in the all factors, obviously. This causes due to the peripheral devices of machine tool. This factor is also proportional to machining time. That is to say the high speed milling in dry machining method may be superior machining from the view point of CO_2 emission because of the short machining time, although detailed analysis will be required.

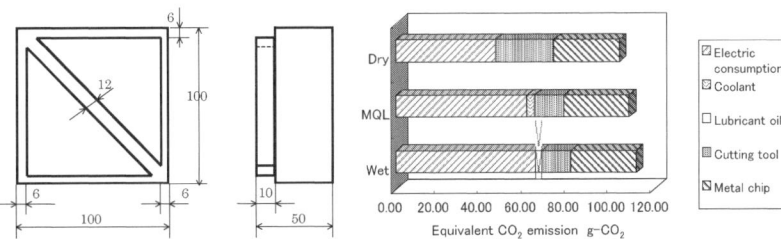

Figure2: Product shape   Figure 3: Analyzed environmental burden results

Also, the equivalent CO_2 emission of the MQL machining and the wet machining is larger than the dry machining. As shown in the Figure 3, the equivalent CO_2 emission of cutting tool is smaller due to the mitigation of tool wear, but one of peripheral devices operated by coolant usage and one of coolant are added and total one becomes larger. It is found, however, one of peripheral devices operated by coolant usage is larger than one of coolant effect. Furthermore, equivalent CO_2 emission of CH_4 and N_2O is calculated using analyzed results of wet machining. These are related to environmental burden of cutting fluid. Equivalent CO_2 emission of them is less than 0.001 g-CO_2. In other word, CO_2 is a dominant environmental burden in machining operation about the global warming.

Here, the production cost is evaluated using cost data. This analysis can be realized that equivalent CO_2 emission intensity data in equations (2)-(9) is changed to cost data. These equations are constructed by considering the activities related to machine tool operation, hence this cost accounting method correspond to activity-based costing (ABC) (Brimson, 1997). The cost due to electric consumption has to be changed a little, because the basic rate of the electricity is considered. The equation of the cost due to electric consumption is following. In this research, JPY (Japanese Yen) is used as currency.

$Ec = Ebc \times MT + ER \times CE$ (10)

Ebc: Basic rate of electricity [JPY/min] MT: Machining time [min]
ER: Electricity bill [JPY/kWh] CE: Electric consumption [kWh]

Cost data are searched by hearing the related companies. In these dates, the metal chip processing value is minus and cutting tool disposal cost is 0, because metal chip becomes profit and cutting tool disposal is carried out free fee in Japan, respectively. Using these data, same machining operations are compared. Figure 4 shows the analyzed results of cost evaluation. As shown in the Figure, the dry machining is largest, and the MQL machining and the Wet machining are almost same value. The dry machining is best from the view point of environmental burden, but this is worst from the view point of cost. So, adequate machining strategy has to be decided according to the situations. It is also found that the reduction of electric consumption of the machine tool peripheral device and the cutting tool consumption is effective from the view point of cost down and mitigation of global warming.

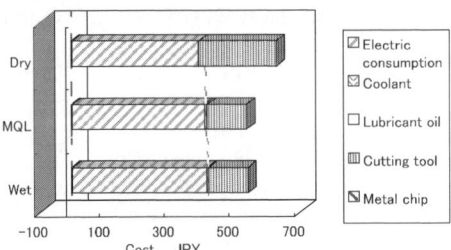

Figure 4: Analyzed cost results of case study

CONCLUSIONS

1. The evaluation model of environmental burden for machining operation has been proposed and evaluation system has been developed. The feasibility of the developed system is also demonstrated through case studies.
2. CO_2 is a dominant environmental burden in machining operation about the global warming by comparing with equivalent CO_2 emission of CH_4 and N_2O.
3. It is found that relationship of the emission factor of global warming and the cost for the machine tool operation isn't always the proportional through the analysis.

REFERENCES

Society of Environmental Toxicology and Chemistry (SETAC) (1993), *Guidelines for Life-Cycle Assessment: A code of Practice*, SETAC.

Narita, H., Shirase, K., Wakamatsu, H., Tsumaya, A. and Arai, E. (2002) Real-Time Cutting Simulation System of a Milling Operation for Autonomous and Intelligent Machine Tools, *International Journal of Production Research*, **40:15**, 3791-3805.

Tokyo Waterworks (2003) Environmental report of Tokyo waterworks 2002, (in Japanese) <http://www.waterworks.metro.tokyo.jp/pp/kh14/index.html> (accessed Mar 25, 2004)

Tokyo electric power company (2004) The Earth, People & Energy TEPCO Sustainability Report 2003, <http://www.tepco.co.jp/index-e.html> (accessed Mar 25, 2004)

Nansai, K., Moriguchi, Y., Tohno, S. (2002), Embodied Energy and Emission Intensity Data for Japan Using Input-Output Tables (3EID) – Inventory Data for LCA -, Center for Global Environment Research, National Institute of Environmental Studies, Japan,.

Mizukami, H., Yamaguchi, R., Nakayama, T., Maki, T. (2002) Off-gas Treatment Technology of ECOARC, NKK Technical Report, **176**, 1-5. (in Japanese)

The Intergovernmental Panel on Climate Change (IPCC) (1995) *Second Assessment Report: Climate Change 1995*

Brimson, J. A. (1997) *Activity Accounting: An Activity-Based Costing Approach*, John Wiley & Sons Inc.

ASSEMBLY SYSTEM BY USING PROTOTYPE OF ACTIVE FLEXIBLE FIXTURE

T. Yamaguchi[1], M. Higuchi[2] and K. Nagai[3]

[1]Department of Mechanical Engineering, Kansai University
3-3-35 Yamate-cho, Suita, Osaka 564-8680 JAPAN
[2]Department of Mechanical System Engineering, Kansai University
3-3-35 Yamate-cho, Suita, Osaka 564-8680 JAPAN
[3]Department of Robotics, Ritsumeikan University,
1-1-1 Nojihigashi, Kusatsu, Shiga 525-8577 JAPAN

ABSTRACT

Our goal is the development of fixture with the function of handling of various works with practicability in automated assembly system for job shop type production. This paper describes the "active flexible fixture (AFLEF)" on plane level as a prototype of the goal. The AFLEF is an active and practical fixture, and it can fix any work rigidly and position the work at a few millimeters to correct the position error after holding. It is multi-fingered hand type, but it is not more dexterous than general hands of this type but more practical than those. As results of the experiments in rigid fixing and short positioning, the fixture rigidity to external force was within about 0.031 mm/N and 0.88 deg./N·m and the maximum error in positioning of a fixed work at ±3.0 mm or ±3.0 deg. was within about 0.3 mm and 0.3 deg..

KEYWORDS

Fixture, Job shop type production, Automated assembly, Peg-in-hole task, Multi-fingered hand

1. INTRODUCTION

The function of handling various works with practicability has been required for automated assembly system in job shop type production. A usual automated assembly unit is composed of a manipulator with a robot-hand and a fixture. In order to equip the manipulator with the above function, many researchers, e.g. Rapela *et al.* (2002), tried assembly task by using a multi-fingered robot-hand. On the other hand, in order to equip the fixture with the function, some researchers, e.g. Asada and By (1985), Lee and Cutkosky (1991), Brost and Goldberg (1996) and Cai *et al.* (1997), applied the modular fixture like T-slot type, dowel type, or pin-array type. However, since the positioning of the tool like dowel, pin, etc. is passive, it is hard to rearrange the fixture layout immediately for the change of work, i.e. this type of fixture is not suitable for practical assembly. Thus the active function is also needed for the fixture suitable for practical assembly. Some active fixtures were developed by Grippo *et al.* (1988), Hazen and Wright (1990), Chan and Lin (1996), and Kimura and Yashima (1996). However, any fixture except for Kimura and Yashima's one can fix any work rigidly but cannot correct the position error occurring at the contact with the work because it is adaptive surface-fitting type. On the other hand, Kimura and Yashima's fixture can correct the position error but is hard to fix a work rigidly.

Therefore we developed "active flexible fixture (AFLEF)" that can fix any work rigidly and actively by only position control and also position the fixed work at a few millimeters in order to correct the location of fixing point into the assembling point. This paper describes the AFLEF on plane level as a prototype and the performance of each function.

2. PROTOTYPE OF THE AFLEF

The prototype of AFLEF is composed of four contact-fingers. The schematic diagrams of the prototype and the contact-finger are shown in Fig. 1. Each contact-finger touches the side of a work to grip the work. The contact-finger has the probe, the contact-tip and two driving joints: translational and rotatory driving joints. As shown in Fig. 1, the contact-tip is joined to the probe and can rotate freely around the vertical axis, and the probe is equipped with the force sensor to measure the contact force and the potentiometer to measure the angle to the contact-tip. Moreover, the contact-tip is equipped with a rubber-slab to cause large friction in the contact point. Each driving joint is controlled by inputting the individual reference position data from a computer simultaneously. The movable range of translational driving joint is from 85 mm to 105 mm and that of rotatory joint is from 0 deg. to 360 deg.. Both the driving joints are usually rigid because of the reduction gear, but the only translational joint can be made elastic by the feedback of a force sensor's signal in addition to a displacement sensor's signal.

The AFLEF needs to have practically the functions both of the rigid fixing realized usually by setting all joints rigid and of the short positioning realized by setting some joints elastic. Osumi and Arai (1994) reported the necessary and sufficient condition where the rigid fixing is compatible with the short positioning, i.e. the positioning accuracy is maintained against the arbitrary external force without generating the excessive internal force. Figure 2(a) shows the characteristic of each joint in the prototype of AFLEF determined under satisfying the condition. Here, the rubber-slab in the contact-finger can be regarded as a passive and elastic joint.

3. EVALUATION OF THE PROTOTYPE OF AFLEF

3.1 Rigid Fixing

We evaluated the function of rigid fixing in the prototype of AFLEF by experiment. The work was a

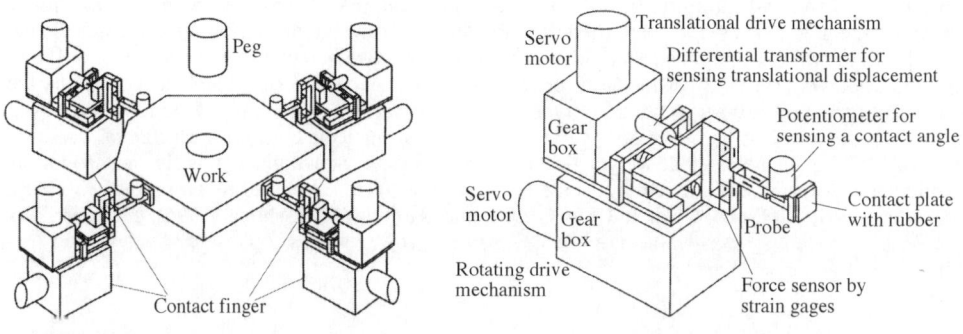

(a) Prototype of active flexible fixture (AFLEF)   (b) Contact finger

Figure 1: Schematic diagrams of prototype of active flexible fixture and contact finger

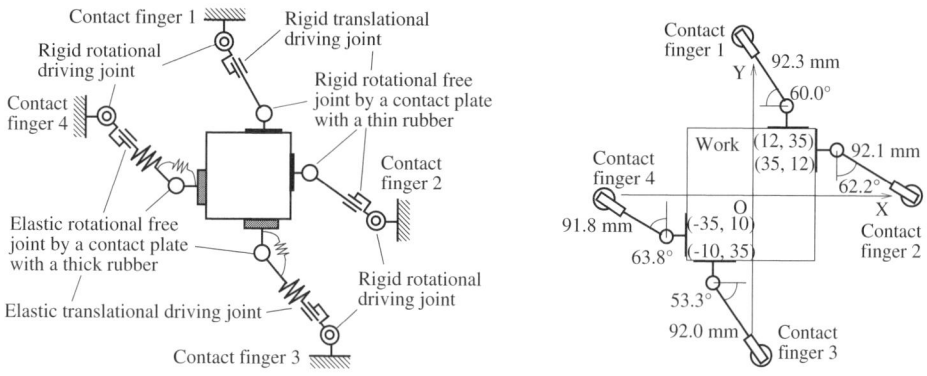

(a) Characteristic of each joint   (b) Coordinates of each contact point in experiments

Figure 2: Prototype of AFLEF fixing a work

rectangular parallelepiped whose size was 70 by 70 by 30 mm and made of hard plastic. Its weight was 0.78 N. The maximum coefficient of static friction between the contact-tip and the work was 0.5. The spring constant of the linear driving joints in the contact-finger 3 and 4 was 5.0 N/mm. At first the prototype fixed the work whose side was parallel to the axis of global coordinates as shown in Fig. 2(b). The coordinates of contact points and the contact angle are also indicated in Fig. 2(b). Then an external force: 9.8 N was added to the side of the work in the +X, -X, +Y or -Y direction. Moreover, an external moment: 0.34 N·m was also added to the side of the work in the +θ or -θ direction. The displacement of work in each direction caused by the external force was measured with a CCD camera (resolution = 0.03 mm/pix.).

The displacement in each direction is shown in Table 1. As can be seen from Table 1, any displacement is within ±0.3 mm (translation) or ±0.3 deg. (rotation). These real displacements are slightly larger than theoretical those founded from the rigidity of each mechanism composing the contact finger. We presume that the result is caused by the unexpected deformation of rubber-slab.

3.2 Short Positioning

We also evaluated the function of short positioning in the prototype by experiment. The experimental conditions were identical to those in the experiment in rigid fixing. The position-control of fixed work was as follows: the reference input to each driving joint can be found from the geometrical relation between the coordinates of each contact point before positioning and that after positioning, and each driving joint was positioned with this reference input. The reference input was set at ±3.0 mm in the X- or Y-direction and at ±3.0 deg. in the θ-direction, because the maximum displacement of the work was 2.6 mm and 2.5 deg. in trying to grip it at twenty times.

The real positioning in each direction for each reference positioning is shown in Table 2. These

TABLE 1

RIGIDITY OF THE WORK FIXED BY THE PROTOTYPE OF AFLEF FOR EXTERNAL FORCE

Direction of external force: 9.8N		+X	-X	+Y	-Y	+θ	-θ
Displacement of each axis	X-axis mm	0.24	-0.29	-0.09	0.00	0.08	-0.11
	Y axis mm	0.03	-0.01	0.19	-0.28	-0.11	0.05
	θ-axis deg.	0.2	-0.1	-0.1	0.1	0.3	-0.2

TABLE 2
REAL POSITIONING IN EACH DIRECTION FOR SHORT POSITIONING BY USING THE PROTOTYPE OF AFLEF

Reference displacement		X-direction mm		Y-direction mm		θ-direction deg.	
		+3.0	-3.0	+3.0	-3.0	+3.0	-3.0
Real displacement at each axis	X-axis mm	0.24	-0.29	-0.09	0.00	0.08	-0.11
	Y axis mm	0.03	-0.01	0.19	-0.28	-0.11	0.05
	θ-axis deg.	0.2	-0.1	-0.1	0.1	0.3	-0.2

displacements were measured with the same CCD camera. As can be seen from Table 2, the positioning error in each positioning is within ±0.3 mm (translation) and ±0.3 deg. (rotation). These real errors agree with theoretical those founded from the positioning performance of each driving joint.

4. CONCLUSIONS

We have developed the 2-dimensional active flexible fixture (AFLEF) with the generally conflicting functions of rigid fixing and short positioning. As a result of experiment, the fixture rigidity to external force was within about 0.031 mm/N and 0.88 deg./N·m, and the maximum error in positioning of a fixed work at ±3.0 mm or ±3.0 deg. was within about ±0.3 mm and ±0.3 deg.. Thus, in the prototype of AFLEF, the function of rigid fixing was compatible with that of short positioning. We have tried to realize the 3-dimensional AFELF by the improvement of the contact-tip and the additional of a vertical translational joint.

REFERENCES

Asada H. and By A. B. (1985). Kinematic Analysis of Workpart Fixturing for Flexible Assembly with Automatically Reconfigurable Fixtures. *IEEE Journal of Robotics and Automation* **RA-1:2**, 86-94.
Brost R. C. and Goldberg K. Y. (1996). A Complete Algorithm for Designing Planar Fixtures Using Modular Components. *IEEE Transactions on Robotic and Automation* **12:1**, 31-46.
Cai W., Hu S. J. and Yuan J. X. (1997). A Variational Method of Robust Fixture Configuration Design for 3-D Workpieces. *Transactions of the ASME, Journal of Manufacturing Science and Engineering* **119:4A**, 593-602.
Chan K. C. and Lin C. S. (1996). Development of a Computer Numerical Control (CNC) Modular Fixture - Machine Design of a Standard Multifinger Module. *International Journal of Advanced Manufacturing Technology* **11:1**, 18-26.
Grippo P. M., Thompson B. S. and Gandhi M. V. (1988). A Review of Flexible Fixture Systems for Computer-integrated Manufacturing. *International Journal of Computer-Integrated Manufacturing* **1:2**, 124-135.
Hazen F. B. and Wright P. K. (1990). Workholding Automation: Innovations in Analysis, Design and Planning. *Manufacturing Review* **3:4**, 224-237.
Kimura H. and Yashima M. (1996). Dynamics and control of intelligent jig with function of manipulation. *JSME International Journal Series C, Dynamics Control Robotics Design and Manufacturing* **39:3**, 549-559.
Lee S. H. and Cutkosky M. R. (1991). Fixture Planning with Friction. *Transactions of the ASME, Journal of Engineering for Industry* **113:3**, 320-327.
Osumi H. and Arai T. (1994). A Method for Introducing Industrial Robots to Cooperative Tasks. *Journal of the Robotics Society of Japan* **12:8**, 1192-1197 *(In Japanese)*.
Rapela D. R., Rembold U. and Kuchen B. (2002). Planning of Regrasping Operations for a Dexterous Hand in Assembly Tasks. *Journal Intelligent & Robotic Systems* **33:3**, 231-266.

ASSEMBLY SEQUENCE PLANNING USING K-NEAREST-NEIGHBOR RULE

T. Murayama[1], T. Eguchi[2], and F. Oba[2]

[1]Division of Oral Health Engineering, Faculty of Dentistry, Hiroshima University
1-2-3 Kasumi, Minami-ku, Hiroshima, 734-8553, Japan
[2]Dept. of Mechanical Systems Engineering, Hiroshima University
1-4-1 Kagamiyama, Higashi-hiroshima, 739-8527, Japan

ABSTRACT

This paper describes an approach to the efficient planning of assembly sequences. K-nearest-neighbor rule reduces the search space for the assembly sequences by using sample data on products, of which assembly sequences are known. Additional sample data are made from the assembly sequences generated by this approach. As the assembly sequence planning and the addition of the sample data are executed more times, the assembly sequences can be generated more efficiently. Some experiments are carried out to show: the effectiveness and efficiency of the approach; and the superiority of the k-nearest-neighbor rule over the heuristics that were used in our previous work.

KEY WORDS

Assembly Sequences, Assembly Planning, K-nearest-neighbor Rule, CAPP, CAD/CAM

INTRODUCTION

Recently, many research efforts have been made to plan assembly sequences automatically and efficiently. Most of the existing approaches generate a disassembly sequence by identifying a part or subassembly to be removed from a product repeatedly, and then generate an assembly sequence by reversing the disassembly sequence (Lambert, 2003.) In order to identify a part or subassembly to be removed, the approaches test which parts and/or subassemblies can be removed from the product. The tests for all the parts and/or subassemblies are computationally very expensive, especially in the case that paths to remove them are searched for at the tests. Therefore some of the approaches focus on reducing the number of the tests. Bourjault (1984) proposed superset and subset rules that can avoid the unnecessary tests; however, the number of the remainder (i.e., the necessary tests) is still large especially for the products composed of many parts. Subassembly extraction (Lee & Yi, 1993) and heuristics (Murayama & Oba, 1993) are effective to reduce the number of the tests further.

This paper describes a method of reducing the number of the tests, in which k-nearest-neighbor rule is used instead of the heuristics. In this method the k-nearest-neighbor rule extracts some parts and/or subassemblies whose possibilities of being removed without any interference are strong, and then the tests for only them are performed by using CAD data.

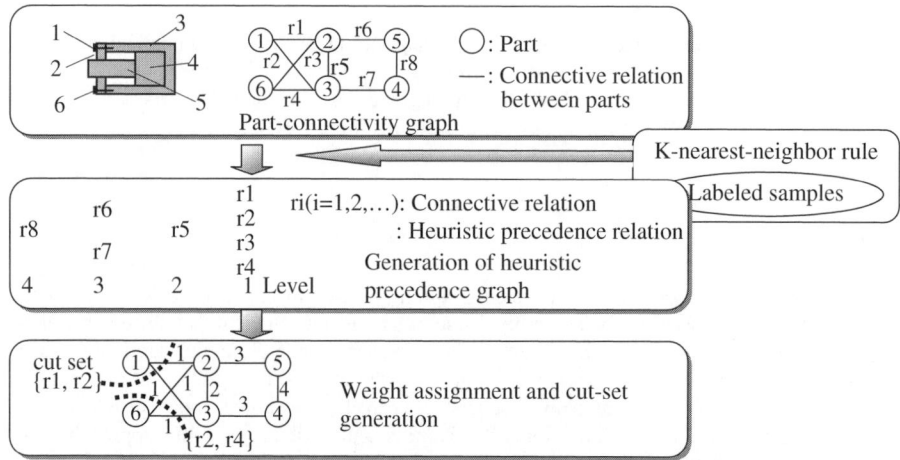

Figure 1: Identification of parts and subassemblies to be removed

IDENTIFICATION OF PARTS AND SUBASSEMBLIES TO BE REMOVED

As shown in Fig.1, first, our method generates a heuristic precedence graph for a given product by using the k-nearest-neighbor rule. Each node of the heuristic precedence graph expresses a connective relation between two parts, and each arc expresses a heuristic precedence relation between two connective relations. This heuristic precedence relation means that the connective relation represented by its terminal node very probably emerges earlier than that represented by its starting node when the product is assembled (conversely, the connective relation represented by the starting node is very probably released earlier than that represented by the terminal node when the product is disassembled). Our previous work used the heuristics to generate such a heuristic precedence graph.

Next, by using the heuristic precedence graph, we assign weights to the connective relations in a part-connectivity graph, each of which nodes expresses a part and each of which arcs expresses a connective relation between parts. The weights are assigned according to the levels in the heuristic precedence graph. For example, connective relation $r5$ shown in Fig.1 is in level 2 of the heuristic precedence graph, and therefore 2 is assigned to it in the part-connectivity graph. The larger weight the connective relation has, the earlier it very probably emerges in the assembly stage (conversely, the smaller weight the connective relation has, the earlier it is very probably released in the disassembly stage). Then, some cut sets composed of the arcs with small weights in the part-connectivity graph are generated, based on a genetic algorithm. A part or subassembly cut off by each of such cut sets has strong possibility of being removed from a product. It is tested by using CAD data whether such a part or subassembly cut off by each of the generated cut sets can be removed without any interference. This approach can avoid the tests for parts and subassemblies that have little possibilities of being removed. This brings about efficient planning of assembly sequences since this means the reduction of the search space for assembly sequences.

GENERATION OF PRECEDENCE GRAPH BY K-NEAREST-NEIGHBOR RULE

Generation of Initial Labeled Samples

To generate the heuristic precedence graph by using the k-nearest-neighbor rule, we collect in advance initial labeled samples obtained from products whose assembly sequences are known. In this study, each sample is an ordered set of three parts, $<p_i, p_j, p_k>$, among which there is a relationship that p_i is connected with p_j and p_k. The class label of each sample is either of the followings:
- Class P expressing the case that p_i and p_j are joined together before p_k is joined to them.
- Class N expressing the other cases.

We generate all the combinations of parts for the products, and then give the class label to every combination, $<p_i, p_j, p_k>$, according to their known assembly sequences.

Assigning a Class Label to an Unlabeled Sample

By using the labeled samples, the k-nearest-neighbor rule assigns a class label to every unlabeled sample, $x=<xp_i, xp_j, xp_k>$, obtained from a given product whose assembly sequence is unknown and needs to be planned. The algorithm of the k-nearest-neighbor rule is as follows.

Step 1: Fix k, which is the number of the closest neighbors of x. Let the labeled samples be $y_1, y_2, ..., y_h, ..., y_n$, where n is the number of the labeled samples.
Step 2: Set $h=1$ and $E=\phi$ (i.e., null set).
Step 3: Calculate the distance from x to a labeled sample y_h, $d(x, y_h)$.
Step 4: If $h \leq k$, add y_h in the set E and go to Step 6. Otherwise go to Step 5.
Step 5: If y_h is closer to x than any member in E, delete the farthest in the set E and include y_h in E.
Step 6: If $h=n$, go to Step 7. Otherwise set $h=h+1$ and go to Step 3.
Step 7: Determine the majority class represented in the set E and classify x in the majority class.

If Class P is assigned to the sample $x=<xp_i, xp_j, xp_k>$, a heuristic precedence relation is generated, which represents that a connective relation between xp_i and xp_j precedes a connective relation between xp_i and xp_k. By applying this method to all the combinations of parts included in the given product whose assembly sequence is unknown, we can generate a heuristic precedence graph for the product.

Calculation of Distance

In Step 3, the distance from x to a labeled sample y_h, $d(x, y_h)$, is calculated by:

$$d(x, y_h) = d_p(xp_i, yp_i) + d_p(xp_j, yp_j) + d_p(xp_k, yp_k) \tag{1}$$

where yp_i, yp_j, and yp_k are the parts included in y_h (i.e., $y_h=<yp_i, yp_j, yp_k>$), and $d_p(*,*)$ is the degree of similarity between two parts. To calculate $d_p(*,*)$, we use is-a hierarchy of part types, which is a relationship among super- and sub-classes of parts.

Addition of Labeled Samples

Additional labeled samples are made from the information on the assembly sequences planned newly by the assembly planning method (Murayama & Oba, 1993) incorporated with the method described in this paper. As the assembly sequence planning and the addition of the labeled samples are executed more times, the assembly sequences can be generated more efficiently.

EXPERIMENTS

Some experiments were carried out to show: the efficiency of the approach; and the superiority of the k-nearest-neighbor rule over the heuristics. First, we made the data on eight types of products. Next we selected one product out of them and we made the initial labeled samples from the data on the selected product. Then, we repeated: the selection of one product out of the remainder; the assembly sequence planning for it by the proposed method; and the addition of the labeled samples, until all the products were selected. We carried out the process mentioned above ten times with changing the order of the selection of products. Figure 2(a) shows how the system inefficiency is improved through the repeated process. In this figure, system inefficiency SI is defined by Nf/Np, where Nf is the average number of the times when a part or subassembly identified by the proposed method can not be removed, and Np is the average number of the parts included in the given product. As shown in this figure, the method can generate the assembly sequences more efficiently as the process is repeated. Figure 2(b) shows the comparison of the computational time by three methods: the proposed method; the method using the heuristics; and the method searching whole space. This figure shows that the proposed method and the method using the heuristics considerably reduce the computational time since these two methods reduce the search space for assembly sequences. The computational time by the proposed method is the shortest for any of the products.

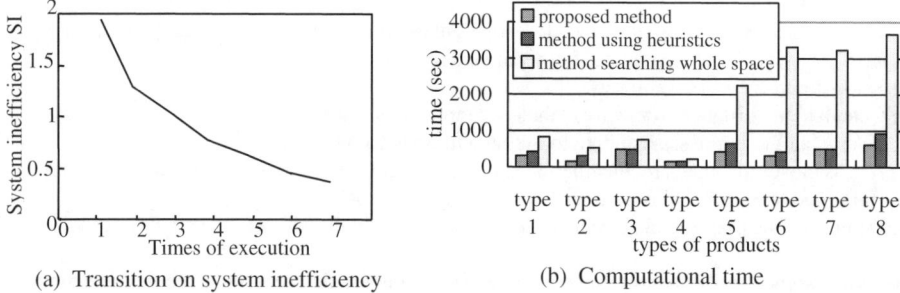

(a) Transition on system inefficiency   (b) Computational time

Figure 2: Result of experiments

CONCLUSIONS

We proposed an approach to the efficient planning of the assembly sequences, in which k-nearest-neighbor rule is used to reduce the search space for the assembly sequences. We carried out some experiments and showed the effectiveness and efficiency of the approach. We can conclude that the proposed approach will be able to contribute much to the efficient planning of assembly sequences.

References

Bourjault, A. (1984). Contribution a une approche methodologique de l'assemblage automatise: Elaboration automatique des sequences operatories, Ph. D. dissertation, Universite de Franche-Comte.

Lambert A. J. D. (2003). Disassembly sequencing: a survey. *International Journal of Production research* **41:16**, 3721-3759.

Lee S. and Yi C. (1993). Subassembly Stability and Reorientation. *Proc. of IEEE Robotics and Automation*, 521-526.

Murayama T. and Oba F. (1993). An Efficient Method for Generating Assembly Sequences in Product Design stages. *Proc. of IEEE International Conference on Industrial Electronics, Control and Instrumentation (IECON'93)*, 564-569.

PROPOSAL OF UBIQUITOUS DISASSEMBLY SYSTEM FOR REALIZING REUSE AND RECYCLING IN COOPERATIVE DISTRIBUTED FACILITIES

T. Tateno[1] and S. Kondoh[2]

[1]Tokyo Metropolitan University, Graduate School of Mechanical Engineering
1-1 Minami-Ohsawa, Hachioji, Tokyo, 192-0397, Japan
[2]National Institute of Advanced Industrial Science and Technology
1-2-1 Namiki, Tsukuba, Ibaraki, 305-8564, Japan

ABSTRACT

In product recovery (reuse and recycle) processes, transportation costs and the lead-time for reuse can be reduced by starting recovery operations near the user's site. This study proposes an operation and information system that is termed a ubiquitous disassembly system, to realize recovery operations in distributed facilities. First, the requirements for this system are considered, and the system architecture is proposed. Second, a realization model of the system using RFID (Radio Frequency Identification) and an agent-based implementation approach is introduced. Finally, a prototype of the ubiquitous disassembly system is developed, and effects of the prototype developed with RFID and the agent-based implementation are verified through case studies.

KEYWORDS

Product recovery, Disassembly system, Ubiquitous, RFID, Agent system

INTRODUCTION

Environmental issues have become more and more important recently, and the establishment of a closed-loop manufacturing system with an effective product recovery process (reuse and recycle) is urgently required (Thierry, et. al. (1995)). The first operation in the product recovery process is a disassembly operation. In many cases, return products are collected at a dedicated facility from over a wide area with high transportation costs, and then disassembled. This causes unnecessary transportation and a probabilistic lead-time for component reuse, and results in raising the costs and environmental load of product recovery. To solve this problem, disassembly operations should be executed in multiple distributed facilities (i.e., carrying out disassembly where products break down). Figure 1 shows a conceptual image of the logistics reduction. A product returned from users is usually transported to a dedicated factory and disassembled into its components, then each component is transported to a second process factory for material recycling, component reuse or landfill. On the other hand, if a product is disassembled and its condition is checked at the user's site or the nearest factory, and each component is then transported directly to the second process factory, the transportation cost and lead-time will be reduced.

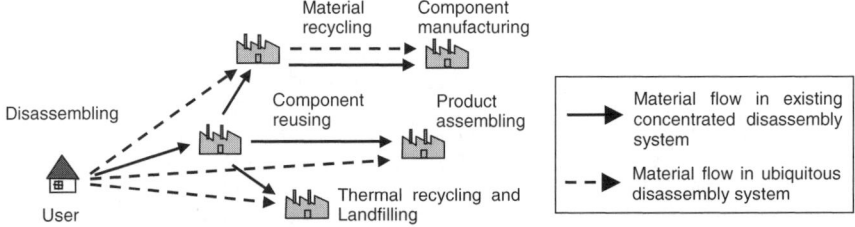

Figure 1 Differences between material flows of the concentrated and ubiquitous disassembly systems

INFORMATION SYSTEM ARCHITECTURE FOR THE UBIQUITOUS DISASSEMBLY SYSTEM

Logistics planning to minimize transportation costs and lead-time seems to be solvable with an conventional planning method, but it is not so simple. The product recovery process contains many uncertainties, such as what, when and where products will be returned.

- What will be returned?

There are sometimes unknown components in a returned product because users have customized it. A product identification method is required and, if possible, information about the use conditions of the product should be recorded.

- When will products be returned?

We cannot estimate accurately the amount of returned products. However, the reuse plan should be decided upon before the product is returned. Sometimes the reuse plan will change after a product is returned. Rapid matching of demand and supply is needed.

- Where will products be returned?

We cannot predict where a returned product will appear because the users are distributed worldwide. Even if there is only a small-scale factory near the returned product, the recovery process should be started there.

To cope with the uncertainties of the product recovery process, three functional requirements are arranged for the ubiquitous disassembly system. Each of the following requirements corresponds to the relevant uncertainty condition written above.

- Sharing information on target products throughout all life cycle stages

All products should have a unique ID number, and their life-cycle information, which includes historical records of their use conditions and assembly structure, should be recorded and managed for each component individually throughout its life. In this paper, RFID will be introduced as a realization method.

- Rapid matching of demand and supply for recovered components and materials

The demand and supply for reusable components are adjusted. In this work, this function is realized as a blackboard system among product agents.

- Operation with inexpensive and flexible equipment

The disassembly operations are assigned to appropriate workers and/or robots for the situation. In this work, this function is realized as a blackboard system among operation agents.

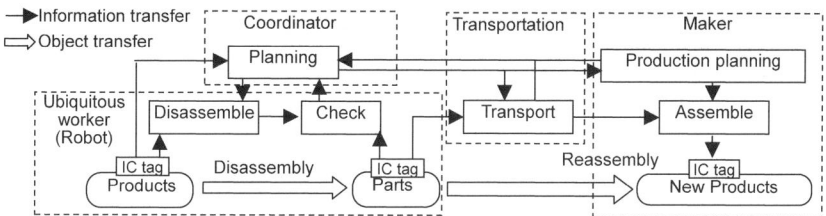

Figure 2 Conceptual architecture of the ubiquitous disassembly system

Figure 2 illustrates the conceptual architecture of the ubiquitous disassembly system. Returned products are transported to the nearest ubiquitous worker, and a worker reads the ID number of the product and sends the number to the coordinator. The tag ID number is coupled with the corresponding component information in the database. Makers send requests for amounts of components corresponding to their production plan. The coordinator decides which components should be reused, recycled to materials or disposed, taking into consideration the real time demands from makers and the historical records of all components. The worker executes the disassembly operations and condition checks according to instructions from the coordinator. The transporter receives request messages from the coordinator and transports products to makers.

However, these recovery processes are not simple because the object and information flows are governed by the factors of malfunction, reuse demand, available disassembly facilities and other factors that change dynamically. This process flow is too complex and too variable to be managed by the conventional centralized system. The proposed architecture provides an intelligible and flexible system enough for the process flow.

REALIZATION APPROACH

Realization Approach with RFID and Mobile Agent System

Three functional requirements for the ubiquitous disassembly system means that decisions should be made dynamically and individually for each component. If these decisions could be made uniformly, the software could be realized easily. However, to realize a system corresponding to the dynamic situation, the software tends to be large and complex, and it must sometimes be modified to adapt to unexpected changes. Therefore, we propose the adoption of new technologies, namely, RFID(Radio Frequency Identification) and mobile agent.

Prototype System

A prototype system is implemented with the mobile agent platform Aglets (Lange and Oshima (1998)) to test the behavior of the system. This system is an approach to realization of two parts of the system proposed in Figure 2, namely, the coordinator and the worker. The coordinator coordinates demand and supply by using agent technology. The worker performs disassembly operations and corresponding checking operations. The operation system is constructed on the basis of assumptions that the facility is a small company specialized in disassembly, that human workers do not have expertise knowledge about products, and that intelligent but inexpensive robots can be used for the disassembly operation. In the case of disassembly operation by a human worker, the operation system includes a worker support system that provides intellectual support for the disassembly operation. In the case of robot disassembly operation, on the other hand, human workers perform simple tasks such as loading a product onto a pallet, and robots execute the disassembly operations and checking operations.

Figure 3 shows the system configuration. This figure is not a process flow. The process flow is not described explicitly but determined by the relations among existing agents. If there is a different agent, a different process flow may be executed. The product agent and operation agent are defined as mobile agents, while the others are defined as stationary agents. These agents are described in the following scenario.

(1) One of the RFID tags on the product is detected by a RFID reader, and a product agent corresponding to the ID number is created.
(2) The product agent moves to a product database and retrieves information about the use conditions and assembly structures of all components in the product.
(3) The product agent moves to the demand and supply blackboard and retrieves demand information for all components in the product.
(4) The product agent moves to the facility database, searches the facilities and generates a list of all operation agents available to work.
(5) The product agent moves to the operation blackboard and writes an operation plan for the extraction of components.
(6) Operation agents move to the operation blackboard and assign each task to an appropriate agent.
(7) Operation agents move to the operation site and execute the assigned task.

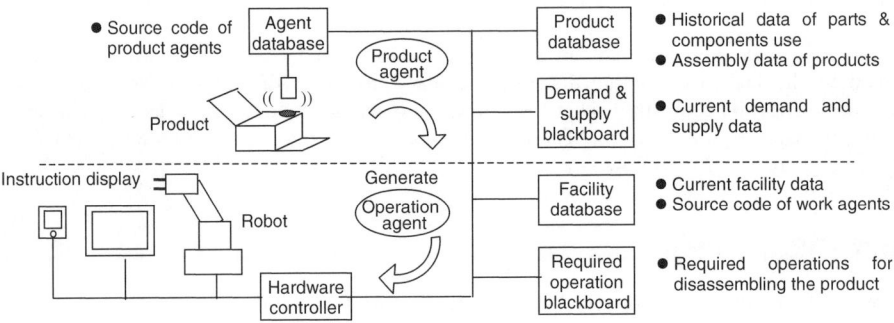

Figure 3 Prototype system using RFID and agent-based implementation

CASE STUDY

Disassembly of a Printer

A laser printer is tested to examine the behaviors of the prototype system. The work object consists of three components, which are a base, a toner cartridge and a photoconductor unit, as shown in Figure 4. Every component has an IC tag attached to its surface. The product assembly structure is described as an and/or graph in Figure 5. This graph is used for disassembly planning.

Here, we assume that a toner cartridge and a photoconductor unit have been requested by different makers, and these requests are listed on the demand and supply blackboard. When a worker checks the IC tag on the base by applying a RFID antenna, a product agent corresponding to the printer is loaded. At this moment, the product agent has its own program but it has no data on the components. The product agent retrieves these data from the product database. Figure 6 shows the product agent window that presents the retrieved data on the assembly structure and the demands for components.

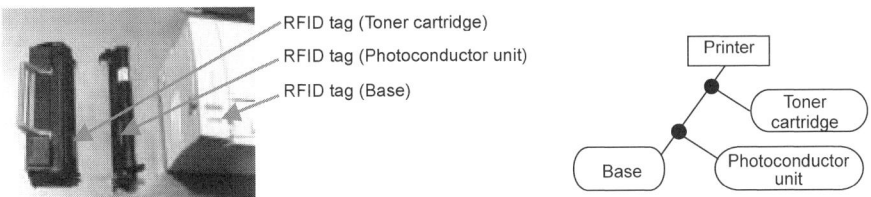

Figure 4 Components used for case study   Figure 5 And/or graph of the product

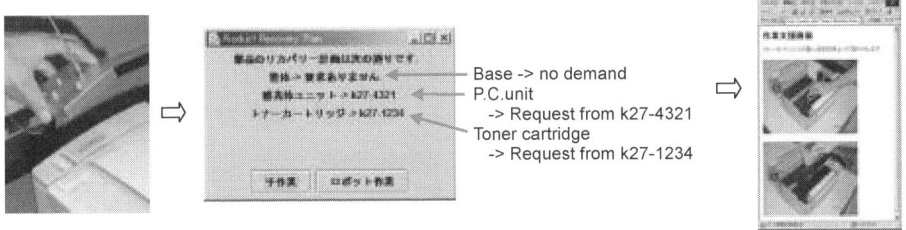

(a) RFID detection   (b) product agent window showing reuse plan   (c) work instruction
Figure 6 Case study (Extraction of photoconductor unit and toner cartridge by a human worker)

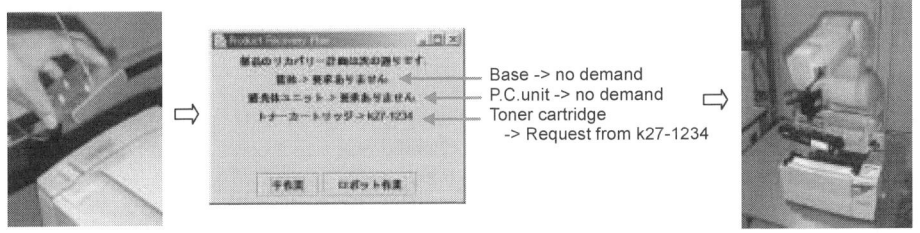

(a) RFID detection   (b) product agent window showing reuse plan   (c) robot operation
Figure 7 Case study (Extraction of toner cartridge by a robot)

Then the worker selects the human worker button in the window. Normally, the product agent retrieves the available operation agents from the facility database. However, in this case, there is only one operation agent, that presents instructions to a human worker. Then, the operation agent opens a web browser and presents a web page for an URL address. The web pages are presented in order with respect to the disassembly. These pages are not hyperlinked. The operation agent arranges the URL addresses appropriately to correspond to the operation sequence.

As another case, we assume only a toner cartridge is demanded by a maker, and a robot executes the disassembly operations along with a human worker. In the trial, after instruction for opening a lid of the printer is given to a worker, the robot replaces the toner cartridge. Figure 7 shows the robot performing the replacing operation.

Through these case studies, the agents performed as expected and the realization of the agent-based system was confirmed.

Effects of Agent-based Implementation

As for the case studies described in above section, even a non-agent system seems to be able to achieve it. However, the important effects of agent-based implementation will become apparent in system reconfiguration. For example, in the case that we change a program in order to refer to an additional database, in which not only the product data but also the processing program must be modified, the agent-based system allows in-process modification in intelligible programming. Moreover, the rum time processing load can be optionally distributed by modification of the agent work place.

In this section, two procedures, namely, the modification of an agent-based system and that of a conventional system, are compared as a case study. We assume that a new printer is released and a new product agent is defined. This printer has an ink cartridge and the product agent must refer to an ink-cartridge database that is different from the laser printer's database. Figure 8 shows each step in the procedure of system modification.

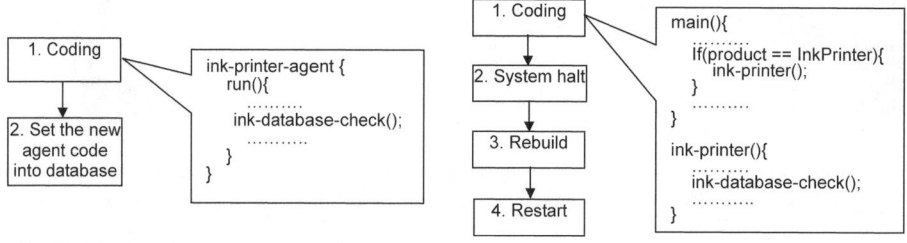

(a) agent system　　　　　　　　　　　　　(b) conventional system
Figure 8 Difference between modification of agent system and that of conventional system

We can see that, in the conventional system, an "if" statement must be added to the main process every time a new process function is defined. On the other hand, in the agent-based system, the modification is described as the definition of a new agent, and other agents are not affected by this modification process. Even halting of the system for related maintenance is not necessary. Moreover, the additional database system helps to distribute the processing load. Therefore, we have confirmed the effects of agent-based implementation through this case study.

CONCLUSIONS

(1) A ubiquitous disassembly system that reduces the logistic costs and lead-time required for product recovery is proposed.
(2) The architecture of the ubiquitous disassembly system is presented, and a model realizing the RFID and agent-based implementation approach is proposed.
(3) A prototype system for disassembly operation using distributed facilities is developed. Through case studies using the prototype system, the realization of the ubiquitous disassembly system is verified.

REFERENCES

Thierry M., Salomon M., Nunen J.V. and Wassenhove L.V. (1995) Strategic Issues in Product Recovery Management, *California Management Review*, 37:2, 114-135.

Lange B.D and Oshima M. (1998) *Programming and Deploying Java Mobile Agents with Aglets*, Addison Wesley.

DEVELOPMENT OF A MICRO TACTILE SENSOR UTILIZING PIEZORESISTORS AND CHARACTERIZATION OF ITS PERFORMANCE

J. Izutani, Y. Maeda and S. Aoyagi

Systems Management Engineering, Kansai University
3-3-35, Yamate-cho, Suita, Osaka 564-8680, Japan

ABSTRACT

Many types of tactile sensor have been proposed and developed. They are becoming miniaturized and more precise at the present state. Micro tactile sensors of high performance equal to a human being are now desired for robot application, in which the skillful and dexterous motion like a human being is necessary. In this research, piezoresistors are made on a diaphragm to detect the distortion of it, which is caused by a force input to a pillar on the diaphragm. Three components of the force in x, y and z direction can be simultaneously detected in this sensor. The concept is proposed and its measuring principle is confirmed by using FEM simulation. Also a practical sensor chip is fabricated by micromachining process and characterization of its performance is reported.

KEYWORDS

Tactile sensor, Piezoresistor, Microstructure, Micromachining, Gauge factor

INTRODUCTION

An advanced tactile sensor is strongly desired now for the purpose of realizing complicated assembly tasks of a robot, recognizing objects in the space where vision sensor cannot be used (in the darkness, etc.), and so on [1, 2]. Besides industry, development of a robot hand will become more important to realize human-like robots, such as a humanoid. In order to give a tactile sense like human to a robot's fingertip, development of the tactile sensor with high performance would be required in the near future. Many tactile sensors have been proposed until now; however, limited by fabrication process a tactile sensor compatible to human's one has not been achieved yet. On the other hand, micromachining process based on semiconductor manufacturing process is hot research area and available now. Using this technology, many tactile sensors are proposed and developed now [3-7]. By this technology many arrayed sensing elements with uniform performance characteristics can be fabricated on a silicon wafer with fine resolution of several microns. Authors are also now developing a tactile sensor comprising many arrayed sensing elements by this technology. The schematic view of concept of arrayed tactile sensor for robotic finger is shown in **Fig. 1**.

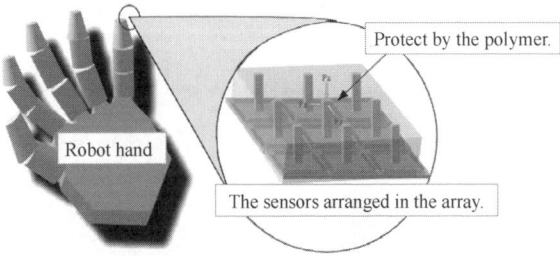

Figure 1: Schematic view of concept of arrayed tactile sensor for robotic finger (future work)

In this paper, a microstructure having a pillar and a diaphragm is fabricated. The schematic structure of one sensing element is shown in **Fig. 2 [8]**. In near future, by arranging many of this structure, the development of a micro tactile sensor which can be used to realize a robot's fingertip is aimed at. Piezoresistors are fabricated on a silicon diaphragm to detect the distortion which is caused by a force input to a pillar on the diaphragm. Three components of force in x, y, z direction can be simultaneously detected in this sensing element. The principle of measurement is shown in **Fig. 3**. Piezoresistors are formed by boron ion-implantation on n-type Si substrate. In order to determine a piezoresistors arrangement, FEM analysis is carried out. This device has four features as follows: 1) It has three-dimensional structure at the front and back side of SOI substrate. 2) It is able to be miniaturized by using a semiconductor process. 3) This sensor utilizes sensitive semiconducting piezoresistors. 4) This sensor is able to detect three components of the force in x, y and z direction by arrangement of four piezoresistors.

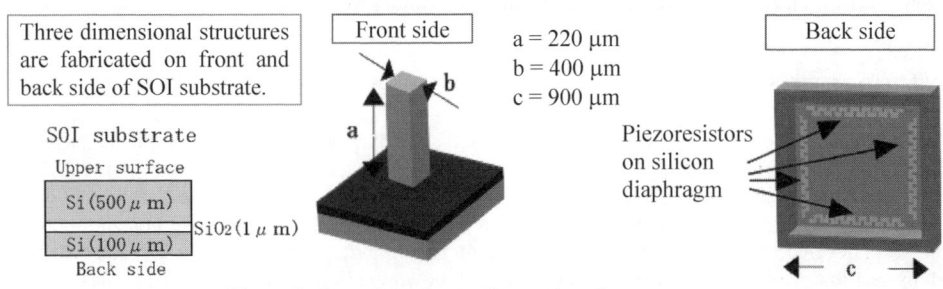

Figure 2: Structure of a tactile sensing element

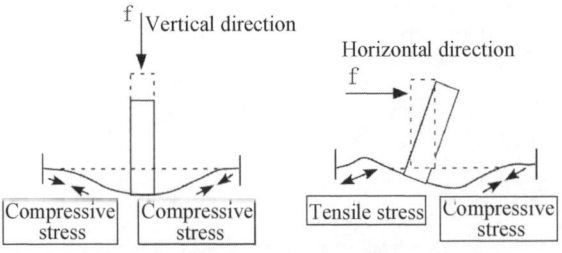

Figure 3: Principle of measurement

FEM (FINITE ELEMENT METHOD) ANALYSIS

In order to determine the position of piezoresistor, FEM analysis is carried out. When the force of 10 gf is applied to the pillar tip of the sensing element, the results of distortion of a diaphragm is shown in **Fig. 4**. **Figure 4 (a)** shows the distribution of strain in the horizontal direction, when the force of 10gf is applied in the vertical direction. **Figure 4 (b)** shows the distribution of strain in the horizontal direction, when the force of 10 gf is applied in the horizontal direction. It is proved that the strain is maximal at the edge of the diaphragm. Therefore, the four piezoresistors are designed to be located as close as possible to the edge of the diaphragm.

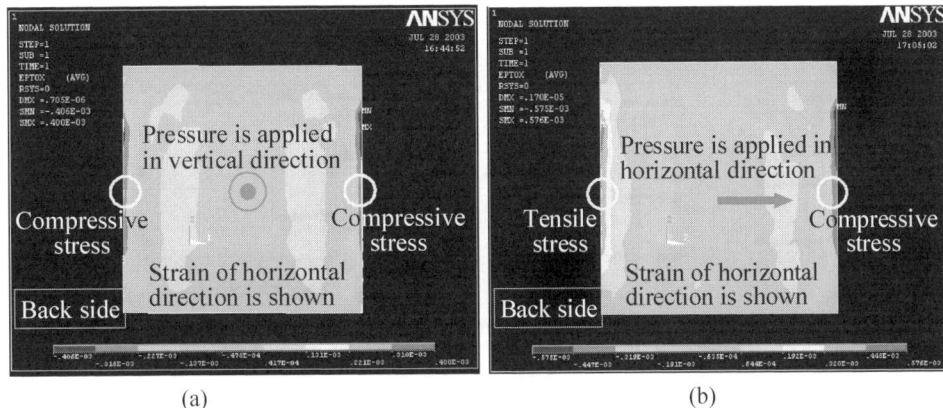

(a)   (b)

Figure 4: FEM result of distortion of a diaphragm.

FABRICATION PROCESS

The micro-machining fabrication process of a tactile sensing element is shown in **Fig. 5**. The microstructure detecting a force is practically fabricated as follows: a SOI wafer is prepared, which consists of a silicon layer (called as active layer) of 100 μm, a silicon dioxide layer of 1μm (called as box layer), and a silicon layer of 500 μm (called as support layer) (see Fig. 5①). A diaphragm is fabricated by anisotropic wet etching of the active layer using KOH solution (see Fig. 5⑤). Piezoresistors are produced by implanting p-type boron ions into the n-type silicon of the diaphragm using an ion implantation apparatus (see Fig. 5⑧). A pillar is fabricated by dry etching the support layer using a deep ICP-RIE apparatus (see Fig. 5⑬). ICP-RIE was performed by Bosch process and their condition are shown in **Table 1** [9]. Aluminum is evaporated and patterned for electrodes, which connect the piezoresistors to the bonding pads. The wafer is diced to square chips, and each chip is set on a print board. The bonding pads of the chip are connected to the print board pads by aluminum wires using a wire bonding apparatus.

THE DESIGN OF EVALUATION CIRCUIT

The direction of applied forces and the position of piezoresistors are shown in **Fig. 6**. When force is applied to the pillar in the x direction, the distortion appears as shown in the upper right of **Fig. 6**. When force is applied to the pillar in the z direction, the distortion will appear as shown in the lower right of **Fig. 6**. This distortion can be detected by four piezoresistors arranged as shown in **Fig. 6** [8].

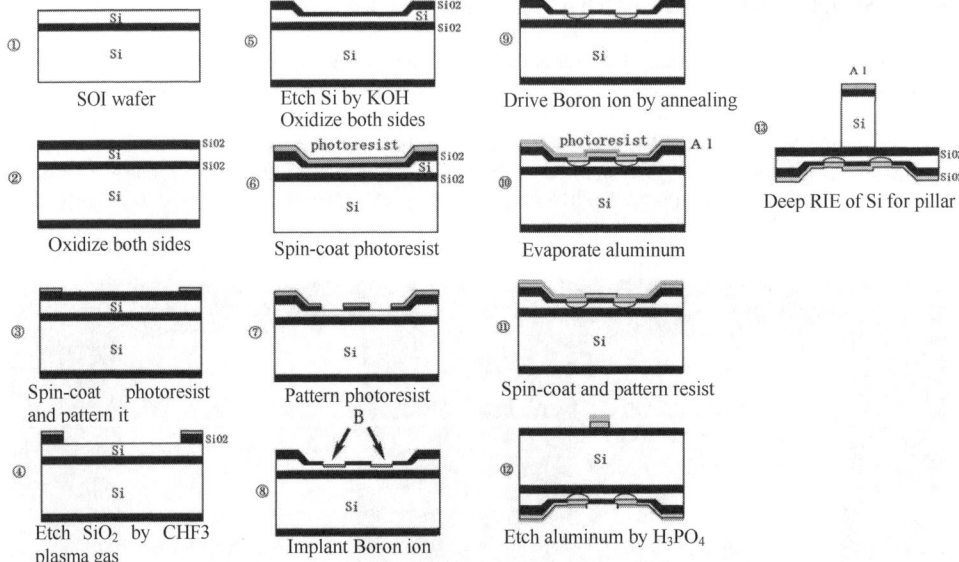

Figure 5: The micromachining fabrication process of a tactile sensing element

TABLE 1
The conditions of the used Bosch process

	Etching	Deposition
Time[s]	4	3
SF_6[sccm]	100	0.5
C_4F_8[sccm]	0.5	100
Ar[sccm]	0.5	0.5
BIAS[w]	25	15
ICP[w]	500	600
Pressure[Pa]	5	5

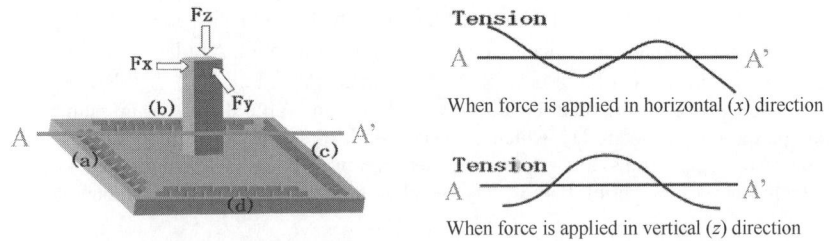

Figure 6: Direction of applied forces and the position of piezoresistors

The change of each resistance is able to be detected as voltage V(a), V(b), V(c), V(d). The output voltage (Vx) corresponding to force (Fx) is calculated using Eq. (1). Similarly, the voltage (Vy) corresponding to force (Fy) is calculated using Eq. (2), and the voltage (Vz) corresponding to force (Fz) is calculated using Eq. (3). These operations were carried out with accumulator and subtractor by using operational amplifiers as shown in **Fig. 7**.

$$V_X = V(a)-V(c) \quad (1)$$
$$V_Y = V(b)-V(d) \quad (2)$$
$$V_Z = V(a)+V(b)+V(c)+V(d) \quad (3)$$

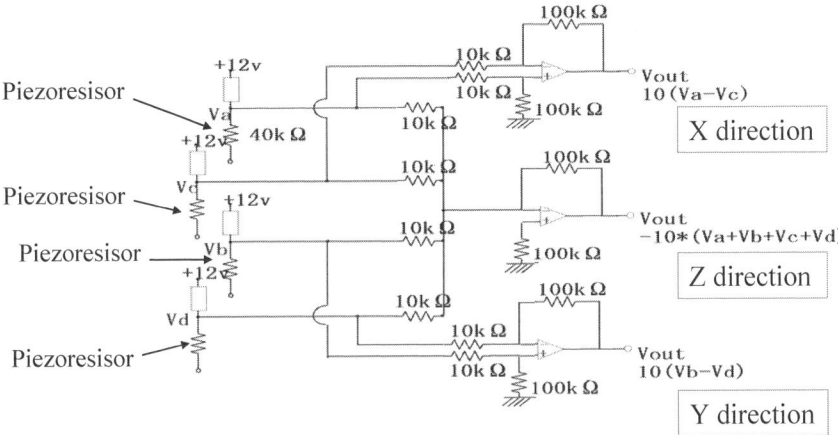

Figure 7: Evaluation circuit using operational amplifiers

CHARACTERISTICS OF SENSOR

SEM image of fabricated tactile sensing element in both sides is shown in **Fig. 8**. Pillar exists on the upper surface. Diaphragm, piezoresistors and aluminum wiring exist on the back side. The produced piezoresistor is measured and it is 0.5 kΩ. The performance of force detection in z direction is experimentally characterized. The known weight is put on the pillar vertically by using a jig, and the resistance change is detected. The relationship between the input weight and the resistance change has good linearity within the range from 0 to 200 gf as shown in **Fig. 9**. By using FEM method, the strain at the resistor is simulated when the weight is input. From the relationship between this strain and the resistance change, the gauge factor of the piezoresistor is proved to be about 133, which is almost equal to the common experimental value of other references.

From these experimental results, it is proved that this microstructure has good potential to detect a force. Characterization of performance of force detecting in x and y direction, and fabrication of an arrayed type micro tactile sensor by using many microstructures are ongoing.

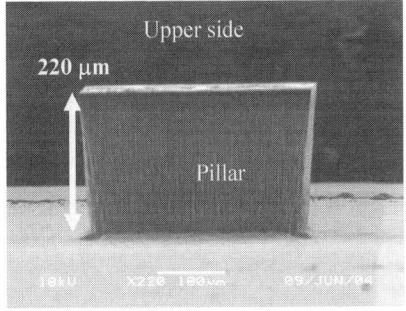
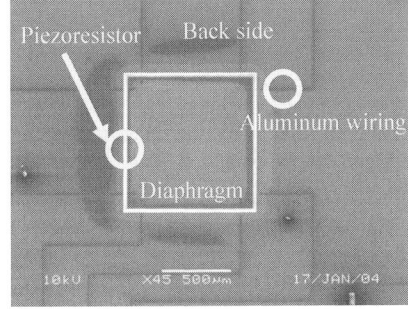

Figure 8: SEM image of fabricated tactile sensing element (upper and back side)

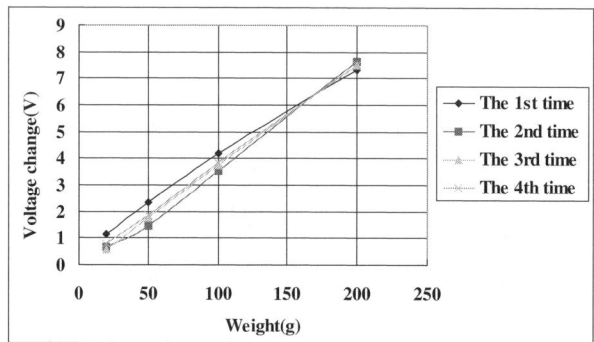

Figure 9: The voltage change when pressurized using weight.

CONCLUSINS

A micromachined force sensing element having a pillar and a diaphragm is proposed and fabricated. It can detect three components of the force in x, y and z direction by using four piezoresistors located four edges of the diaphragm. The performance of force detection in z direction is experimentally characterized. The relationship between the input weight and the resistance change has good linearity within the range from 0 to 200 gf.

ACKNOWLEDGEMENT

This work was mainly supported by MEXT (Ministry of Education, Culture, Sports, Science and Technology).KAKENHI (17656090). This work was also partially supported by JSPS (Japan Society for the Promotion of Science).KAKENHI (16310103), "High-Tech Research Center" Project for Private Universities: Matching Fund Subsidy from MEXT, 2000-2004 and 2005-2009, the Kansai University Special Research Fund, 2004 and 2005.

REFERENCES

[1] Lee M. H. and Nicholls H. R. (1999). Tactile Sensing for Mechatronics - A State of the Art Survey -. *Mechatronics* **9**, 1-31.
[2] Shinoda H. (2000). Tactile Sensing for Dexterous Hand. *J. The Robotics Society of Japan* **18:6**, 772-775.
[3] Kovacs G. T. A. (1998). *Micromachined Transducers Sourcebook*, McGraw-Hill, USA, 268-275.
[4] Kobayashi M. and Sagisawa S. (1991). Three Direction Sensing Silicon Tactile Sensors. *Trans. Institute Electronics, Information and Communication Engineers* **J74-C-II:5**, 427-433.
[5] Esashi M., Shoji S. Yamamoto A. and Nakamura K. (1990). Fabrication of Semiconductor Tactile Imager. *Trans. Institute Electronics, Information and Communication Engineers* **J73-C-II:1**, 31-37.
[6] Kane B. J., Cutkosky M. R. and Kovacs G. A. (2000). A Tactile Stress Sensor Array for Use in High-Resolution Robotic Tactile Imaging. *J. Microelectromechanical Systems*, **9:4**, 425-434.
[7] Suzuki K., Najafi K. and Wise K. D. (1990). A 1024-Element High-Performance Silicon Tactile Imager. *IEEE Trans. Electron Devices* **37:8**, 1852-1860.
[8] Ohka M., Kobayashi M., Shinokura T. and Sagisawa S. (1991). Tactile Expert System Using a Parallel Fingered Hand Fitted with Three-Axis Tactile Sensors. *JSME Int. J., Series C*, **.37-1:138**, 427-433.
[9] Chen K. (2002). Effect of Process Parameters on the Surface Morphology and Mechanical Performance of Silocon Structures after Deep Reactive Ion Etching. *J. Microelectromechanical Systems*, **11:3**, 264-275.

DEVELOPMENT OF SENSORS
BASED ON THE FIXED STEWART PLATFORM

K. Irie, J. Kurata and H. Uchiyama

Department of Mechanical Systems Engineering, Kansai University
3-3-35, Yamate-cho, Suita, Osaka 564-8680, Japan

ABSTRACT

We propose new type of spatial vector sensor based on the Fixed Stewart Platform. Since six measuring units are arranged in periodic and represented on the links of Stewart platform, the errors accompanying each measurement axis are not accumulated. Our aim is focused to measure six components of spatial vector, and we propose the structure composed without movable links. We described the constructing method and the calculating solution from link parameters, which resulted ease of the calculation. In order to confirm the validity of our proposal, the acceleration and angular-acceleration sensor was manufactured. As the results of the triaxial acceleration measurement, the validity of our sensor was confirmed as comparing with the performance of typical commercial product.

KEY WORDS

Sensor, Accelerometer, Angular accelerometer, 6DOF, Method of measurement, Stewart platform

INTRODUCTION

Recently, many machines need much information on motion with six degrees of freedom more and more. In these machines, the measuring instruments that can measure 6DOF motion are included, and many multi-axis measuring sensors have been developed. However, it is generally very difficult to measure 6DOF motion individually at once. A multi-axis measuring sensor measures each component simultaneously although the influence of component to the others is curbed as much as possible. Because of the reduction of this disadvantage, the structure of such kind of sensor seems to be complicated and a measurement axis is restricted to a certain direction. Some of multiple sensors, which can measure 6DOF motion, employ two kinds of sensors, they are three acceleration sensors arranged according to the orthogonal coordinates and three angular-acceleration sensors put in center of rotation. In our proposed sensor, six sensors of the same kind are employed and arranged according to the special structure of sensor body like Stewart platform, for example Stewart (1965). Six measurement sensor units are arranged along the parallel structure represented on the Stewart platform so that the errors in each measurement axis are not accumulated. In our proposed structure, there were no movable links, and this structure resulted the ease of calculation of six components from six measured values. We described the constructing method and the calculating solution on each link parameters. In order to confirm the validity of this method of measurement, the acceleration and angular acceleration sensor system was manufactured.

MEASUREMENT ALGORITHM

The calculating solution was worked out by thinking that the upper plate was moving as six links were expanding and/or contracting, and that the motion of links were measured by single axis accelerometer. The calculating algorithm could be resolved as follows by using points and vectors shown in Figure 1. When a vector is described in one of the two plate, the superscripts written on the left of each vector indicate the coordinate. Superscript 'b' means bottom plate and 'p' upper plate. The matrix '$^b R_p$' is coordinate transformation matrix from the upper coordinate to the bottom coordinate. When a position and posture of upper plate was given, the vector l_i could be shown by the following equation.

$$^b l_i = {}^b a - {}^b b_i + {}^b R_p {}^p p_i \tag{1}$$

The vectors '$^p p_i$' and '$^b b_i$' were constant vector determined by the structural specimen. By differentiating equation 1 with respect to time, the following equation can be obtained.

$$\frac{d^b l_i}{dt} = \frac{d^b a}{dt} + \left\{ \frac{Rot(k_r, d\phi) - E}{dt} \right\} {}^b R_p {}^p p_i \tag{2}$$

Since all links would not expand and contract, the infinitesimal deformation caused by the motion of upper plate would return to zero in a very short time. Therefore, the velocity of links can be expanded by introducing next equation.

$$\frac{d|^b l_i|}{dt} = \frac{d^b l_i}{dt} \cdot \frac{^b l_i}{|^b l_i|} \tag{3}$$

Since the components of each terms $\frac{d|^b l_i|}{dt}$, $\frac{Rot(k_r, d\phi) - E}{dt}$, $\frac{d^b a}{dt}$ include the vector v and angular velocity vector ω, next equation can be obtained from above equations.

$$\begin{pmatrix} v \\ \omega \end{pmatrix} = C^{-1} V_L \tag{4}$$

Here, the vector V_L is composed of link's expanding velocities and the matrix C is coefficient matrix about components of v and ω. In our proposal, all links would not expand and contract, therefore the coefficient matrix C should be constant. In same manner, the following equation can be obtained.

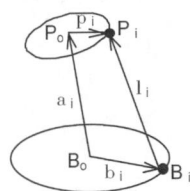

Points and vectors
B₀ : Origin of coordinates in bottom plate
P₀ : Origin of coordinates in upper plate
B ᵢ : Node of i-th link to bottom plate
P ᵢ : Node of i-th link to upper plate
b ᵢ : Vector from B₀ to B ᵢ
p ᵢ : Vector from P₀ to P ᵢ
a ᵢ : Vector from B₀ to P₀
l ᵢ : Vector from B ᵢ to P ᵢ

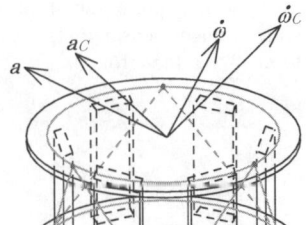

Figure 1: Model of Stewart Platform   Figure 2: Acceleration vector

$$\begin{pmatrix} \mathbf{a} \\ \dot{\boldsymbol{\omega}} \end{pmatrix} = \frac{d}{dt}\begin{pmatrix} \mathbf{v} \\ \boldsymbol{\omega} \end{pmatrix} = \mathbf{C}^{-1}\frac{d\mathbf{V}_L}{dt} = \mathbf{C}^{-1}\mathbf{A}_L \qquad (5)$$

As mentioned above, six components of acceleration and angular acceleration can be calculated from measured accelerations along the direction of each link by using constant coefficient matrix C in advance.

CALCULATING OPTIMUM STRUCTURAL PARAMETER

The coefficient matrix C should be nonsingular matrix, and the calculation results tend to come under the influence of misalignment and measurement errors of each sensors when the matrix C is near singular point. Since we use that platform as not an actuator but a base structure of measuring instrument, we found the optimal structure based on Stewart Platform to reduce the influence of misalignment. Two plates are in a direction parallel each other. The centerline, which connects centers of plate, is vertical to both plates. And nodes are placed evenly spaced apart (120degrees interval). In this time, we calculated normalized radius of upper plate 'R' and normalized distance between two plates 'H' according the centerline, when bottom plate radius is fixed to 1. By adding virtual error to the accelerations of (a, ω) up to 10%, the set of calculated accelerations (a_C, ω_C) from equation 5 and the average of evaluation value S calculated from equation 6 were obtained. The optimal radius of upper plate R and the optimum distance between both plates H were found out by making average value of S minimum.

$$S = \frac{|a_C - a|}{|a|} + \frac{|\dot{\omega}_C - \dot{\omega}|}{|\dot{\omega}|} \qquad (6)$$

Calculated results were shown in Figure 3. As R and H increased or decreased from the optimum value, the average of evaluation value S increased. Because the coefficient matrix became close to the singular point, the calculation results tended to come under the influence of added error. After searching optimum values, the optimum radius of upper plate R should be 0.83 and the optimum distance between two plates H should be 0.93. On the optimum structure with these parameters, angle made by each link and each plate was 43degrees. However, when the detectors are in a manufacturing process, the more simple of manufacture and the reliability of processing would be our prior attention. Therefore, the angle made by each link and each plate should be 45degrees, we decided. Under this condition, the semi-optimum parameters of the structure were R=0.81 and H=0.92.

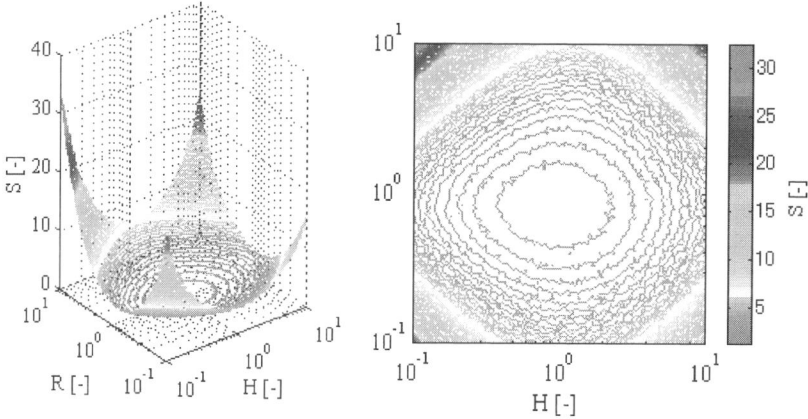

Figure 3: Simulation results on the evaluate function S to fix the optimum structure

MEASUREMENT INSTRUMENT

The picture of the manufactured sensor system was shown in Figure 4. This structure of sensor body had two plates with same diameter and six pillars with same size. The each pillar has single axis accelerometer (Analog Devices Inc., ADXL105) in the central part of the pillar. And, the angle made by measurement axis of each sensor and plates made 45 degrees each other.

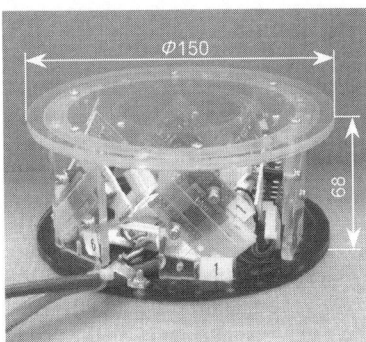

Figure 4: Manufactured device

EXPERIMENTAL RESULT

In order to confirm the validity, the acceleration and angular acceleration were measured while reciprocating the manufactured detector. The X-Y plane was set on the upper plate, and Z-axis was vertical to X-Y plane. From the various experimental results, the detected values out of the main direction of movement were about 1% on the peak value in main direction. Although it could not be measured strictly by this data, the cross talk could be -35dB at least. From the experimental results of measurement in translational and rotational reciprocation simultaneously, the error of measured acceleration value was 15% on the calculated value, and the error of angular acceleration value was 6%. The error of angular acceleration was similar to the value in only rotational motion. Due to the scatter in measured performance of each accelerometer, the error of acceleration was increased, we considered.

CONCLUSION

In this detector, we use only one kind of sensor (ADXL105 in this report) as single axis detector. The proposed sensing device could measure six components of special motion at once. The maximum cross axis sensitivity of sensing device is 5%, and it is almost equal to the specifications of each sensor tips. Totally, the cross talk value is about -35dB. The acceleration and angular acceleration could be measured by this method in translational and rotational motion respectively. In experimental confirmation, the amplitude of acceleration was about $0.01 m/s^2$ (about 5% on the peak value). Assuming that this value would be electrical noise of acceleration sensor tips, the acceleration and angular acceleration could be measured in translational and rotational motion simultaneously without calculating errors. From the results of experimental confirmation, it has been clear that new type of sensor device, which was designed based on the fixed Stewart Platform by us, would be essential way to construct the various kind of six component sensing device.

Reference
   D. Stewart (1965), A Platform with Six Degrees of Freedom, *UK Institution of Mechanical Engineers Proceedings 1965-66*, **180:Pt 1:15**

MICROFABRICATION OF
A PARYLENE SUSPENDED STRUCTURE
AND INVESTIGATION OF ITS RESONANT FREQUENCY

D. Yoshikawa[1], S. Aoyagi[1] and Y. C. Tai[2]

[1]Systems Mangement Engineering, Kansai University
3-3-35, Yamate-cho, Suita, Osaka 564-8680, Japan
[2]California Institute of Technology
136-93, Pasadena, CA 9112, USA

ABSTRACT

Polymer material of Parylene has intrinsic tensile stress on account of mismatch of thermal coefficient of expansion (TCE) between the substrate and the deposited film. Therefore, the stiffness k of the Parylene suspended structure under tensile stress is much higher than that under no stress, which also leads to its higher resonant frequency f_r. These mechanical characteristics are investigated in this study. First, FEM simulation is employed, and it is proved that $k \propto 1/l$ holds true under tensile stress, while $k \propto 1/l^3$ holds true under no tensile stress according to the theory of strength of materials, where l is beam length. This means a relatively long beam is necessary under tensile stress for the purpose of lowering f_r, which leads to obtaining high sensitivity in case that the suspended structure is applied to a sensor such as an accelerometer. Considering this, a structure with spiral beams is proposed. Second, Parylene suspended structures are practically fabricated. Their experimental resonant frequencies are obtained by a LDV. They coincide well with simulated ones. As the result, it is proved that the structure with spiral beams is effective for lowering f_r.

KEYWORDS

Parylene, Resonant frequency, Stiffness, Tensile stress, Spiral beam, Accelerometer

INTRODUCTION

Parylene is polymer material expected to be applied in micromachine field and many sensors and actuators using Parylene has been investigated and reported [1]. For example, Parylene accelerometer as shown in **Fig.1** is being developed by authors [2]. Parylene has intrinsic tensile stress on account of mismatch of thermal coefficient of expansion (TCE, α) between substrate and Parylene deposited on it [3]. The stiffness k of this accelerometer structure changes according to the tensile stress of Parylene. In this study, the mechanical characteristics of suspended microstructures are investigated by using FEM (Finite Element Method) analysis. From the result of FEM simulation, it is proved that $k \propto 1/l$

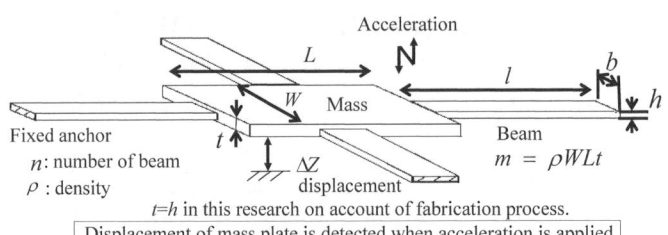

Figure1: Accelerometer comprising a proof mass plate and support beams

holds true under tensile stress, while $k \propto 1/l^3$ holds true under no tensile stress according to the theory of strength of materials. Lowering the stiffness k, which means lowering the resonant frequency f_r, is important in order to increase the sensitivity of the accelerometer, since the sensitivity is $1/(2\pi f_r)^2$. Therefore, spiral shaped long beam structure is proposed in this paper. And free standing Parylene suspended structures are practically fabricated. Vibrations of them are observed by a LDV (Laser Doppler Vibrometer) and their resonant frequencies are obtained experimentally. These results have good agreement with simulated ones. This means large l is necessary for lowering resonant frequency. As the result, it is proved that the structure with spiral shaped beam is effective for lowering the resonant frequency.

FEM ANALYSIS

Mechanical characteristics under tensile stress are numerically simulated by using FEM. FEMLAB produced by Comsol, Inc. is adopted as FEM software. In the case of the structure shown in **Fig.1,** the stiffness k is analytically calculated according to the theory of strength of materials as follows:

$$k = \frac{nEbh^3}{l^3} \propto b\left(\frac{h}{l}\right)^3 \qquad (1)$$

where l is beam length, b is beam width, h is beam thickness, n is the number of beams, and E is Young's modulus. However, these equations are derived under no tensile stress. In order to estimate these mechanical characteristics under severe tensile stress, FEM simulation is carried out. In this simulation, it is assumed $t=h$ since it is difficult to fabricate the structure of which t and h are different, where t is plate thickness.

Dependence of deflection ΔZ on the beam size of l, b, $h(=t)$ are simulated. The results are shown in **Figs.2-4**. In **Fig.2**, ΔZ is increased in proportion to the first power of the beam length under tensile stress. In **Fig.3**, ΔZ is decreased in proportion to the first power of the beam width b. In **Fig.4**, ΔZ has no dependence on the beam thickness h. From the results of **Figs.2-4** totally, the relation holds true as follows:

$$\Delta Z = \frac{F}{k} = \frac{\rho W L t \cdot \alpha}{k} \propto \frac{l}{b} \qquad (2)$$

where ρ is density, W is plate width, L is plate length and α is the input acceleration. Taking account that ΔZ is proportional to t/k as shown in the former part in Eq.(2), and taking account that the condition of $t=h$ holds true, it is concluded that k is proportion to h, since ΔZ is irrespective of h as shown in **Fig.4**. Eventually, the relationship holds true as follows:

$$k \propto b\frac{h}{l} \qquad (3)$$

Eq.(3) under tensile stress is derived from FEM simulation, and it is different from that of Eq.(1) under no intrinsic stress derived from the theory of strength of materials. It means a rather longer beam is necessary for lowering the stiffness k, which also leads to lowering the resonant frequency f_r. When the length of beam is longer, larger space is required. Considering space efficiency, spiral shaped

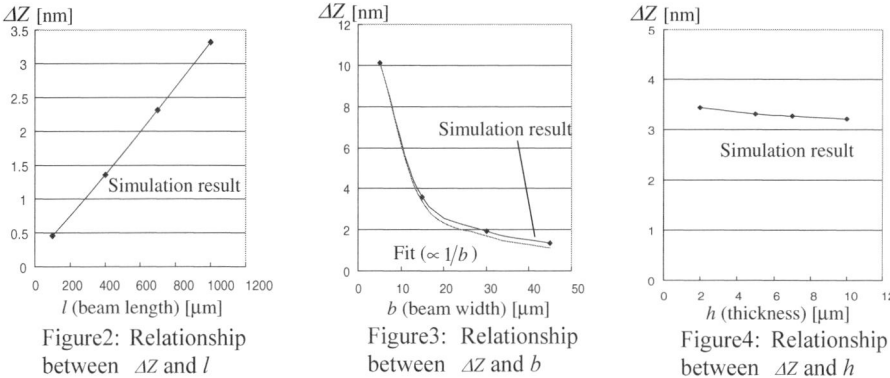

Figure2: Relationship between ΔZ and l

Figure3: Relationship between ΔZ and b

Figure4: Relationship between ΔZ and h

beam is efficient as compared with other beam shapes in order to form a long beam structure in a limited space.

RESONAT FREAQUENCY MEASUREMENT

Next, the resonant frequencies of Parylene suspended structures are investigated experimentally. Free standing Parylene suspended structures are fabricated. The process flow is shown in **Fig. 5**. A SEM images of fabricated structures are shown in **Fig. 6**. And, a rotation tip is also employed in order to check the actual tensile stress as shown in **Fig. 7 [4]**, and tensile stress is proved to be about 30 MPa.

Resonant frequencies of the fabricated structures are measured. The structures are shaken by a piezoelectric actuator and the out-of-plane vibrations of them are observed by a LDV (Laser Doppler Vibrometer). A vacuum chamber is specially developed in order to decrease the influence of air damping. This vacuum pressure is about 0.8 Pa during measurement. The measured structures are the same as shown in **Fig. 6**. Changing the driving frequency of the piezoelectric actuator, the amplitude of the center of the plate is measured. The result of the frequency response is shown in **Fig. 8**. From the results of this figure, it is found that the resonant frequency of normal straight beam structure is 24 kHz and that of spiral shaped beam structure is 12 kHz. From simulation results (omitted from the want of space), the resonant frequencies of these structures under tensile stress of 30 MPa are 26 kHz and 11 kHz respectively. Considering that the experimental resonant frequencies agree with simulated

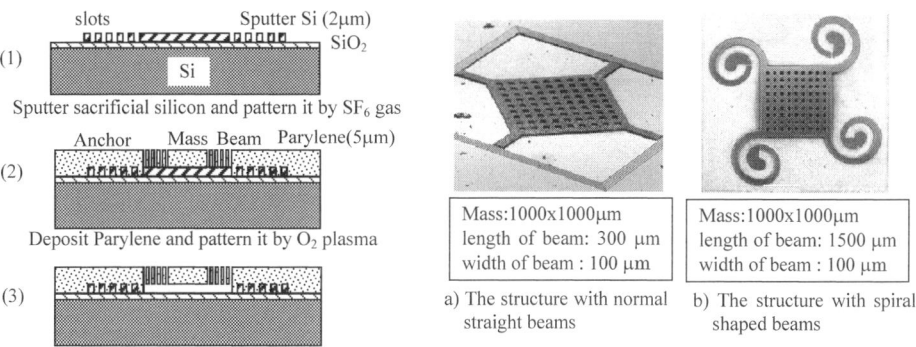

Figure 5: Fabrication process flow of suspended structure

Figure 6: SEM image of fabricated structure

results, the FEM simulation taking account of intrinsic tensile stress in this study is thought to have good validity. It is surely confirmed theoretically and experimentally that the a spiral beam is effective for lowering the resonant frequency of f_r, which leads to the sensitivity of accelerometer of $1/(2\pi f_r)^2$.

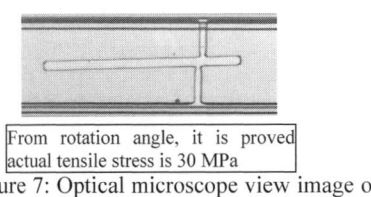

Figure 7: Optical microscope view image of Rotation Tip

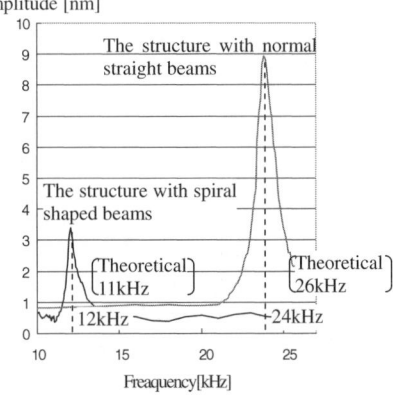

Figure 8: Result of frequency response

CONCULUSION

An accelerometer made of Parylene, which comprises a proof mass and support beam, has been developed now. In this study, the stiffness and the resonant frequency of suspended microstructures under tensile stress are investigated. The summary is as follows:
1) It is proved by FEM simulation that the stiffness is decreased in proportion to the first power of the beam length, while it is decreased in proportional to the third power of it under no stress according to the theory of strength of materials. Therefore, the structure with spiral beam is proposed.
2) Free standing Parylene suspended structures are fabricated by a micromachining process. The vibrations of these structures are observed by using a LDV and resonant frequencies of them are obtained. It is found that the resonant frequency of the structure with spiral beams is lower than that with straight beams, which shows the effectiveness of spiral beams for obtaining high sensitivity of accelerometer.

ACKNOWLEDGEMENT

This work was mainly supported by JSPS (Japan Society for the Promotion of Science).KAKENHI (16310103). This work was also partially supported by MEXT (Ministry of Education, Culture, Sports, Science and Technology). KAKENHI (17656090), "High-Tech Research Center" Project for Private Universities: Matching Fund Subsidy from MEXT, 2000-2004 and 2005-2009, the Kansai University Special Research Fund, 2004 and 2005.

REFERENCE

[1] Tai Y. C. (2003). Parylene MEMS: Material, Technology and Application. *Proc. 20th Sensor Symposium*, 1-8.
[2] Aoyagi S. and Tai Y. C. (2003). Development of Surface Micromachinable Capacitive Accelerometer Using Fringe Electrical Field. *Proc. Transducers'03*, 1383-1386.
[3] Harder T. A., Yao T. J., He Q., Shih C. Y. and Tai Y. C. (2002). Residual Stress in Thin-Film Parylene-C. *Proc. MEMS'02*, 435-438.

DIRECT PREDICTION OF CUTTING ERROR IN FINISH ENDMILLING BASED ON SEQUENCE-FREE ALGORITHM

J. Kaneko[1], K. Teramoto[2], K. Horio[1] and Y. Takeuchi[2]

[1] Department of Mechanical Engineering, Faculty of Engineering, Saitama University,
Saitama, Saitama, Sakura-ku, Shimo-Ohkubo, 255, Japan
[2] Department of Computer Controlled Mechanical systems, Graduate School of Engineering,
Osaka University
Osaka, Suita, Yamadaoka, 2-1, Japan

ABSTRACT

This study deals with a new estimation method of cutting error distribution on workpiece surface, which is caused by cutting force and tool deflection. The proposed procedure is based on "sequence-free" algorithm of cutting force prediction, which makes it possible to predict directly cutting error in an arbitrary tool position regardless of the order of tool movement in NC program. By applying the proposed procedure, quick estimation of cutting error distribution is realized. As a result, it is expected that NC operators can collect easily cutting conditions and cutter location in NC program with consideration of cutting error.

KEYWORDS

End mill, Machining, Instantaneous Cutting Force, Prediction, Cutting error, Tool swept volume

INTRODUCTION

Today, verification process for NC program plays very important roles. Especially, estimation of cutting error distribution on workpiece surface in finishing process is earnestly required. In conventional studies, many verification methods have been proposed. They are designed to verify geometric errors in NC program and already widely used. On the other hand, prediction of cutting error caused by instantaneous cutting force and tool deflection is not yet put in to practical use.

As a reason the error prediction about tool deflection does not spread, we focus following problems.
- In order to predict instantaneous cutting force, accuracy of estimated cutting depth is needed.
- Estimation process of cutting depth requires accurate explicit information of workpiece shape.
- The workpiece shape is usually changed by each tool movement step and complicated.

Usually, each part of workpiece surface is generated at different moment in machining. These facts mean that the prediction process for cutting error distribution caused by tool deflection requires vast amount of geometric calculation to estimate the explicit workpiece shape information.

So, in order to solve these problems and realize an efficient estimation, we propose a new prediction procedure. In the proposed procedure, the cutting depth on cutting edge in machining is calculated by a new "sequence-free" algorithm, which is based on the idea of tool swept volume (Wang W.P. 1986). Because the new algorithm does not require the explicit information of workpiece shape, it is thought that immediate and accurate prediction of cutting error is attained regardless of both the complexity of workpiece shape and sequence of tool moving in NC program.

NEW PREDICIOTN PROCEDURE OF CUTTING ERROR DISTRIBUTION

As mentioned above, difficultness of the prediction is caused by repetition of workpiece shape estimation process. So, in the new proposed procedure, the cutting depth is directly estimated using NC program and workpiece initial shape. This process can be performed regardless sequence of change of workpiece shape in machining, as shown in Figure 1.

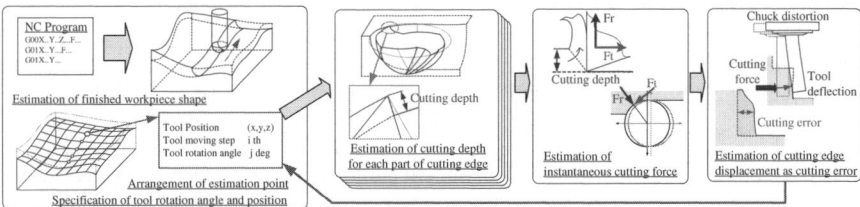

Figure 1: Proposed procedure for prediction of cutting error distribution

The proposed procedure consists of the following four estimation steps.
1. Arrangement of estimation point on nominal surface of finished workpiece. Nominal surface is workpiece shape estimated under assumption that the tool deflection did not happen.
2. Specification of tool rotation angle and position at the moment each point was generated.
3. Estimation of cutting depth on each part of cutting edge and prediction of total cutting force.
4. Prediction of displacement on the part of cutting edge.

These steps are repeated for each estimation point on the nominal surface of finished workpiece. In the following sections, we explain the details from 1st to 3rd step in case of 3-axis controlled machining with ball end mill.

Arrangement of Estimation Point and Specification of Tool Rotation Angle and Position

In this study, we regard the cutting error as the distance between the nominal surface and actual machined workpiece surface. In order to estimate the distance, we arrange estimation points on the nominal workpiece surface and specify both tool rotation angle and position at the moment the estimation points was appeared. So, we introduce Z-map representation (Takeuchi Y. 1989) and the idea of tool swept volume for estimation of workpiece shape at the time when machining is finished. As illustrated in Figure 2, By finding the tool moving step which distance $|p_{ij}-p0_{ij}|$ is the smallest, we can specify the coordinate of estimation point p_{ij}, tool rotation angle θ_{ij} and tool position tc_{ij}.

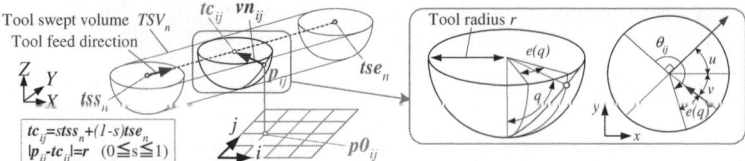

Figure 2: Arrangement of estimation point and specification of tool rotation angle and position

Estimation of Cutting Depth using Existence Evaluation of Workpiece Volume

In order to estimate the cutting depth without repetition of workpiece shape estimation, we introduce some estimation models proposed in former studies (Takata S. 1989). In these models, it is assumed that cutting edge is regarded as a set of finite flutes and cutting force can be estimated as the sum total of the force loaded on each finite flute as Figure 3. Furthermore, if both finite flute and tool feed is sufficiently small, the cutting depth at can be calculated from tool feed in each cutting edge passing f and the result of existence evaluation EE, as shown Equation 1 and 2.

$$at = EE \cdot \left(r + fb - \sqrt{f^2(b^2-1) + r^2} \right) \qquad (1)$$
$$b = \sin\theta \sin q \cos\phi + \cos q \sin\phi \qquad (2)$$

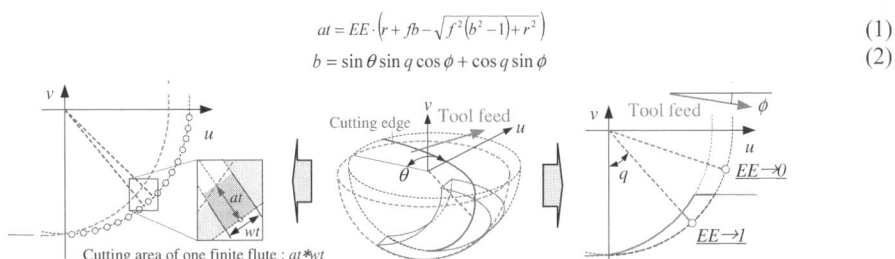

Figure 3: Cutting depth estimation based on the idea of finite flute and existence evaluation

This equation means that cutting force estimation can be realized by referring results of the existence evaluation for each finite flute. So, following section, we propose a new method of existence evaluation without the explicit information about workpiece shape in machining.

Efficient solution of Existence Evaluation based on the Idea of Tool Swept Volumes

Existence evaluation requires only judgment of workpiece volume existence where the finite flute is located. It does not surely need the explicit information of workpiece shape. So, we introduce the idea of tool swept volume and set operation between volumes. As shown in Equation 3, workpiece volume in machining of nth tool moving step $MWV_n{'}$ can be described by volume of initial workpiece MWV_0, ith tool swept volume TSV_i, a part of nth tool swept volume $SubTSV_n$ and set operations.

$$MWV_n' = MWV_0 \cap \left(\bigcap_{i=1}^{n-1} \overline{TSV_i} \right) \cap \overline{SubTSV_n} \qquad (3)$$

Then, we define function of existence evaluation $EE(p,V)$. If point p is located in the inside of volume V, the value of $EE(p,V)$ is 1. In the case of others, $EE(p,V)$ is 0. Applying this function to Equation 3, the existence evaluation for volumes performed set operations can be achieved by multiplication of result about the existence evaluation for each volume, as illustrated in Figure 4.

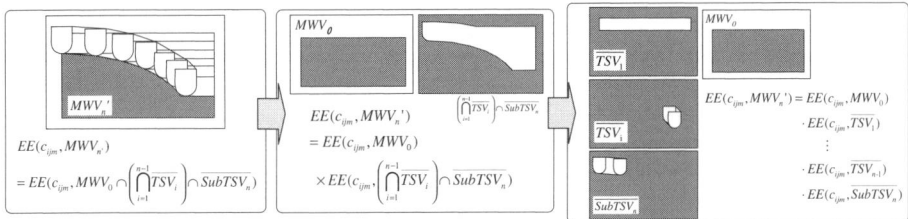

Figure 4: Decomposition of existence evaluation process for workpiece volume in machining

By using this relation, we can judge whether workpiece volume exists on the finite flute c_{ijm} with the result of existence evaluation between c_{ijm} and each tool swept volume. Because the tool swept volume usually has simple shape as shown in Figure 2, we can estimate accurately interference between c_{ijm} and the tool swept volume. As a result, the existence evaluation for workpiece volume in machining is realized without the estimation of worikpiece shape. It means the instantaneous cutting force can be predicted directly without the estimation of explicit workpiece shape. By using the proposed algorithm, we can predict the cutting error on the nominal workpiece surface regardless of the order of tool movement.

EVALUATION OF THE PROPOSED PROCEDURE AND CONCLUSION

In order to evaluate the proposed procedure, we develop a prototype system and conduct an experiment using NC program simulates finish machining. In the prototype system, we introduced P-Voxel representation method (Kaneko J. 2002) in order to accelerate the existence evaluation.

(a) Using cutting tools with helix flutes of 30 degrees   (b) Using cutting tools with no-helix flutes

Figure 5: Estimated results of cutting error distribution by the developed prototype system

Figure 5 shows the estimated results by the developed system. The NC program is created by commercial CAM system. Workpice is sculptured by contour milling of 7600 steps with square end mill and profile milling of 10600 steps by ball end mill. Figure 5(a) shows the estimated result when helix angle of flutes on cutting tool is 30 degrees. And, Figure 5(b) shows the result when the helix angle is 0 degree. The difference of each result is caused by changes of loaded cutting force resulted from the helix angle of cutting edge and removal process of workpiece volume.

The required time for estimation of case (a) was about 130 seconds, and was 133 seconds in case (b). The total number of estimation points on sculptured surface is about 44000. The prototype system can calculate the cutting error on finished workpiece surface in about 0.003 seconds per one estimation point. As a result, it is thought that the proposed procedure realizes the estimation of cutting error distribution with sufficient performance.

References

Kaneko J., Teramoto K. and Onosato M. (2002). An implicit shape representation method for sequence-free force estimation in end-milling. *Proceedings of ICMT2002*, 260-265.

Takata S., Tsai M.D., Inui M. and Sata T. (1989). A Cutting simulation System for Machinability Evaluation Using a Workpiece Model. *Annals of the CIRP* **38:1**, 417-420.

Takeuchi Y., Sakamoto M., Abe Y. and Orita R. (1989). Development of a Personal CAD/CAM System for Mold Manufacture Based on Solid Modeling Techniques. *Annals of the CIRP* **38:1**, 429-432.

Wang W.P. and Wang K.K. (1986). Geometric Modeling for Swept Volume of Moving Solids. *IEEE Computer Graphics and Applications* **6:12**, 8-17.

DEVELOPMENT OF CURVED HOLE MACHINING METHOD
- SIZE REDUCTION OF HOLE DIAMETER -

T. Nakajima[1], T. Ishida[1], M. Kita[2], K. Teramoto[1] and Y. Takeuchi[1]

[1] Dept. of Mechanical Eng., Graduate School of Eng., Osaka University
Yamadaoka 2-1, Suita, Osaka 565-0871 JAPAN
[2] Dept. of Machinery System Production Technolgy, Kinki Polytechnic College
Inabatyou 1778, Kishiwada, Osaka 596-0103 JAPAN

ABSTRACT

This study deals with a diameter reduction of curved holes that can be machined by the method developed by the authors. In order to improve the productivity of molding, it is necessary to increase the efficiency of the cooling stage in a molding cycle. It depends on the shape and the arrangement of water channels, i.e., pipelines built in molds. However, water channels consist of a series of straight holes due to the fabrication by drilling. Accordingly, it is strongly required to develop a machining method of curved holes since curved water channels are desirable. To meet the requirement, the device has been developed, which can make an electrode move along a curved trajectory with electrical discharge machining. The device can fabricate curved holes. However, the fabricated curved holes have a problem that their diameter is too large to employ them as a water channel. In the study, thus, the size reduction of the curved holes is tried by improving the electrode and its peripheral parts.

KEYWORDS

curved hole, electrical discharge machining, size reduction, helical compression spring, servomotor, water channel, wire feeding

INTRODUCTION

Injection molding is one of very important manufacturing methods and is employed to create various products in a variety of industries. Therefore, an innovation in an injection molding technologies has a very strong impact on our society. In general, injection molding has the cycle which is composed of melting material, injecting the melted material to a mold, solidifying the material and taking out a desired-shaped material from the mold. Accordingly, the productivity of injection molding is improved if the time for the cycle can be shortened. More than a half of the cycle time is wasted in the stage to solidify the injected material. Consequently, the shortening of the solidifying time results in improvement of molding productivity. The solidification is generally accomplished by cooling a mold by means of coolant flow which runs through water channels built in mold. Water channels are pipelines fabricated in a mold. They are usually made by drilling. Namely, they consist of a straight hole or a series of straight holes. As a result, the shape of a water channel is polygonal line. This causes the restriction of the degree of freedom in their position and shape.

To achieve the optimal position and shape of water channels, it is demanded to develop a curved hole machining method, Goto et al. (2002), Ichiyasu et al. (1997), Uchiyama & Shibasaki (2004). This leads to the reduction of the cycle time in molding since the solidifying time can be shortened by the optimal water channels. As a result, it will be possible to improve productivity of injection molding. To meet the requirement, the authors have also developed the devices which can machine curved holes, Ishida & Takeuchi (2002), (2004). The devices can control the moving trajectory of a tool electrode attached to an electrical discharge machine (EDM) and simultaneously make the electrode perform electrical discharge machining. If the electrode moving trajectory is curved one, the curved hole can be machined, which has the identical shape with the envelope of electrode moving locus. Additionally, the device is able to fabricate various-shaped curved holes since the electrode moving trajectory can be controlled by a software.

However, the device has a problem that the diameter of the machined curved holes is too large to employ them as water channels. To solve the problem, in the study, the diameter of the former electrode, 20mm, is reduced to a half of it. According to the size reduction of the electrode, the parts constituting the electrode and the peripheral parts around the electrode are redesigned. Concretely, some parts are omitted, the size of some other parts is reduced, or assembling method is changed. From the results in the motion and machining experiments, it is found that the redesigned device is effective and can machine the curved holes of half size diameter.

CURVED HOLE MACHINING DEVICE

Structure and Motion of the Device

Figure 1 illustrates a schematic view of the developed device. The device is installed on an EDM and consists of a helical compression spring, an electrode for electrical discharge machining, wires, pulleys, three ball screws with servomotors, motor drivers, a linear scale, and a personal computer (PC). The electrode is mounted on the end of the spring, which is connected to a head of the EDM through a shaft and a tabular jig. Three wires are fastened in equal angles of 120° on the end of the electrode side of the spring and are respectively led to the nuts of the ball screws through the pulleys on the tabular jig. Each servomotor is connected with the PC through the motor driver. On the other hand, a L-shaped jig rests on the bottom of a working tank of the EDM. On the wall of the L-shaped jig, the linear scale is mounted so that it can measure the position of the EDM head. Additionally, it is also connected to the PC. In summary, the PC can measure the EDM head position and can control respective servomotors, i.e., respective feeds of the wires at the same time. Consequently, this device can independently control the wire feeds according to the EDM head feed. In the study, the PC controls the wire feeds so that the feeding amounts of two wires on the left side are identical and that they are different from the feeding amount of the wire on the right side and so that the relationship between the EDM head feed and the wire feeds can be expressed as follows:

$$L_{s1} = N_1 h, \quad L_{s2} = N_2 h \tag{1}$$

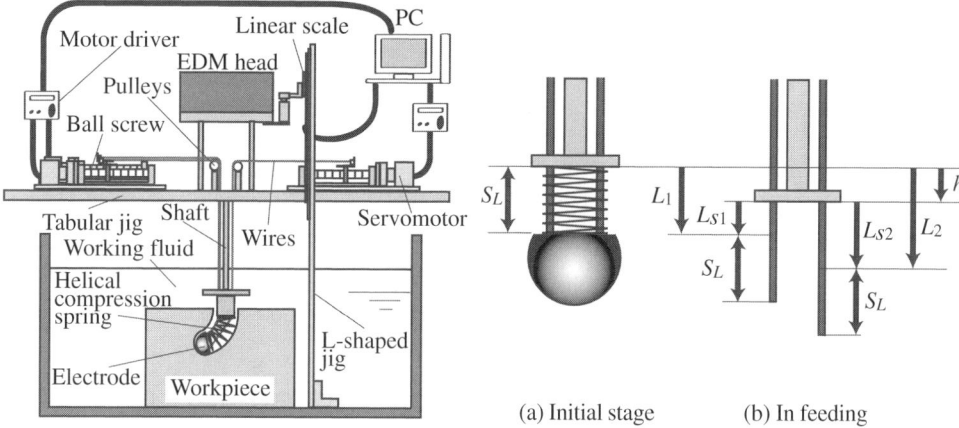

Figure 1: Schematic view of experimental device

(a) Initial stage (b) In feeding
Figure 2: Wire feed from the viewpoint of observers

, where h is EDM head feed, L_{S1}, L_{S2} are the feeding amounts of two wires on the left side and that of the wire on the right side, and N_1, N_2 are arbitrary values given in the software in the PC, respectively.

Figure 2 illustrates the wire feeds obtained by setting L_{S1}, L_{S2}. Figure 2(a) represents the initial stage of a curved hole machining. In the stage, the spring is deeply compressed. S_L is the length of the spring in the initial stage. When the EDM head moves down, as illustrated in Figure 2(b), the device moves down with the EDM head and the wires are fed. In the situation, the wire feeds from the viewpoint on the working tank and their ratio are expressed as follows:

$$L_1 = L_{S1}+h,\ L_2 = L_{S2}+h,\ L_1 : L_2 = (L_{S1}+h):(L_{S2}+h) \tag{2}$$

, where L_1, L_2 are the feeding amounts of two wires on the left side and that of the wire on the right side from the viewpoint on the working tank, and $L_1 : L_2$ is called wire feed ratio.

These processes make the spring stretch with its posture bent due to the difference of three wire feeds. This results in a curved motion of the electrode. Additionally, discharge gap control is realized at the electrode since the electrode motion reflects the EDM head motion. This means that electrical discharge machining is performed on the electrode, thus allowing a curved hole machining. Moreover, the electrode can move along various kinds of curved trajectories and various kinds of curved holes can be fabricated since the relationship between the EDM head feed and the wire feeds can be variously set by means of the software in the PC.

Size Reduced Electrode

Figure 3 shows the former electrode and the parts around the electrode. To the electrode, a flexible tube is attached through a joint in order to provide a working fluid to a discharge gap to remove chips. The diameter of the electrode is 20mm. This results in fabrication of curved holes with a diameter of about 20mm, which is too large as water channels. Therefore, the size reduction of electrode has been required. In the study, it is intended to reduce the diameter of the electrode to 10mm. For the reduction, the parts around the electrode must be omitted. Concretely, screws used for assembling the parts and the flexible tube for working fluid supply are not used. Instead of the screws, transition fit is adopted for fixing the parts. In addition, the flexible tube is not used since working fluid supply is not necessarily indispensable for electrical discharge machining. As a result, the number of the parts

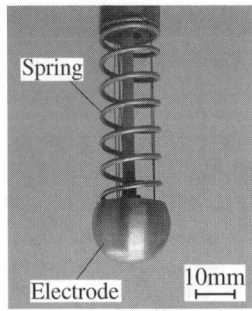 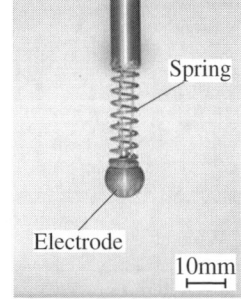

Figure 3: Former electrode and its peripherals   Figure 4: Size reduced electrode and its peripherals

around the electrode can be reduced. According to the electrode diameter reduction, moreover, the size reduction of the spring and the wires is required. The specifications of the spring and the wires are determined in consideration of the diameter of the electrode, buckling limit of the spring and the relationship between broken load of the wires and the maximum force generated by the spring. Figure 4 shows the electrode of 10mm in diameter and the parts around the electrode. Since Figure 3 and 4 are depicted in an identical scale, it is seen that the diameter of the newly developed electrode is half of that of the former electrode.

MOTION EXPERIMENT

To confirm the behavior of the size reduced electrode, motion experiments are conducted. In the initial stage of the experiment, the spring is deeply compressed. The relationship between h and L_{s1} and L_{s2} is set as follows. In case that $0 \leq h \leq P$, $L_{s1} = L_{s2} = 0$, and in case that $P \leq h \leq Q$, $L_{s1} = N_1 h$, $L_{s2} = N_2 h$. That is to say, when $0 \leq h \leq P$, the spring moves down and straight, same as the EDM head, with its posture deeply compressed, and when $P \leq h \leq Q$, the spring stretches with its posture bent. As a consequence, the electrode moves along a curved trajectory after a straight one whose length is P.

Letting $P = 20$mm and $(N_1, N_2) = (1, 1.5), (1, 2)$ when $P \leq h \leq Q$, actual motion experiment is carried out. From Equation 2, wire feed ratio in each case is 4:5 and 4:6. Figure 5 depicts actual behavior of the spring and the electrode according to EDM head feed in each case. Additionally, Figure 6 illustrates the actually obtained electrode trajectories in the cases, respectively. The trajectories are expressed by the change in the electrode position. The electrode positions are obtained at every 5mm movement of the EDM head in the straight section and at every 0.5mm in the curved section, respectively. As can be seen from Figure 5 and 6, the electrode moves along a straight trajectory until the EDM head goes down to 20mm and continuously moves along a curved trajectory in both cases. The total EDM head feeds, i.e., Q, in the cases of the wire feed ratios of 4:5 and 4:6 are 25.5mm and 24mm, respectively. This is because the bent spring deviates from the envelope of the electrode trajectory if the EDM head goes down more than the values.

MACHINING EXPERIMENT

From the results of the motion experiments, it will be possible to machine two types of curved holes. In order to verify the expectation, machining experiments are carried out in the identical settings of the motion experiments. Machining condition is set as follows. Materials of employed electrode and workpiece are oxygen-free copper and aluminum alloy (A5052), respectively. Working fluid is oil. Working current, pulse duration and duty factor

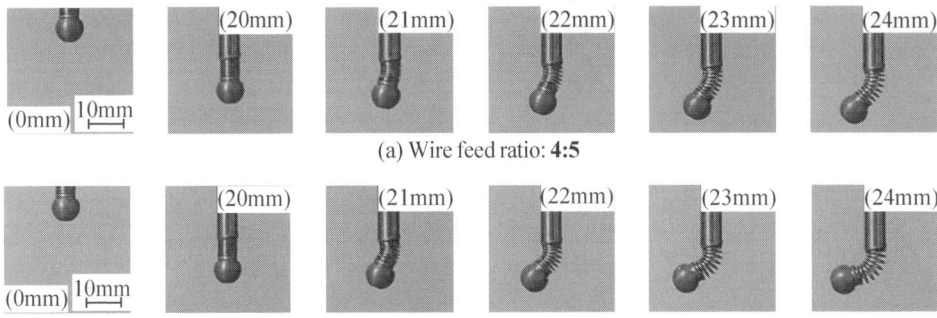

(a) Wire feed ratio: **4:5**

(b) Wire feed ratio: **4:6**

Figure 5: Actual behavior of spring and electrode according to EDM head feed in case of wire feed ratio of 4:5 and 4:6

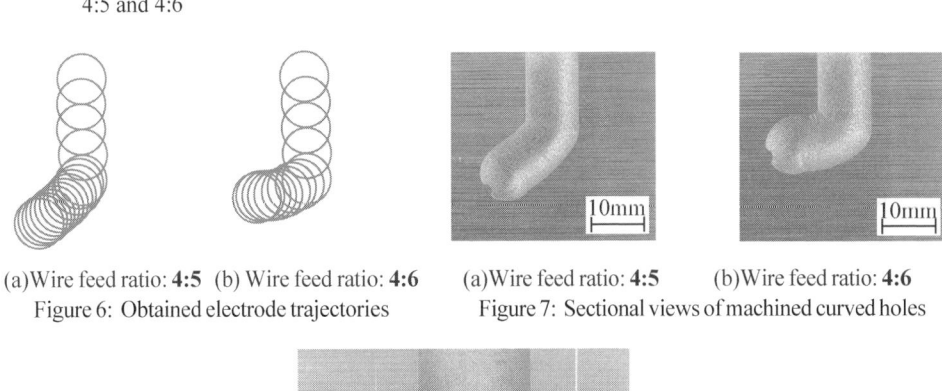

(a) Wire feed ratio: **4:5** (b) Wire feed ratio: **4:6**
Figure 6: Obtained electrode trajectories

(a) Wire feed ratio: **4:5** (b) Wire feed ratio: **4:6**
Figure 7: Sectional views of machined curved holes

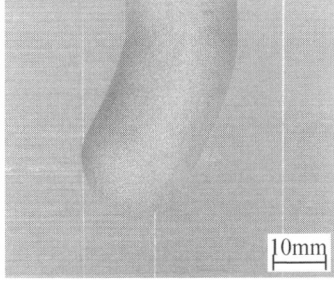

Figure 8: Sectional view of workpiece machined by former electrode (wire feed ratio: **4:5**)

are respectively set in 8A, 135μs and 30% under the polarity of electrode(+)/workpiece(-). Jump and orbital motion of the EDM head is not used.

Figure 7 shows sectional views of the workpieces obtained by the size reduced electrode in two cases of the wire feed ratios. As shown in the figure, it is found that two-different-shaped curved holes with a straight hole can be machined. The diameter of the hole is 10.2mm in both straight and curved parts. In these machining, stable electrical discharge machining continues from start to finish even without supplying a working fluid to the discharge gap. Machining times in the cases are 97min. and 91 min., respectively. Namely, each machining speed is 26.9mm^3/min and 27.1mm^3/min. Additionally, it can be seen from Figure 6 and 7, each curved hole section is identical with the shape with which each electrode trajectory is covered. Consequently, it is found that the electrode moves accurately along the trajectories obtained in the motion experiments even in machining. Figure 8 shows a sectional view of the workpiece machined by the former electrode in case of wire feed ratio of 4:5. Figure 7(a) and 8 are depicted in an identical scale. From the comparison of them, it is seen that a curved hole with half diameter can be machined.

CONCLUSIONS

Aiming at size reduction of curved hole diameter, the device is developed, which can machine a curved hole of about 10mm in diameter. From the motion and machining experiments, the obtained conclusions are summarized as follows:

(1) The diameter of the electrode is reduced to 10mm, which is a half of the former one, by redesigning the electrode, the parts constituting the electrode and the peripheral parts of the electrode.

(2) The size reduced electrode and its peripheral parts for the curved hole machining device being well provided, the electrode can move along a straight and curved trajectory as well as the former electrode.

(3) By means of the curved hole machining device equipped with the redesigned parts, the movement of an EDM head can be transmitted to the size reduced electrode, which means that electrical discharge machining by the electrode works well. As a result, it is achieved to create curved holes with an approximately half diameter, compared with those by the former electrode.

ACKNOWLEDGEMENT

The study is partly supported by the Ministry of Education, Culture, Sports, Science and Technology, Grant-in-Aid for Young Scientists, (B)14750083

REFERENCES

Goto A., Watanabe K. and Takeuchi A. (2002). A Method to Machine a Curved Tunnel with EDM. *International Journal of Electrical Machining* **7**, 43-46.

Ichiyasu S., Takeuchi A. and Watanabe K. (1997). Machining Curved Tunnel for Coolant with Mole EDM. *Proceedings of 4th International Conference on Die & Mould Technology*, 224-330.

Ishida T. and Takeuchi Y. (2000). Curved Hole Machining by Means of Electrical Discharge Phenomena and Electrode Feed Mechanism. *Proceedings of the 2000 Japan USA Flexible Automation Conference*, 1283-1288.

Ishida T. and Takeuchi Y. (2002). L-shaped Curved Hole Creation by Means of Electrical Discharge Machining and an Electrode Curved Motion Generator. *International Journal of Advanced Manufacturing Technology* **19:4**, 260-265.

Uchiyama M. and Shibasaki T. (2004). Development of an Electromachining Method for Machining Curved Holes. *Journal of Materials Processing Technology* **149:1**, 453-459.

MICROCHANNEL ARRAY CREATION BY MEANS OF ULTRAPRECISION MACHINING

F. Andou[1], A. Yamamoto[2], T. Kawai[3], H. Ohmori[4], T. Ishida[1] and Y. Takeuchi[1]

[1] Dept. of Mechanical Eng., Graduate School of Eng., Osaka University,
Yamadaoka 2-1, Suita, Osaka 565-0871, JAPAN
[2] Reconstitution Materials Group, Biomaterials Center,
National Institute for Materials Science (NIMS),
1-2-1 Sengen, Tsukuba, Ibaraki 305-0047, JAPAN
[3] Robomachine Laboratory, FANUC Ltd.,
Oshino, Yamanashi 401-0597, JAPAN
[4] Materials Fabrication Laboratory,
The Institute of Physical and Chemical Research (RIKEN),
2-1 Hirosawa, Wako, Saitama 351-0198, JAPAN

ABSTRACT

The study deals with ultraprecision machining of microchannel array chips made of several metals to evaluate the compatibility between blood and metallic materials which are buried in human bodies as parts of artificial internal organs, etc. The blood-compatibility of the metallic materials is very important since their influences for human bodies are not completely clear. Therefore, it is essential to estimate their bio- and blood-compatibilities. To evaluate them efficiently, it is planned to employ the blood fluidity measurement by a microchannel array chip with a micro-rheology device. However, the chips are made of silicon and the shape of their microchannels is limited since they are generally fabricated by photo-lithographic technologies. To solve the problem, it is required to fabricate the chips with various-shaped microchannels made of several metals. In the study, consequently, ultraprecision cutting is applied to the fabrication of the microchannel array. From the experimental results, it is found that ultraprecision cutting has potential of fabricating arbitrary-shaped microchannel array made of various kinds of metals with high accuracy.

KEYWORDS

ultraprecision cutting, metallic microchannel array, microgroove, rotational tool, non-rotational tool, blood-compatibilities, blood fluidity evaluation, multi-axis control ultraprecision machining center

INTRODUCTION

In recent years, many kinds of metals are applied to medical usages instead of ceramics, high polymer and so on. Metals have the advantage in terms of strength, elasticity and stiffness. Usually employed metals are stainless steel, cobalt-chromium alloy, titanium, gold and so forth. Naturally, these metals are widely employed as materials of such medical implements as are buried in human bodies, for example, fixture for fracture, artificial joints, tooth implants, and others. Accordingly, it is important to investigate the influences or toxicities of the metals for human bodies. For satisfactory selection of metals used in the medical implements, therefore, it is essential to evaluate bio- and blood- compatibilities of the metals. Conventionally, the evaluation has been done by making experiments on living animals, which consumes a lot of money and time. To save the cost, it is required to develop a new evaluating method.

On the other hand, micro-rheology device to measure blood-fluidity has been developed to investigate flow mechanism of blood. The device allows human blood flow to pass through microchannel array built on a chip, which is a model of capillary vessels due to its shape in which many microgrooves are arranged in parallel. At the same time, the blood flow through the microchannel array can be visually observed, which can evaluate its fluidity.

Consequently, the employment of microchannel array chips made of various metals is expected to evaluate the compatibility between blood and metals. However, the microgrooves constituting a microchannel array is generally built on silicon by photolithographic techniques, which do not have high abilities to control the shape of the microgrooves and to increase the accuracy of the shape. Their shape and accuracy are extremely important to measure blood-fluidity with a microchannel array chip.

Accordingly, the study aims at fabrication of the microchannel array chip by ultraprecision cutting. Cutting can make complicated microgroove shapes with high degree of freedom and high accuracy, and have no choice of materials to be fabricated, Takeuchi et al., (2001) and (2002), Kumon et al., (2002). As a result of actual machining experiments, it is succeeded to fabricate chips with two-kinds-shaped microchannel array made of some metals by means of ultraprecision cutting.

ULTRAPRECISION MACHINING CENTER AND MACHINING METHOD

Figure 1 illustrates the setups in cutting with the ultraprecision machining center used for the experiments. The utilized machining center is ROBONANO make by FANUC Ltd., and has five axes, i.e., X, Y and Z axis as translational axes, and B and C axis as rotational ones. The positioning resolutions of the translational axes and the rotational axes are 1nm and 0.00001 degree, respectively. The machining center is designed based on the concept of friction-free servo structures. As illustrated in the figure, the machining center has two type cutting methods according to the employed tool, viz., rotational tool or

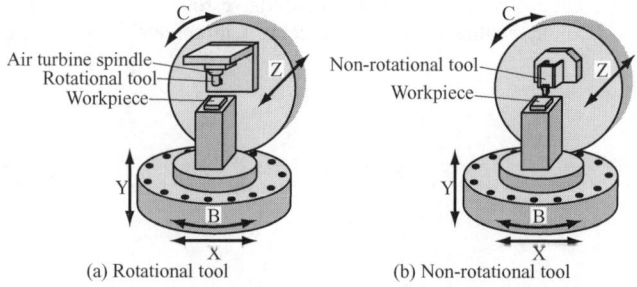

Figure 1: Two kinds of setups of ultraprecision cutting

non-rotational tool. The former is attached to a high speed air turbine spindle mounted on C table. The latter is directly fixed on C table through a jig. A workpiece is mounted on B table in both cases.

CREATION OF V-SHAPED MICROCHANNEL ARRAY CHIP

Figure 2 illustrates schematic views and dimensions of V-shaped microchannel array chip. The chip has a glass contact surface on its outside circumference, a shape like a bank in its center, hollows in both sides of the bank and a through hole on the bottom of each hollow, which are an entrance and exit of blood. V-shaped microchannel array, i.e., parallel-arranged V-shaped microgrooves, is fabricated on the bank. One of the microgrooves is 10μm in width, 5μm in depth and 100μm in length. They are arranged at intervals of 10μm, and the total number of them is 250. The top surface of the array has the same height as the glass contact surface. The shapes to be machined are the microgrooves and the glass contact surface.

Fluidity of blood, viz., compatibility between blood and metal, is evaluated as follows. A cover glass is attached to the top surface of the chip, and blood flow comes in and out of the holes through the microchannel array. The blood flow through it is observed over the cover glass. Consequently, the top surface of the chip, namely the glass contact surface and the top surface of the array, must be a mirror surface to prevent blood from leaking.

Figure 3 illustrates the employed machining manner of the V-shaped microchannel array chip in the study. First, the top surface of the chip is machined with a large-diameter rotational tool so as to be a mirror surface. Secondly, the bank is formed with a small-diameter rotational tool so that the width of its top shape can be 100μm. Lastly, the V-shaped microchannel array, i.e., the V-shaped microgrooves, are fabricated with two kinds of methods using a rotational tool or a non-rotational tool. Each tool has a diamond tip with the cutting edge of 90°. The former and the latter are respectively applied to the workpiece made of gold and aluminum due to the results of the basic experiments that V-shaped microgrooving by

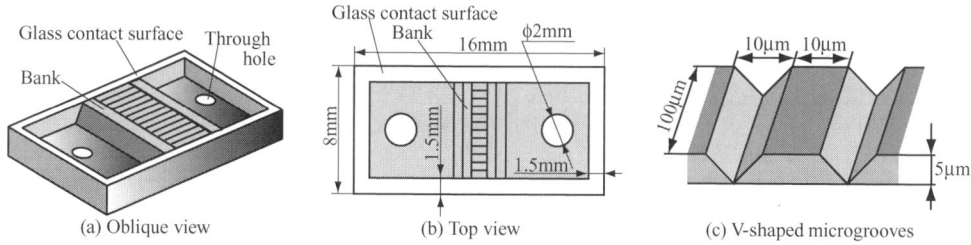

Figure 2: Schematic views and dimensions of V-shaped microchannel array chip

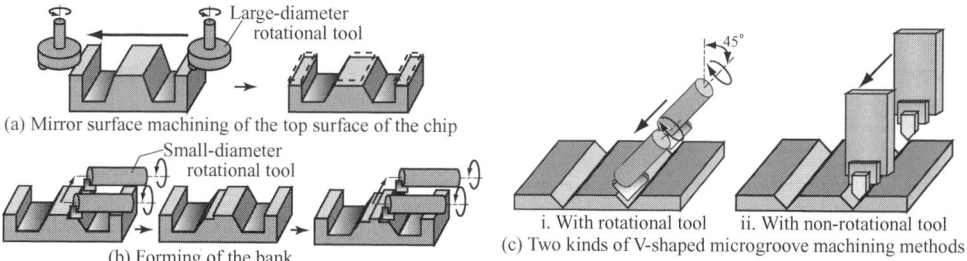

Figure 3: Machining manner of V-shaped microchannel array

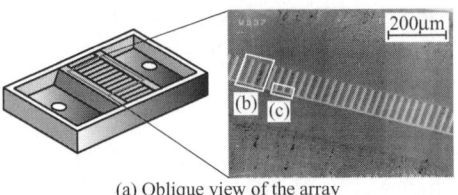 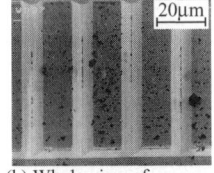 

(a) Oblique view of the array (b) Whole view of V-shaped microgrooves (c) Enlarged view of edges of V-shaped microgrooves

Figure 4: Machined V-shaped microchannel array made of gold with rotational cutting

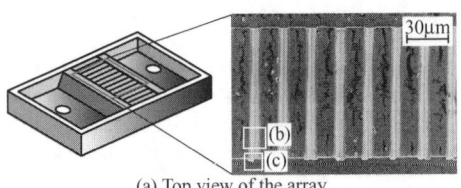 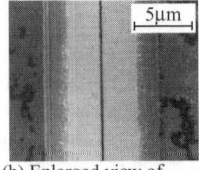 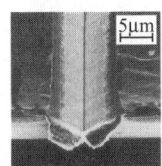

(a) Top view of the array (b) Enlarged view of V-shaped microgroove (c) Enlarged view of edge of V-shaped microgroove

Figure 5: Machined V-shaped microchannel array made of aluminum with non-rotational cutting

the tools has been tested to the workpieces made of various metals.

Figure 4 shows the V-shaped microchannel array machined with the rotational tool under the cutting conditions that cutting speed is 14.7 m/s, tool feed speed is 50.0mm/min., depth of cut is 2.0μm in roughing and 1.0μm in finishing and the workpiece is sprayed with cutting fluid of kerosene. As can be seen from the figures, it is found that the microchannel array has good surfaces, accurate shapes, and sharp edges without any burr.

Figure 5 shows the V-shaped microchannel array fabricated with the non-rotational tool under the cutting conditions that cutting speed (= tool feed speed) is 40.0mm/min. in roughing and 1.0mm/min. in finishing, depth of cut is 0.5μm in both roughing and finishing and the workpiece is submerged in cutting fluid of kerosene. From the figures, it is seen that the microchannel array can be almost machined well, similarly to that with the rotational cutting. However, burr is formed on the edge of the V-shaped microgrooves. The blood flow in the blood fluidity evaluation will be affected by the burr. Consequently, it is required to remove the burr or to improve the tool path not to generate the burr.

The V-shaped microchannel array chip made of gold machined with the rotational tool is actually used for evaluating the blood fluidity. However, the V-shaped microchannel array is clogged with the ingredients contained in blood at its entrance in only 3 minutes after starting to make blood flow into the chip. After all, the chip is not available for the evaluation of the blood fluidity. Consequently, it is necessary to redesign the shape of the microgrooves constituting the microchannel array.

CREATION OF SQUARE-SHAPED MICROCHANNEL ARRAY CHIP

Figure 6 illustrates schematic view and dimensions of the redesigned microchannel array, i.e., parallel-arranged square-shaped microgrooves. Changing the view point, the redesigned array is a row of slender rectangular-prism-shaped objects with diamond-shaped ends. The object is 10μm in width, 5μm in height and 100μm in length. The objects are arranged at several intervals of 25μm, 50μm, 100μm and 150μm, and each interval is repeated 8 times. The gaps between the objects play a role of the square-shaped microgrooves. Accordingly, the interval, height and length of the objects are respectively equal to the width, height and length of the square-shaped microgrooves. In addition, the both sides of the microgroove are gradually open due to the diamond-shaped ends of the objects. The other dimensions of the square-shaped microchannel array chip are identical with that of the V-shaped one.

Figure 7 illustrates the adopted machining manner of the square-shaped microchannel array chip. In the initial stage, the top surface of the chip is machined with the same method as the V-shaped one. In the next stage, the bank is formed. In the final stage, the square-shaped microchannel array, i.e., the square-shaped microgrooves, is fabricated. In the last two stages, a same non-rotational tool is employed, as illustrated in the figure. The utilized non-rotational tool is depicted in Figure 8. First reason is because the square-shaped microgrooves cannot be machined with a rotational tool since the revolving radius of the diamond cutting edge is so large that the shapes to be left have been cut, and second reason is because the positioning error of the tool is suppressed which occurs in exchanging the tool. The array machining is done under the identical cutting conditions with those in machining the V-shaped microgrooves with the non-rotational tool except that depth of cut is 1.0μm in roughing and that the workpiece material is gold.

Figure 9 (a) and (b) show the actually machined square-shaped microgrooves whose width is 25μm. As seen from the figure, it is found that the microchannel array is well machined as designed and has very good surface. Figure 9 (c) depicts the profile of the cross section that is represented as A-A in Figure 9 (b). The depth of the object, i.e., the height of the microgrooves, is 4.95μm. This proves that the microchannel array is precisely fabricated. Figure 9 (d) shows an enlarged view of the end of the object between the microgrooves. From the figure, it is seen that the diamond shape of the object is sharply fabricated though its edges are a little wavelike shape with burr in nanometer order. This is due to the ductility of gold. However, they do not affect the evaluation of blood fluidity.

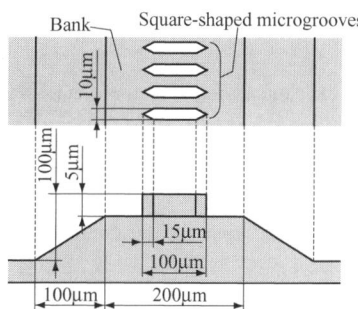

Figure 6: Schematic view and dimensions of square-shaped microchannel array

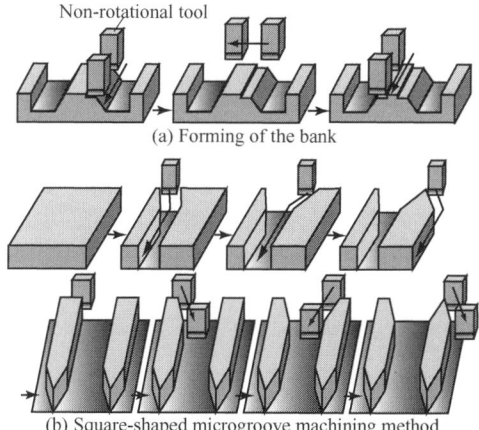

Figure 7: Machining manner of square-shaped microchannel array

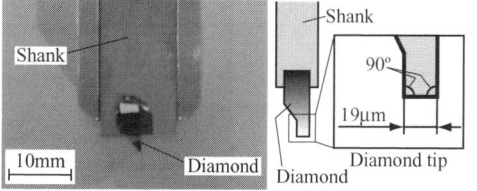

Figure 8: Non-rotational tool employed to machine square-shaped microgrooves

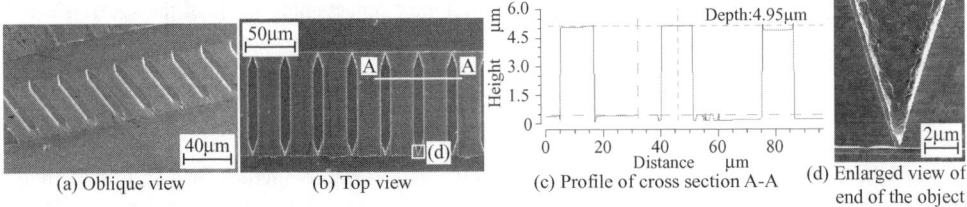

Figure 9: Several views and measurements of machined square-shaped microchannel array made of gold

The microchannel array is actually used for the evaluation of the blood fluidity. The cover glass is well fitted with the chip and the blood flows smoothly. It is found that the chip is valid for the evaluation.

CONCLUSIONS

Microchannel array chip is available for evaluation of blood fluidity. This chip is generally built on silicon with photolithographic techniques. Therefore, the study aims at creation of metallic microchannel array chips by means of an ultraprecision machining center and diamond cutting tools. The reason to employ the traditional cutting technology is the high possibility of selecting various kinds of metals and fabricating complicated shapes. The conclusions obtained in the study are summarized as follows:

(1) V-shape microchannel arrays made of gold and aluminum are well fabricated with rotational and non-rotational cutting tools.

(2) Square-shaped microchannel array made of gold is finely created with a non-rotational cutting tool.

(3) Blood flow can be observed by use of metallic chips with the square-shaped microchannel array.

ACKNOWLEDGEMENT

This study is partly supported by the Ministry of Education, Culture, Sports, Science and Technology, Grant-in-Aid for Scientific Research, B(2)16360069.

REFERENCES

Kumon T., Takeuchi Y., Yoshinari M., Kawai T. and Sawada K. (2002). Ultraprecision Compound V-shaped Micro Grooving and Application to Dental Implants. *Proc. of 3rd Int. Conf. and 4th General Meeting of EUSPEN* 313-316.

Takeuchi Y., Maeda S., Kawai T. and Sawada K. (2002). Manufacture of Multiple-focus Micro Fresnel Lenses by Means of Nonrotational Diamond Grooving. *Annals of the CIRP* **50:1,** 343-346.

Takeuchi Y., Miyagawa O., Kawai T., Sawada K. and Sata T. (2001). Non-adhesive Direct Bonding of Tiny Parts by Means of Ultraprecision Trapezoid Microgrooves. *J. of Microsystem Technologies* **7:1,** 6-10.

AUTOMATION OF CHAMFERING BY AN INDUSTRIAL ROBOT (DEVELOPMENT OF POSITIONING SYSTEM TO COPE WITH DIMENSIONAL ERROR)

Hidetake TANAKA[1], Naoki ASAKAWA[1], Tomoya KIYOSHIGE[2] and Masatoshi HIRAO[1]

[1] Graduate School of Natural Science and Technology Kanazawa University
2-40-2, Kodatsuno, Kanazawa City, Ishikawa, Japan
[2] Honda Engineering Co., Ltd.
Haga-dai 16-1, Haga Town, Tochigi, Japan

ABSTRACT

The study deals with an automation of chamfering by an industrial robot. The study focused on the automation of chamfering without influence of dimensional error piece by piece. In general, products made by casting have dimensional error. A cast impeller, used in water pump, is treated in the study as an example of the casting product. The impeller is usually chamfered with handwork since it has individual dimensional errors. In the system, a diamond file driven by air reciprocating actuator is used as a chamfering tool and image processing is used to compensate the dimensional error of the workpiece. The robot hand carries a workpiece instead of a chamfering tool both for machining and for material handling. From the experimental result, the system is found to have an ability to chamfer a workpiece has the dimensional error automatically.

KEYWORDS

Industrial robot, Chamfering, Image processing, Impeller, Error compensation

INTRODUCTION

Chamfering is essential processes after machining for almost all machined workpieces to control products appearance. Usually, workpieces, which having simple shapes can be chamfered by an automatic chamfering machine. However, complicated shaped workpieces are obliged to chamfer with handwork because of their intricacy. Especially, products made by sand mold casting basically have dimensional errors. A cast impeller, used in water pump, is treated in the study as an example of the workpiece with individual dimensional error. The objective chamfering part is an edge of outlet of the impeller. The part is usually chamfered by human handwork because it is located in narrow space and its dimension is largely influenced by individual dimensional errors piece by piece. Figure 1 shows the appearance and dimension of the workpiece. The objective chamfering part is an edge of outlet of the impeller between front and rear shroud as shown in Fig. 1. The impeller has 6 parts to be chamfered. In the study, y-z plane is defined as tangent plane on the chamfering part. The dimensional errors occurred in y-z plane and θ, rotating error around the normal direction on tangent plane are considered.

Since the industrial robot has a large number of degrees of freedom, it provides a good mimic of a human handwork. Formerly, some studies to automate such contaminated workings by use of industrial robots. [1][2][3]

To automate the chamfering, an industrial robot is used to handle and hold the impeller in front of a "tool station" our own developed in our study. The tool station fixed on a worktable has positioning actuators and a file driven by air reciprocating actuator as a chamfering tool. To detect positioning and dimensional errors of the workpiece based on an image of the objective part taken by a camera. The tool station can compensate the errors and chamfer the objective edge based on the calculated positioning information. In the article, implementation of the chamfering system and experiments are reported.

SYSTEM CONFIGURATION

The system configuration is illustrated in Fig. 2. Workpiece shapes are defined with 3D-CAD system (Ricoh Co. Ltd. :DESIGNBASE) on EWS (Sun Microsystems Inc.: UltraSPARC-II 296MHz). Tool path for material handling is generated with our own developed CAM system on the EWS and a PC (AT compatible, OS: FreeBSD) on the basis of CAD data followed by conversion to the robot control command. A 6-DOF industrial robot (Matsushita Electric Co. Ltd: AW-8060), 2840mm in height, the positioning accuracy is 0.2mm and the load capacity is 600N, is used. Robot control command generated on the PC is transferred to the robot through a RS-232-C. A 3-finger parallel style air gripper attached to the end of the robot hand holds the workpiece. The robot carries the workpiece in front of a CCD camera to take the image of chamfering part. Positioning and dimensional errors of the workpiece are detected based on an image of objective part taken by the CCD camera on a PC. The tool station can compensate the errors and chamfer the objective edge based on the calculated positioning information using three liner actuators (axis X, Y, Z) and a rotary actuator (axis A) to rotate the file. In the study, the industrial robot handles the workpieces instead of the chamfering tools. The method has following two advantages.
(1) The workpiece can be chamfered while transferring to reduce lead-time.
(2) No additional transferring/handling equipment is required.

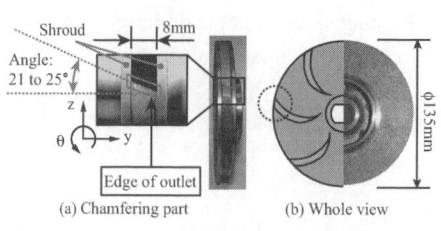

Figure 1: Shapes and dimension of the workpiece   Figure 2: System configuration

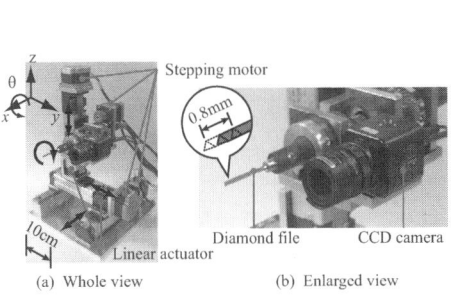

Figure 3: Tool station

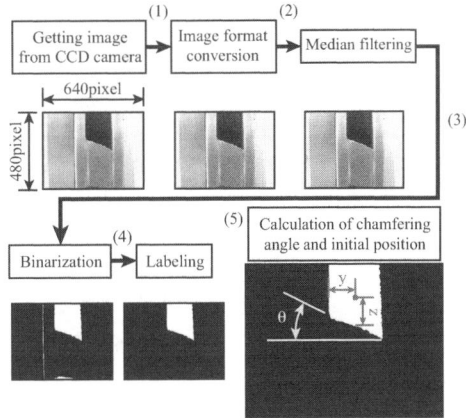

Figure 4: Outline of the image processing

TOOLSTATION

In order to compensate positioning error of the robot and dimensional error of the workpiece, the tool station is developed. The whole view of the tool station is shown in Fig.3. The tool station consists of 4-DOF actuators to compensate the positioning and dimensional errors, a diamond file driven by air reciprocating actuator is attached as a chamfering tool and CCD-camera for image acquisition. The 4-DOF actuators consist of three liner actuators to compensate translational errors about x, y and z axes and one rotary actuator to compensate angular error about θ as illustrated in Fig. 3. Both of them are driven by stepping motor. The maximum strokes of the liner actuators are 50mm. The maximum resolutions of the liner actuators are 0.03mm and that of rotary actuator is 0.1 degree. Although the objective chamfering part is too narrow to chamfer with rotational tools, the tool station adopt a diamond file driven by air reciprocating actuator.

IMAGE PROCESSING

The tool station can compensate the errors and chamfer the objective edge based on the calculated positioning information using three liner actuators (axis X, Y, Z) and a rotary actuator (axis A) to rotate the file. Relative distance and angle between the file and workpiece are calculated by processing the taken image. Outline of the image processing is explained as follows and illustrated in Fig. 4.
(1) The color image (ppm image: 640 x 480 pixel) is taken and converted to gray scale image (pgm image).[5]
(2) Apply median filtering to remove noise.
(3) Binarize the image.
(4) Apply labeling to extract the edge to be chamfered.
(5) Calculate the positioning information (y,z and θ).
Method of image bi-linear is used to enlarge the image and method of least squares is used to calculate the angle θ.

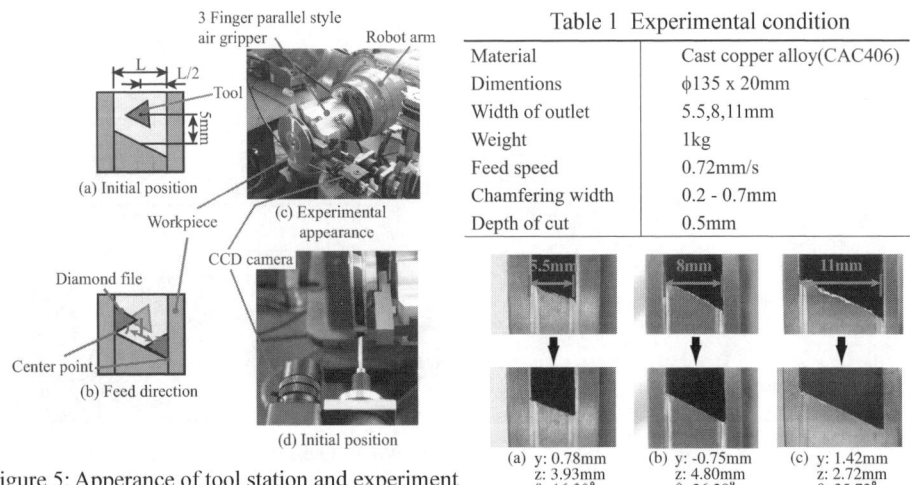

Figure 5: Apperance of tool station and experiment

Table 1 Experimental condition

Material	Cast copper alloy(CAC406)
Dimentions	φ135 x 20mm
Width of outlet	5.5,8,11mm
Weight	1kg
Feed speed	0.72mm/s
Chamfering width	0.2 - 0.7mm
Depth of cut	0.5mm

Figure 6: Experimental result

EXPERIMENT

In order to evaluate the ability of the developed chamfering system with the tool station, the chamfering experiments on the different type of impellers are carried out. The material of the workpiece cast copper alloy (CAC406). The conditions of the experiment are shown in Table 1. Figure 5 (a) illustrates the initial position of the tool on chamfering, Fig. 5 (b) illustrates movement of the tool path on the chamfering part, Fig. 5 (c) shows the appearance of the system under chamfering and Fig. 5 (d) shows the tool at the initial position in front of the impeller. The initial position of the tool is located at mid point of inner side of shrouds for y-direction and having offset from the edge to be chamfered for x-direction to avoid interference between the tool and the shrouds. As shown in Fig. 5 (b), the tool sways from side to side at first and next rotates up to the file face becomes parallel to the shroud in order to completely chamfer at the corner. The appearances after chamfering and measured dimensions are shown in Fig. 6. Upper and lower pictures show workpieces before and after chamfering respectively. Smooth finishing are seen at the chamfered part respectively.

CONCLUSION

The system to automate chamfering to cope with dimensional error by industrial robot is developed. From the experimental result, the system is found to have an ability to chamfer the workpieces without influence of dimensional error automatically.

REFERENCES

[1] Asakawa,N., Mizumoto, Y., Takeuchi,Y., 2002, Automation of Chamfering by an Industrial Robot; Improvement of a System with Reference to Tool Application Direction, Proc. of the 35th CIRP Int. Seminer on Manufacturing Systems :529-534.
[2] Hidetake,T., Naoki, A., Masatoshi, H., 2002, Control of Chamfering Quality by an Industrial Robot, Proc. of ICMA2002 : 399-346.
[3] Takayuki, N., Seiji, A., Masaharu, T., 2002, Automation of Personal Computer Disassembling Process Based on RECS, Proc. of ICMA2002 : 139-146

INTERACTIVE BEHAVIORAL DESIGN BETWEEN AUTONOMOUS BEHAVIORAL CRITERIA LEARNING SYSTEM AND HUMAN

Min An and Toshiharu Taura

Graduate School of Science and Technology, Kobe University,
1-1, Rokkodai, Nada Kobe, 657-8501, Japan

ABSTRACT

Conventional robotic behaviors are directly programmed depending on programmer's personal experience. On the other hand, an artist cannot easily convey their interesting behavioral patterns to the programmers due to difficulty in expressing such behaviors. Therefore, interesting behavioral patterns can hardly be produced at present. It is necessary to develop an effective method of designing robotic behavior. In this study, the authors propose a method of designing robotic behavior though interaction with a computer and establish a design system with the method. For demonstrating the design system, we invited both engineering students and art students to use this design system and value it in our survey. The survey results showed that the design system could not only help a user present the behavioral pattern through an interface with the computer, but could also expand the user's creativity from the interface with the computer.

KEYWORDS

Robotics, genetic algorithm, genetic programming, behavioral design, interactive design

INTERODUCTION

A variety of robots are created all over the world. However, there has been little research focusing on robotic behavioral design. It is necessary to develop an effective method of designing robotic behavior. In this study, the authors aim to establish a method of designing robotic behaviors by operating behavioral criteria, because one of the most effective techniques in design is the operation of multiple information or knowledge. For example, we can combine the action of moving a leg forward with the action of rotating it at the hip into a kicking behavior. Here, the behavioral criteria of a computer program are used to bring the behavior candidates into an optimum behavior. The behavioral criteria measure the behavior candidates in terms of the error produced by the computer program. The closer this error is to a minimum value, the better the behavior is.

One of the characteristics of the method in this study is that the proposed design system operates the behavioral criteria that evaluate behaviors and creating novel behaviors with the operated behavioral criteria. Another characteristic of the method is that it can obtain novel behavioral criteria from novel behaviors.

DEFINITION

When designing behavioral patterns for a robot, the designer focuses on the coordinates of the robot's fingertip, the joint of its elbow, and the joint of its shoulder, and their angles; otherwise, M. An and T. Taura (2003) suggested that the designer may only pay attention to behavioral criteria such as 'smoothly', 'quickly' and so forth, which describe the whole movement from the start point of the movement to the goal point.

Definition of Behavioral Pattern

In this study, we have defined behavioral patterns as trajectories drawn by an effector of the robot . Figure 1 shows the elements of the effector of the robot. The coordinates of a fingertip, a wrist and an elbow are expressed as (x_finger, y_finger & z_finger), (x_wrist, y_wrist, & z_finger) and (x_elbow, y_elbow& z_elbow), and the angles of motion are θfinger, θwrist,θelbow, φfinger, φwrist, φelbow, λfinger, λwrist, and λelbow, respectively.

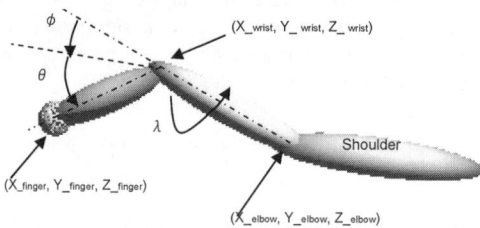

Figure 1: Robotic effector

Definition of Behavioral Criteria

In this study, behavioral criteria are defined as criteria for evaluating whether a robot performs behaviors as what the robot is expected to do. The behavioral criteria are treated as mathematic forms in this study. For example, equation 1 shows a behavioral criterion that is for evaluating whether the robot fingertip reaches a target.

$$E_I = (X - x_T)^2 + (Y - y_T)^2 + (Z - z_T)^2 = 0 \qquad (1)$$

Here, T indicates the numbers of steps needed to reach the target, x_T, y_T and z_T are the coordinates of the fingertip of the robot, and X, Y and Z are the coordinates of the target.

DESIGN SYSTEM

The design system is proposed as shown in Figure 2. In step 1, we let the system acquire several basic behavioral criteria of evaluating a model behavioral pattern. In step 2, the system reproduces behavioral patterns based on the acquired behavioral criteria, and then the reproduced behavioral patterns are shown on a computer screen. In step 3, a designer selects two preferred behavioral patterns from the computer screen. Finally, in step 4, the system combines behavioral criteria of the behaviors selected by the designer into a new behavioral criterion, and then creates a new behavioral pattern based on the newly combined behavioral criterion and shows the behavioral pattern, again.

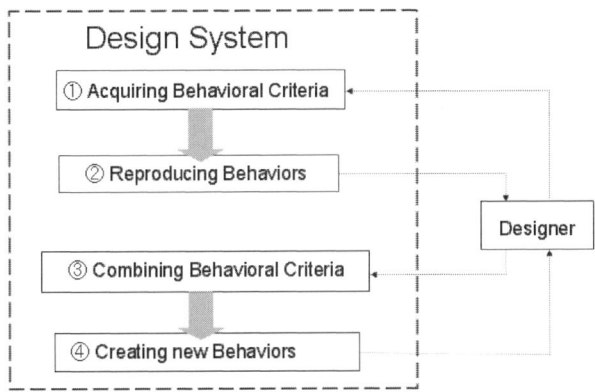

Figure 2: Design system

Individual of GA

In our study, the behavioral patterns are produced by Genetic Algorithms (GAs). The variation of each angle is presented as a GA gene, so that a series of variations from the start point to the target point is replaced by one individual in the GA. The behavioral patterns are evaluated by the behavioral criteria prepared or combined in the design system.

Behavioral criteria acquisition from new behaviors

In addition to the existing behavioral criteria, we aimed to construct forms of novel behavioral criteria from behaviors by Genetic Programming (GP). The set of functions appearing at the internal points of the GP tree includes "+", "-", "*" and "/". The set of terminals appearing at the external points includes "x_t", "y_t", "x_{t+1}", "y_{t+1}", "x_{t+2}", and "y_{t+2}".

EXPERIMENTS

Demonstrating the proposed design system, we invited both design students and engineering students to use and evaluate it through 2 experiments. 10 students participated in the experiments including 5 design students who are family with art but do not have any programming experience and 5 engineering students who are good in engineering but not good in art. The participants filled in a questionnaire to evaluate the design system, after they had used the design system.

System interface

Figure 3 shows the windows presented by the implemented system. The number of individual is set to 200 at each generation. 6 individuals of the 200 individuals are shown on these windows. 2 selected individuals are shown on the top two windows.

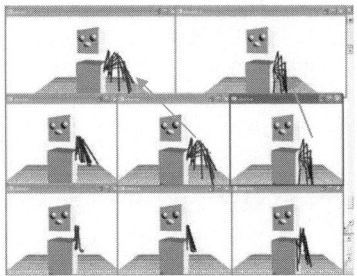

Figure 3: System interface

Experiment

In our experiment, we provide participants our design system to design behaviors of pitching a baseball. The participants evaluated the design system in a questionnaire, in which there are two questions for evaluating the methods: one is whether the behaviors produced by the methods have creativity; the other is whether the software created by the methods can be regarded as a design tool. The questions are ranked from 1 to 5. The answers from Design Students (DS) and Engineering Students (ES) are arranged in table 1.

TABLE 1
DATA FROM EXPERIMENT

	Creativity	Possibility as tools
Answers from DS	3.5	3.7
Answers from ES	4.0	3.4

Results analysis

We compared the data of answers from design students with those of engineering students, and we found that the scores from design students for evaluating creativity is lower than those from engineering students, while the scores for evaluating possibility as tools is higher than those from engineering students. Probably, the reason of the difference is that the design system helped design students who are good at creating novel items but not good at programming techniques to program behaviors; and it helped engineering students expand their creativity.

CONCLUSIONS AND FUTURE TASKS

We have described a prototype of behavioral design system using evolutionary techniques. New robotic behavioral patterns have been created by the design system. As a result of the interaction between the user and the system, it becomes possible to help the users who do not have any experience in programming to produce interesting behavioral patterns with computer.

REFERENCES

An Min, Kagawa Kenichi and Taura Toshiharu, 2003, A study on acquiring model's criterion focusing on learning efficiency, proceedings of the 12th IASTED International Conference on Applied Simulation and Modeling, 2003, pp. 163-168.

HUMAN BEHAVIOR BASED OBSTACLE AVOIDANCE FOR HUMAN-ROBOT COOPERATIVE TRANSPORTATION

Y. Aiyama[1], Y. Ishiwatari[1] and T.Seki[2]

[1] Department of Intelligent Interaction Technologies, University of Tsukuba,
Tsukuba, Ibaraki, 305-8573, Japan
[2] Graduate School of Science and Engineering, University of Tsukuba,
Tsukuba, Ibaraki, 305-8573, Japan

ABSTRACT

In this paper, we propose a new method to compensate for lack of robot abilities of environment recognition and global path planning which are very important abilities to use robots at general environment such as homes or offices. Robots lack these abilities in unstructured environment, but human beings have great abilities of them. We pay attention that human behavior is a result of their recognition and path planning. Robots should use this information if it can easily sense human motion with like as human-robot cooperation transportation task. When a robot transports an object with a human, it senses human motion, recognize obstacles by the human behavior, and plan a local path to follow the human with avoidance the obstacles.

KEYWORDS

Cooperative transportation, Human-robot interaction, Human interactive manipulation, Environment recognition

INTRODUCTION

Recently, many researches aim to use mobile robots in "general environment" such as houses or offices. In these cases, obstacle recognition and global path planning are large problem for robots. However, human beings have very high ability for this recognition. At a glance, human can find obstacles to be avoided. With this recognition, human can find a global path to a goal very easily. It is useful to combine abilities of robots and human; robots do works which require force, and human does obstacle recognition and global path planning. This combination will bring immediately a practical application with current robot technology.

In this research, we pay attention to the information which exists in human behavior and use it for robot to recognize obstacles and to generate its path. For this purpose, we introduce cooperative transportation by human and robots. In this task, human and robots bring one object. So it is easy for robots to sense the human behavior.

In this paper, we introduce two methods for this research. One is for a case that robots do not have any outer sensors and then know only its internal information. In this case, we do not use global information of environment but use local one, which is described by potential of probability. The other is for a case that robots can sense its position and orientation in its environment by some kind of landmark method or so. In this case, robots can use global information of environment.

OBSTACLE RECOGNITION FROM HUMAN BEHAVIOR

When human and robots cooperatively transport one object, the robots can sense the human motion by sensing the object motion. Then robots can sense human behavior, which is result of human's environment recognition and path planning. So, by observation of this human motion, robots can recognize obstacles without any observation of outer environment by themselves. For example, if human who has been moving towards goal position changes its motion direction, robot can recognize that there exist some obstacles in front of the direction. Then robot can generate following path not to collide with the recognized obstacles.

The structure of this system is as shown in Figure 1. Here, there exists a very important assumption. ``When human recognize obstacles around, the human acts avoidance motion in according to a certain behavior model.'' With this assumption, robots can recognize obstacles from the human motion by using inverse model of the human behavior model.

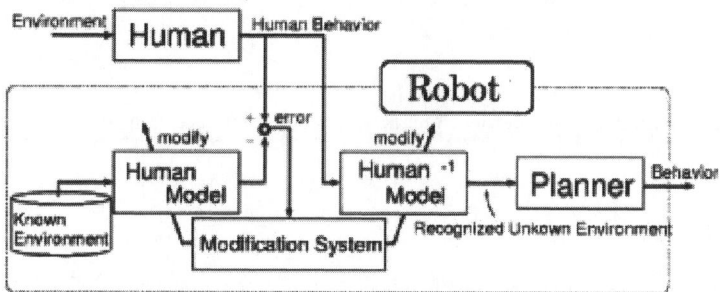

Figure 1: Human-model based obstacle recognition

With this structure, robots can achieve recognition of obstacles by observation of human behavior and can achieve cooperative transportation with human.

PROBLEM SETTINGS

For the cooperative transportation task, we have some assumptions which are common for both two methods; Human and robots support an object at one point respectively. At each point, the object can change its pose, so robots can move any position with keeping relative distance to the human. Human leads the object and robots. When human finds obstacles within the area of radius r_p, human acts to avoid the obstacles with keeping the distance. Robots recognize the object and environment in 2-D space. C-Obstacle is a set of convex polygons. Robots know the shape and their support position of the object.

OBSTACLE RECOGNITION AND PATH PLANNING WITH GLOBAL INFORMATION

In this section, we introduce an obstacle recognition method in the case that robots have their global information in environment by some way like as sensing landmarks. In this case we have additional assumptions as followings; Robots can sense its position and orientation in the environment. Robots have a map of the environment with some known obstacles and goal position. Human tries to move straight towards its goal position.

When human does not move towards the goal, robots recognize that there exists an obstacle on the goal direction. The area around the measured human position with radius r_p must be safe area where no obstacles exist. And the point where the distance from the human is r_p towards the goal is a point where an obstacle exists. Small circle marks in the Figure 2 show recognized obstacles. Robots add these recognized obstacles on their environment map.

With the information in the environment map, robots decide their path to move. There are some conditions for their path; Robots must keep their relative distance to the human. Robots and the object must not collide with both of known obstacles and recognized obstacle points. Robots and the object should have large surface within the safe area.

With these conditions, robots decide their following path. So they make a path which bypaths obstacles as shown in Figure 2.

According to the algorithm, we did experiment. As a robot for the experiment, we use a TITAN-VIII, a four-legged robot. Since this four-legged robot can move omni-direction which is differ with normal wheeled mobile robots, we do not need additional condition to the path planning algorithm.

Figure 3 shows the result of the experiment. The human moves keeping the distance from the obstacle as r_p=500[mm]. However, as shown in the figure, the robot moves to bypath the obstacle to avoid collision between the object and the obstacle.

Figure 2: Recognition of safe area and obstacles and following path plan

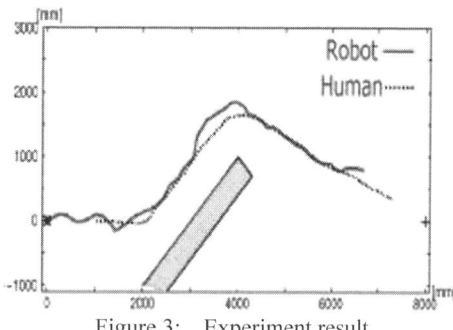

Figure 3: Experiment result

OBSTACLE RECOGNITION AND PATH PLANNING WITH LOCAL INFORMATION

We introduce another method in the case that robots do not have any information about the environment and then cannot use global information. In this case, human does not need to move towards its goal, but need to go straight where no obstacles exist. Robots cannot sense its position, orientation nor any information of its surroundings. Robots do not have any map of environment.

Different from the previous method, robots cannot recognize obstacles from the fact that human does not move towards the goal. In this case, robots cannot recognize obstacles correctly. Then we make a strategy for this method. During human moves straight, it must be the safest way to follow human's behind. When human turns, there is high possibility that there exists obstacle at the corner. So robot should bypath the corner. With this strategy, we adopt "local potential map" which describes possibility of obstacle existence locally around robots.

Local potential map is generated as shown in Figure 4. Robots modify the map by adding this potential according to its motion. Robots decide their motion to lower the sum of the potential. Figure 5 and Figure 6 show the result of an experiment. The potential value at the corner is higher and the robot moves to bypath the corner. Finally the robot has large error in motion direction, but it correctly generates following path since it depends only on local information.

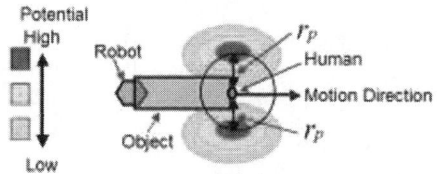

Figure 4:   Obstacle potential

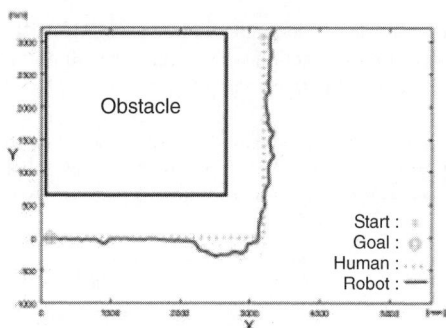

Figure 5:   Experiment result

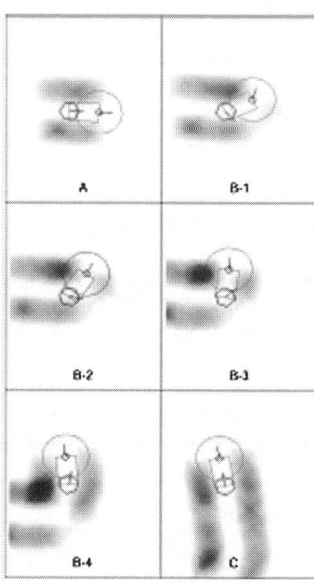

Figure 6:   Local potential map

CONCLUSION

We propose two methods that robots recognize obstacles by observation of human motion when they cooperatively transport an object. Each experiment uses just one robot, but the idea is expandable to multiple robot transportation. Further, there must exist other applications that use the human ability of sensing and global path planning. Power assist system may be another type of the application.

REFERENCES

Hirata Y. et al. (2002). Motion control of multiple DR helpers transporting a single object in cooperation with a human based on map information. Proc. IEEE International Conf. on Robotics and Automation, 995-1000.

Takubo T. et al. (2001). Human-robot cooperative handling using virtual nonholonomic constraint in 3-D space.   Proc. IEEE International Conf. on Robotics and Automation, 2680-2685.

EVALUATION METHODS FOR DRIVING PERFORMANCE USING A DRIVING SIMULATOR UNDER THE CONDITION OF DRUNK DRIVING OR TALKING DRIVING WITH A CELL PHONE

Y.Azuma[1], T.Kawano[1] and T.Moriwaki[2]

[1] Department of Industrial and Systems Engineering, Setsunan University,
Neyagawa, Osaka 572-8508, JAPAN
[2] Department of Mechanical Engineering, Kobe University,
Kobe, Hyogo 657-8501, JAPAN

ABSTRACT

The purpose of this study is to fabricate a driving simulator and establish the methods to evaluate the driving performance using the simulator under the condition of drunk driving or talking driving with a cell phone. Two indices are proposed to evaluate the driving performance. One is the degree of unsteadiness of the driving path and the other is the reaction time in pressing the brake pedal with a foot. The degree of unsteadiness is defined as composition of the degree of weaving from side to side and the degree of fluctuating of the distance between cars. Using the driving simulator experiments were carried out for six subjects. As a result it is demonstrated that the drunk driving or the talking driving with a cell phone are evaluated appropriately.

KEYWORDS

Driving Simulator, Driving Performance, Safe Driving, Drunk Driving, Talking Driving, Cell Phone, Human behavior

1. INTRODUCTION

Numerous driving simulators have been already developed for many applications[1] [2] [3]. Using a driving simulator Contardi et al.[4] analyzed mean and standard deviation of lane position according to the circadian variation of alertness. Reed and Green[5] recorded driving speed and steering-wheel angle while periodically dialing simulated phone calls. Gawron and Ranney[6] examined the driving performances including lateral acceleration on the approach and negotiation of horizontal curves of varying length and curvature when sober or alcohol-dosed. In those studies various evaluation methods were adopted for driving performances. However, those methods varied depending on the researchers. Particularly, adequate and uniformalized evaluation methods of drunk driving or talking driving with a cell phone have not quite established.

The purpose of this study is to fabricate a driving simulator and establish the methods to evaluate the driving performance with the simulator under the condition of drunk driving or talking driving with a cell phone.

2. DRIVING SIMULATOR

The driving simulator is rebuilt as an automatic shift car from the components of a car taken apart. The simulated driving is assumed to be conducted on a one-way highway in the suburb, therefore no traffic signals and no intersections appear. The road patterns of straight and curve are designed to appear at random. The other car, which runs with speed increased and decreased in the range of 40 to 60km/h in front of the car simulator, is displayed on the same lane. If the distance between the car simulator and the preceding car in front of it becomes more than 70m, another following car is designed to cut in 10m ahead.

3. EVALUATION METHODS FOR DRIVING PERFORMANCE

In this study two indices are proposed to evaluate the driving performance under the conditions of drunk driving or talking driving with a cell phone. One is the degree of unsteadiness of the driving path and the other is the reaction time in pressing the brake pedal with a foot.

3.1 *Unsteadiness of Driving Path*

The degree of unsteadiness of the driving path is defined newly in this study as composition of the degree of weaving from side to side per unit time (Δw_i) and the degree of fluctuating of the distance between cars per unit time (Δf_i). The unit time is defined as 0.2s. The unit of Δw_i and Δf_i is meter.
In this study the degree of unsteadiness U of the driving path is defined as the composition of Δw_i and Δf_i as follows:

$$U = \frac{\sum_{i=1}^{n} \log_{10}(\Delta u_i)}{n} = \frac{\sum_{i=1}^{n} \log_{10}\{\Delta f_i^2 + (a \cdot \Delta w_i)^2\}}{n} \tag{1}$$

where n=300 for one minute drive. Δu_i is unsteadiness of the driving path per unit time. The weight ($a = 6$) was obtained as the ratio of $|\Delta f|$ to $|\Delta w|$.

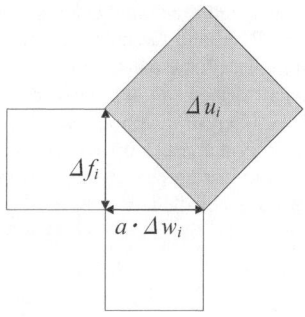

Figure 1: Geometric illustration of unsteadiness of the driving path per unit time

As shown in Figure 1 it corresponds to square measure of the hypotenuse in right triangle which includes Δf_i and $a \cdot \Delta w_i$ of two sides. The logarithm in the equation is applied since the square measure Δu_i varies more widely as the value becomes larger. From the results of various driving simulations it is found that the driving performance is classified under the following five qualitative assessments. That is stable vs. $U < 0.1$, somewhat unstable vs. $0.1 \leq U < 0.3$, unstable vs. $0.3 \leq U < 0.5$, rather unstable vs. $0.5 \leq U < 0.7$, and much unstable vs. $0.7 \leq U$.

3.2 Reaction time

Driver perception reaction time is one of the essential factors for the drunk driving or talking driving with a cell phone. The time lag of pressing the brake pedal with a foot is measured. Drivers do not perform the driving task but only press the brake pedal during watching a colored circle(ϕ 300) displayed on the screen. Subjects are asked to press the brake pedal with a right foot immediately when the color of the circle is changed.

4. FEASIBILITY TEST

Using the driving simulator experiments were carried out to demonstrate that the evaluations of drunk driving or talking driving with a cell phone were appropriate. Six male subjects participated in this study. They were all right handed and were aged between 20 and 40 years. Firstly, the degree of unsteadiness U was assessed. The talking tasks through the cell phone were arithmetic questions. The subjects were asked to reply the number added 1 to each figure of a certain number; e.g. 8 for 7, 73 for 62, and 397 for 286. The number of the figures corresponds to the talking task level 1, 2, and 3. On the other hand, under the condition of DUI(Driving Under the Influence of alcohol), two drunken levels i.e. above 0.15mg/l and above 0.25mg/l were adopted.

Figure 2 shows the degree of unsteadiness U under the condition of drunk driving or talking driving with a cell phone. Each bar was averaged by 3 times by a subject and then was grand averaged by six subjects. The degree of unsteadiness increased as the drunken level and the talking task level came up. In addition, the degree of unsteadiness under the drunk driving was similar to that under the talking driving over the level 2. The correlation between the degree of unsteadiness U and the subjective scores asked after every talking driving was 0.93(p<0.05).

Secondly, the reaction time of pressing brake pedal under drunk driving or talking driving was assessed. Figure 3 shows the results. The reaction time of pressing brake pedal increased as

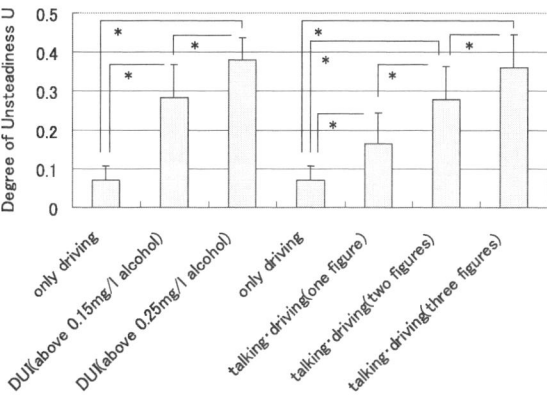

Figure 2: Degree of Unsteadiness U (drunk driving and talking driving) (*:p<0.05)

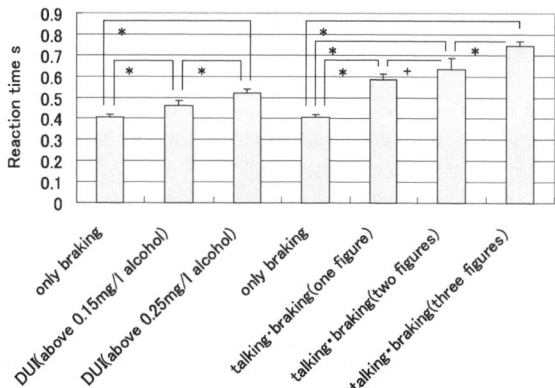

Figure 3: Reaction time of braking (drunk braking and talking braking) (*:p<0.05,+:p<0.1)

the drunken level and the talking task level came up. The reaction time under the talking driving was larger than that under the drunk driving.

5. CONCLUSION

A driving simulator and evaluation methods of the driving performance were established. Feasibility tests of the simulator and the evaluation methods were carried out under the conditions of drunk driving and talking driving with a cell phone. The results are summarized as follows:

(1) The driving simulator rebuilt from a real car is assumed to run on a one-way highway in the suburb. The road scene and the preceding car are displayed with computer graphics.
(2) The degree of unsteadiness of the driving path is defined newly in this study as composition of the degree of weaving from side to side and the degree of fluctuating of the distance between cars.
(3) The reaction time is defined as the time from when the color of the circle displayed on the screen is changed to when the brake pedal is pressed.
(4) The degree of unsteadiness of the driving path and the reaction time of pressing brake pedal both increased as the talking task level through a cell phone came up. The results were in close agreement with the subjective evaluations.
(5) The degree of unsteadiness and the reaction time similarly increased as the drunken level came up.
(6) The driving simulator and the evaluation methods developed in this study can be utilized to evaluate the drunk driving or the talking driving appropriately.

6. REFERENCES

[1] Kading W. and Hoffmeyer F. (1995). The Advanced Daimler- Benz Driving Simulator. *SAE Technical Paper Series* **950175**, 91-98.
[2] Papelis Y., Brown T., Watson G., Holtz D. and Pan W. (2004). Study of ESC Assisted Driver Performance Using a Driving Simulator. **N04-003-PR** *The University of IOWA*,1-35.
[3] Shiiba T. and Suda Y. (2002). Development of Driving Simulator with Full Model of Multibody Dynamics. *JSAE Review* **23**, 223-230.
[4] Contardi S., Pizza F., Sancisi E., Mondini S. and Cirignotta F. (2004). Reliability of a Driving Simulation Task for Evaluation of Sleepiness. *Brain Research Bulletin* **63**, 427-431.
[5] Reed P. and Green A. (1999). Comparison of Driving Performance On-Road and in a Low-Cost Simulator Using a Concurrent Telephone Dialing Task. *Ergonomics* **42**, 1015-1037.
[6] Gawron J. and Ranney A. (1990). The Effects of Spot Treatments on Performance in a Driving Simulator under Sober and Alcohol-Dosed Conditions. *Accid. Anal. & Prev.* **22:3**, 263-279.

COMPUTATIONAL MODEL AND ALGORITHM OF HUMAN PLANNING

H. Fujimoto, B. I. Vladimirov, and H. Mochiyama

Robotics and Automation Laboratory, Nagoya Institute of Technology
Gokiso-cho, Showa-ku, Nagoya 466-8555, Japan

ABSTRACT

In this paper, we investigate an application of a working memory model to learning robot behaviors. We implement an extension that allows learning from model-based experience to reduce the costs associated with learning the desired robot behaviors and to provide a base for exploring neural network based human-like planning with grounded representations. A simulation of applying the approach to a random walk task was performed and a basic plan was obtained in the working memory.

KEYWORDS

Human mimetics, Human behavior, Mobile robot, Planning

INTRODUCTION

Using neural networks, it is relatively easy to learn separately simple mobile robot behaviors like approaching, wall following, etc., and with appropriate network architectures, combinations of such behaviors can be learned too. However, since these combinations are encoded into the network weights, switching from one combination to another often requires retraining. An interesting approach addressing the problem of switching among different mappings is presented in a working memory model proposed recently in O'Reilly & Frank (2004). It comes from the field of computational neuroscience and is a computational model of the working memory based on the prefrontal cortex (PFC) and basal ganglia. An important aspect of applying this model to learn a combination of behaviors is that the information for that combination is maintained explicitly as activation patterns in the PFC. Compared to a weights based encoding, these activation patterns can be updated faster and thus switching among possible combinations becomes easier.

In this paper, an implementation of that working memory model is applied to a five-state random walk task. Furthermore, an environment model is added to provide model-based learning, motivated by the fact that reinforcement learning based only on real experience is associated with high costs (in terms of time, energy, etc.) when applied to real robots. Using additional model-generated experience helps to decrease the associated costs and also provides a link to planning, since, as argued in Sutton & Barto (1998), planning can also be interpreted as learning from simulated experience. In light of this interpretation, the information (about the learned specific combination of behaviors) maintained in the working memory can be viewed as a simple plan to achieve the rewarded goal state.

RELATED WORKS

While simple mobile robot behaviors can be learned with feed-forward neural networks, combinations of behaviors, where sometimes identical sensory inputs should trigger different actions, require additional coordinating mechanisms. For example, in Calabretta, Nolfi, Parisi, & Wagner (1998) a Khepera robot is trained to perform a garbage collecting task and the authors find a correspondence between specific behaviors and the evolved neural network modules. The interaction among these modules is controlled by selector neurons that give precedence of a given module over the others.

In contrast to the above work, where the modules are physically separate entities, Ziemke (2000) interprets the trained Recurrent Neural Network (RNN) as a *diachronically* structured controller. In this case, instead of modules existing separately at the same time, a monolithic neural network instantiates different input-output mappings at various time points. An important aspect of the mechanism by which RNN achieve modularity is discussed in Cohen, Dunbar, & McClellandl (1990), where the switching between two input-output mappings is achieved by attentional control (attention is viewed as "an additional source of input that provides contextual support for the processing of signals within a selected pathway" (p. 335)). In RNN, the source that provides contextual support favoring one of the competing input-output mappings is the context layer. The state maintained in the context layer disambiguates the inputs and thus different outputs can be obtained for similar inputs.

Since, in RNN, the internal state plays a central role in switching between the alternative input-output mappings, the flexibility of updating and maintaining this internal state affects directly the flexibility of the resulting robot behaviors implemented by the network. The potential of the computational model of working memory based on the PFC and basal ganglia (PBWM model), proposed in O'Reilly & Frank (2004), to provide such flexibility motivated us to investigate its application to learning combinations of robot behaviors.

APPROACH

In the presented approach, the PBWM model is used to implement several possible input output mappings and then to learn specific combinations. Also, a model of the environment is added to provide model-generated experience. We are interested in two consequences of using an environment model: lowering the costs associated with actually performing the actions and extending the neural network model to a planning system supporting grounded representations.

Working Memory Model

Here we present an outline of the PBWM model (refer to O'Reilly & Frank (2004) for details). The model implementation is based on the Leabra framework (O'Reilly & Munakata, 2000), uses point neuron activation function for modelling the neurons, k-Winners-Take-All inhibition to model competition among the neurons in a layer, and a combination of Hebbian and error-driven learning.

The neural network structure (Figure 1c) consists of two groups of layers. The first group includes the *Input*, *Hidden*, *Output*, *NextInput*, and *PFC* layers. The *NextInput* layer is used for the environment model and will be explained later. The *Input*, *Hidden*, and *Output* layers form a standard three-layer neural network structure. The *PFC* layer is an improved context layer, which is bi-directionally connected with the *Hidden* layer, and influences the input-output pathways. The *PFC* layer is divided into stripes to allow independent control over the updating and maintenance of parts of the activation state. The rest of the layers form the second group, which implements a gating mechanism for control over the updating and maintenance of the *PFC* activation state. Generally, a positive reward leads to stabilizing of the current *PFC* activation state, while a negative reward results in updating (a part of it) and establishing of another state.

Model of the Environment

Under the reinforcement learning framework (Figure 1a), an *Agent* performs an action *a* based on the current sensory input and the policy formed so far. The *Environment* (or the *Environment Model*) responds with a new sensory input *s* and an external reward *r*. The *Agent* adjusts its policy based on the reward and completes the cycle by performing a new action.

The two parts of the environment model are implemented as follows. The model of the next input is implemented as an additional output layer, trained to predict the next input based on information from the current network state. The model of the external reward at this stage is implemented outside of the network as a simple lookup table keeping the last reward received for each input-output pair.

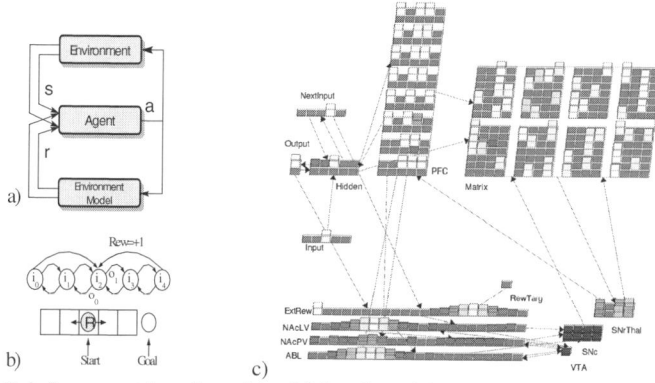

Figure 1. a) Reinforcement learning with additional model-generated experience. b) Random walk task settings. c) Neural network structure.

SIMULATION

A five-state random walk task was used to test the approach. In this task, there are five squares in a row, and an agent that moves one square left or right. The start position is the middle square and a move outside from the leftmost and rightmost squares sends the agent back to the start position. Two goals were used: moving right from the rightmost square and moving left from the leftmost square. Figure 1b shows the settings and a finite state automaton describing the states and the transitions (inputs *i* and outputs *o* in the network). The reward value corresponds to goal set to the right side.

For this simulation, we used the PDP++ neural network simulator (PDP++ software package, ver. 3.2a, http://www.cnbc.cmu.edu/PDP++/PDP++.html). The network input (see Figure 1c) is the current position of the agent. The network outputs are the current action in the *Output* layer and the prediction of the next input in the *NextInput* layer. The Hidden layer has one neuron for each state-action combination. The top row encodes *move-right* and the bottom row encodes *move-left*. A restriction is imposed through the k-Winners-Take-All function to allow only one active neuron. The weights between the *Input*, *Hidden*, *Output*, and *NextInput* layers are hand-coded (in a separate experiment we have confirmed that these weights can be learned too) so that from each state the two possible actions are equally probable. The *PFC* has 8 stripes, each one with the same size as the *Hidden* layer. The *Hidden* layer has one-to-one connections with each stripe in the *PFC* layer.

The training process, inspired by the Dyna algorithm (Sutton & Barto, 1998), is an interleaving execution of two loops. One for the real experience, receiving the next input and the external reward from the environment and the other, for the model-generated experience, obtaining the input from the *NextInput* layer and the external reward from the lookup table.

Two groups of simulations were performed: with and without model-generated experience. In each group, there were two simulations: with the goal on the right and on the left. After training for 300 sequences of real experience, a test consisting of 10 trials, 50 sequences each, was performed. The test results are summarized in Figure 2.

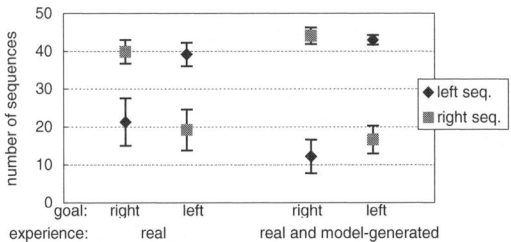

Figure 2. Plot of the average number and standard deviation of left and right sequences over the 10 test trials. The horizontal axis shows the settings for the four simulations.

DISCUSSION

From the simulation results in Figure 2, it can be seen that the neural network learned to achieve the goal state. Also, the neural network trained with additional model-generated experience performs better than the one trained only with real experience. These results were obtained using only the reward as a teaching signal (using supervised learning as in the original PBWM model leads to better results but is not suitable for experiments with planning). Another result is evident from the obtained activation patterns in the *PFC* layer. The neural network shown in Figure 1c, has been trained to achieve the goal state on the right side. As can be seen, mostly active are the units in the top row of the *PFC* stripes. They correspond to the units for *move-right* in the *Hidden* layer and consequently, bias the neural network output to prefer this action in each state. Thus, the contents of the *PFC* layer can be interpreted as a simple plan (a combination of actions) leading to the goal state. The future work is directed toward using distributed representations in the network and more complex tasks.

REFERENCES

Calabretta R., Nolfi S., Parisi D., and Wagner G. (1998). Emergence of functional modularity in robots. In From Animals to Animats 5, Edited by Blumberg B., Meyer J.A., Pfeifer R., and Wilson S.W., MIT Press, Cambridge. pp 497-504.

Cohen J.D., Dunbar K., and McClelland J.L. (1990). On the control of automatic processes: A parallel-distributed processing account of the stroop effect. *Psychological Review*, **97:3**, 332-361.

O'Reilly R.C. and Frank M.J. (2004). Making working memory work: A computational model of learning in the prefrontal cortex and basal ganglia. Technical Report 03-03 (Revised-Version Aug. 2, 2004). University of Colorado Institute of Cognitive Science.

O'Reilly R.C. and Munakata Yuko. (2000). *Computational explorations in cognitive neuroscience: Understanding the mind by simulating the brain.* MIT Press, Cambridge.

Sutton R.S. and Barto A.G. (1998). *Reinforcement Learning: An Introduction.* MIT Press, Cambridge.

Ziemke Tom. (2000). On 'parts' and 'wholes' of adaptive behavior: Functional modularity and diachronic structure in recurrent neural robot controllers. In From Animals to Animats 6 – Proceedings of the Sixth International Conference on the Simulation of Adaptive Behavior. MIT Press, Cambridge.

SAFETY DESIGN FOR SMALL BIPED-WALKING HOME-ENTERTAINMENT ROBOT SDR-4XII

Masatsugu Iribe, Tomohisa Moridaira, Tetsuharu Fukushima, Yoshihiro Kuroki

Motion Dynamics Reseach Lab., Information Technologies Labs, Sony Corporation,
6-7-35 Kita-Shinagawa, Shinagawa-ku, Tokyo, 141-0001, Japan

ABSTRACT

In March 2003, we proposed a small biped-walking home-entertainment robot SDR-4XII (Sony Dream Robot -4XII, a prototype), which was developed for home use and entertainment. As SDR-4XII is developed for home use, it is designed to be safe. And the important roles of the safety features depend on the functions of newly developed robotic actuator "ISA-4" (Intelligent Servo Actuator 4th). The actuator contributes SDR-4XII's motion control and safety management very much. In this paper we introduce the safety functions which are achieved in the development of SDR-4XII and ISA in detail.

KEYWORDS

home-entertainment robot, biped-walking, safety design, motion control, SDR-4XII, Intelligent Servo Actuator (ISA).

INTRODUCTION

In March 2003, we proposed a small biped-walking home-entertainment robot SDR-4XII (Sony Dream Robot -4XII, a prototype), which was developed for home use and entertainment (Kuroki 2003-2). For that purpose, SDR-4XII is designed to be safe, i.e. not to hurt human users, itself, or any other beings in the environment. The robot has 38 DOF (degrees of freedom) in its joints, and the 22 joints important for its motion control are driven by newly developed robotic actuator ISA-4 (Intelligent Servo Actuator 4th). The robotic actuator ISA-4 consists of three major parts; a motor which contains originally designed electromagnetic circuits, high precision gears, and electrical circuits (Fukushima 2004). By this construction, we achieved to develop a compact and high performance robotic actuator. And in addition, ISA-4 detects unusual situations such as overloads by signals from the sensors both in its own body and in the SDR-4XII's parts. We achieved important safety features for human user's safety and the robot's body protection with the ISA-4.

In this paper we introduce the safety features which are achieved in the development of SDR-4XII and ISA-4, and also describe the methods of the functions in detail.

DESIGN AND CONFIGURATION OF SDR-4XII

Figure 1 shows the appearance and basic specifications of SDR-4XII. SDR-4XII is 58 [cm] tall and weights 7.0 [kg], and has 28 DOF in its major joints. Each leg has 6 DOF, the trunk has 2 DOF, each arm has 5 DOF and the neck has 4 DOF. In addition, 5 fingers are attached to each hand.

SDR-4XII adopts the OPEN-R system as its control architecture which has scalability not only in software but also in hardware, and adopts the proprietary operating system Aperios which is suitable for real-time control (Fujita 1997).

CPU	64 bit RISC Processor (×3)
Memory	64MB DRAM (×3)
Operating Sys. & Architecture	Aperios & OPEN-R
Robot Control Supplying Media	Memory Stick
Input/output	PC Card Slot (Type II)/MS Slot
Image Input (color/stereo)	110,000 pixels CCD
Sound Input/output	7 Microphones/Speaker
Walking Speed	6m/min. max (unleveled surface)
Weight (including battery)	Approximately 7Kg
Dimensions (height/width/depth)	Approximately 580/270/190mm

Figure 1: Appearance and basic specifications of SDR-4XII

DEVELOPMENT OF THE ROBOTIC ACTUATOR ISA-4

For the development of SDR-4XII, we developed 3 types of robotic actuators ISA-4 (Intelligent Servo Actuator 4th prototype) which consist of a motor with electrical circuits and a precise gear unit. Figure 2 shows their appearances. ISA-4MH is used for the body parts which need high-power output such as knee joints, ISA-4S is used for the parts which need high angular velocity but not high power output such as arm joints, and ISA-4M is for general use. The actuators are applied to the 22 driving joints except joints in the head, wrists, and fingers, and contribute to the robot's motion control and safety operation. In this chapter we describe the control mechanism and functions of ISA-4 (Iribe 2004-1).

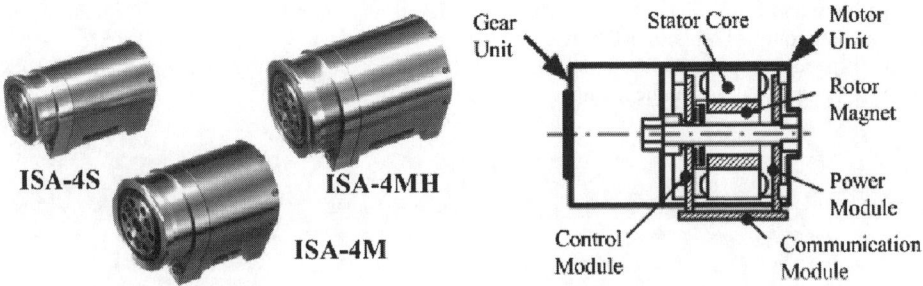

Figure 2: Robotic Actuator "ISA-4"   Figure 3: Construction of ISA-4

Configuration of ISA-4

ISA-4 is largely composed of a Motor unit and a Gear unit as shown in Figure 3. The Gear unit has precise gears, and the Motor unit consists of Rotor Magnet, Stator Core, and three circuit modules. Each circuit module has its own function as Control module, Communication module, or Power module.

The Control module manages all ISA-4 functions such as servo control or safety management, the Communication module manages the communication between the Control module and the CPU (Central Processing Unit) in the upper layer which controls the robot's whole body motion, and the Power module works as motor drive circuit and is very compact and efficient.

Functions of ISA-4

As described above, we achieved to give various functions to ISA-4 by means of its built-in control system. Figure 4 shows the block diagram of ISA's control system, and we describe the functions thereinafter.

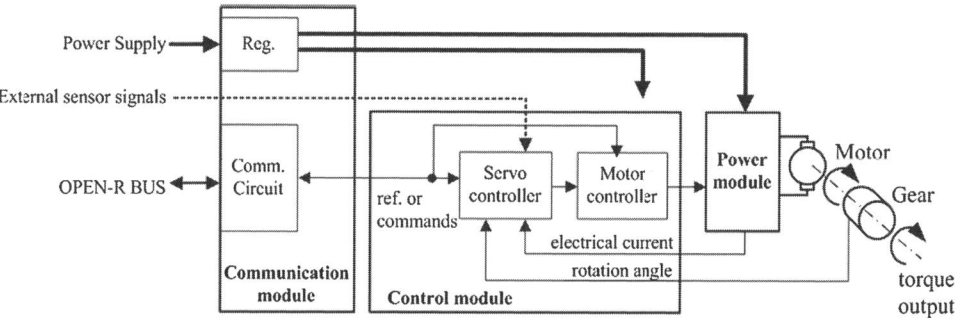

Figure 4: Block diagram of Control system in ISA-4

The main function of ISA-4 is the ***Angular position servo control***. ISA-4 achieves over 10 Hz frequency responses by the Servo controller shown in Figure 4. Its mechanism is that; Communication module receives the reference signal from the motion control CPU in the upper layer via OPEN-R BUS, and then the signal is sent to the servo controller in the Control module. By this signal, ISA-4 controls its servo control gain and angular position of Gear unit's output axis.

ISA-4's characteristic functions are the ***Electrical current monitor*** and the ***Sensor processing***. ISA-4 possesses a sensing circuit which monitors the electrical current consumption of the Power module in order to detect the overloads of ISA-4. As the electrical current consumption of the Power module is in proportion to the added loads to ISA-4, the Control module is able to sense the overload of the ISA-4 by monitoring the electrical current. In addition, there are some built-in sensors such as the angular position sensor and some extra input ports for external sensors such as pinching detection sensors set in the body parts which tend to clamp foreign objects. The Control module senses and processes those sensor signals to send the motion control CPU in the upper layer useful information for its motion control.

Other noteworthy functions are the ***Viscosity friction control for joint characteristics*** and the ***Electrical Shock Absorbing System***. It is a very important problem for robots which do contact with the environment to absorb impact shocks of the contact. Especially, it has been a serious problem for walking robots to absorb the impact shock when landing their feet on the floor surface as their legs form a closed link system with the floor. For this reason, some research works which adopted new foot mechanisms applied cushioning material or springs to settle the problem (Yamaguchi 1996) (Takenaka 2001) (Collins 2001). On the other hand, ISA-4 achieved to settle the problem via software control and the design of an electrical circuit.

The Motor controller, which is shown in Figure 4, is able to control the coefficient of viscosity of ISA-4's output axis dynamically by controlling the induced electromotive force of the Motor unit. Figure 5 shows the property of the controlled coefficient of viscosity. And as shown in the figure, the values of the coefficient of viscosity are monotone increasing for the normalized input which is sent to the Motor controller. By applying this function, ISA-4 achieved to control SDR-4XII's joints' stiffness in a soft or solid way.

Furthermore, some kind of shock absorbing mechanism against the loads by impact shock is needed for the robot safe design. For this purpose, it is popular to apply torque limit mechanisms between mechanical links and actuators. However those kinds of mechanisms sometimes cause a deterioration of the robot's control performance such as a deviation increase of angular position. Therefore SDR-4XII adopted a new shock absorbing system in ISA-4 instead of a mechanical method to settle the problem.

As the magnitude of the impact shock which causes the plastic deformation is able to be detected by monitoring the kinetic energy, the Control module monitors the 2nd order differential term of the kinetic energy to detect quickly by calculating the approximate expression Eqn.1. And if the detected value is large enough to break the robot's body, the Control module immediately cuts off ISA-4's output torque and makes the stiffness of the output axis soft by the viscosity friction control function described above.

Figure 6 shows the advantageous effect of this shock absorbing system. We added the impact shock which is equal to 24-joules kinetic energy to the mechanical link attached to ISA-4M for one hundred times by dropping an 8kg weight from 30cm height, and compared the values of Gear unit's backlashes. The changes of backlashes with the shock absorbing system are about 15% at the maximum, though the changes without the system are about 75% at the maximum as shown in Figure 6.

$$f(I,\theta,t) = K \frac{d}{dt}I(t) \frac{d}{dt}\theta(t) \tag{1}$$

K: Torque constant [N-m/A], $I(t)$: Electrical current [A], $\theta(t)$: Angular position of output axis [rad]

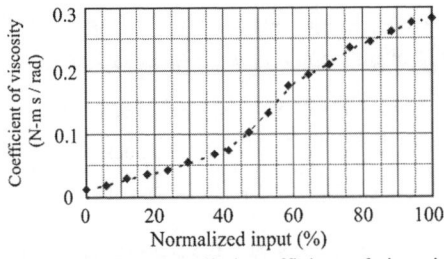
Figure 5: The controlled coefficient of viscosity

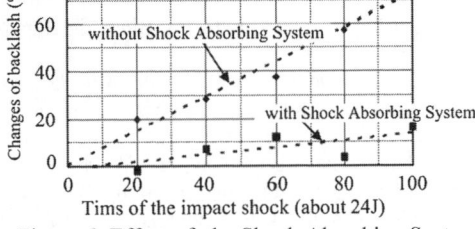
Figure 6: Effect of the Shock Absorbing System

SAFETY FUNCTIONS ON SDR-4XII

As SDR-4XII is supposed to be used in a home environment, the robot has several capabilities for self-protection and user-protection (Iribe 2004-2). For the safety operation of home robots, they must be able to fall over softly by controlling their posture to protect themselves and the environment from the damage of their falling over, and of course, home robots must stand up by themselves (Kuroki 2003-1) (Fujiwara 2003). Therefore SDR-4XII makes its joints soft and loose to soften the damage when it falls over. And when the robot stands up, it makes its joint solid and stable to get high accuracy position control. The behavior of these features, the *Falling over motion control* and the *Standing up motion Control* are shown in Figure 7 and 8, and were achieved world first on the previous prototype SDR-4X.

And SDR-4XII has capabilities of detecting added loads and its internal temperature rise to protect the robot itself and users from its overloads and overheats. When constant loads or impact loads are added to the robot, it measures the strength of the loads by ISA-4 which is used in each joint. If the loads are large enough to break the body of SDR-4XII, the robot cuts off the power of the joints partly or in whole. And if the robot's inner temperature becomes high enough to break itself or to burn the users who touch it, it cuts off all the power of its body. These features, the ***Overload protection*** and the ***Overheat protection***, also contribute to SDR-4XII's safety operation.

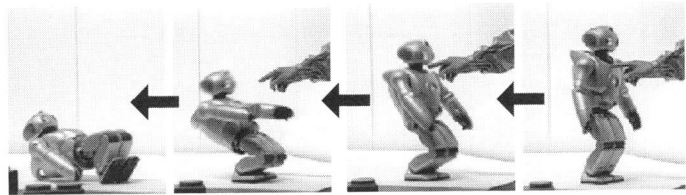

Figure 7: Falling over Motion Control

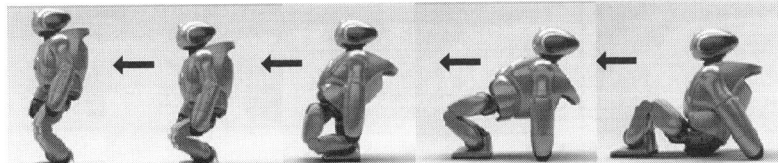

Figure 8: Standing up Motion Control

In addition to the above, SDR-4XII possesses the ***External pinching detection sensors*** in its joints of axillas, elbows, waists, knees, and ankles as shown in Figure 9. Signals of those sensors are sent to the motion control CPU via ISA-4 which is put into the nearest part of the body, and then the CPU gets the condition of the pinching detection parts. If the robot clamps something in the above-mentioned joints, it is able to detect the condition and release them. By this means the robot protects users not to hurt their hands or fingers from pinching. The functions of ISA-4 contribute those five above-mentioned safety features.

And at last, as shown in Figure 10 and Figure 11, the ***Cover design*** obviously contributes to the safety operation of SDR-4XII. Users can touch the cover parts of the robot, and as the cover parts are designed rounded, they do not hurt or scratch user's body. In addition, Figure 10 shows the covering mechanism of its waist to guard the user's body, and Figure 11 shows the rounded shapes of the elbow part which tend to release the pinched object.

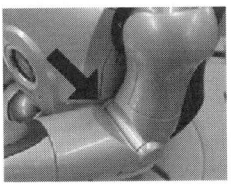
in the elbow

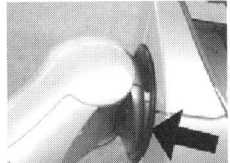
in the axilla

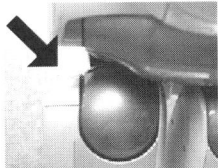
in the waist

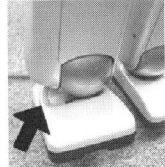
in the ankle

Figure 9: Extra Pinching detection sensors

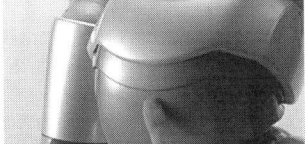

A pinched finger is easy to be released by the shapes of the elbow

Figure 10: Backside covering mechanism   Figure 11: Shape of the elbow joint

CONCLUTION

In this paper, we described the safety design for the small biped-walking home-entertainment robot SDR-4XII and the outline of the robotic actuator ISA-4 which contributes to the safety management of the robot. The cross relationships between functions and features are shown in Figure 12.

As SDR-4XII is designed to be used in a home environment, we had to encounter several problems for its safe operation. Therefore we developed new ingenious functions described in this paper and settled the problems.

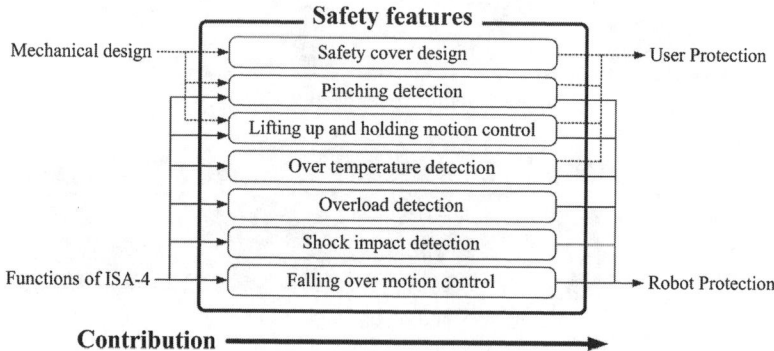

Figure 12: Cross relationship between functions and features

REFERENCES

Collins, H.S., Wisse, M., Ruina, A. (2001), "A Three-Dimensional Passive-Dynamic Walking Robot with Two Legs and Knees", *Int. Journal of Robotics Research*, **Vol.20, No.7**, pp.607-615.

Fujita, M., Kageyama, K. (1997), "An Open Architecture for Robot Entertainment", *Proc. Int. Conference on Autonomous Agents 1997*, pp.435-450.

Fujiwara, K., Kanehiro, F., Kajita, S., Yokoi, K., et al. (2003), "The First Human-size Humanoid that can Fall Over Safely and Stand-up Again", *Proc. IEEE/RSJ Int. Conference of Intelligent Robotics and Systems 2003*, pp.1920-1926.

Fukushima, T., Kuroki, Y., Ishida, T. (2004), "Development of a New Actuator for a Small Biped Walking Entertainment Robot – Using the optimization technology of Electromagnetic Field Analysis", *Proc. ISR 2004*.

Iribe, M., Fukushima, T., Yamaguchi, J., Kuroki, Y. (2004), "Development of a New Actuator for a Small Biped Entertainment Robot Which has Suitable Functions for Humanoid Robots", *Proc. The 30th Annual Conference of the IEEE Industrial Electronics Society 2004*.

Iribe, M., Moridaira, T., Fukushima, T., Kuroki, Y. (2004), "Safety design for small biped walking home entertainment robot SDR-4XII", *Proc. The 5th Int. Conference on Machine Automation 2004*, pp.303-308.

Kuroki, Y., Fujita, M., Ishida, T., Nagasaka, K., Yamaguchi, J. (2003), "A Small Biped Entertainment Robot Exploring Attractive Applications", *Proc. of the IEEE Int. Conference on Robotics & Automation 2003*.

Kuroki, Y., Fukushima, T., Nagasaka, K., Moridaira, T., Doi, T., Yamaguchi, J. (2003), "A small Biped Entertainment Robot Exploring Human-Robot Interactive Applications", *Proc. The 12th Int. IEEE Workshop on Robot and Human Interactive Communication 2003*, 303.

Takenaka, T. (2001), "Honda humanoid robot "ASIMO" ", *Report of Honda foundation*, **No.99**

Yamaguchi, J., Takanishi, A., Kato, I. (1996), "Stabilization of Biped Walking and Acquisition of Landing Surface Position Information Using Foot Mechanism with Shock Absorbing Material", *Journal of the Robotics Society of Japan*, **Vol.14 No.1**, pp.67-74.

A STUDY ON A REAL-TIME SCHEDULING OF HOLONIC MANUFACTURING SYSTEM
- COORDINATION AMONG HOLONS BASED ON MULTI-OBJECTIVE OPTIMIZATION PROBLEM -

Koji IWAMURA[1], Yota SEKI[1], Yoshitaka TANIMIZU[1], Nobuhiro SUGIMURA[1]

[1] Graduate School of Engineering, Osaka Prefecture University,
1-1, Gakuen-cho, Sakai, Osaka 599-8531, Japan

ABSTRACT

This paper deals with a real-time scheduling system for HMS (Holonic Manufacturing System). A new real-time scheduling method for HMS is proposed, in the paper, to consider both the objective functions of the individual holons and the whole HMS. In this method, all the pareto optimal combinations of the resource holons and the job holons for the machining processes are generated based on the objective functions of the individual holons. Following this, a most suitable combination is selected from the pareto optimal ones, based on the objective functions of the whole HMS, such as the total make span and the total tardiness.

KEYWORDS

Holonic Manufacturing System, Real-time scheduling, Multi-objective optimization, Coordination

INTRODUCTION

Recently, automation of manufacturing systems has been much developed aimed at realizing flexible small volume batch productions. New distributed architectures of manufacturing systems have been proposed to realize more flexible control structures of the manufacturing systems, in order to cope with the dynamic changes in the volume and the variety of the products and also the unforeseen disruptions, such as malfunction of manufacturing equipment and interruption by high priority jobs. They are so called as autonomous distributed manufacturing systems, biological manufacturing systems, and holonic manufacturing systems [1]-[6].

In the previous report [6], decision making processes using effectiveness values have been proposed and applied to the real-time scheduling problems of the HMS (Holonic Manufacturing System), and it was shown, through case studies, that the proposed methods generate suitable schedules from the view point of the objective functions of the individual holons. New systematic methods for the individual holons in the HMS are proposed, in the paper, to consider both the objective functions of the individual holons and the whole HMS. The proposed methods are verified through case studies.

REAL-TIME SCHEDULING PROCESSES OF HOLONS

Real-time Scheduling of Holons

New real-time scheduling process of the individual holons is proposed to select a suitable combination of the resource holons and the job holons which can carry out the machining processes in the next time period. The resource holons and the job holons mean here the equipment carrying out the machining processes and the work-pieces to be machined, respectively.

At the time t when some machining processes are finished, and some resource holons and job holons become 'idling' status, all the 'idling' holons select their machining schedules in the next time period. The real-time scheduling processes consist of following five steps.

(1) Collection of status data
The individual 'idling' holons firstly gather the status data from the other holons.
(2) Selection of candidate holons
The individual 'idling' holons select all the candidate holons for the machining processes in the next time period.
(3) Evaluation of objective function values of individual holons
The individual 'idling' holons evaluate the objective function values for the cases where a holon selects candidate holons for the next machining process.
(4) Generation of all pareto optimal combinations based on objective functions of individual holons
The individual holons send the selected candidates and their objective function values to the coordination holon. The coordination holon generates all pareto optimal combinations of the job holons and the resource holons which can carry out the machining processes in the next time period, based on their objective function values. The pareto optimal combinations means that there are no feasible combination which will improve the objective function value of one holon without degrading the objective function value of at least one another holon [7].
(5) Determination of suitable combination based on objective functions of whole HMS
The coordination holon selects a most suitable combination of the job holons and the resource holons from the pareto optimal combinations, from the view point of the objective functions of the whole HMS.

Evaluation of Objective Functions of Individual Holons

The objective functions of the individual holons were proposed in the previous research [6], as shown in Table 1. The individual holons have one of the objective functions. The objective functions are evaluated by referring to the following technological information representing the machining process and machining capability of all the job holons and the resource holons.

M_{ik}: k-th machining process of the job holon i ($i = 1, \cdots, \alpha$), ($k = 1, \cdots, \beta$).
R_{ikm}: m-th candidate of resource holon, which can carry out the machining process M_{ik} ($m = 1, \cdots, \gamma$).
T_{ikm}: Machining time in the case where the resource holon R_{ikm} carries out the machining process M_{ik}.
W_i: Waiting time until the job holon i becomes idle if it is under machining status.
AC_{ik}: Required machining accuracy of machining process M_{ik}. It is assumed that the machining accuracy is represented by the levels of accuracy indicated by 1, 2, and 3, which mean rough, medium high, and high accuracy, individually.

The individual resource holons have the following technological information representing the machining capability of the resource holons for the machining process M_{ik}.

W_m: Waiting time until the resource holon R_{ikm} becomes idle if it is under machining status.
MAC_{ikm}: Machining accuracy in the case where the resource holon R_{ikm} carries out the machining process M_{ik}. MAC_{ikm} is also represented by the levels of 1, 2 and 3.
MCO_{ikm}: Machining cost in the case where the resource holon R_{ikm} carries out the machining process M_{ik}.

TABLE 1
OBJECTIVE FUNCTIONS OF HOLONS

Objective functions		Objective function values
Resource Holon	Efficiency	Σ Machining Time / Total Time
	Machining Accuracy	Σ (Machining Accuracy of Resources – Required Machining Accuracy of Jobs)
Job Holon	Flow Time	Σ (Machining Time + Waiting Time)
	Machining Cost	Σ (Machining Cost of Resources)

The following procedures are provided for the job holons to evaluate the objective functions. Let us consider a job holon i at time t. It is assumed that $JT_{i \cdot t}$ and $JC_{i \cdot t}$ give the total time after the job holon i is inputted to the HMS and the machining cost, respectively. If the job holon i selects a candidate resource holon j ($= R_{ikm}$) for carrying out the machining process M_{ik}, the flow time $JT_{i \cdot t+1}(j)$ and the machining costs $JC_{i \cdot t+1}(j)$ are estimated by the following equations.

$$JT_{i \cdot t+1}(j) = JT_{i \cdot t} + T_{ikj} + W_j \tag{1}$$

$$JC_{i \cdot t+1}(j) = JC_{i \cdot t} + MCO_{ikj} \tag{2}$$

As regards the resource holons, the following equations are applied to evaluate the efficiency $ME_{j \cdot t+1}(i)$ and the machining accuracy $MA_{j \cdot t+1}(i)$, for the case where a resource holon j ($= R_{ikm}$) selects a candidate job holon i for carrying out the machining process M_{ik}.

$$ME_{j \cdot t+1}(i) = -(ME_{j \cdot t} TT_{j \cdot t} + T_{ikj})/(TT_{j \cdot t} + T_{ikj} + W_i) \tag{3}$$

$$MA_{j \cdot t+1}(i) = MA_{j \cdot t} + (MAC_{ikj} - AC_{ik}) \tag{4}$$

where, $TT_{j \cdot t}$, $ME_{j \cdot t}$, and $MA_{j \cdot t}$ show the total time after the resource holon j starts its operations, the efficiency, and the evaluated value of machining accuracy of the resource holon j, respectively. Eqn. 3 contains the minus sign in order to evaluate the efficiency as the minimization problem.

The holons may select to wait in the next time period without executing any machining processes. In this case, the objective functions of the individual holons are evaluated by the following equation.

$$JT_{i \cdot t+1}(0) = \max_{j=1,\cdots,\gamma} \{ JT_{i \cdot t+1}(j) \} \tag{5}$$

$$JC_{i \cdot t+1}(0) = \max_{j=1,\cdots,\gamma} \{ JC_{i \cdot t+1}(j) \} \tag{6}$$

$$ME_{j \cdot t+1}(0) = \max_{i=1,\cdots,\delta} \{ ME_{j \cdot t+1}(i) \} \tag{7}$$

$$MA_{j \cdot t+1}(0) = \max_{i=1,\cdots,\delta} \{ MA_{j \cdot t+1}(i) \} \tag{8}$$

where, γ and δ are the number of candidate resource holons for the job holon i, and the number of candidate job holons for the resource holon j, respectively. Eqn. 5 to 8 mean that these objective function values are defined by the worst values of all the candidate resource holons, if they select waiting.

COORDINATION AMONG HOLONS BASED ON MULTI-OBJECTIVE OPTIMIZATION PROBLEM

Pareto Optimal Combination of Holons

After the individual holons evaluate the objective functions, the coordination holon generates all pareto optimal combinations of the job holons and the resource holons, which carry out the next machining processes. The

TABLE 2
COMBINATION OF RESOURCE AND JOB HOLONS

	wait	Resouce1	Resource2	...	Resourceγ
wait		a_{01}	a_{02}	...	$a_{0\gamma}$
Job1	a_{10}	a_{11}	a_{12}	...	$a_{1\gamma}$
Job2	a_{20}	a_{21}	a_{22}	...	$a_{2\gamma}$
...	...	...	...	...	...
Jobδ	$a_{\delta 0}$	$a_{\delta 1}$	$a_{\delta 2}$	...	$a_{\delta\gamma}$

procedure for generating all pareto optimal combinations is formalized as a multi-objective optimization problem, and the pareto optimal combinations of the job holons and the resource holons are defined as follows.

A matrix $A = \{a_{ij} \mid (i = 0, 1, \cdots, \delta, j = 0, 1, \cdots, \gamma)\}$ gives the combinations of job holons and resource holons, as shown in Table 2. Where $a_{ij} = 1$, if the job holon i is machined by the resource holon j in the next time period. Otherwise, $a_{ij} = 0$. If the job holon i or the resource holon j waits in the next time period, $a_{i0} = 1$ or $a_{0j} = 1$. Otherwise, $a_{i0} = 0$ or $a_{0j} = 0$. Only one job holon is machined by one resource holon, therefore, the following equations shall be satisfied.

$$\sum_{j=0}^{\gamma} a_{ij} = 1 \qquad i = 1, 2, \cdots, \delta \qquad (9)$$

$$\sum_{i=0}^{\delta} a_{ij} = 1 \qquad j = 1, 2, \cdots, \gamma \qquad (10)$$

If A is determined, the objective function values $x_{J_i}(A)$ of the job holon i and the ones $x_{R_j}(A)$ of the resource holon j are given by following equations, respectively.

$$x_{J_i}(A) = \sum_{j=0}^{\gamma} a_{ij} \cdot JOF_i(j) \qquad i = 1, 2, \cdots, \delta \qquad (11)$$

$$x_{R_j}(A) = \sum_{i=0}^{\delta} a_{ij} \cdot ROF_j(i) \qquad j = 1, 2, \cdots, \gamma \qquad (12)$$

where, $JOF_i(j)$ and $ROF_j(i)$ are the objective function values of the job holon i and the resource holon j given by following equations.

$$JOF_i(j) = JT_{i \cdot t+1}(j) \text{ or } JC_{i \cdot t+1}(j) \qquad (13)$$

$$ROF_j(i) = ME_{j \cdot t+1}(i) \text{ or } MA_{j \cdot t+1}(i) \qquad (14)$$

The objectives of the individual holons are to minimize their objective function values, therefore, the objective functions for coordination among holons are given by following equations as the multi-objective optimization problem.

$$\text{minimize } X(A) \qquad X(A) = [x_{J_1}(A), \cdots, x_{J_\delta}(A), x_{R_1}(A), \cdots, x_{R_\gamma}(A)] \qquad (15)$$

A^* is a pareto optimal combination, if there is no A such that the following equation is satisfied.

$$x_k(A) \leq x_k(A^*) \quad \text{for all } k, k = J_1, J_2, \cdots, J_\delta, R_1, R_2, \cdots, R_\gamma \qquad (16)$$

$$x_l(A) < x_l(A^*) \quad \text{for any } l, l = J_1, J_2, \cdots, J_\delta, R_1, R_2, \cdots, R_\gamma \qquad (17)$$

The coordination holon firstly generates all the candidates of A, which represent all the combinations of the job holons and the resource holons. This process does not take long time, since the number of 'idling' holons is limited at the time t. A set of pareto optimal combinations $\{A_p\}$ are secondly obtained based on Eqn. 16 and Eqn. 17.

Determination of Combination of Next Machining Processes

The coordination holon selects a suitable combination of the job holons and the resource holons from all the pareto combinations, based on the objective functions of the whole HMS. The following two performance indices of the whole HMS are considered in this research.

(1) Total slack
The total slack is given by the following equation.

$$SLACK = \sum_{i=1}^{\alpha} (d_i - t - TWKR_i) \tag{18}$$

where, α, d_i and t are the number of the job holon in the HMS, the due date of the job holon i, and the current time, respectively. $TWKR_i$ is the average of the total processing time of the remaining machining processes of the job holon i which is given by following equation.

$$TWKR_i = \sum_{k=\xi+1}^{\beta} (\sum_{m=1}^{\gamma} T_{ikm}/\gamma) \tag{19}$$

where, T_{ikm} is the machining time in the case where the m-th ($m = 1, \cdots, \gamma$) candidate resource holon carries out the k-th machining process of the job holon i. β and ξ are the total number of the machining processes of the job holon i, and the number of the machining processes finished by the current time t.

(2) Sum of the ratio of the next processing time and the remaining processing time
The sum of the ratio of the next processing time and the remaining processing time is given by the following equation.

$$PT/TWKR = \sum_{i=1}^{\delta} (T_{i(\xi+1)m}/TWKR_i) \tag{20}$$

where, δ and $TWKR_i$ are the number of the candidate job holons in the HMS, and the average of the total processing time of the remaining machining processes of the job holon i, respectively. $T_{i(\xi+1)m}$ means the machining time of the next machining process of the job holon i.

The coordination holon calculates the total slack $SLACK$ or the sum of the ratio of the next processing time and the remaining processing time $PT/TWKR$ for all the pareto combinations $\{A_p\}$. Following this, the coordination holon selects the combination of the job holons and the resource holons, which minimizes the $SLACK$ or $PT/TWKR$. That is, the coordination holon applies one of the rules called 'minimum $SLACK$' and 'minimum $PT/TWKR$'.

CASE STUDY

Some case studies have been carried out to verify the effectiveness of the proposed methods. The HMS model consisting of 10 machining centers (MC) is considered for the case study. The individual machining center holons have the different objective functions and the different machining capacities, such as the machining time T_{ikm}, the machining accuracy MAC_{ikm}, and the machining cost MCO_{ikm}. As regards the job holons, 24 job holons are considered in the case study, which have the different objective functions and the machining process. 8 cases are considered in the case study by changing the machining capacities of the individual resource holons.

Figure 1 shows the verification of the objective functions of the individual holons and the whole HMS. The vertical axis and the horizontal axis in the figures of the left and middle are the average of the objective function values of all the holons and the type of the objective functions, respectively. It is found that the proposed method keeps the objective function values of the individual holons in almost same as the ones obtained by the previous method. The figures in the right give the average values of the total tardiness and the total make span of all the job holons. It is shown that the proposed method improves the total tardiness and the total make span which are the objective functions of the whole HMS.

CONCLUSIONS

(1) A new real-time scheduling method for the HMS is proposed, in order to generate a suitable schedule of holons considering both the objective functions of the individual holons and the whole HMS.
(2) The proposed method is applied to the real-time scheduling problems of the HMS, and the scheduling results are compared with the ones by the previous method. It was shown, through case studies, that the proposed method is effective to improve the production schedules from the viewpoint of the objective functions of the whole HMS.

REFERENCES

1. Ueda, K. (1992). An approach to bionic manufacturing systems based on DNA-type information. *Proc. of the ICOOMS '92*, 303-308.
2. Moriwaki, T. and Sugimura, N. (1992). Object-oriented modeling of autonomous distributed manufacturing system and its application to real-time scheduling. *Proc. of the ICOOMS '92*, 207-212.
3. Iwata, K., et al. (1994). Random manufacturing system: A new concept of manufacturing systems for production to order. *Annals of the CIRP* **43:1**, 379-384
4. Wiendahl, H.P. and Garlichs, R. (1994). Decentral production scheduling of assembly systems with genetic algorithm. *Annals of the CIRP* **43:1**, 389-396
5. Wyns, J., et al. (1996). Workstation architecture in holonic manufacturing systems. *Proc. of the 28th Int. Seminar on Manufacturing Systems*, 220-231
6. Iwamura, K. et al. (2003). A study on simulation system for real-time scheduling of holonic manufacturing system. *Proc. of The 7th World Multiconference on Systemics, Cybernetics and Informatics* **8,** 261-266
7. Vira C. et al. (1983). *Multi-objective decision making: theory and methodology*, North Holland

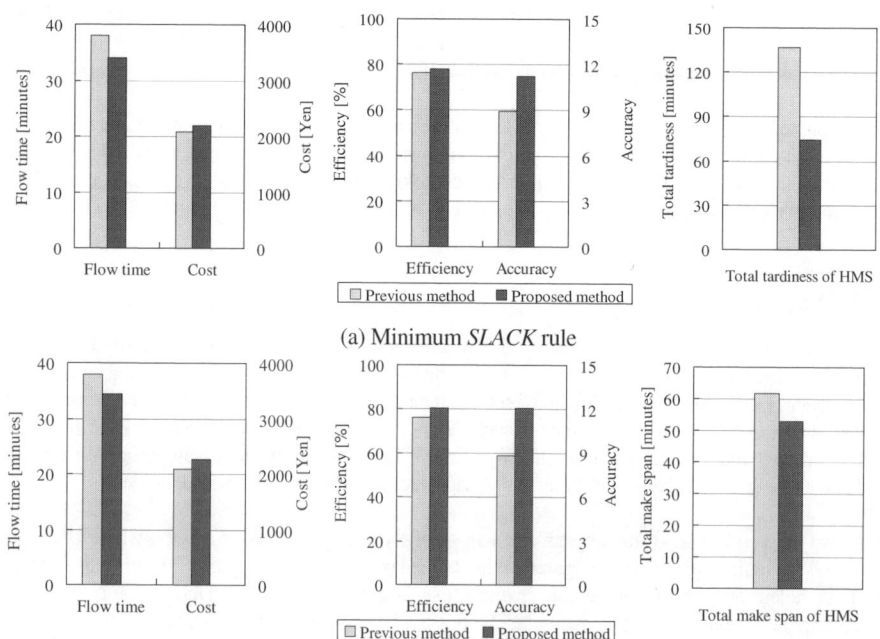

(a) Minimum *SLACK* rule

(b) Minimum *PT/TWKR* rule

Figure 1: Comparison of objective function values

A STUDY ON INTEGRATION OF PROCESS PLANNING AND SCHEDULING SYSTEM FOR HOLONIC MANUFACTURING SYSTEM - SCHEDULER DRIVEN MODIFICATION OF PROCESS PLANS-

Rajesh SHRESTHA[1], Toshihiro TAKEMOTO[1], Nobuhiro SUGIMURA[1]

[1] Graduate School of Engineering, Osaka Prefecture University,
1-1, Gakuen-cho, Sakai, Osaka 599-8531, Japan

ABSTRACT

In case of small batch productions with dynamic changes in volumes and varieties of products, the conventional manufacturing systems are not adaptable and thus, new architectures of manufacturing system known as autonomous distributed manufacturing system has been proposed, which can cope with dynamic changes in volume and variety of products, and also with unscheduled disruptions. Holonic manufacturing system is one of the autonomous distributed manufacturing systems. The purpose of the present research is to develop an integrated process planning and scheduling system, which is applicable to the HMS. In this research, the process plans of the individual product are modified with the help of the feedback information of the generated schedule. A systematic method based on the DP and the heuristic rule is proposed to modify the predetermined process plans, based on the load balancing of the machining equipment.

KEYWORDS

Holonic Manufacturing, Scheduling, Process Planning, Dynamic Programming, Heuristic Rule

INTRODUCTION

In case of small batch productions with dynamic changes in volumes and varieties of products, the conventional manufacturing systems are not adaptable, and thus, new architectures of manufacturing system have been proposed. The new architectures known as autonomous distributed manufacturing systems cope not only with the dynamic changes but also with the unscheduled disruptions such as the breakdown of equipment and the interruption of high priority jobs. Holonic manufacturing system is one of the autonomous distributed manufacturing systems besides biological manufacturing systems, fractal manufacturing systems and agile manufacturing systems. [1]-[4]

The objective of the present research is to develop an integrated process planning and scheduling system applicable to the holonic manufacturing system. In the previous papers [5],[6], integration of process planning and scheduling was carried out, wherein the scheduling system for multi-products as a whole uses the process plan information of a set of individual products to generate a suitable schedule. But, there is not any feedback information from the scheduling system to the process planning system. This paper deals with the integration of the process planning and the scheduling systems where there is a scheduler driven modification of the process plans of the products. A systematic method is proposed to generate modified sequences of machining equipment for the individual products based on the feedback information of the scheduling results, and to generate a modified production schedule for the whole manufacturing system.

PROCESS PLANNING AND SCHEDULING

The process planning system generates suitable process plans for the individual products to be manufactured. The process plans give suitable sequences of manufacturing equipment needed to manufacture the machining features of the products, and machining time of the machining features. The scheduling system determines suitable production schedules of manufacturing equipment in the HMS for manufacturing a set of products. The production schedules give the loading sequences of the products to the manufacturing equipment and the starting times of the individual machining processes of the products. The production schedules are verified based on the objective functions such as the make span and the tardiness against due date.

SCHEDULING BY SCHEDULING HOLON

Input Information

The input information of the scheduling holon is summarized here. The following production management information is the requirements to the scheduling process.
(1) Starting time and due time of job holons.
(2) Candidate machining sequence of machining features and candidate sequences of machining equipment.
(3) Machining time of machining features.
(4) Alternative machining equipment for each machining feature.
(5) Machining time by alternative machining equipment.

Objective Functions

This research deals concurrently with both the process planning of the individual jobs and the scheduling of all the jobs to be manufactured in the HMS. The following objective functions are considered for the scheduling task of the HMS[5].
(1) Make span: MS
(2) Total machining cost: TMC
(3) Weighted tardiness cost: WT

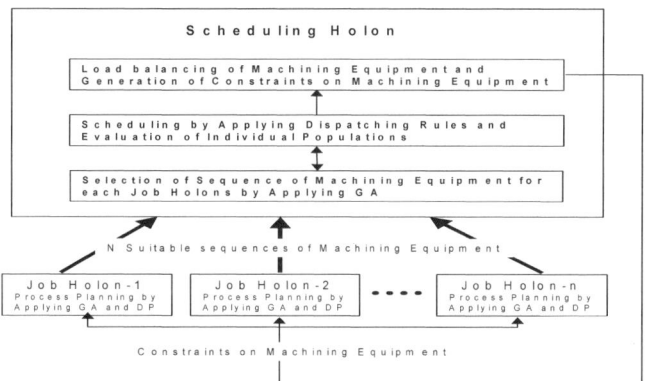

Figure 1: Scheduler driven modification of process plans

SCHEDULING BASED ON GA AND DISPATCHING RULES[6]

A procedure shown in Figure 1 is proposed to generate suitable production schedules for all the jobs. All the job holons firstly select suitable process plans based on their objective functions and send the candidate process plans to the scheduling holon. Following this, the scheduling holon selects a combination of the process plans of all the jobs and generates a production schedules for the selected combination. The procedure of the scheduling holon is summarized in the followings.

Selection of a combination of process plans

A genetic algorithm (GA) based method is adopted for selecting a combination of process plans. The individual job holon send N candidate process plans to the scheduling holon. The scheduling holon finally obtains both a suitable combination of the process plans of all the jobs and a suitable schedule of the HMS.

Scheduling based on dispatching rules

A set of dispatching rules is adopted, in the research, for solving the scheduling problems. The dispatching rules give the priority to one job against all the candidate jobs that are waiting for the machining process of the manufacturing equipment. Let the j-th process of the i-th waiting job be denoted by $OP_{ij}^{(k)}$ ($i = 1, 2, ..., m$) and its processing time of the machining process be $MAT_{ij}^{(k)}$ ($j = 1, 2, ..., n_i$). Three different dispatching rules are applied to the waiting jobs. These rules have been widely used for the large scale job shop scheduling problems. The followings give the dispatching rules considered in the research[7].
(1) SPT (Shortest Processing Time).
(2) SPT/TWKR (Shortest Processing Time / Total Work Remaining).
(3) Apparent Tardiness Cost (ATC).

SCHEDULER DRIVEN MODIFICATION OF PROCESS PLANS

Modification Process of Process Plans

In the newly proposed method, the constraints on the machining equipment are sent to the job holons as the feed back information of the scheduling results, to generate the modified sequences of the machining equipment for the individual job holons. It was found that the machining process in some of the machining equipment are concentrated where as the other machining equipment is remaining idle. Therefore, the global objective functions, such as the total make span and the weighted tardiness cost, can be improved, if the scheduling holon redistributes the concentrated load of the machining processes to the other machining equipment and reduces the waiting time. The process plan modification procedure basically consists of two stages, they are, the load balancing of the machine equipment by the scheduling holon and the modification of the sequence of the machining equipment by the job holons.

Load Balancing

The load balancing means here to reallocate all the machining features and their machining processes to the suitable machining equipment, in order that the load of all the machining equipment is well balanced, taking into consideration of the entire alternative machining equipment MEA_{ijp} for the machining features.

The following steps are being taken during the load balancing.
STEP 1 Generation of load chart: The load chart of all the machining equipment is drawn based on the scheduling results.

STEP 2 Calculation of average balanced load: The average balanced load (ABL) is estimated from the load chart, based on the following equation.

$$ABL = \Sigma \Sigma MAT_{ij}^{(k)} / N \tag{1}$$

where, i is ID of the job holon, j is ID of the machining features machined by the j-th position in the machining sequence, k is ID of the process plans of the job holon i, which is selected in the scheduling process and N is total number of machining equipments.

STEP 3 Selection of machining equipment to be reallocated: The machining equipment with the maximum load is selected, which is reallocated first. The reallocation process is carried out step-by-step from the machining equipment with large load in the load chart.

STEP 4 Reallocation of machining features to selected machining equipment: The machining features are reallocated to the machining equipment selected in the STEP 3. The LPT (Longest Processing Time) rule is used in the research to determine the machining features to be loaded to the selected machining equipment. By the LPT rule, the highest priority is given to the machining features with the maximum value of the machining time $MAT_{ij}^{(k)}$. Therefore, the machining features with the high priorities are allocated to the selected machining equipment according to the priority.

STEP5 Termination of reallocation process: The reallocation process is terminated, just before the load of the selected machining equipment crosses the average balanced load (*ABL*).

After STEP 1 to STEP 5, some of the machining features are loaded to the selected machining equipment. The machining equipment, which carries out these machining features, is fixed. On the other hand, the remaining machining features shall be loaded to the machining equipment except the selected one. The procedures in the next section are applied for selecting the suitable machining equipment for the remaining machining features.

Selection of suitable machining equipment

Figure 2 shows an example of the status of the alternative machining equipment of the machining features of the job holon *i*, after the reallocation process is completed. In this case, the machining equipment ME_2 is reallocated and balanced, therefore, the machining feature MF_{12} is fixed to ME_2, and the other alternative machining equipment for MF_{12} are deleted. As regards to other machining features, if they have ME_2 as the alternative machining equipment, ME_2 is deleted from the alternative.

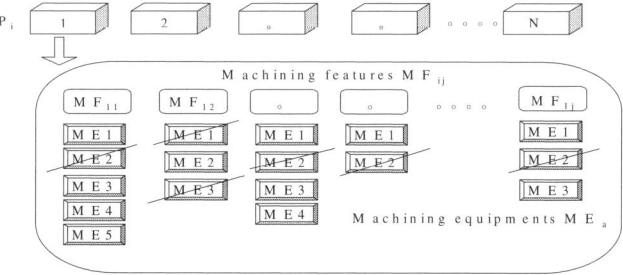

Figure 2 : Modified process plans with alternative machine equipment

Following this, all the job holons regenerate new sequences of the machining equipment under the constraints determined in the load balancing process.

CASE STUDY

The algorithm has been constructed based on the load balancing method and the dynamic programming method and a prototype of the process planning and scheduling system has been implemented using C++ language. One of the case result is summarized in Figures. 3 and 4, which show that the make span has been reduced from 28561.5 sec. before load balancing to 19335.7 sec. after load balancing. Balancing of the machining equipment is carried out in the sequence of most busy machining equipment to the least busy machining equipment, and the balancing sequence of the machining equipment is MT12, MT3, MT6, MT17, MT14, MT9 and finally MT15, in this case.

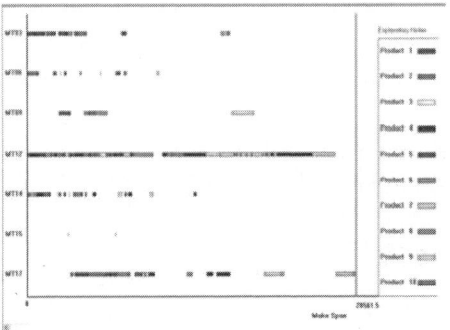
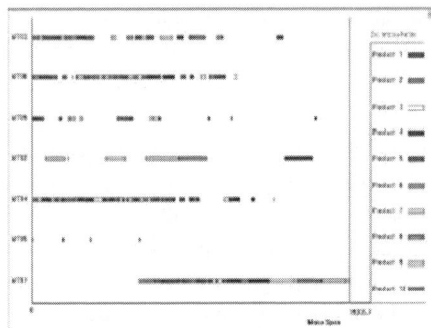

Figure 3: Gantt chart before load balancing   Figure 4: Gantt chart after load balancing.

CONCLUSIONS

This paper dealt with the integration of process planning and scheduling systems using process plan modification system. Following are the conclusions:
(1) Systematic methods for load balancing of the machining equipment and for modifying the process plans are proposed in order to obtain a modified processed plan based on the feedback information from the scheduling results.
(2) A prototype of the process planning and scheduling systems has been implemented. Some case studies show that the total make span can be improved from the modified process plans obtained after the feed back information from the scheduling results.

REFERENCES

1. Moriwaki, T. and Sugimura, N. (1992). Object-oriented modeling of autonomous distributed manufacturing system and its application to real-time scheduling. *Proc. of the ICOOMS'92*, 207-212.
2. Ueda, K. (1992). An approach to bionic manufacturing systems based on DNA-type information. *Proc. Of the ICOOMS'92*, 303-308.
3. Warnecke, H. J. (1993). *The Fractal Enterprise*, SpringerVerlag, New York
4. Sugimura, N. et. al. (1996). Modeling of holonic manufacturing system and its application to real-time scheduling. *Manufacturing Systems* **25:4**,1-8.
5. Shrestha, R. et.al. (2003). A study on process planning system for Holonic manufacturing – Process planning considering both machining time and machining cost –. *Proc. of LEM21*, 753-758.
6. Shrestha, R. et.al. (2004). A study on Integration of Process Planning and Scheduling Systems for Holonic Manufacturing – Manufacturing multi-products-. *Proc. of 2004 Japan-USA Symposium on Flexible Automation*, 1-8.
7. Vepsalainen, A. P. J. and Morton, T. E. (1987). Priority rules for job shops with weighted tardiness costs. *Management Science*.33:8, 1035-1047.

GENETIC ALGORITHM BASED REACTIVE SCHEDULING IN MANUFACTURING SYSTEM – ADVANCED CROSSOVER METHOD FOR TARDINESS MINIMIZATION PROBLEMS –

T. Sakaguchi[1], Y. Tanimizu[2], K. Harada[3], K. Iwamura[2] and N. Sugimura[2]

[1]Graduate School of Science and Technology, Kobe University,
1-1 Rokkodai, Nada-ku, Kobe 657-8501, JAPAN
[2]Graduate School of Engineering, Osaka Prefecture University,
1-1 Gakuen-cho, Sakai, Osaka 599-8531, JAPAN
[3]Manufacturing Engineering Service Dev., Toyota Motor Corporation,
1 Shimoyama, Uchikoshi, Miyoshi-cho, Nishikamo-gun, Aichi 470-0213, JAPAN

ABSTRACT

Recently, flexible scheduling systems are required to cope with dynamic changes of market requirements and manufacturing environments. A reactive scheduling method based on Genetic Algorithm (GA) was proposed, in the previous research, in order to improve an initial production schedule delayed due to unscheduled disruptions, such as delays of manufacturing processes. The objective of the research is to propose a new GA based reactive scheduling method for tardiness minimization scheduling problems, aiming at improving the disturbed production schedule efficiently and generating suitable production schedules faster than the previous reactive scheduling method. A prototype of reactive scheduling system is developed and applied to computational experiments.

KEYWORDS

Scheduling, Genetic algorithm, Flexible system, Tardiness of job, Recovery, Object-oriented

INTRODUCTION

Unscheduled disruptions, such as delays of manufacturing processes, addition of emergent jobs and failures in manufacturing equipment, often occur in the actual manufacturing systems. However, most of the traditional scheduling researches assume that manufacturing environments are well stabilized. The manufacturing system becomes impossible to satisfy the constraints on the due dates and the make-span, when the initial schedules are delayed due to the unscheduled disruptions.

The reactive scheduling method (Smith 1995) is defined here as the method that modifies and improves the predetermined initial production schedules, when some unscheduled disruptions of manufacturing processes occur in the manufacturing systems. A reactive scheduling method for delays of manufacturing processes was proposed in the previous research papers (Tanimizu 2002). This method used Genetic Algorithm (GA) to generate new feasible production schedules. The previous paper showed that the initial production schedule is modified and improved through the GA based reactive scheduling processes.

The objective of the research is to propose a new GA based reactive scheduling method for tardiness minimization scheduling problems, aiming at improving the disturbed production schedule efficiently and generating suitable production schedules faster than the previous reactive scheduling method. A prototype of reactive scheduling system is developed and applied to computational experiments.

CURRENT REACTIVE SCHEDULING METHOD

Reactive scheduling process is activated, only when the initial production schedule cannot satisfy the constraint on the make-span, due to the unscheduled disruptions. It is necessary to consider the progress of the manufacturing process in the reactive scheduling process.

Figure 1 shows the whole reactive scheduling process. The reactive scheduling process is activated at the present time T_1, only when the delay of the make-span occurs and the predetermined initial production schedule does not satisfy the given constraint on the make-span. The reactive scheduling process takes computation time dt to generate a new feasible schedule. The time dt is the time in which GA creates a new generation of the population representing the modified production schedules. The computation time dt is estimated based on the time needed to generate a new population of the feasible production schedules by applying GA. Therefore, the schedule of the operations starting after $(T_1 + dt)$ can be modified in the reactive scheduling process. If the make-span of the newly generated schedule is shorter than the make-span of the current schedule, the current schedule is substituted by the newly generated one. The reactive scheduling process is repeated, until the newly generated schedule satisfies the constraint on the make-span, or until all the manufacturing operations have already started.

If new operations start during the reactive scheduling process, the next reactive scheduling process inherits only the individuals that are consistent with the schedule of the operations starting between T_x and $(T_x + dt)$. It is because that the schedule of these operations should be fixed in the reactive scheduling process. The other individuals are deleted, and new individuals are randomly created. Therefore, the proposed GA based reactive scheduling method can continuously modify and improve the production schedule, taking into consideration of the progress of the manufacturing processes.

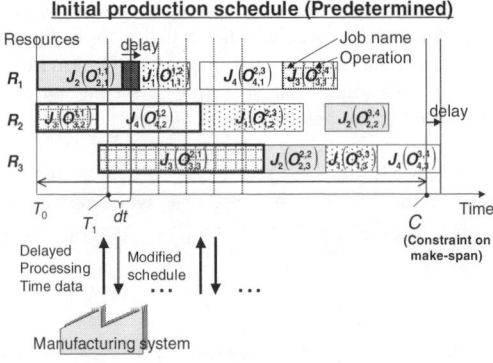

Figure 1: Reactive scheduling process

REACTIVE SCHEDULING METHOD FOR TARDINESS MINIMIZATION PROBLEMS

Tardiness minimization problems

In this paper, the main concern is total tardiness, which is a criterion based on job due dates, and is defined by Eqn. 1.

$$\sum_{i=1}^{n} \max(0, C_i - d_i) \quad (1)$$

Where C_i and d_i are the completion time and the due date of the job J_i respectively, and n is the total number of the jobs under consideration.

The reactive scheduling process using GA is a time-consuming process in order to generate a good solution. However, the reactive scheduling modifies the schedule in parallel to the production activity, therefore, it is required to find a good solution in the limited time. An advanced crossover method is discussed in the followings, in order to modify and improve the schedule in a short time.

Advanced crossover method

Each gene in an individual of the proposed GA method corresponds to a manufacturing operation to be executed in the manufacturing system, and the list of the genes in the individual represents the priorities for the execution of manufacturing operations in the production schedule. The lower bound of the tardiness is estimated for the y-th gene of the individual, by applying Eqn. 2.

$$LT_y = \left(ft_i^{(h)} + \sum_{s=h+1}^{n} pt_i^{(s)} \right) - dd_i \quad (2)$$

where,
LT_y: lower bound of tardiness for the y-th gene, which corresponds to the h-th operation of job J_i.
$ft_i^{(h)}$: finishing time of the h-th operation of job J_i.
$pt_i^{(s)}$ ($s=h+1, \ldots, n$): processing time of remaining operations of job J_i.
dd_i: due-date of job J_i.

If the LT_y is more than zero, it is impossible for job J_i to finish the remaining operations by its due date.

Two parent individuals and their crossover points are randomly selected in the first step of the crossover operation. After that, only the genes having positive number of LT_y between two crossover points are exchanged with the genes of another parent individual, by the newly proposed crossover method. The other genes of the parent individuals are survived to the offspring individuals, as shown in Figure 2.

Reactive scheduling process

The reactive scheduling process is carried out by the following steps.

STEP1 Initialization
The present time T_x ($x = 1, 2, \ldots$) is set up. Computation time dt is estimated. It is the time for creating the modified production schedules through STEP2 to STEP4.

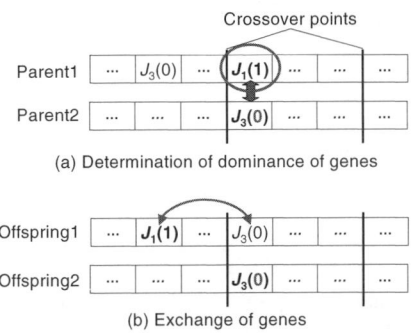

Figure 2: Crossover process

STEP2 Creation of initial population
Two cases are considered in the creation of the initial population including the individuals which represent the production schedules. They are,

(1) First activation of reactive scheduling process at time T_1
(2) Second or later activations of reactive scheduling process at time T_2 or later.

For the cases of 1 and 2, the reactive scheduling process creates the initial population through the STEP2-1 and STEP2-2, respectively.

STEP2-1 First activation of the reactive scheduling process
The reactive scheduling process generates the initial population randomly. The initial population created here should satisfy the constraint on the schedule of the operations starting before $(T_1 + dt)$.

STEP2-2 Second or later activations of the reactive scheduling process
In the case 2, the reactive scheduling process can inherit the population created in the previous reactive scheduling process. Two cases are considered for the inheritance process of the population as shown in the followings.

Case-A No operations start between T_x and $(T_x + dt)$
If no operations start between T_x and $(T_x + dt)$, all the individuals of the last population of the previous reactive scheduling process are inherited to a new reactive scheduling process between T_x and $(T_x + dt)$.

Case-B Some operations start between T_x and $(T_x + dt)$
If some operations start between T_x and $(T_x + dt)$, the production schedules of these operations should be fixed. Therefore, a new reactive scheduling process can inherit only the individuals, which are consistent with the schedules of the fixed operations, from the last population created in the previous reactive scheduling process. The other individuals are deleted, and new individuals are created from the inherited ones randomly.

STEP3 Application of genetic operators to the population
The fitness value of each individual is calculated. The total tardiness of the production schedule which has to be minimized is selected as the fitness value. Based on the fitness value, genetic operators, such as selection, crossover and mutation, are applied to the individuals of the population created in STEP2, in order to create new individuals of the next population. Crossover is carried out by the following steps.

STEP 3-1: Selection of crossover points
STEP 3-2: Calculation of lower bound of tardiness
STEP 3-3: Determination of dominance of gene
STEP 3-4: Exchange of genes

STEP4 Evaluation of modified production schedule
If the shortest total tardiness of all the new individuals created in STEP3 is shorter than the total tardiness of the current production schedule, the new modified production schedule is substituted for the current production schedule. If the total tardiness of the new production schedule is shorter than the constraint, the reactive scheduling process is terminated.

All the steps from STEP1 to STEP4 are repeated, until the created production schedule satisfies the given constraint on the tardiness or all the manufacturing operations have started in the manufacturing system.

COMPUTATIONAL EXPERIMENTS

Prototype of reactive scheduling system

A prototype of reactive scheduling system was implemented by using an object-oriented language, Smalltalk. It was developed on a personal computer operating under the Windows system. The prototype system was applied to some reactive scheduling problems for the tardiness minimization problems in order to verify the effectiveness of the proposed method. The following experimental conditions are based on the test cases proposed by Storer, Wu and Vaccari (Storer 1992).

- Job-shop type production scheduling problem
- Number of resources: 10
- Number of jobs: 50
- Parameters of GA: population size, crossover rate and mutation rate were 30, 0.5 and 0.1, respectively. The values of these parameters were determined based on some case studies of the job-shop type production scheduling problems.
- Interruptions: Some operations were randomly selected, and their operation times were enlarged.

Experimental results

A prototype of reactive scheduling system was applied to computational experiments for the tardiness minimization scheduling problems. Some delays of manufacturing processes occurred, while the manufacturing processes were in progress. The prototype system activated the reactive scheduling process, in order to modify and to improve the disturbed initial production schedule.

Figure 3 shows the experiment results for the previous reactive scheduling method and the newly proposed reactive scheduling method. The horizontal axis and the vertical axis show the time and the total tardiness, respectively. The lines show that the new reactive scheduling method improves the delayed initial production schedule faster than the previous reactive scheduling method.

Ten experimental results of the new reactive scheduling method were also compared with the results of four types of rule based real-time scheduling methods, as shown in Figure 4. Through the comparison, it was shown that the proposed reactive scheduling method improves the total tardiness shorter than the real-time scheduling methods.

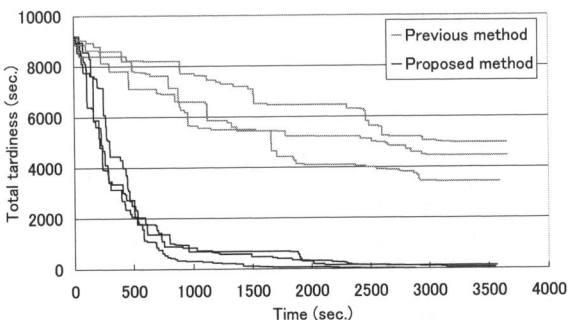

Figure 3: Experimental results

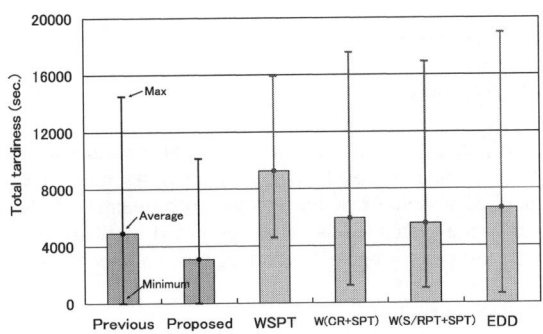

Figure 4: Comparison of 10 cases of experimental results

CONCLUSIONS

This research proposed a new reactive scheduling method in order to improve the performance of the GA based reactive scheduling method for tardiness minimization scheduling problems. A new crossover method was proposed, in this research, to exchange the genes between the parent individuals efficiently, aiming at generating suitable offspring individuals effectively. The effectiveness of the proposed method was verified through some computational experiments.

REFERENCES

Smith S. F. (1995). Reactive scheduling systems. Intelligent Scheduling System, Kluwer Academic, 155-192.

Storer R. H., Wu D. D. and Vaccari R. (1992). New search spaces for sequencing instances with application to job shop scheduling. Management science **38**, 1495-1509.

Tanimizu Y. and Sugimura N. (2002). A study on reactive scheduling based on genetic algorithm. Proc. of the 35th CIRP-ISMS, 219-224.

A BASIC STUDY ON COST BASED SCHEDULING

Kentarou Sashio[1], Susumu Fujii[1], Toshiya Kaihara[2]

[1]Faculty of Eng. Kobe University, Rokkodai 1-1, Nada, Kobe, Japan
[2]Graduate School of Science and Technology, Kobe University, Rokkodai 1-1, Nada, Kobe, Japan
ADDRESS : Rokkodai 1-1, Nada, Kobe, Japan

ABSTRACT

In most of the studies on manufacturing scheduling, only time and quantity based criteria, such as, queue length, average inventory level, and so on, have been evaluated for measuring their performance. However, it is difficult to find the most important criterion, since there are many criteria and it is varying every moment. In addition, some of the criteria are forming trade-off relationship. Therefore, we focus on the product cost as a criterion of scheduling performance. They are possible to estimate the product cost with accounting methods and to reduce the product cost directly. In this paper, we propose two kinds of cost based scheduling, such as, Activity Based Costing approach and Genetic Algorithm based approach. And their performance is evaluated through experiments with a Distributed Virtual Factory.

KEYWORDS

Cost, Scheduling, Distributed Virtual Factory, Activity Based Costing

1. INTRODUCTION

To deal with the diversification of consumers' needs and to survive in severe competitions, manufacturers are facing problems of shortening lead time, cutting indirect cost and so on. For these problems, Information Technology (IT) has been fully utilized in manufacturing systems. On the other hand, many studies on manufacturing scheduling have been achieved to provide solutions for these problems. In most of the studies, however, only time and quantity based criteria, such as, queue length, average inventory level, and so on, have been evaluated for measuring their performance. It is difficult to find the most important criterion even for veteran engineers, since there are many criteria based on time and quantity and it is varying every moment. In addition, some of the criteria are forming trade-off relationship.

Therefore, we focus on the product cost as a criterion for measuring scheduling performance. They are possible to estimate the product cost with accounting methods and to reduce the product cost directly by applying the product cost as a criterion. In this paper, we propose two kinds of cost based scheduling approaches, such as, Activity Based Costing (ABC)[1] approach and Genetic

Algorithm based approach. And their performance and characteristics are investigated through experiments with a Distributed Virtual Factory[2,3].

2. PRODUCT COST ANALYSIS

To evaluate and estimate the product cost and its composition, variety of accounting method has been proposed. In recent manufacturing systems, the share of indirect costs in the total cost is relatively increasing due to the development of the automation and IT. And it becomes more and more important to reduce and control the indirect costs. Respecting the back ground, we employ ABC, since the indirect costs are reasonably distributed with ABC compared with the other accounting methods.

As shown in Figure 1, ABC firstly pools the indirect costs to the objects whose activities consume economical resources. The pooled costs are called Activity Costs. At the second step of ABC, the activity costs are distributed to each product by cost driver. The indirect costs are reasonably distributed to products, since the indirect costs are distributed to products or facilities in proportion to the cost drivers which are carefully selected as reasonable criteria. In this study, the product cost is obtained by summarizing eleven costs listed in Table 1. Cost drivers for each cost and charge rate are also listed in the table. Costs without cost driver 1, such as Depreciation Cost and Direct Energy are directly charged on each facility, therefore, it is not necessary to calculate their activity cost. And the costs with neither cost drivers are direct costs.

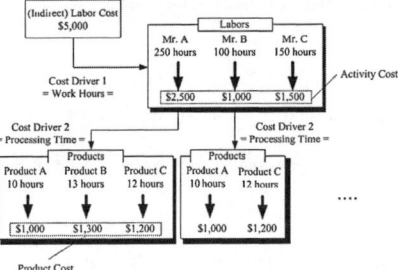

Figure 1 Concept of Activity Based Costing

3. DISTRIBUTED VIRTUAL FACTORY

As a simulation environment for global manufacturing system simulation, Distributed Virtual Factory (DVF) has been proposed[2,3]. DVF is constructed by integrating local area simulation systems via the internet based on the concept of distributed simulation. In this study, we developed a DVF as shown in Figure 2 and process histories of each product are obtained as simulation logs.

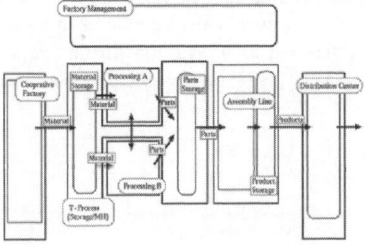

Figure 2 Overview of DVF model

4. ABC BASED APPROACH

As a first algorithm of cost based scheduling, we propose an ABC based approach. We tend to the practicality and global improvement of cost structure rather than the optimality.

We focus on three types of simple dispatching rules, named EDD (Earliest Due Date), SPT (Shortest Processing Time) and HC (Highest Commonality). HC is our original rule. Under this rule, materials which have higher commonality for the products are granted higher priority. Simulation for term t_1 is performed three times applying each rule (shown as Step 1 in Figure 3). Then, product costs are estimated for each trial with ABC and the rule which produces with the minimum cost is selected as the scheduling rule for term t_1 (shown as Step 2 and 3 in Figure 3). Those procedures are iterated for all over the scheduling term.

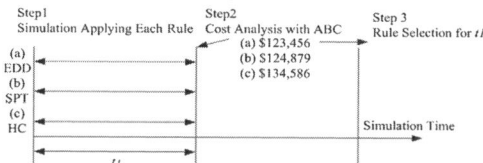

Figure 3 Concept of Proposed Method

Table 1 List of Objective Costs

Cost Name	Cost Driver 1	Cost Driver 2	Charge Rate
Depreciation Cost A		Utilization Time	100,000(Yen/Month)
Depreciation Cost B		Utilization Time	150,000(Yen/Month)
Depreciation Cost C		Utilization Time	500,000(Yen/Month)
Stock Cost A			3,000(Yen/Month)
Stock Cost B			8,000(Yen/Month)
Indirect Energy Cost	Utilization Time	Processing Time	100,000(Yen/Month)
Indirect Labor Cost	Processing Time	Processing Time	750,000(Yen/Month)
Material Cost			100(Yen/Each)
Penalty for Tardiness			10(Yen/Minute)
Direct Energy		Processing Time	10(Yen/Minute)
Setup Cost			50(Yen/Each)

4.1. Experiments 1

We have simple experiments with the DVF to investigate the effectiveness of the proposed method and the effectiveness of dispatching rules for product cost. The proposed method is implemented in processing A and B. Experiments conditions are following.
1. Total scheduling period is 6 days.
2. Term of rule selection (shown as $t1$ in Figure 3) is 2 days.
3. Order amount for day 1 and 2 makes utilizations of facilities about 85%.
4. Order amount for day 3 and 4 is 10% higher than that of day 1 and 2.
5. Order amount for day 5 and 6 is 10% lower than that of day 1 and 2.

4.2. Results of Experiments 1

Product cost at 4 areas, such as, Processing A, B, Assembly Line and Storages, for day 1 and 2 are shown in Table 2. At first, we decided HC as the dispatching rule for day 1 and 2, since HC produces in the lowest cost. However, HC is the best rule only for Processing B and it is also the worst rule for storages. That means the best rule from view point of total manufacturing system might not be the best for each area. Conversely, the best rule at an area might not be the global best rule. It is important to evaluate the global manufacturing system for cutting total product cost. All costs in assembly line are the same. We consider that the initial inventory level of parts storage is enough to absorb the fluctuation of parts arrival from processing areas. Those facts also suggest the importance of evaluating total manufacturing system.

Product costs of 4 days are listed in Table 3 and 6 days are listed in Table 4. As shown in these tables, total best rule is varied at end of 4 days. This fact shows the difficulty to select the most important criterion from time and quantity based criteria. In other words, the advantage of cost criterion is shown through the experiments.

Table 2   Product Cost of Day 1 and 2

	HC	EDD	SPT
Processing A	566,149	566,149	566,149
Processing B	1,210,868	1,231,775	1,231,775
Assembly Line	527,166	527,166	527,166
Storages	2,933,024	2,931,088	2,932,017
Total	5,237,207	5,256,178	5,257,107

Table 3   Accumulated Product Cost of 4 days

	HC	EDD	SPT
Processing A	1,286,396	1,259,266	1,259,277
Processing B	2,897,404	2,836,145	2,927,244
Assembly Line	976,833	979,033	978,333
Storages	4,808,301	4,820,264	4,813,092
Total	9,968,934	9,894,708	9,977,946

Table 4   Accumulated Product Cost of 6 days

	HC	EDD	SPT
Processing A	1,959,670	1,959,649	1,959,670
Processing B	3,943,778	3,943,778	3,943,778
Assembly Line	1,748,700	1,748,700	1,748,700
Storages	6,936,707	6,943,103	6,943,182
Total	14,588,855	14,595,730	14,596,330

5. GENETIC ALGORITHM BASED APPROACH

To reduce product cost more aggressively, we propose another algorithm based on Genetic Algorithm. In this approach, product cost is estimated as the fitness value.

5.1. Objective System

In this experiment, we implement the algorithm to Processing A. As shown in Figure 4, there are four HMCs and three VMCs in Processing A. We assume that MRP system sends order messages to this area every day and the detail schedules are composed in this area. All materials for order messages are stocked in Material Storage. In this area, 20 kinds of materials are processed. Material 1-5 are processed only on HMCs, material 6-10 are processed only on VMCs and material 11-20 are processed on HMCs and VMCs.

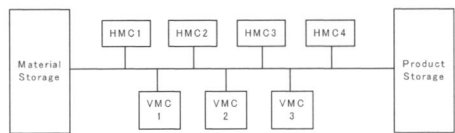

Figure 4   Overview of Processing A

5.2. Target Costs
In this experiment, product cost is estimated by summarizing following costs as equation (1).
- Stock Cost A   (Material Storage): $CSTA$
- Stock Cost B   (Product Storage): $CSTB$
- Energy Cost : CE
- Setup Cost   : CS
- Late Penalty : CL

$$C = \sum_{i \in \Phi} \left(CSTA \cdot TSTA_i + CSTB \cdot TSTB_i + CE \cdot TE_i + CS \cdot TS_i + CL \cdot TL_i \right) \quad (1)$$

Here, Φ is a set of all materials, stock time in material storage of material i is $TSTA_i$, stock time in product storage of material i is $TSTB_i$, total processing time on MCs of material i is TE_i, total setup time on MCs of material i is TS_i and late time of material i is TL_i.

5.3. Gene Structure
Gene is represented as process sequences of each MCs (Figure 5). Each gene consists of seven arrays.

5.4. Crossover
MC arrays of each parent are combined into one array with ordered crossover to preserve consistency as shown in Figure 6.

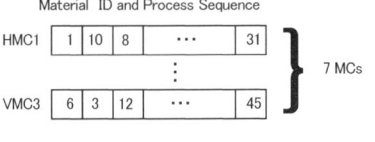
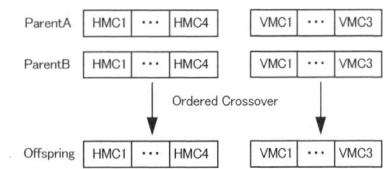

Figure 5   Gene Structure   Figure 6 Concept of Crossover

5.5. Experiments 2
In addition to the product cost, make span and setup time are also applied as the fitness value, to evaluate the performance of the proposed algorithm. Cost settings and GA settings are following.
- Cost Settings
  - Stock Cost A:CSTA = 100Yen/minute
  - Stock Cost B:CSTB = 100Yen/minute
  - Energy Cost :CE   = 100Yen/minute
  - Setup Cost   : CS   = 100Yen/minute
  - Late Penalty :CL   =100Yen/minute
- GA Settings
  - Gene number / Generation : 10
  - Number of Elite   :6
  - Number of Crossover   :4
  - Probability of Mutation   :5%
  - Number of Generation   :5000

The result of experiment 2 is shown in Table 5. If make span is applied as fitness value, the shortest make span is achieved. In the case of setup time, same result is observed. However, these fitness values tend to make only one objective function better. On the other hand, if product cost is applied as fitness value, good make span and setup time are obtained.

Table5 Result of Experiment 2

Fitness Value	Make Span	Setup Time	Cost
Make Span	1,389	1,384	1,558
Setup Time	2,485	1,935	2,870
Tardiness Time	0	0	796
Total Cost	87,110,200	87,794,800	80,785,400
Stock Cost A	26,299,300	25,559,700	32,901,400
Stock Cost B	59,963,400	61,442,600	46,918,400
Energy Cost	599,000	599,000	599,000
Setup Cost	248,500	193,500	287,000
Late Penalty	0	0	79,600

As the next experiment, we changed setting of Stock Cost B (*CSTB*) from 100 Yen/minute to 200 Yen/minute. The results of this experiment are listed in Table 6. In this condition, late penalty is lower than that of Stock Cost B. Thus, the tardiness of materials makes total cost better. Focusing on the results of proposed algorithm, total cost is considerably lower than the others. Efficiency of our proposed algorithm is shown through this experiment.

Table 6   Result of Experiments 2-2

Fitness Value	Make Span	Setup Time	Cost
Make Span	1,378	1,470	1,390
Setup Time	2,435	2,050	2,000
Tardiness Time	0	35	0
Total Cost	57,123,500	57,092,000	57,080,000
Stock Cost A	26,311,300	25,980,200	26,051,100
Stock Cost B	29,969,700	30,304,300	30,229,900
Energy Cost	599,000	599,000	599,000
Setup Cost	243,500	205,000	200,000
Late Penalty	0	3,500	0

6. CONCLUSION

In this paper, we discussed the importance of the cost as a criterion for measuring performance of schedules and two kinds of cost based scheduling algorithms are proposed. Due to the simplicity, ABC based algorithm is applicable to the large scale manufacturing systems. GA based algorithm is not applicable to the large scale manufacturing systems, however, performance of this algorithm is much better than the simple rules.

In the future research works, we are planning to improve these algorithms to achieve much better cost globally.

REFERENCES

[1] Cooper,R. and Kaplan, R.S.,Activity-based systems: measuring the costs of resource usage. Accounting Horizons, September, pp.1-13.
[2] Susumu Fujii, Toshiya Kaihara, Hiroshi Morita, Masaya Tanaka, *A Distributed Virtual FactoryIn Agile Manufacturing Environment*, The 15thInternational Conference on Production Research,1999, pp. 1551-1554
[3] Kentarou Sashio, Susumu Fujii, Toshiya Kaihara, *A Study on Push and Pull Production Control Systems –Experiments on a Distributed Virtual Factory-*, 2003, International Conference on Production Research (ICPR) – 17. Electronic Proceedings, No. 0276.

SEARCH AND POSE RECOGNITION OF INDUSTRIAL COMPONENTS USING CURVATURE OF OPTIMIZED EDGE PIXELS

Kunihiro Goto[1] and Fumihiko Saitoh[1]

[1] Department of Information Science, Faculty of Engineering, Gifu University, 1-1 yanagido, Gifu-shi, Gifu, 501-1193, Japan

ABSTRACT

We propose a method to recognize a target image area that has a free location and a free inclination in an objective image. This method uses curvatures that are measured in two sizes of areas as the matching key in order to improve the reliability of matching. Besides, if the only effective edge pixels can be used for matching, the higher reliability and the low computational cost may be expected because the number of registered template data decreases. To realize this function, the proposed method uses the genetic algorithm (GA) to determine the optimal combination of effective edge pixels for matching from the huge combinations of selected edge pixels.

KEYWORDS

search and pose recognition, pair of curvatures, edge pixel selection, voting process, genetic algorithm

IMAGE MATCHING METHOD BASED ON PAIR OF CURVATURES

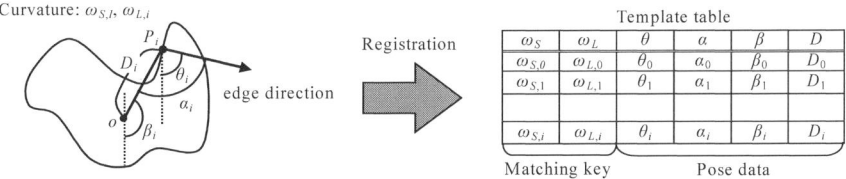

Figure 1: Geometrical parameters to register a template table

The proposed matching method consists of the template registration phase and the matching phase. In the template registration phase, edge pixels are extracted in the template image. Fig. 1 shows the geometrical relation between parameters that are stored in a template table. The variance of the edge directions around the neighboring edge pixels is defined as the curvature of the edge pixel[Saitoh03]. If many curvatures that have similar values to ones registered in the template table are included in an objective image, some wrong votes may be executed and the reliability of matching may decrease in the case of by using a single curvature. To solve this problem, the proposed method uses two kinds of curvature ω_S and ω_L that are measured in two sizes of areas in order to improve the reliability. In this method, ω_S and ω_L is called pair of curvatures. The parameters $\omega_{S,i}$ and $\omega_{L,i}$ are invariant to change of two-dimensional inclination and treated as the matching key. The other parameters are treated as pose date to determine a position and an inclination of a target image.

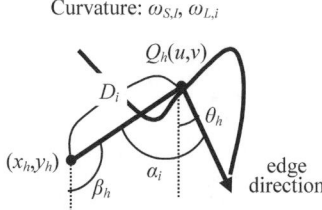

Figure 2: Relation between edge pixel and candidate of base pixels

In the matching phase, the edge directions and the curvatures of all edge pixels detected in an objective image are measured. The template table is referred by using the curvature of the edge pixels in the objective image. When a pair of curvatures between in the template table and in the objective image are similar, the address (x_h, y_h) to be vote is calculated using the geometrical parameters in the template table as shown in Fig. 2. The value of the address in the voting space is incremented as the voting process and the candidate for inclination is stored into the vote log. This process is repeated at all edge pixels in the objective image and the position of the target image area is determined by the address with the maximum voted value in the voting space. The inclination of the target image area is obtained from the peak in the histogram that is generated by the vote log.

EDGE PIXEL SELECTION BY GA

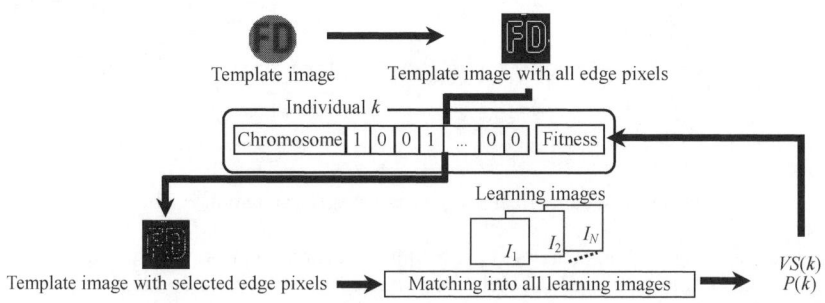

Figure 3: Evaluation of fitness

All edge pixels in a template image are used for matching described above. But, all edge pixels may be not useful for matching. The reliability may be improved by using only effective edge pixels. Additionally, the size of a template table becomes smaller and the computational cost is less by decreasing the number of edge pixels. To realize this function, the proposed method uses the GA[Holland75] to determine the optimal combination of effective edge pixels for matching from the huge combinations of edge pixels.

An individual chromosome is represented by the arrayed bits whose length is same as the number of edge pixels included in the template image. Each bit corresponds to each edge pixel in the template image. The bit 1 shows that the corresponding edge pixel is used for matching, and the bit 0 shows that the corresponding edge pixel is not used. Fig. 3 shows the evaluation method of an individual fitness in the GA. The proposed method generates a new template table only the selected edge pixels in the template image. Edge pixels are selected by an individual k whose fitness is evaluated by matching results for learning images.

The fitness is evaluated by two parameters. The one parameter $VS(k)$ means the ratio between the maximum voted value and the second voted value. This parameter evaluates the reliability of matching. The another parameter $P(k)$ means the reduced ratio of the number of edge pixels for matching. This parameter evaluates the possibility of high-speed matching. In the GA, the total number of individual is set to 100, the continuation rate of all individuals is set to 50% and the mutation rate is set to 1%.

EXPERIMENTAL RESULTS

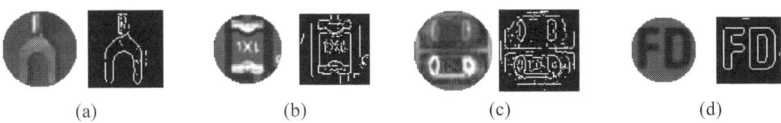

(a) (b) (c) (d)

Figure 4: Template images and template images with all edge pixels

Fig. 4 shows the four kinds of template images with gray scale and template images with all edge pixels. The size of the objective images were 192x256 pixels. The experiments were executed by three kinds of method described as follows in order to demonstrate the effectiveness of the pair of curvatures as the matching key and the selection of edge pixels.
(Method1) the matching method with the single curvature that was measured in a local area. (Using the template image with all edge pixels)
(Method2) the matching method with the pair of curvatures. (Using the template image with all edge pixels)
(Proposed method) the matching method with the pair of curvatures. (Using the template image with selected edge pixels by GA)

TABLE 1
COMPARISON OF RECOGNITION RATE BY MATCHING KEY

Template	(a)	(b)	(c)	(d)	Ave.
Method1	84%	78%	76%	100%	84.5%
Method2	100%	94%	78%	100%	93.0%

The results of the method1 and the method2 were compared in order to evaluate effectiveness of the pair of curvatures as the matching key. Table 2 shows the recognition rate that was obtained from the results of matching by using 50 objective images for each template image. This result shows that the superior result in the recognition rate was obtained by the method2 that was used the pair of curvatures as the matching key.

Fig. 5 shows the generated template images with selected edge pixels by GA. The number of edge pixels was reduced to about 28.1% on the average. It is expected that high-speed matching can be performed because the number of edge pixels was reduced.

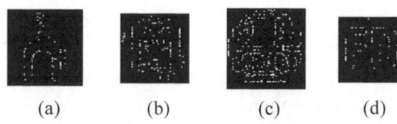

(a)   (b)   (c)   (d)

Figure 5: Generated template images with selected edge pixels

(a)   (b)   (c)   (d)

Figure 6: Matching results by the proposed method

TABLE 2
RECOGNITION RATE BY THE PROPOSED METHOD

Template	(a)	(b)	(c)	(d)	Ave.
Proposed method	100%	96%	90%	100%	96.5%

The proposed matching method was executed by using the generated template images with selected edge pixels. Fig. 6 shows examples of matching results by the proposed method. It is found the inclined target image areas were sought successfully. Over 90% recognition rates were recorded as shown in Table 2. The proposed method reduced the large number of edge pixels for matching by using the selected template table and the method recorded the equal or higher recognition rates than using the template image with all edge pixels.

TABLE 3
MATCHING TIME [ms]

	(a)	(b)	(c)	(d)	Ave.
Method1	75	115	296	98	146.0
Method2	71	121	223	99	128.5
Proposed method	67	110	142	89	102.0

Table 3 shows the processing time to search for the target image area. The averaged processing time of the proposed method was 102ms. The proposed method that was used a template image with selected edge pixels was recorded the equal or faster processing time than the method1 and method2. The proposed method can search a target component even if the component has a free location and an inclination and the method is supposed to be realized the high-speed matching for industrial use.

REFERENCES

Hollond J.H. (1975). *Adaptation in Natural and Artificial Systems*, The Univ. Michigan Press

Saitoh F. (2003). Rotation Invariant Image Matching Based on Correlation of Curvature Distribution. *Electrical Engineering in Japan* **145:4,** 975-981.

VISION-BASED NAVIGATION OF AN OUTDOOR MOBILE ROBOT USING A ROUGH MAP

Jooseop Yun, Jun Miura and Yoshiaki Shirai

Department of Mechanical Engineering, Osaka University,
Suita, Osaka, 565-0871, Japan

ABSTRACT

We describe a method of mobile robot navigation based on a rough map using stereo vision, which uses multiple visual features to detect and segment the buildings in the robot's field of view. The rough map is a map with large uncertainties in the shapes and locations of objects so that it can be built easily. The robot fuses odometry and vision information using an extended Kalman filter to update the robot pose and the associated uncertainty based on the detection of buildings in the map. An experimental result shows the potential feasibility of our localization method in an outdoor environment.

KEYWORDS

Outdoor mobile robot, Vision-based navigation, Rough map.

INTRODUCTION

In this paper, we deal with the case that the robot has an environment map to be represented as a set of 2D segments. The map approximates the outlines of buildings except for feature information to be used as landmarks (Georgiev, et al. 2002). We propose a method to robustly estimate the robot pose using multiple visual features: walls of buildings, vanishing points, and corners of buildings. The walls of buildings are extracted from the stereo vision observation. The vanishing points are calculated from the non-vertical skylines of buildings. And the corners of buildings are the vertical skylines. The visual features are matched to the given map and the results are integrated into the odometry information for the estimation of the robot pose using an Extended Kalman Filter.

FEATURE DETECTION

For the matching process, we use multiple visual features: walls of buildings from disparity image, vanishing points from non-vertical skylines, and corners of buildings from vertical skylines.

Walls of Buildings

We use the SAD (Sum of Absolute Difference) algorithm for the area-based stereo matching in order to extract disparity image (Moon, et al. 2002). In this study, the walls of buildings are extracted from the regions with a same value in the disparity image. The Building regions are extracted using the height information from the disparity information with a priori knowledge of the one-floor height of building.

Vanishing Points

A non-vertical skyline caused by the roof of a building can provide information on the relative orientation between the robot and the building. What is necessary for estimating the relative orientation is the vanishing point. We first calculate the vanishing points of the non-vertical skylines with the horizontal scene axis. And we estimate an angle between the image plane and the line from the camera center to a vanishing point which is parallel to the direction of a visible wall in the building.

Corners of Buildings

The boundary lines are the vertical skylines of buildings adjoining to the sky regions (Katsura, et al. 2003). The boundary lines correspond to the corners of buildings on the given map.

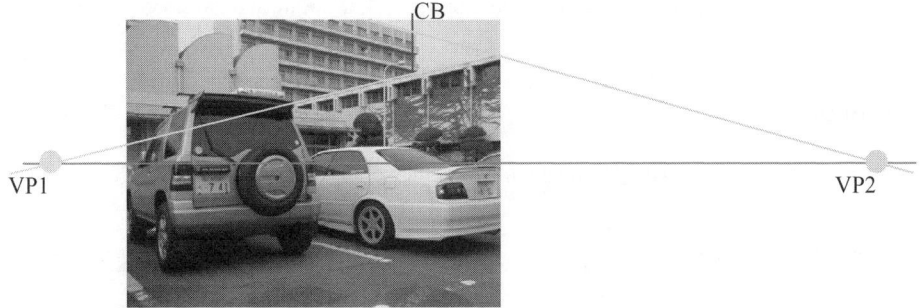

Figure 1: A boundary line and two vanishing points.

Figure 1 shows an extraction result of a corner of building (CB) from a vertical skyline and two vanishing points (VP1 and VP2) from two non-vertical skylines, respectively. The vertical and non-vertical skylines are adjoining to the sky region at the top right of the image.

ROUGH MAP

Although an accurate map provides accurate and efficient localization, it needs a lot of cost to build and update (Tomono, et al. 2001). A solution to this problem would be to allow a map to be defined roughly since a rough map is much easier to build. The rough map is defined as a 2D segment-based map that contains approximate metric information about the poses and dimensions of buildings. It also has rough metric information about the distances and the relative directions between the buildings present in the environment.

The map may carry a characteristic of the initial position as a current position and the goal position on the map. The approximate outlines of the buildings can be also represented in the map and thus used for recognizing the buildings in the environment during the navigation. And besides, we can arrange the route of robot on the map (Chronis, et al. 2003). Figure 2 shows a guide map for visitors to our university campus and an example of rough map. We use this map as a rough map representation for our localization experiments. We approximate the buildings on the map to the polygons and compute the uncertainties of their poses and dimensions for estimating the uncertainty of robot pose from the map matching.

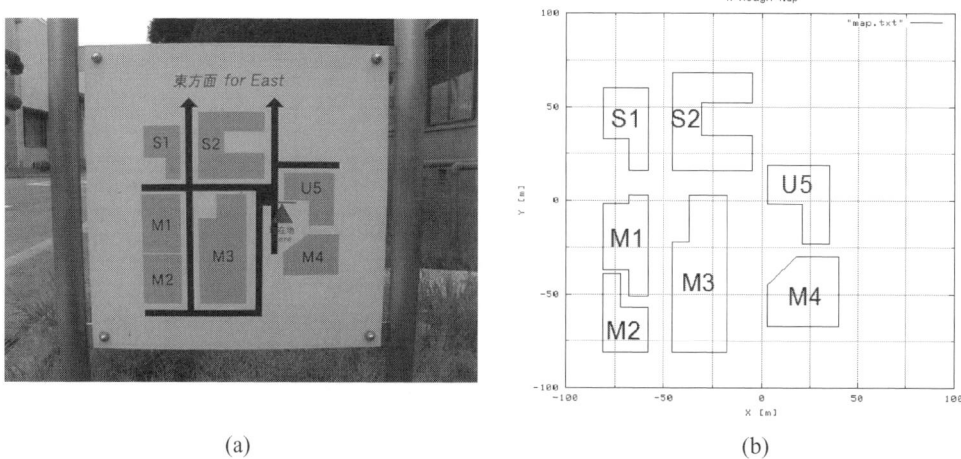

(a) (b)

Figure 2: A guide map of our university campus (a) and an example of rough map (b).

LOCALIZATION

The robot matches the extracted planar surfaces from the disparity image to the building walls on the map using the *Mahalanobis* distance criterion. Note that the distance is a quantity which is computed in the *disparity space*. The *disparity space* is constructed such that the x-y plane coincides with the image plane and the disparity axis d is perpendicular to the image plane.

The map matching provides for a correction of the estimated pose of the robot that must be integrated with odometry information. We use an extended Kalman filter for the estimation of the robot pose from the result of the map matching and this integration (DeSouza, et al. 2002).

Kalman Filter Framework

Prediction

The state prediction $X(k+1|k)$ and its associated covariance $\Sigma_X(k+1|k)$ is determined from odometry based on the previous state $X(k|k)$ and $\Sigma_X(k|k)$. The modeled features in the map, M, get transformed into the observation frame. The measurement prediction $z(k+1) = H(X(k+1|k), M)$, where H is the non-linear measurement model. Error Propagation is done by a first-order approximation which requires the Jacobian J_X of H with respect to the state prediction $X(k+1|k)$.

Observation

The parameters of features constitute the vector of observation $Z(k+1)$. Their associated covariance estimates constitute the observation covariance matrix $R(k+1)$. Successfully matched observation and predictions yield the innovations

$$V(k+1) = Z(k+1) - z(k+1), \qquad (1)$$

and their innovation covariance

$$S(k+1) = J_X \Sigma_X(k+1|k) J_X^T + R(k+1). \qquad (2)$$

Update

Finally, with the filter equations

$$W(k+1) = \Sigma_X(k+1|k) J_X^T S^{-1}(k+1), \qquad (3)$$

$$X(k+1|k+1) = X(k+1|k) + W(k+1)V(k+1), \qquad (4)$$

$$\Sigma_X(k+1|k+1) = \Sigma_X(k+1|k) - W(k+1)S(k+1)W^T(k+1). \qquad (5)$$

the posterior estimates of the robot pose and associated covariance are computed.

Filter Setup for the Walls of Buildings

We formulate the walls of buildings by $y = A + Bx$ in the map and those transformed into the *disparity space* by $y = \alpha + \beta x$ with $z = (\alpha, \beta)^T$. The observation equation $Z = (\hat{\alpha}, \hat{\beta})^T$ of the walls of buildings in disparity image is described as follows:

$$Z = F(X,L) + v$$
$$= \begin{pmatrix} fl \dfrac{\sin\theta_p - B\cos\theta_p}{A + Bx_p - y_p} \\ -l \dfrac{\cos\theta_p + B\sin\theta_p}{A + Bx_p - y_p} \end{pmatrix} + v, \qquad (6)$$

where $X = (x_p, y_p, \theta_p)^T$ is the robot pose, $L = (A, B)^T$ the map parameter, and v the random observation error. The filter setup for this feature is as follows:

$$S = J_X \Sigma_X J_X^T + J_L \Sigma_L J_L^T + \Sigma_V, \qquad (2)'$$

$$W = \Sigma_X J_X^T S^{-1}, \qquad (3)'$$

where J_X and J_L are the respective Jacobians of F w.r.t. X and L, and Σ_X, Σ_L and Σ_v are the uncertainty covariance matrices of X, L and v, respectively.

Filter Setup from the Vanishing Points

We can directly observe the robot orientation using the angle from vanishing point and the direction of building. Thus the observation $Z = \pi/2 + \theta_b - \theta_{vp}$, where θ_b is the direction angle of a wall of building and θ_{vp}, the angle from the vanishing point, and the prediction $z = \theta_r$ is the robot orientation of the last step. The filter setup for this feature is as follows:

$$S = \begin{bmatrix} 0 & 0 & 1 \end{bmatrix} \Sigma_X \begin{bmatrix} 0 & 0 & 1 \end{bmatrix}^T, \qquad (2)''$$

$$W = \Sigma_X \begin{bmatrix} 0 & 0 & 1 \end{bmatrix}^T S^{-1}. \qquad (3)''$$

Filter Setup for the Corners of Buildings

After we found the corresponding corner of building to the boundary line $Z = (x, d)^T$ in the *disparity space*, the observation equation can be described like the following equation:

$$Z = H(X,M) + v$$
$$= \begin{pmatrix} f \dfrac{(m_x - x_p)\sin\theta_p - (m_y - y_p)\cos\theta_p}{(m_x - x_p)\cos\theta_p + (m_y - y_p)\sin\theta_p} \\ \dfrac{fl}{(m_x - x_p)\cos\theta_p + (m_y - y_p)\sin\theta_p} \end{pmatrix} + v, \quad (7)$$

where $M = (m_x, m_y)^T$ is the coordinates of the building corner on the map. The filter setup for this feature is as follows:

$$S = J_X \Sigma_X J_X^T + J_M \Sigma_M J_M^T + \Sigma_V, \quad (2)'''$$

$$W = \Sigma_X J_X^T S^{-1}. \quad (3)'''$$

EXPERIMENTAL RESULTS

The experimental results are shown in figure 3 magnified from figure 2 which shows the estimation of the robot pose by the localization algorithms using the EKF. The color ellipses with 1σ uncertainty are the estimated uncertainties of the robot poses by matching the features to the map.

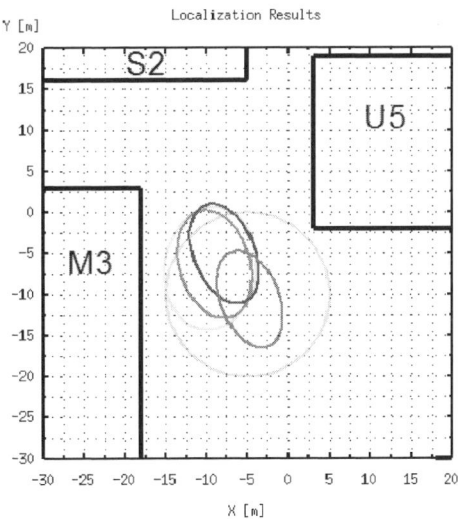

Figure 3: Localization results with uncertainty ellipses.

Table 1 shows the estimates of the robot pose in each feature used by the localization algorithm. The left figures of each table row represent the estimate of the localization method. The right parenthesized figures of the same row represent the standard deviation of the robot pose.

TABLE 1
ROBOT POSE AND STANDARD DEVIATION BY EACH FEATURE

Color	Feature	x (m) (std. dev.)	y (m) (std. dev.)	θ (°) (std. dev.)
green	Odometry	-5.0 (10.0)	-10.0 (10.0)	140.0 (10.0)
cyan	Disparity	-9.5 (7.5)	-6.8 (5.6)	141.8 (7.5)
magenta	Vanishing Point 1	-9.0 (6.7)	-6.3 (4.3)	142.7 (4.7)
blue	Vanishing Point 2	-7.8 (6.4)	-5.0 (3.7)	144.8 (3.1)
red	Corner	-4.7 (6.2)	-10.5 (3.6)	148.8 (3.0)

The table clearly demonstrates the improvements achieved by integrating several visual features using the proposed algorithm.

CONCLUSION AND FUTURE WORK

In this paper, an approach to determine the robot pose was presented in an urban area where GPS can not work since the satellite signals are often blocked by buildings. We tested the method with real data and the obtained results show that the method is potentially applicable even in the presence of errors in feature detection of the visual features and incomplete model description of the rough map. This method is a part of an ongoing research aiming autonomous outdoor navigation of a mobile robot. The system depends on the stereo vision and the rough map to compensate for the long-term unreliability of the robot odometry. No environmental modifications are needed.

Future works include performing experiments at other various places in our campus to test the robustness of the proposed approach in more detail. And finally, we will apply the approach described in this research to the autonomous navigation of a mobile robot in an outdoor urban, man-made environment consisting of polyhedral buildings.

REFERENCES

Chronis G. and Skubic M. (2003). Sketch-Based Navigation for Mobile Robots. *Proc. of IEEE Int. Conf. on Fuzzy Systems* 284-289.

DeSouza G.N. and Kak A.C. (2002). Vision for Mobile Robot Navigation: A Survey. *IEEE Trans. on Pattern Analysis and Machine Intelligence* **24:2,** 237-267.

Georgiev A. and Allen P.K. (2002). Vision for Mobile Robot Localization in Urban Environments. *Proc. of IEEE/RSJ Int. Conf. on Intelligent Robots and Systems* 472-477.

Katsura H. Miura J. Hild M. and Shirai Y. (2003). A View-Based Outdoor Navigation using Object Recognition Robust to Changes of Weather and Seasons. *Proc. of IEEE/RSJ Int. Conf. on Intelligent Robots and Systems* 2974-2979.

Moon I. Miura J. and Shirai Y. (2002). On-line Extraction of Stable Visual Landmarks for a Mobile Robot with Stereo Vision. *Advanced Robotics* **16:8,** 701-719.

Tomono M. and Yuta S. (2001). Mobile Robot Localization based on an Inaccurate Map. *Proc. of IEEE/RSJ Int. Conf. on Intelligent Robots and Systems* 399-405.

TEACHING A MOBILE ROBOT TO TAKE ELEVATORS

Koji Iwase, Jun Miura, and Yoshiaki Shirai

Department of Mechanical Engineering, Osaka University
Suita, Osaka 565-0871, Japan

ABSTRACT

The ability of moving between floors by using elevators is indispensable for mobile robots operating in office environments to expand their work areas. This paper describes a method of interactively teaching the task of taking elevators for making it easier for the user to use such robots for various elevators. The necessary knowledge of the task is organized as the *task model*. The robot examines the task model and determines what are missing in the model, and then asks the user to teach them. This enables the user to teach the necessary knowledge easily and efficiently. Experimental results show the potential usefulness of our approach.

KEYWORDS

Mobile robots, Interactive teaching, Task models, Take an elevator, Visual navigation.

INTRODUCTION

The ability of moving between floors by using elevators is indispensable for mobile robots performing service tasks in office environments to extend their working areas. We have developed a mobile robot that can take elevators, but we had to give the robot in advance the necessary knowledge such as the shape of the elevator and the positions of the buttons. Since the necessary knowledge of the task of taking elevators is different from place to place, it is desirable that the user can easily teach such knowledge on-site.

We have been developing a teaching framework called *task model-based interactive teaching* (Miura et al. 2004), in which the robot examines the description of a task, called *task model*, to determine missing pieces of necessary knowledge, and actively asks the user to teach them. We apply this framework to the task of taking elevators (*take-an-elevator* task) by our robot (see Fig. 1). This paper describes the task models and the interactive teaching method with several teaching examples.

TASK MODEL-BASED INTERACTIVE TEACHING

Interaction between the user and a robot is useful for an efficient and easy teaching of task knowledge. Without interaction, the user has to think by himself/herself about what to teach to the robot. This is difficult for the user partly because he/she does not have enough knowledge of the robot's ability (i.e., what the robot can (or cannot) do), and partly because the user's knowledge may not be well-structured. If the robot knows of *what are needed* for achieving the task, then the robot can ask the user to teach them; this enables the user to easily give necessary knowledge to the robot. This section explains the representations for task models and the teaching strategy.

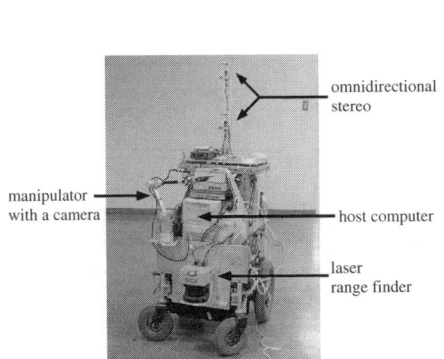

Figure 1: Our mobile robot.

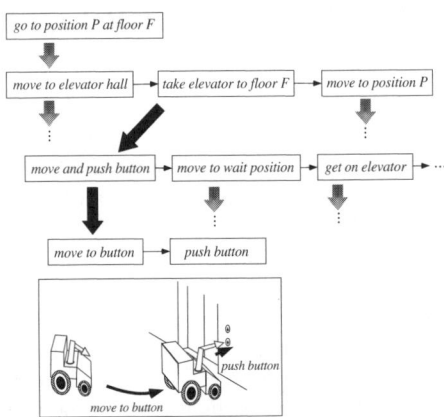

Figure 2: A hierarchical structure of the *take-an-elevator* task.

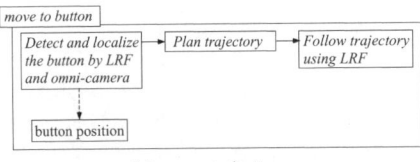

(a) move to button.

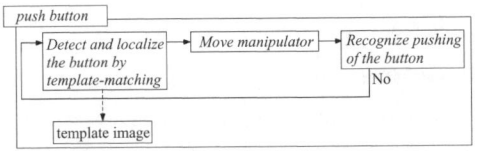

(b) push the button.

Figure 3: Diagrams for example primitives. Dashed lines indicate dependencies.

Task Model

In our interactive teaching framework, the knowledge of a task is organized in a *task model*, in which necessary pieces of knowledge and their relationships are described. Some pieces of knowledge require other ones; for example, a procedure for detecting an object may need the shape or the color of the object. Such dependencies are represented by the network of knowledge pieces. The robot examines what are given and what are missing in the task model, and asks the user to teach the missing pieces of knowledge.

Hierarchical Task Structure Robotic tasks usually have hierarchical structures. Fig. 2 shows a hierarchy of robot motions for the *take-an-elevator* task. For example, a subtask, *move and push button*, is further decomposed into two steps (see the bottom of the figure): moving to the position where the robot can push the button, and actually pushing the button by the manipulator using visual feedback. Such a hierarchical task structure is the most basic representation in the task model.

Non-terminal nodes in a hierarchical task structure are *macros*, which are further decomposed into more specific subtasks. Terminal nodes are *primitives*, the achievement of which requires actual robot motion and sensing operations.

Robot and Object Models The robot model describes knowledge of the robot system such as the size and the mechanism of components (e.g., a mobile base and an arm) and the function and the position of sensors (e.g., cameras and range finders). Object models describe object properties including geometric ones, such as size, shape, and pose, and photometric ones related to visual recognition.

Movements The robot has two types of movements: free movement and guarded movement. A free movement is the one that the robot is required to a given destination without colliding with obstacles; the robot does not need to follow a specific trajectory. On the other hand, in a guarded movement, the robot has to follow some trajectory, which is usually generated from the configuration of surrounding obstacles; movements of this type are basically used for reaching a specific pose (position and orientation) or for passing through a narrow space. Fig. 3(a) shows the diagram for the subtask of moving to the position where the robot can push a button.

Hand Motions   Hand motions are described by its trajectory. They are usually implemented as sensor-feedback motions. Fig. 3(b) shows the diagram for the subtask of pushing a button.

Sensing Skills   A sensing operation is represented by a *sensing skill*. Sensing skills are used in various situations such as detecting and recognizing objects, measuring properties of objects, and verifying conditions on the geometric relationship between the robot and the objects.

Interactive Teaching Using Task Model

The robot tries to perform a task in the same way even in the case where some pieces of knowledge are missing. When the robot cannot execute a motion because of a missing piece of knowledge, the robot pauses and generates a query to the user for obtaining it. By repeating this process, the robot completes the task model with leading the interaction with the user. It could be possible to examine the whole task model before execution and to generate a set of queries for missing pieces of knowledge.

ANALYSIS OF TAKE-AN-ELEVATOR TASK

The *take-an-elevator* task is decomposed into the following steps:

(1) Move to the elevator hall from the current position. This step can be achieved by the free space recognition and the motion planning ability of the robot (Negishi, Miura, and Shirai 2004), provided that the route to the elevator hall is given.
(2) Move to the place in front of the button outside the elevator, where the manipulator can reach the button. The robot recognizes the elevator and localizes itself with respect to the elevator's local coordinates. For the movement, the robot sets a trajectory from the current position to the target position, and follows it by a sensory-feedback control.
(3) Localize the button and push it using the manipulator. The robot detects that the button is pushed by recognizing that the light of the button turns on.
(4) Move to the position in front of the elevator door where the robot waits for the door to open.
(5) Get on the elevator after recognizing the door's opening.
(6) Localize and push the button of the destination floor inside the elevator, as the same as (3).
(7) Get off the elevator after recognizing that the door opens (currently, the arrival at the target floor is not verified using floor signs inside the elevator).
(8) Move to the destination position at the target destination floor, as the same as (1).

Based on this analysis, we developed the task model for the *take-an-elevator* task. Fig. 4 shows that the robot can take an elevator autonomously by following the task model.

TEACHING EXAMPLES

The robot examines the task model, and if there are missing pieces of knowledge in it, the robot acquires them through the interaction with the user. Each missing piece of knowledge needs the corresponding teaching procedure.

The above steps of the *take-an-elevator* task are divided into the following two parts. Steps (1) and (8) are composed of free movements. The other steps are composed of guarded movements near the elevator and hand motions. The following two subsections explain the teaching methods for the first and the second parts, respectively.

approach an elevator   push the button   wait for the opening

get on the elevator   push the button inside   get off the elevator

Figure 4: The mobile robot is taking an elevator.

Route Teaching

The robot needs a free space map and a destination or a route to perform a *free movement*. The free space map is generated by the map generation capability of the robot, which is already embedded (Miura, Negishi, and Shirai 2002). The destination may be given by some coordinate values, but they are not intuitive for the user to teach. So we take the following "teaching by guiding" approach (Katsura et al. 2003, Kidono, Miura, and Shirai 2002).

In route teaching, we first take the robot to a destination. During this *guided* movement, the robot learns the route. Then the robot can reach the destination by localizing itself with respect to the learned route. Such two-phase methods have been developed for both indoor and outdoor mobile robots; some of them are map-based (Kidono, Miura, and Shirai 2002, Maeyama, Oya, and Yuta 1997) and some are view-based (Katsura et al. 2003, Matsumoto, Inaba, and Inoue 1996).

In this work, the robot simply memorizes the trace of its guided movement. Although the estimated trace suffers from accumulated errors, the robot can safely follow the learned route because of the reliable map generation; the robot moves to the direction of the destination within the recognized free space.

The next problem is how to guide the robot. In Katsura et al. (2003) and Kidono, Miura, and Shirai (2002), we used a joystick to control the robot; but this requires the user to know the mechanism of the robot. A user-friendly way is to implement a person-following function to the robot (Huber and Kortenkamp 1995, Sawano, Miura, and Shirai 2000). For a simple and reliable person detection, we use a teaching device which has red LEDs; the user shows the device to the robot while he/she guides it to the destination (see Fig. 5). The robot repeatedly detects the device in both of the two omnidirectional camera by using a simple color-based detection algorithm, and calculates its relative position in the robot coordinates. The calculated position is input to our path planning method (Negishi, Miura, and Shirai 2004) as a temporary destination. Fig. 6 shows a snapshot of person tracking during a guided movement.

Teaching of Vision-Based Operation

This section describes the methods for teaching the position of an elevator, the positions of buttons, and the views of them.

Teaching the Elevator Position   Suppose that the robot has already be taken to the elevator hall, using the method described above. The robot then asks about the position of the elevator. The user indicates it by pointing the door of the elevator (see Fig. 7). The robot has a general model of elevator shape, which is mainly composed of two parallel lines corresponding to the wall and the elevator door projected onto the floor. Using this model and the LRF (laser range finder) data, the robot searches the indicated area for the elevator and sets the origin of the elevator local coordinates at the center of the gap of the wall in front of the door (see Fig. 8).

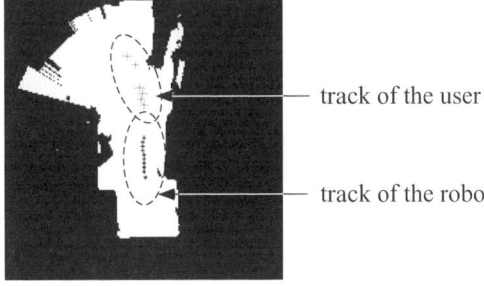

Figure 5: Taking the robot to the destination.

Figure 6: Tracking the user. The white area is the detected free space.

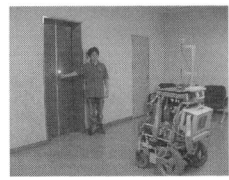

Figure 7: Teaching the elevator position to the robot.

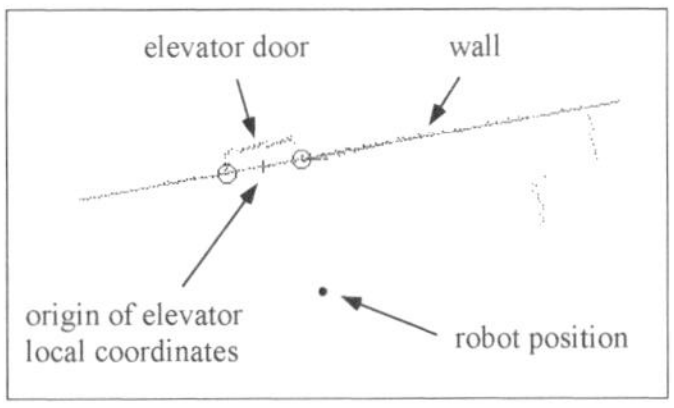

Figure 8: Elevator detection from the LRF data.

Figure 9: A detected button outside the elevator.

Teaching the Button Position The robot then asks where the buttons are, and the user indicates their rough position. The robot searches the indicated area on the wall for image patterns which match the given button models (e.g., circular or rectangular). Fig. 9 shows an example of detected button. The position of the button with respect to the elevator coordinates and the button view, which is used as an image template, are recorded after the verification by the user. The robot learns the buttons inside the elevator in a similar way; the user indicates the position of the button box, and the robot searches there for buttons.

CONCLUSION

This paper has described a method of interactively teaching the task of taking elevators to a mobile robot. The method uses *task models* for describing the necessary pieces of knowledge for each task and their dependencies. Task models include the following three kinds of robot-specific knowledge: object models, motion models, and sensing skills. Using the task model, the robot can determine what pieces of knowledge are further needed, and plans necessary interactions with users to obtaining them. By this method, the user can teach only the important pieces of task knowledge easily and efficiently. We have shown the preliminary implementation and experimental results on the *take-an-elevator* task.

Currently the task model is manually designed for the specific, *take-an-elevator* task from scratch. It would be desirable, however, that a part of existing task models can be reused for describing another. Since reusable parts are in general commonly-used, typical operations, a future work is to develop a repertoire of typical operations by, for example, using an inductive learning-based approach (Dufay and Latombe 1984, Tsuda, Ogata, and Nanjo 1998). By using the repertoire, the user's effort for task modeling is expected to be reduced drastically.

Another issue is the development of teaching procedures. Although the mechanism of determining missing pieces of knowledge in a dependency network is general, for each missing piece, the corresponding procedure for obtaining it from the user should be provided. Such teaching procedures are also designed manually at present and, therefore, the kinds of pieces of knowledge that can be taught are limited. Implementing the procedures for various pieces of knowledge requires much user's effort, especially for non-symbolic (e.g., geometric or photometric) knowledge. Another future work is thus to develop interfaces that can be used for teaching a variety of non-symbolic knowledge. Graphical user interfaces (GUIs) (e.g., Saito and Suehiro 2002) or multi-modal interfaces (MMIs) (e.g., Iba, Paredis, and Khosla 2002) are suitable for this purpose.

Acknowledgments

This research is supported in part by Grant-in-Aid for Scientific Research from Ministry of Eduction, Culture, Sports, Science and Technology, the Kayamori Foundation of Informational Science Advancement, Nagoya, Japan, and the Artificial Intelligence Research Promotion Foundation, Nagoya, Japan.

REFERENCES

B. Dufay and J.C. Latombe. (1984). An Approach to Automatic Robot Programming Based on Inductive Learning. *Int. J. of Robotics Research*, **3:4**, 3–20.

E. Huber and D. Kortenkamp. (1995). Using Stereo Vision to Pursue Moving Agents with a Mobile Robot. In *Proceedings of 1995 IEEE Int. Conf. on Robotics and Automation*, 2340–2346.

S. Iba, C.J. Paredis, and P.K. Khosla. (2002). Interactive Multi-Modal Robot Programming. In *Proceedings of 2002 IEEE Int. Conf. on Robotics and Automation*, 161–168.

H. Katsura, J. Miura, M. Hild, and Y. Shirai. (2003). A View-Based Outdoor Navigation Using Object Recognition Robust to Changes of Weather and Seasons. In *Proceedings of 2003 IEEE/RSJ Int. Conf. on Intelligent Robots and Systems*, 2974–2979.

K. Kidono, J. Miura, and Y. Shirai. (2002). Autonomous Visual Navigation of a Mobile Robot Using a Human-Guided Experience. *Robotics and Autonomous Systems*, **40:2-3**, 121–130.

S. Maeyama, A. Ohya, and S. Yuta. Autonomous Mobile Robot System for Long Distance Outdoor Navigation in University Campus. *J. of Robotics and Mechatronics*, Vol. 9, No. 5, pp. 348–353, 1997.

Y. Matsumoto, M. Inaba, and H. Inoue. (1996). Visual Navigation using View-Sequenced Route Representation. In *Proceedings of 1996 IEEE Int. Conf. on Robotics and Automation*, 83–88.

J. Miura, Y. Negishi, and Y. Shirai. (2002) Mobile Robot Map Generation by Integrating Omnidirectional Stereo and Laser Range Finder. In *Proceedings of 2002 IEEE/RSJ Int. Conf. on Intelligent Robots and Systems*, 250–255.

J. Miura, Y. Yano, K. Iwase, and Y. Shirai. (2004). Task Model-Based Interactive Teaching. In *Proceedings of IROS2004 Workshop on Issues and Approaches on Task Level Control*, 4–11.

Y. Negishi, J. Miura, and Y. Shirai. (2004). Adaptive Robot Speed Control by Considering Map and Localization Uncertainty. In *Proceedings of the 8th Int. Conf. on Intelligent Autonomous Systems*, 873–880.

F. Saito and T. Suehiro. (2002). Toward Telemanipulation via 2-D Interface – Concept and First Result of Titi. In *Proceedings of IECON 02*, 2243–2248.

Y. Sawano, J. Miura, and Y. Shirai. (2000). Man Chasing Robot by an Environment Recognition Using Stereo Vision. In *Proceedings of the 2000 Int. Conf. on Machine Automation*, 389–394.

M. Tsuda, H. Ogata, and Y. Nanjo. (1998). Programming Groups of Local Models from Human Demonstration to Create a Model for Robotic Assmebly. In *Proceedings of 1998 IEEE Int. Conf. on Robotics and Automation*, 530–537.

GENERATED IMAGE FEATURE BASED SELECTIVE ATTENTION MECHANISM BY VISUO-MOTOR LEARNING

T. Minato[1] and M. Asada[1]

[1] Department of Adaptive Machine Systems, Graduate School of Engineering, Osaka University,
2-1 Yamada-oka, Suita, Osaka 565-0871 Japan

ABSTRACT

Visual attention is an essential mechanism of an intelligent robot. Existing research typically specifies in advance the attention control scheme required for a given robot to perform a specific task. However, a robot should be able to adapt its own attention control to varied tasks. In our previous work, we proposed a method of generating a filter to extract an image feature by visuo-motor learning. The generated image feature extractor is considered to be generalized knowledge to accomplish a task of a certain class. We propose an attention mechanism, by which the robot selects the generated feature extractors based on its task-oriented criterion.

KEYWORDS

Mobile Robot, Selective attention, Image feature generation, Image feature selection, Task-oriented

INTRODUCTION

Attention control is an essential mechanism for an intelligent robot to avoid processing enormous amounts of data. It is a data reduction process to facilitate decision making. With regard to visual attention control, it involves selection of focus, image features, and so on. Existing research typically specifies in advance the attention control scheme required for a given robot to perform a specific task. However, a robot should be able to adapt its own attention control to varied tasks and environments.

We have focused on visual attention control related to a robot's actions to accomplish a given task and proposed a method in which a robot generates an image feature extractor (i.e., image filter) which is necessary for the selection of actions through visuo-motor map learning (Minato & Asada, 2003). The robot's learning depends on the experience gathered while performing a task. In this method, the robot uses only one feature extractor for a given task. For more complex tasks, however, multiple feature extractors are necessary to accomplish the tasks and a method of selecting them should be addressed.

Some research has focused on a method of feature selection based on task-relevant criteria. McCallum (1996) proposed a method in which a robot learns not only its action but feature selection using

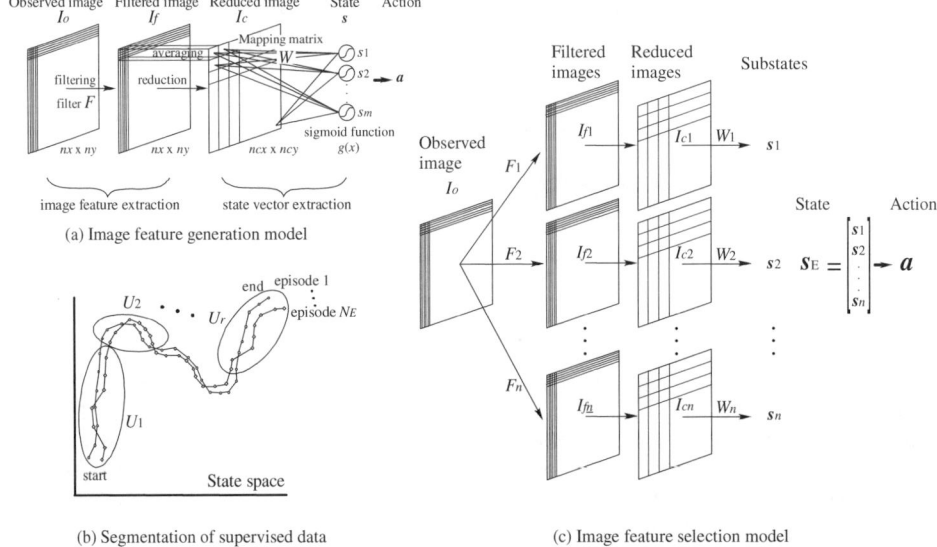

Figure 1: Image feature generation and selection models

reinforcement learning. Mitsunaga and Asada (2000) proposed a method to select a landmark according to the information gain on action selection. In these methods, however, the image features to detect the landmarks from the observed image are given a priori. It is desirable that the image feature adapts to environmental changes.

This paper proposes a method in which a robot learns to select image feature extractors generated by itself according to a task-relevant criterion. The generated feature extractors are not always suitable for new tasks. The robot must learn to select them to accomplish the task. The criterion of selection is the information gain calculated from given task instances (supervised data). Furthermore, a part of supervised data which gives the local information of the task makes the selective mechanism more effective. The method is applied to indoor navigation.

THE BASIC IDEA

In the proposed method, a robot generates an image feature extractor that is necessary for the action selection through visuo-motor map learning (Minato & Asada, 2003). The state calculation process is decomposed into feature extraction and state extraction (Figure 1(a)). A robot learns the effective feature extractor and state mapping matrix for a given task through a mapping from observed images to supervised actions. During feature extraction, the interactions between raw data are limited to local areas, while the connections between the filtered image and the state spread over the entire space to represent non-local interactions. It is, therefore, expected that the feature extractors are more general and could be generalized knowledge to accomplish a task of a certain class.

The robot calculates the filtered image I_f from the observed image I_o using the feature extractor F. The state $s \in \Re^m$ is calculated from a compressed image I_c by the sum of weighted pixel values. The robot decides the appropriate action for the current state s. The function model of the feature extractor is given, and the robot learns its parameters and the mapping matrix W by maximizing the information gain of s with respect to action a.

The robot, which generates one feature extractor for a given task, obviously needs multiple feature extractors for more complex tasks. It is unnecessary to learn a feature extractor for every given task. The generated feature extractor must be generalized to make the robot more adaptable.

In this method, the robot reuses a number of generated feature extractors from past experiences and selects effective ones for action decision. The system is shown in Figure 1(c). The robot is given a number of different feature extractors, but must select those which are appropriate for the given task. The robot, therefore, learns the state mapping matrix using the supervised data and evaluates which feature extractor is appropriate from the distribution of supervised data. If the robot uses all of the supervised data in the evaluation, optimality in a local part of the task is lost. To evaluate the effectiveness in the local task, the robot estimates which local task it is performing from the history of observations and selects the feature extractor using a portion of the supervised data corresponding to the local task.

SELECTIVE ATTENTION MECHANISM BASED ON GENERATED IMAGE FEATURE EXTRACTORS

The System Overview

The robot is given n different feature extractors ($F_i, i=1,...,n$) and calculates the substate $s_i \in \Re^m$ using the mapping matrix W_i corresponding to F_i. Each mapping matrix is learned by maximizing the information gain of s_E (direct product of $s_1,...,s_n$) with respect to the supervised action $a \in A$.

The robot selects the feature extractor which has a maximum expected information gain and decides the appropriate action for the substate calculated using the selected feature extractor. It cannot always decide the appropriate action using one feature extractor. It, therefore, estimates the reliability of selected feature extractors and selects repeatedly until the reliability exceeds a given threshold.

For evaluation in the local task, the supervised data is segmented by temporal order. The robot selects a sub-supervised data according to the history of observation and selects feature extractors to decide an action using the selected one.

State learning

First, the robot collects supervised successful instances of the given task for N_E episodes. An episode ends when the robot accomplishes the task. An instance u consists of an observed image I_o^u and a given action a^u. Next, the robot learns the mapping matrices. The state s_E^u consists of substates s_i^u which are calculated from I_o^u using F_i and W_i (the superscript denotes the corresponding instance). The evaluation function used to learn W_i is to maximize the information gain of s_E with respect to a. It is equivalent to minimizing the following risk function R (see Vlassis, Bunschoten, and Kröse (2001)).

$$R = -\frac{1}{N}\sum_{u \in U} \log p(a^u \mid s_E^u). \tag{1}$$

In Eqn. 1 U denotes a set of all instances and N denotes the number of instances. The probability density functions are computed using kernel smoothing. Using the gradient method, the mapping matrices W_i, which minimize R, are obtained.

Feature Extractor Selection

The set of instances U is divided into r subsets $U_j, j = 1,...,r$ before performing the task (Figure 1(b)). The subsets are arranged by temporal order. The choice of r includes a trade-off between the locality of the evaluation and the reliability of the action decision. To evaluate it, U is divided so that instances of similar state and action are included into the same subset. The vector $c^u = (s_E^u, a^u, \tau^u / L)$ is defined from the instance u, and U is divided by applying the ISODATA algorithm for the set $\{c^u\}$. Here, L is the time taken to accomplish the task and τ is the time when the instance u is observed. The value of each component is normalized to the range [0,1]. To avoid aliasing problems, the robot always uses two neighbouring subsets to evaluate the effectiveness of a feature extractor.

The robot executes the following process at every interval.
1) Selecting subsets of instances: Select subsets of instances $\mathcal{V}$ according to a procedure shown in the next section. $k = 0$.
2) Calculating a reliability of action decision: Calculate substate s_{ok} corresponding to the k-th selected feature extractors F_{ok} and the entropy $H_\mathcal{V}(A | S_o)$ using the instances in $\mathcal{V}$.

$$H_\mathcal{V}(A | S_o) = -\sum_{u \in \mathcal{V}} P_\mathcal{V}(a^u | S_o) \log P_\mathcal{V}(a^u | S_o), \qquad (2)$$

where $S_o = \{s_{o1},..., s_{ok}\}$ ($S_o = \phi$, if $k = 0$) and $P_\mathcal{V}$ denotes a probability calculated on the set $\mathcal{V}$. $H_\mathcal{V}(A | S_o)$ means an uncertainty of the action decision. Evaluate the uncertainty using a threshold H_{th}.
- If $H_\mathcal{V}(A | S_o) \leq H_{th}$, then go to 4.
- Otherwise, $k = n$ and $\mathcal{V} = U$, then go to 4.
- Otherwise, $k = n$ and $\mathcal{V} \neq U$, then go to 2 with $\mathcal{V} = U$ and $k = 0$.
- Otherwise, go to 3.
3) Selecting a feature extractor: Let the set of unselected feature extractors be $\mathcal{F}$. Calculate an expected entropy for each unselected feature extractor $F_z \in \mathcal{F}$. The expected entropy is:

$$\sum_{u \in \mathcal{V}} P_\mathcal{V}(s_z^u) H_\mathcal{V}(A | S_o, s_z^u), \qquad (3)$$

where s_z is a substate corresponding to F_z. Select the feature extractor F_{ok+1} which has the minimum entropy, that is, has the maximum information gain. $k \leftarrow k+1$. go to 2.
4) Deciding an action: Execute the action a which maximizes $P_\mathcal{V}(a | S_o)$.

Selecting Subsets of Instance

The robot selects subsets of instances $\mathcal{V}$ in order to calculate a probability and an entropy according to the states $S_o(\tau - 1),..., S_o(\tau - h)$ observed in the past h steps. For each subset U_j the robot counts the number of substates which satisfy $P_{U_j}(S_o(\cdot)) > 0$ in h substates. If the count C_j is greater than a threshold C_{th}, U_j and U_{j+1} are added to $\mathcal{V}$. If $C_j = 0$, the robot uses all instances ($\mathcal{V} = U$).

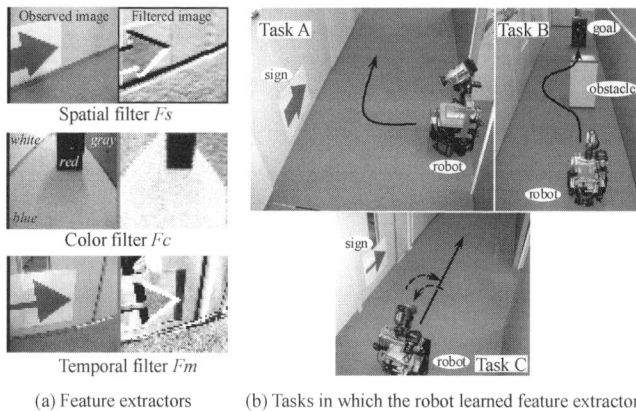

Figure 2: Feature extractors which are generated in the past tasks

EXPERIMENT

Experimental Setting

We used a small mobile robot which is about 40 cm high and has a camera with a fixed orientation to look ahead at the floor. The task is to move along a given path to a destination. The size of I_o and I_f in pixels is 64 x 54 and that of I_c is 8 x 6. We defined the dimension of a substate as $m = 1$. The robot can move at a translational speed v and a steering speed ω independently. To reduce the computation cost, we discretized the state and action space and calculated the probabilities. We set the history length to $h = 10$. The thresholds are set to $H_{th} = 0.4$ and $C_{th} = 0.8h$.

The robot was given three feature extractors shown in Figure 2. Fs, Fc, and Fm are generated in tasks A, B, and C, respectively. The feature extractors have the following characteristics.
- 3 x 3 spatial filter Fs: This type of filter calculates sum of weighted brightness values of nine neighbouring pixels. The generated filter emphasizes and inhibits horizontal edge.
- Color filter Fc: This type of filter calculates sum of weighted red, green, and blue. The generated filter inhibits red.
- Spatial filter Fm: This type of filter calculates sum of weighted past five images. The generated filter emphasizes the current image and inhibits the past image.

Feature Extractor Selection

The task given to the robot is shown in Figure 3(a). The robot moves to the front of the door and waits for it to open. It moves to the destination after the door opens. The environment is the same as that of tasks A, B, and C. We gave three episodes of successful instances ($L = 234, 254, 233$). After learning, the robot divided all instances into 13 subsets using ISODATA algorithm.

Figure 3(b) shows the learned behavior and Figure 3(c) shows the selected feature extractors at each time step. The selected feature extractors differ depending on the situation. The average number of selected feature extractors per step is 1.57. Figure 3(d) shows the selected subsets of instances at each step. When the robot could not choose an action from the selected subsets because of low reliability, it used all instances to decide again. $\mathcal{V}$ in the figure shows the step when the robot could choose an action from the selected subsets. It is verified that the robot accomplishes the task while selecting

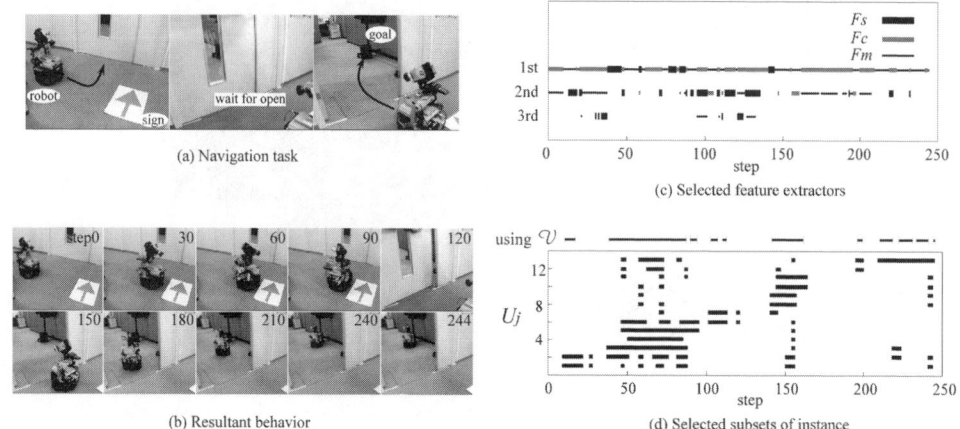

Figure 3: Task and experimental results

effective feature extractors.

To verify the subset of instances, we performed the same experiment except for the procedure to select the subsets. In this experiment, the robot always selected all instances. In the result, the average number of selected feature extractors per step is 1.97, which is larger than the result of Figure 3. This means that the robot spent much more time for action decision at each step. Hence, the robot effectively decides the action using a portion of the instances.

CONCLUSION

This paper has proposed a method in which a robot learns to select image feature extractors generated by itself according to a task-relevant criterion. A portion of supervised data which gives the local information of the task makes the selection of feature extractors more effective. In the proposed method, a robot can accomplish more complicated tasks using multiple feature extractors. Our future work is to verify the extent of effectiveness of the proposed method.

References

McCallum, A. (1996). Learning to Use Selective Attention and Short-Term Memory in Sequential Tasks. *Proceedings of International Conference on Simulation of Adaptive Behavior*, 315-324.

Minato, T. and Asada, M. (2003). Towards Selective Attention: Generating Image Features by Learning a Visuo-Motor Map. *Robotics and Autonomous Systems* **45**, 211-221.

Mitsunaga, N. and Asada, M. (2000). Observation Strategy for Decision Making based on Information Criterion. *Proceedings of International Conference on Intelligent Robots and Systems*, 1038-1043.

Vlassis, N., Bunschoten, R., and Kröse, B. (2001). Learning Task-Relevant Features from Robot Data. *Proceedings of International Conference on Robotics and Automation*, 499-504.

PRECISE MICRO ROBOT BIO CELL MANIPUTATION BASED ON THE MICROSCOPIC IMAGE RECONGIOTION

Daigo Misaki [1], Chiba Naoto [2], Takashi Usuda [3], Ohmi Fuchiwaki [2] and Hisayuki Aoyama [2]

[1] Department of Mechanical Engineering, Shizuoka Institute of Science and Technology, 2200-2, Toyosawa, Fukuroi, Shizuoka 437-8555,Japan
[2] Department of Mechanical Engineering & Intelligent Systems, University of Electro-Communications, Tokyo 1-5-1, Chofu, Tokyo 182-8585 ,Japan
[3] Advanced Industrial Science and Technology,Umezono 1-1, Tsukuba, Ibaraki 305-8563, Japan

ABSTRACT

For these years the demand of microscopic manipulation system has been increased. However, conventional microscopic manipulation systems are expensive, low-flexibility and limited working range, so there are various problems in the actual shop floor of microscopic operations. The micro robot technology has the great potential benefit for microscopic manipulations system because they can provide a high positioning accuracy and flexible operations easily. We have developed various kinds of micro robots and they provided several unique accurate applications such fine surface machining and microscopic bio-cell manipulations. In this paper, we describe the micro robot bio manipulator system based on 3D microscopic image recognition. The basic experimental results show the system has high flexibility and high-precision control is available by a simple mechanism.

KEYWORDS

Piezoelectric Actuator, Micro Robot, Bio Cell Manipulation, Microscopic Image Recognition, Non Contact Rotation Mechanism

1. INTRODUCTION

With the demand for microscopic operation, it becomes much useful to control the micro robots system with nano-tool on the desktop. For these years, several groups have succeeded in developing many smart micro or nano robots system which can apply to the micro-nano factories. In particular the fields of biotechnology such as cell manipulations are one of the most important applications and various researches are reported [1], [2]. However, because theses systems consist of conventional precision positioning devices, they have the problems of poor layout flexibility and limited working range to take care of complex and precise microscopic operations.

Our research group also has provided the various micro robots and they provided several unique accurate applications under microscope such a fine surface machining, micro parts conveying and micro bio cell manipulations in resent year [3].

In order to provide such microscopic operations automatically under the microscope by multiple micro robots, we have investigated the robot navigation strategy and the precise positioning method [4].

However, to develop more advanced microscopic biological operations in the bio cell, the more precise and complicated microscopic operations are required. For example, when a micro injection tool attached the micro robot is guided to a bio cell and microscopic operations such a nuclear transplantation can be executed, then the bio cell should be fixed gently at the specified position to keep the high success rate. For this procedure, the microscopic positioning manner of an egg cell is required. Some precise microscopic positioning mechanisms have been studied in resent year. Those systems using an electromagnetic field may damage bio-samples which are sensitive to environmental change and are not suitable in order to handle a bio cell [5].

Then, we have developed a unique precise microscopic positioning mechanism which can control bio cell in three dimensions. This mechanism can provide non-contact rotation positioning. Using this rotational mechanism, we can get the several cell microscopic images from CCD device at every rotating angle. These images can be reconstructed for a 3D bio cell model. This 3D model geometry provides the precise location of the bio cell elements. This microscopic operation system by using multiple micro robots and microscopic image recognition is reasonable and flexible compared with the conventional system. In this paper, the experimental results of the fundamental function for this operation system are described.

2. SYSTEM CONFIGURATION

Fig.1 (a) shows the proposed flexible microscopic manipulation system by using versatile micro robots under microscope which is ongoing development in our group. These micro robots can be controlled using a PC in real-time analyzing of microscopic images. All micro robots are set on a steel table. In this basic setup, operator of microscopic operations can execute flexible microscopic tasks with easy operation by using this micro robot system. All positioning facilities are given by micro robots' movements so that all mechanical functions are simply divided into each robot. This unique arrangement allows the system good flexibility and high mechanical stability although sophisticated control is required. This may be a good application for micro robots practically. We can easily build in the micro robots to the conventional micro processing instruments. Fig.1 (b) shows the actual basic set up of our proposed system. This system can satisfy a requirement of a bio cell manipulation, we can construct the high flexibility system in reasonable cost.

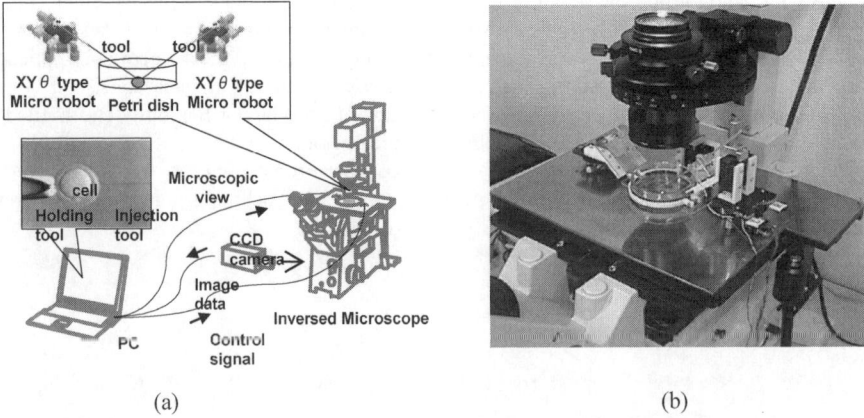

Figure 1: The microscopic operation system using multiple micro robots.

3. VERSATILE MICRO ROBOT

In microscopic operations, an operator needs to operate a tool intricately. Fig.2 (a) shows motion patterns required to carry out the precise microscopic operation. Because there is no surplus space in such a narrow area, such movement was difficult for the previous micro robot [6]. To realize these motion patterns at the same time, a structure was need that could move in XY directions and in rotation in θ independently, namely holonomic behaviour. The structure of the versatile micro robot which is proposed to satisfy the requirements of microscopic operations [7]. Two u-shaped electromagnets which are arranged to cross each other are connected by four piezo elements so that the micro robot can move in any direction like an inchworm. Also we designed the special joint at one of the 4 legs to ensure all legs smooth simultaneous contact on the surface. This layout of actuators can allow it to move precisely in a holonomic locomotion manner. As shown in Fig.2 (b), the length of robot is 35 [mm], width is 35 [mm] and height is 15[mm]. Weight is 34[g]. We use the stacked type PZT elements of 5 [mm] x 5 [mm] x 10 [mm]. Each piezo element is connected to each electromagnet with a plastic insulator. In experiments, we confirmed that the micro robot can move in XY directions. In experiments, we confirmed that the micro robot can move in XY directions as well as in rotation with sub micron positioning performance.

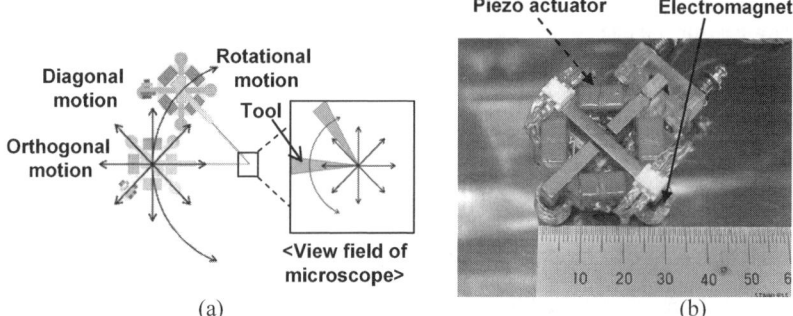

Figure 2: Piezo driven micro robot for microscopic operation

4. PRECISE MICRO ROBOT BIO-CELL MANIPUTAION

To perform bio-cell manipulations, the precise positioning of bio manipulation tool such as holding pipette is required. Furthermore, in order to perform more precise bio-cell manipulations, operation of the direction of Z as well as the operation of XY is required. In our approach, precise positioning of micro pipette is carried out by using microscopic image recognition. In order to construct the 3D information, it is necessary to observe a bio-cell from any direction by rotating a CCD camera around a bio-cell, or rotating bio-cell on a center. In the microscopic manipulation under microscope, because it is difficult to move a camera due to the spatial constraint, the constructing 3D information for precise micro manipulation is carried out by fixing a camera and rotating an object.

In this paper we proposed and used the non contact rotation positioning of an egg cell by pipette vibration, which generates a rotational flow near the pipette. In our system, this rotational flow is generated by using only the versatile micro robots. First, as shown in Fig. 3 (a), the electromagnet 1 is always activated and electromagnet 2 is always not activated. And the position Fig. 3 (a)-(2) and (a)-(3) is repeated by expanding and contracting the piezo-electric element between both electromagnets. By this motion, the pipette on a micro robot can be vibrated with the amplitude of tens of microns. This proposed mechanism can be carried out non contact rotation operation, without using some additional actuators for a micro robot. This is a very important factor, in order to construct a simple and low cost microscopic operation system. Fig.3 (b) shows the microscopic vibration mechanism for the basic experiment. This mechanism can be attached on the micro robot easily.

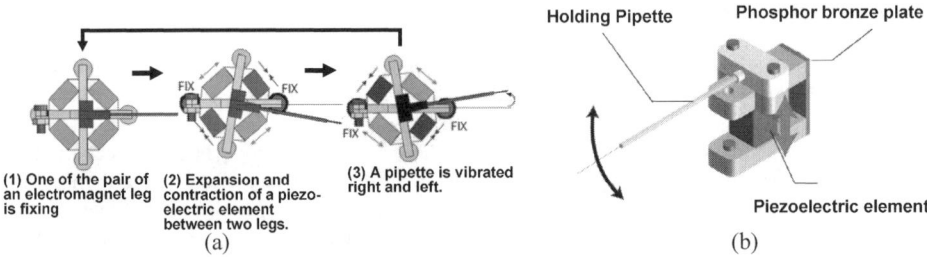

(1) One of the pair of an electromagnet leg is fixing
(2) Expansion and contraction of a piezo-electric element between two legs.
(3) A pipette is vibrated right and left.
(a)

Holding Pipette
Phosphor bronze plate
Piezoelectric element
(b)

Figure 3: Microscopic vibration mechanism

In this experiment, the rotational follow in water of schale was generated using the holding pipette of the bore 15 [μm] and an outside diameter 100 [μm] at a tip of micro pipette. The pipette tip was bent to 45 [°], and it placed in parallel to the schale bottom, and observed from the schale bottom. To tracing partial flows, water mixed with black sumi was used. The flow pattern when vibrating a piezo-electric element by the sine wave of pitch 200 [Hz] and the seal-of-approval voltage±45 [V] is shown in Fig. 4 (a).

These experimental results show that pipette vibration can generate constant flows, which shows a fixed flow pattern near the pipette. When changing the pitch and seal-of-approval voltage of a piezo-electric element, the flow velocity changed, but this flow pattern was almost fixed. Therefore, the proposed mechanism may be effective in order to control rotation of a miniature ball such as an egg cell without any mechanical contact.

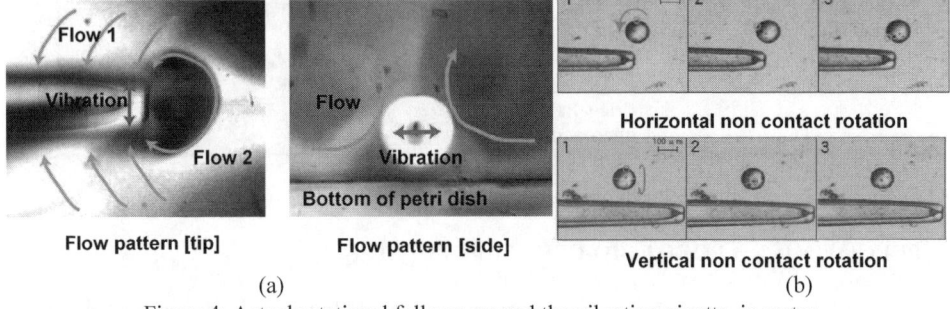

Flow pattern [tip]
Flow pattern [side]
Horizontal non contact rotation
Vertical non contact rotation
(a) (b)
Figure 4: Actual rotational follows around the vibrating pipette in water

As a basic experiment, the non contact rotational positioning of the minute glass ball placed around a pipette was carried out using our proposed mechanism. The glass minute ball of a diameter 130 [μm] (UB-67L Union Co. Ltd) was used for this primary experiment. When vibrating a piezo-electric element by the sine wave of pitch 200 [Hz] and the seal-of-approval voltage±45 [V], a minute ball rotates horizontal and vertical in the same position at the pipette tip as shown in Fig.4 (b). This motion occurs based on the straight micro flow and circle micro flow shown in Fig.4 (a). The direction of rotation of the minute ball in the experimental result of Fig. 4 (b) is the same as the direction of a flow observed in Fig.4 (a). Therefore, it is guessed that rotational positioning is caused out by this rotational flow. From these experimental results, it is shown that the proposed mechanism can be used non-contact rotation positioning in three dimensions.

When the egg cell of the mouse with a diameter 120 [μm] was used as an object of microscopic operations, the non contact rotation positioning of the egg cell by pipette vibration of a micro robot was also available in a similar manner. As show in Fig.5 (a), a general sequence of microscopic operations (rotation positioning of an egg cell in a pipette tip, movement of a pipette and aspiration fixation) required for an operation can be carried out by using only this mechanism without some special additional actuators. The rotational speed of egg cell is 33 [degree/s] and rotational resolution is about 10 [degree] from the experimental result as show in Fig.5 (b).

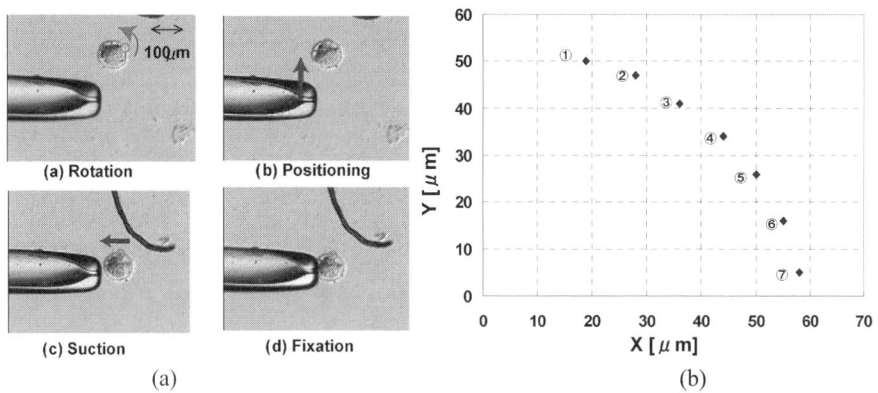

Figure 5: Results of non contact rotational positioning

5. MICROSCOPIC IMAGE RECOGNITION FOR IN BIO-CELL OPERATION

5.1 Precise pipette guiding to the bio cell using microscopic image recognition

The proposed system can carry out the precise positioning and manipulation by using microscopic image feedback. Firstly, the heads of the tools such as pipette set in the microscope view area. In order to carry out the microscopic operations, tools are positioned precisely by using microscope image recognition. In our proposed system as show in Fig.1 (a), a microscopic image is a 512x480-pixel 256 gradation gray scale obtained with a CCD camera. When the objective lens with a magnification of 10 times is used, the width of a field of view is 1mm in every direction. Fig. 6 (a) shows the basic set up of microscopic manipulation. Target of basic experiment is 100 [μm] small steel ball. Diameter of pipette is 100 [μm]. Hough transformation of a circle is used to for detect the position of a target such as a bio-cell from a microscopic image. The retrieval area in Hough transformation is narrowed by applying a differentiation filter to a microscopic image as pre-processing, and the carrying out border-line extraction. The trajectory when guiding a robot to the position (200,400) where a target of a microscopic image was detected. During the navigation, the control signal is calculated based on the position of a fine tool is observed every 10 steps of the micro robot. Fig.6 (b) shows the navigation trajectory from arbitrary start point. In this experiment, a fine tool carried on a micro robot can be guided to the target point from the start point at a speed of 100 [μm /sec] on average.

5.2 Reconstruction 3D model using non contact rotational positioning mechanism

Next basic experiment, using the non-contact rotation positioning mechanism, the miniature object of microscopic operation is recognized in three dimensions. A minute glass ball is used as an object of a basic experiment for bio cell operations. This ball is a diameter 130 [μm] (UB-67L Union Co. Ltd). The diameter of this ball is equivalent to the egg of a mouse. To a preparation of 3D micro image recognition, the 3D sphere model corresponding to the size of this ball is generated on a computer based on the information on the scale factor of a microscope, and the diameter of the circle on a screen. The sphere on a computer and an actual sphere are rotated 360 degrees by the resolution of a unit 10 degree. The circle shape on the surface model of a sphere of the boundary for recognition of front area is extracted. Pre-processing of recognition of the object of circle shape assigns a label number to a continuous pixel area, after calculating the bit map image of a CCD camera. The circular value (Value shows circle-likeness) is calculated to each labelling area. A circular domain is extracted using this value and threshold value. The circular value e is calculated by the following equation using the circumferential length l and the area S of labelling area .

$$e = \frac{4\pi S}{l^2} \quad (0 < e < 1)$$

When a local area is a perfect circle, e is set to 1.0. This value becomes small, so that area is no like a circle. When a circle is extracted, the position and size are measured and the result plotted on the surface-of-a-sphere model is shown in Fig.6 (c) .To generate this 3D surface model in the computer previously is useful to execute more precise operation and this model can be used creating control signal of micro robots.

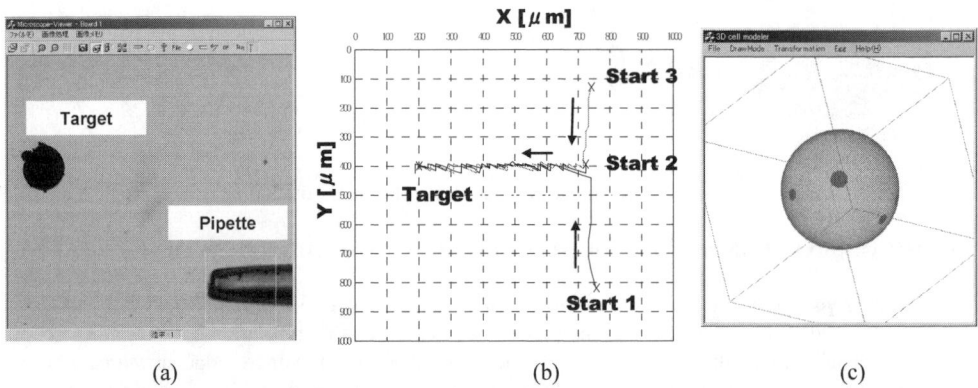

Figure 6: Results of microscopic image recognition for precise bio-cell manipulations

6. CONCLUSIONS

Flexible micro manipulation system organized by versatile micro robots under microscopes was proposed and microscopic image recognition for tool navigation and developed 3D model for microscopic operations based on the non contact rotation positioning mechanism.

In future work, carry out the more precise manipulation by using this 3D model, and then we're going to develop a more precise bio operation system.

REFERENCES

[1] Yu Sun and Bradley J. Nelson. (2002). Biological Cell Injection Using an Autonomous Micro Robotic System . *Journal of Robotics Research* **21:10**, 861-868.

[2] A. Bergander, J.-M. Breguet, C. Schmitt, R. Clavel. (2000). Micropositioners for Microscopy Applications based on the Stick-Slip Effect. *IEEE Int'l Symposium on Micromechanical and Human Sci.* 213-216.

[3] Hisayuki Aoyama and Akira Hayashii.(1999). Multiple Micro Robots for Desktop Precise Production. *Proc. of 1st Conf. of EUSPEN*, 60-63

[4] D. Misaki, S. Kayano, Y. Wakikaido, O. Fuchiwaki, H. Aoyama. (2004). Precise Automatic Guiding and Positioning of Micro Robots with a Fine Tool for Microscopic Operations. *Proc. of 2004 IEEE/RSJ International Conf. on Intelligent Robots and Systems* 213-223.

[5] H. Aoyama, F. Iwata, and A. Sasaki. (1995) Desktop flexible manufacturing system by movable micro robots. *Proc. of Int. Conf .on Robotics and Automation*, 660-665.

[6] F. Arai, A. Ichikawa, M. Ogawa, T. Fukuda, K. Horio and K. Itoigawa. (2001). High Speed Separation System of Single Microbe by Laser Trap and Dielectrophoresis. *Trans. Of JSME* **67:653**, 146-153.

[7] O. Fuchiwaki and H. Aoyama. (2003) Precise manipulation control on three versatile microrobots for flexible micro handling. *Proc. Of 19th ASPE Annual Meeting.* 287-290

SERVICE EXPLORER
- A TOOL FOR SERVICE DESIGN -

Yoshiki SHIMOMURA[1], Tomohiko SAKAO[2], Tatsunori HARA[1],
Tamio ARAI[1] and Tetsuo TOMIYAMA[3]

[1] Research into Artifacts, Center for Engineering, The University of Tokyo,
Komaba 4-6-1, Meguro-ku, Tokyo, Japan
[2] Mitsubishi Research Institute, Inc.,
Otemachi 2-3-6, Chiyoda-ku, Tokyo, Japan
[3] Delft University of Technology, Mekelweg 2, 2628 CD Delft, The Netherlands

ABSTRACT

To design and develop services effectively and efficiently, a methodology of service design and a support by the computer system based on the methodology are needed. The authors are carrying out fundamental research in "service engineering," which deals with services in an engineering manner, and have already proposed a methodology of service modeling. This paper proposes a "service CAD," a computerized tool to support service design, which provides designers with a design environment equipped with knowledge about existing services. To do so, we developed the specifications of a prototype of service CAD and, based on these, a prototype system called Service Explorer. Lastly, the power of the prototype will be demonstrated for an example of hotel service.

KEYWORDS

Service Engineering, Service Design, Service Modeling, Service Design Tool, Service CAD.

INTRODUCTION

Service is attracting more and more attention as manufacturing industries are shifting from a "product seller" toward a "service provider." To design and develop services effectively and efficiently, a methodology of service design and a support by the computer system based on the methodology are needed. However, very few researchers dealt with services from the viewpoint of design, while services have often been traditionally a topic in the field of marketing and management (e.g. Shostack (1981)).

The authors are carrying out fundamental research in "service engineering," which deals with services in an engineering manner (Tomiyama (2001)). A methodology of service modeling has been already

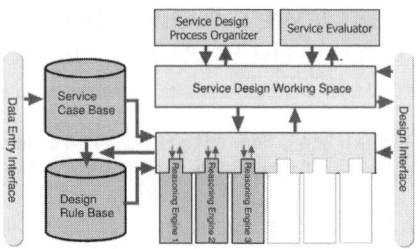

Figure 1: The Conceptual Scheme of Service CAD

proposed (Shimomura (2002)). Based on these, this paper aims at proposing a "service CAD," a computerized tool to support service design, which provides designers with a design environment equipped with knowledge about existing services.

The rest of the paper consists of the following sections. Section 2 proposes the concepts of a service CAD. Section 3 reviews the service modeling method which has been already proposed. Section 4 describes the specifications of a prototype service CAD, and its implementation with an example service.

SERVICE DESIGN AND A SERVICE CAD

In service engineering, service is defined as an action that a provider performs, through which a receiver changes into another state the receiver desires. An analysis of existing designs of services revealed that most of the service designs can be classified into the following three patterns (Shimomura (2002)):

(1) (Re-)design of a new service by enhancing components of and improving existing services,
(2) application of existing service to a different field, and
(3) creative new design.

For the first two classes of service design, the success or failure, the quality, and the efficiency of service design depend to a great extent on the knowledge about service design and existing service cases. However, systematized knowledge about service design hardly exists, while in contrast in mechanical design existing design knowledge can be stored in a reusable form.

Regarding the first pattern of service design, we could observe at least the following three operation patterns (Shimomura (2002)):

(1-1) Substitution of components with something else,
(1-2) removing a part of service, and
(1-3) combination of different existing services.

Pattern (1-1) is an operation to substitute a component of an existing service with another one. Patterns (1-1) and (1-2) are operations to build a new service by changing and modifying the whole or a partial structure of the target service, while Pattern (1-3) comes up with a new combination of services.

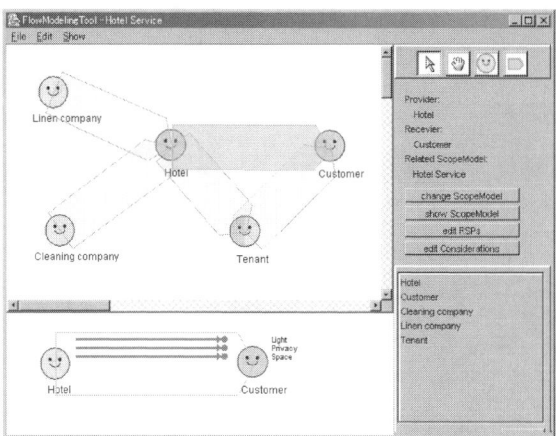

Figure 2: A Screenshot of a Flow Model for a Hotel

Based on the above-mentioned analysis, we have proposed the concepts of service CAD (Computer Aided Design) (Shimomura (2002)) to support engineers in designing services. The service CAD serves as an environment to develop a service by providing the knowledge about existing service cases and various operation rules stored in its database. Figure 1 shows the conceptual scheme of the service CAD, which consists of the followings.

(1) **Service case base:** A database of existing service cases.
(2) **Design rule base:** A database of operation rules for service design.
(3) **Reasoning engines:** Reasoning engines reason about various properties of service such as similarity. A pluggable mechanism is employed so that a necessary reasoning engine is selected based on the designer's request.
(4) **Service evaluator:** An evaluation module to evaluate a service design solution.
(5) **Service design process organizer:** A module to support service design processes based on a specific design methodology by means of calling other components etc.

THE EMPLOYED SERVICE MODEL

In order to develop the service CAD, it is necessary to model a service in a form that can be represented in computer software. A service consists of the following components based on the definition of the service in Section 2 (Shimomura (2002)).

(1) **Provider:** provides the service.
(2) **Receiver:** receives the service.
(3) **Contents:** cause the receiver's state change directly when provided by the provider through the service channel.
(4) **Channel:** what contributes to the receiver's state change indirectly by transmitting, supplying or amplifying the service contents.

In addition, a Receiver State Parameter (RSP) is introduced to represent a receiver's state. The changes of RSPs as a whole represent the receiver's state changes caused when a service is provided.

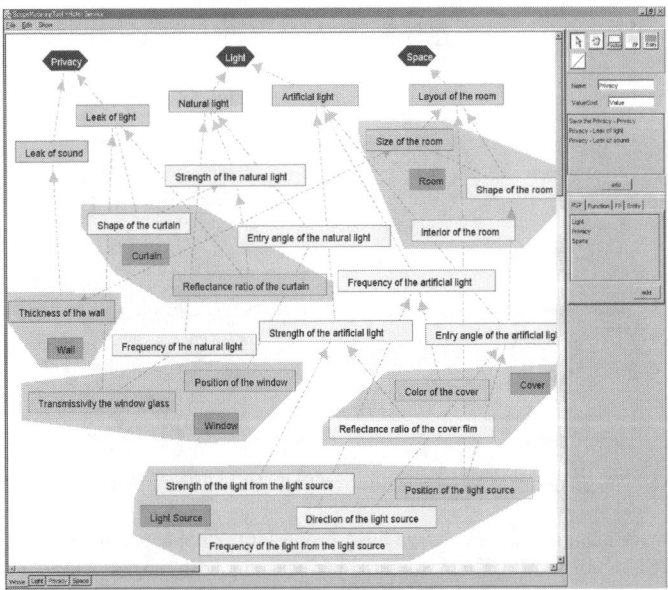

Figure 3: A Screenshot of a Scope Model for a Hotel

An RSP can be a target parameter in service design, but an observable and controllable parameter should be chosen as an RSP.

Then, a Contents Parameter (CoP) is a parameter which causes a receiver's state change directly. Similarly, a Channel Parameter (ChP) is a parameter which indirectly contributes to a receiver's state change through CoP. These parameters are introduced to represent subjective nature of service; even for an identical service, there are recognition differences between service receivers.

In many cases, supplying services requires an environment with a complex structure, in which a provider, a receiver, and several intermediate agents interact each other. A service can involve a chain of subservices. A flow model is a sub model of service and represents those multiple complex structure as a chain of agents who provide/receive a service (Shimomura (2003)). The flow model includes descriptions of a service in terms of a target range which are called scope models. The value and the cost of a service are expressed as the total changes of RSPs contained in the scope model.

A view model represents a concrete method to realize the change of an RSP of a receiver. A view model describes functions of the channels and the contents which realizes a change of an RSP. This functional information includes a function name (FN), a function parameter (FP) that is the main parameter of the function, and a function influence (FI) that is the main influence to FP.

THE PROTOTYPE OF A SERVICE CAD

The prototype system of a service CAD, which is called "Service Explorer," has been developed. It is necessary that Service Explorer has the following five functions;

(1) To allow a user to input and edit a service model,
(2) to display component elements that designers focus on,

251

(3) to register service cases in a service database,
(4) to search in the service database, and
(5) to reuse a service model data stored in the service database.

Required specifications and implemented specifications in details are as follows.

(1) **To allow a user to input and edit a service model**
This is the most fundamental function of Service Explorer. In order to acquire the knowledge of service cases for service design efficiently, an easy graphical interface to describe a service model is needed. A service model is described as a graph structure consisting of nodes and arcs.

(2) **To display component elements that designers focus on**
The system has to be able to (or not to) display component elements selectively depending on designers' demands so that they can understand the structure of the service efficiently. To do so, Service Explorer provides a function to display the function topology and the parameter structure.

(3) **To register service models in the service database**
It is desirable that the service database can store service cases independently of the specific OS/application. For this purpose, Service Explorer employs XML as the database description language.

(4) **To search in the service database**
Service Explorer is equipped with a search function to look up the service database depending on designers' requests. On the current Service Explorer, designers can search for service models with keywords contained in the composition elements (RSP, FN, FP, FI, and Entity).

(5) **To reuse a service model data stored in the service database**
This is the function to reuse composition elements or structure of a service model stored in the service database, when designers inputs and edits a service model.

The Implementation of the Prototype

Based on the above-mentioned functional specifications, Service Explorer was developed in Java SDK 1.4.1 and XML version 1.0. The MVC model (Krasner (1988)), which has been used widely in general GUI applications, was adopted as basic architecture of Service Explorer. By applying the MVC model, high flexibility and reusability of Service Explorer, and robustness of the service model data are achieved.

Below how Service Explorer works is explained using screenshots of a design case of hotel service. Figure 2 shows the user interface to input and edit a flow model, which expresses a service as a chain of agents, depicting a design case of hotel service. Concepts such as "Hotel," "Customer," "Linen company," "Clean company," and "Tenant" are arranged as agents which participate in the service. Multiple scope models which represent provision/receipt relationships among them are defined in this flow model. Furthermore, the interface shown on Figure 2 allows designers to set up RSPs contained in the scope model.

Figure 3 shows the user interface to input and edit a view model for each RSP in the scope model defined on the flow model. Figure 3 depicts a view model about "Light" RSP in the scope model between a hotel and a customer. The RSP is closely related to the customer's demand on brightness. In the view model, "Light" is described as a root element and "Provide Light" is described as a root function to change the RSP. Then, the function is detailed into "Provide artificial light" and "Provide natural light". By detailing gradually the functional structure which realizes the change of the RSP, the RSP is finally linked to entities such as "Light source," "Curtain," and "Window." A view model in the prototype is represented by the graph structure consisting of the following elements.

(1) **RSP nodes**
An RSP is represented by a hexagon node. The color of the node expresses whether it is value or cost. Since "Light" RSP in Figure 3 is a value for the customer, the node is painted blue.

(2) **FN nodes**
An FN (e.g. "Provide natural light") is represented by a square and white node.

(3) **FP nodes**
An FP which influences the RSP directly (i.e. CoP, e.g. "Natural light") is represented by an orange square node, while an FP which influences an RSP indirectly through CoPs (i.e. through ChP such as "Strength of the natural light") is represented by a yellow square node.

(4) **Arcs between FN nodes**
A relationship between functions is represented by an arc between FN nodes. For example, the arc between "Provide light" node and "Provide artificial light" node means that one is developed into the other in Figure 3.

(5) **Arcs between FN nodes and FP nodes**
An FI is represented by an arc between an FN node and an FP node. For example, the arc between "Provide natural light" node and "Natural light" node means "Provide" FI in Figure 3.

(6) **Arcs between FP nodes**
A relationship between FPs is represented by an arc between FP nodes. This relationship is generally defined as a result of the relationship between functions or the effect of the embodiment structure. The designer can set a causal relation on the relationship expressed by a directed arc. For example, the arc between "Strength of the artificial light" node and "Strength of the light from the light source" node means that one affects the other in Figure 3.

References

Krasner G. E. and Pope S. T. (1988). For Using the Model-view-controller User Interface Paradigm in Smalltalk-80. *Journal of Object-Oriented Programing* **1:3**, 26-49.

Shimomura Y. and Tomiyama T. (2002). Service Modeling for Service Engineering. *Proceedings of The 5th International Conference on Design of Information Infrastructure Systems for Manufacturing 2002*, 309-316.

Shimomura Y., Watanabe K., Arai T., Sakao T. and Tomiyama T. (2003). A Proposal for Service Modeling. *Proceeding of the Third International Symposium on Environmentally Conscious Design and Inverse Manufacturing (Eco Design 2003)*, 75-80.

Shostack G.L. (1981). How to Design a Service. in Donnelly, J.H. and W.R. George (eds.), *American Marketing Association*, 221-229.

Tomiyama T. (2001). Service Engineering to Intensify Service Contents in Product Life Cycles *Proceedings of the Second International Symposium on Environmentally Conscious Design and Inverse Manufacturing (EcoDesign 2001)*, 613-618.

A FRAMEWORK FOR SERVICE ENGINEERING BASED ON HIERARCHICAL COLORED PETRI NETS

Guohui TIAN [1,2], Taisuke MIURA [2], Tatsunori HARA [2], Yoshiki SHIMOMURA [2], Tamio ARAI [2]

[1] School of Control Science and Engineering, Shandong University, Jinan, China
[2] Research into Artifacts, Center for Engineering, University of Tokyo, Japan

ABSTRACT

Hierarchical Colored Petri nets (HCPN) method is presented for service engineering. The top level net gives the flow model describing the structure of a target service as a chain of agents existing in the service. The sub pages corresponding to the substitution transitions of the top level net give the scope models determining the sub services. Moreover, the sub pages corresponding to substitution transitions in the scope models give the view models expressing the relationships among the Receiver State Parameters, Content Parameters, and Channel Parameters. The development procedure in this framework is illustrated by studying Consumer Electronics Rental Service using CPN—TOOLS.

KEYWORDS

Artifacts, service engineering, consumer electronics rental service, HCPN, CPN—TOOLS.

1. INTRODUCTION

In the modern society, consumers want to be satisfied with using products rather than with owning them, the industrial production paradigms are changing from product-centered structure to service-centered one considering the whole life cycle of artifacts. The new paradigm should reduce the production and consumption volume of artifacts to an adequate, manageable size and bring this volume into balance with natural and social constraints. Consequently, the aim should be qualitative satisfaction rather than quantitative sufficiency and the decoupling of economic growth from material and energy consumption. To achieve this paradigm, products should have more added values, supplied largely by knowledge and service contents, rather than just materialistic values in order to compensate for volume reduction. Recently, leading scholars were calling for more research in the application of engineering principles to the design and delivery of services, a research field that they called "service engineering" (Tomiyama T., 2001).

A service is well defined in a framework consisting of a service provider, a service receiver, service contents, and service channels. A service model consisting of three sub-models: scope model, view model and flow model, is also presented. A computer-aided design tool, called Service Explorer, is developed to represent a network of the parameters and determines the influence weight one another (Shimomura Y., *et al.* 2003).

In this paper, we present a research framework for service engineering based on a kind of high-level Petri Nets—Hierarchical Colored Petri Nets (Jensen. Kurt, 2004). Firstly we give the flow model in top level net to describe the structure of a target service as a chain of agents existing in the service. Then the sub pages corresponding to the substitution transitions of the top level net give the scope models determining the sub services which include each agent as a receiver. Moreover, there are also substitution transitions in the scope model, the sub pages corresponding to them give the view models expressing the relationships among the RSPs (Receiver State Parameters), CoPs (Content Parameters), and ChPs (Channel Parameters). Under this framework, we can represent material flow information, also deal with RSPs. It will be helpful in intensifying, improving, and automating the whole service, including service creation, service delivery, and service consumption. We illustrate the development procedure by studying Consumer Electronics Rental Service using CPN—TOOLS software.

2. THE WHOLE STRUCTURE OF PRODUCING-CONSUMING SYSTEM AND THE TOP LEVEL PETRI NET

By "leasing" instead of "selling", Consumer Electronics Rental Service can realize a new paradigm from product-selling to function-selling: reducing of cost and trouble of customers (buying, operation, disposal), following customers' situation changes, taking back and renting again, tiding-up with house leasing service with little customization instead of new needs for high functionalities.

The process from producing to consuming is a complicated and large system, obviously it is better to describe it using hierarchical and modular method in order to analyze it clearly. We will deal with it under the 3 sub-model framework and give a realization using Hierarchical Colored Petri nets.

Between the electronic producer (the service provider) and the consumers (the service receivers), there are many intermediate agents, such as wholesalers, lease companies, and so on. They play different roles and carry out relevant activities. Without considering the details about each activity, the providing-receiving service relationship can be represented by flow model which is realized by the top level net of HCPNs depicted as Figure 1. Where, a service provider is a place that has only outgoing arcs; a service receiver is a place that has only incoming arcs; an intermediate agent is a place that has both incoming arcs and outgoing arcs. The places of the top level net are all typed as E. By token e, we can control the progress of the system. Transitions t1~t8 of the top level net are all substitution transitions giving the scope models which determine the sub service activities.

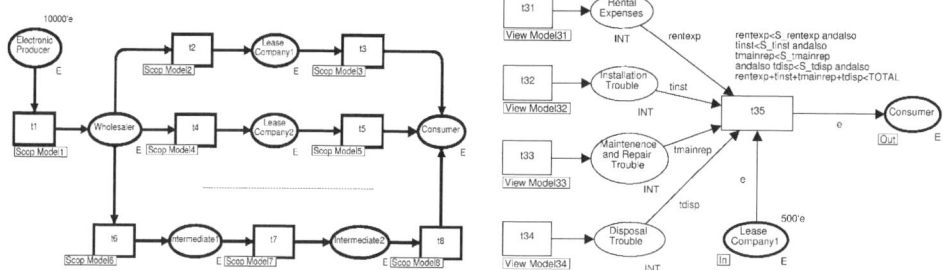

Figure 1: Flow model of producing-consuming System   Figure 2: Scope model of rental service

3. SCOPE MODEL OF THE CONSUMER ELECTRONICS RENTAL SERVICE

In this paper, because we are interested in the sub service bout "leasing", we will pay more attention to the activity caused by Lease company1 to Consumer. The sub page corresponding to substitution transition t3 gives the scope model about it, just depicted as Figure 2. In this activity, Lease company1 will provide the consumers with electronics rental service. The consumers will evaluate the service with 4 RSPs: Rental Expenses, Installation Trouble, Maintenance and Repair Trouble, and Disposal Trouble. So we define 4 places to represent these 4 RSPs respectively. They all are typed as INT, indicating that the color value is integer representing the satisfaction degree corresponding to each RSP. Transition t35 represents the event that the rental service is provided. We give an evaluating criterion on which the consumer judges this service by a transition guard. By using RSPs, the scope model can describe not only whether the consumer can receive the material contents but also whether the consumer is satisfied with the service contents.

4. VIEW MODELS OF THE CONSUMER ELECTRONICS RENTAL SERVICE

In the scope model, we don't consider how lease company1 will manage the rental service, neither how the service management will influence these 4 RSPs. We will describe the details by the view models realized by the relevant sub pages corresponding to the substitution transitions t31, t32, t33, t34 of the scope model respectively. These 4 RSPs are influenced by many aspects respectively. In the view models we will represent these aspects using places which all are typed as INT.

The evaluation about Rental Expenses is described by the view model corresponding to the sub page of the substitution transitions t31 of the scope model, just depicted as Figure 3. The evaluation about Installation Trouble is described by the view model corresponding to the sub page of the substitution transitions t32 of the scope model, just depicted as Figure 4. The evaluation about Maintenance and Repair Trouble is more complicated, and is described by the view model corresponding to the sub page of the substitution transitions t33 of the scope model, just depicted as Figure 5. The evaluation about Disposal Trouble is described by the view model corresponding to the sub page of the substitution transitions t34 of the scope model, just depicted as Figure 6.

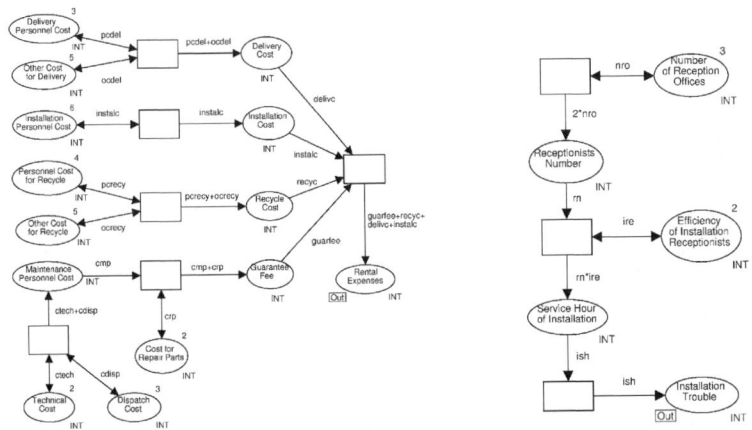

Figure 3: View model about Rental Expenses   Figure 4: View model about Installation Trouble

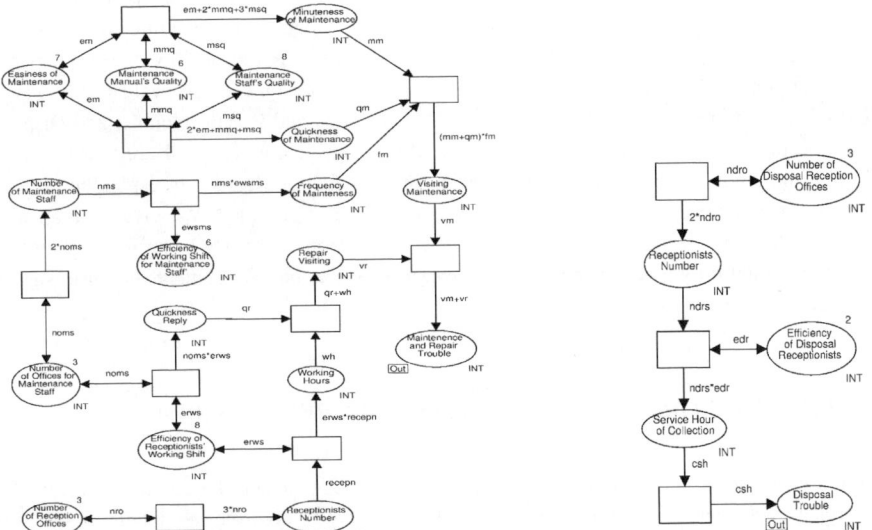

Figure 5: View model for maintenance and repair trouble   Figure 6: View model for disposal trouble

REFERENCES

Jensen Kurt. (2004). *CPN Tools*. Online: http://wiki.daimi.au.dk/cpntools/cpntools.wiki

Tomiyama T. Service Engineering to Intensify Service Contents in Product Life Cycles. *Proceedings of the 2nd International Symposium on Environmentally Conscious Design and Inverse Manufacturing*, IEEE Computer Society, 613-618.

Shimomura Y., et al. (2003). A Proposal for Service Modeling. *Proceeding of 3rd Int. Symposium on Environmentally Conscious Design and Inverse Manufacturing*, IEEE Computer Society, 75-80.

OBSERVABLES OF OPPOSITES ALTERNATIVES IN DECISION MAKING

Junichi Yagi [1], Eiji Arai [2], and Shinji Matsumoto [3]

[1] Institute of Technology, Shimizu Corporation, Tokyo 135-8530, Japan
[2] Osaka University, Graduate School of Engineering, Suita, Osaka 565, Japan
[3] CSP Japan, Tokyo 100-0011, Japan

ABSTRACT

Project management requires a project manager to make a series of hard decisions as his project develops or prior to its commencement. A choice must be made more often rather than otherwise out of the opposite alternatives. This manuscript investigates a proper model for decision mechanism of choice out of the severely contending opposite alternatives, the source of complexity in consequences.

KEYWORDS

dynamic interaction of alternatives, potentials and events, Möbius surface, Verhulst equation, intensional and extensional wholes

INTRODUCTION

A decision maker faces a series of opposite alternatives for choice, seemingly equally valid, from which it is forced to choose one. They are oftentimes under severe contention which way to go may lead to possibly far different consequences or distinct pattern of consequences. It is the norm for decision making, rather than otherwise, to make a choice out of the opposite alternatives.

He confronts with burning potential of opposites at every decision point– to the left or to the right, up or down, metaphorically, A or $\sim A$, speaking most generically, where both opposites *coexist* in acting potential, and both are capable of being, but not yet in existence as event. This mode of existence is called acting potential, whose opposite elements are both rushing toward realization, and only one of which will be realized. In design process, it is the opposite alternatives for almost every parameter that are concerned and stand together under severe contention. This manuscript investigates the peculiar characteristics of acting potential, the logical relation between potentials and events, and the consequent dynamic interaction (Prigogine 1980, Kauffman 1993) among them, which may provide us with better understanding of the underlying mechanism how the opposites influence a decision-making.

MODEL OF ACTING POTENTIALS

The collection of the common attributes which all the elements in a set equally share beyond their own peculiarities is called intensity of the set, while the collection of the members is the extension of the set. The intensity is reciprocal to the extension (Russell and Whitehead 1910). There are two ends in the universe of the set theory, the empty set and the universe. Taking limits towards both ends, the intensity of the empty set is ∞ and that of the universe is 0. The universe has no intensity, i.e. no common ascribable attribute as long as we stay inside the universe (unless from outside, i.e. from a view of a larger whole, it cannot obtain any attribute A, for the attribute requires the existence of its opposite $\sim A$ for A to be defined).

On the other, the empty set could be deemed to contain all the possible pairs of opposite attributes, since $\phi = A \cap \sim A$ for any attribute A relevant in the universe currently dealt with. Any attribute A that predicates the empty set is necessarily cancelled out by its opposite attribute $\sim A$ in the view of extension, whose cancellation does not however evade that the empty set contains both opposite attributes in the view of intensity. The empty set therefore transcends all and contains all – in short, 'it is empty, but plenum'.

The two extremes in the set theory, the empty set and the universe therefore may be deemed as the two opposite wholes, the intensional whole and the extensional whole respectively. The extensional and intensional wholes were shown as two reciprocal modes of the Whole. They are two modes of existence, to which the domains of events and of potentials correspond respectively.

The Whole must thus satisfy the double-fold requirements in its unity; (1) the requirement that the Whole is one, and (2) that there are two distinct reciprocal modes of the Whole. A Möbius strip as shown at Figure 1 can give a plausible model for the Whole so defined to satisfy the double-fold requirements above. The universe Ω yields its copies with different dimidiated partition according to every possible pair of opposite attributes. A series of (infinitely many) copies with different partitioning, Ω_A, Ω_B, Ω_C, ⋯ is thus obtained. Let these copies raise perpendicular to the Möbius surface and align along the surface, whose intersections equally represent the empty set, e.g. $\phi = A \cap \sim A$ *for any attribute A on the surface.*

In this regards, the Möbius model constitutes the null ϕ along its surface as just *one* single surface globally, and rends all the possible opposite attributes across its *two* local faces everywhere. It unifies the reciprocal modes of the Whole, the intensional whole along the null surface and extensional wholes across the surface; (1) The Möbius null surface models the Potential as the intensional whole, pure being of potentials as plenum of attributes. It renders existence to the extensional universe of events and its constituents, (2) an event occurs, when a choice is made out of every attributable opposite. It is because collapsing over the null direction determines the unique accumulation of attributes relevant to a particular event, (3) whenever and wherever an event occurs, holding itself existent extensionally, the Potential acts on the event intensionally to render existence to the event from behind.

DYNAMIC INTERACTION OF OPPOSITES

The innate dynamic interaction of opposites for decision making is thus found well represented by the Möbius model. Given that both wholes, intensional and extensional, are reciprocal opposites, when the one covers the whole surface as it should, there remains no room for the other whole. Immediately after the one whole covers the whole surface, it cannot hold itself, for the one requires its opposite to be well-defined, and thus it inevitably moves to its opposite, the other whole. This cyclic movement never stops or, rather required not to stop on this ceaseless flow of dialectics, in order that the reciprocal wholes both should be defined.

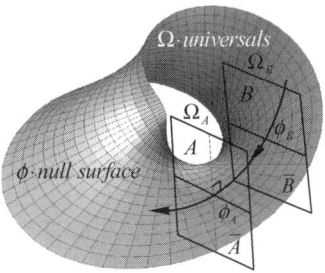

Figure 1: Model for Dynamic Unification of the Acting Potential and Event

The dynamic interaction goes way beyond dynamics of events, physical or otherwise. It is the more fundamental movement between the two wholes that molds both events and potentials with its dynamic framework. It is not just logically anticipated, but governing principle of reality, more akin to Heraclites' proposition in antiquity "*all is in a state of flux*" (Russell 1945). It also gives the substantial ground why the opposite things interact at first place, A and $\sim A$, opposing alternatives which press on decision makers under impending pressure both in the domains of potentials and of events. The potential mode of existence is particularly relevant to decision making, where the opposite potencies are both rushing toward realization as event, but only one of which will be realized exclusively.

One of the simplest equations among possible others which entertains the Möbius model is the Verhulst equation, $x_{n+1} = b \; x_n \cdot \sim x_n$ (Verhulst 1845, Feigenbaum 1978). It is not only relevant to the original application for the growth of populations, but for the rather far-reaching extension of application, that is describing the deterministic interaction of opposites in the process of decision making. The Verhulst equation expresses iterative interplay of reciprocalities of two kinds, additive and multiplicative ($x_n + \sim x_n = 1$ and $x_n \cdot \sim x_n = \xi_{n+1} \; (= x_{n+1}/b\;)$, respectively) at the right hand side of the equation. Both of them equally satisfy the defining relation of reciprocality among the quantities of two or more variables in a way that when one quantity increases, the other decreases or the other way around, though their quite distinct ways of increasing or decreasing.

The Verhulst equation embodies a representation of the iterative fundamental movement between two distinctive wholes, the domains of events and of potentials by capturing the interplay of both types of reciprocals, $x \cdot \sim x = 1$ and $x + \sim x = 1$. Such iterative interplay between both types normally leads to a complex behavior as depicted at Figure2. The equation consists of a series of steps of transformations, where the fundamental movement between two wholes governs along the Möbius surface (Eqn.1); An event x_n at n^{th} generation occurs in the domain of events, and determines it's unrealized opposite 1- x_n (= $\sim x_n$). The opposite then moves to the domain of intension or potential, where both x_n and $\sim x_n$ reside as opposite potentials in the form of $x_n \cdot \sim x_n$. The potential then produces an event of $n+1^{th}$ generation by the dynamic law of Verhulst, $x_{n+1} = bx_n \cdot \sim x_n$. (Note: The additive reciprocality, $x_n + \sim x_n = 1$ expresses the sum of opposites is the whole or "*the whole is the sum of parts*". It is the distinctive characteristic of extension. It does not hold for the intensional whole which completely lacks extension. The multiplicative reciprocality, $x_n \cdot \sim x_n = 1$ is rather "*the intensional whole is the product of parts*", for the intensional parts of attributes are all enfolded in one entangled state of the intensional whole. This entanglement establishes a product as the natural operator for the domain of the intensional whole, where an essential non-linearity reigns.)

$$
\begin{array}{ll}
x_n & [E]_n \\
\to b\,x_n \cdot \sim x_n & [P]_n \\
\to x_{n+1} & [E]_{n+1}
\end{array}
\qquad (1)
$$

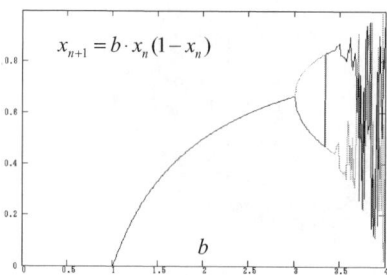

Figure 2: the Verhulst equation

CONCLUSION

A decision maker who manages production process confronts with pouncing disturbance. He must achieve a dynamic equilibrium upon the sweeping waves of both internally and externally oriented disturbance to hold the goal of the whole inviolable at every phase of production, which requires a series of decision making to amend his course of action upon disturbance. However, a decision making itself can be a source of considerable disturbance or, even more than often, it is the primary source, where the opposite alternatives are acting potential for most of decision making. This characteristic state of potential, that is "though neither yet in existence, both opposites are equally capable of being, and contending toward existence" must be properly modeled to understand the mechanism how the real acting potential of opposites undeniably observable in day-to-day human activities, acts on the outcome of choice, and its consequence. The dynamic togetherness of two reciprocal wholes is the primary cause of interference among opposites which produce a complex behavior. The Verhulst equation exemplifies one of the simplest kinds which possibly describe complexity due to interaction between the domains of potentials and of events.

REFERENCES

Feigenbaum M. (1978). Quantitative Universality for a Class of Nonlinear Transformations, *J. Statistical Phys.* **19:25**

Kauffman S. (1993). *The Origins of Order- Self organization and Selection in Evolution*, Oxford University Press, Oxford, ISBN: 0-19-505811-9

Prigogine I. (1980). *From Being to Becoming- Time and Complexity in the Physical Sciences*, W.H. Freeman and Company, New York, ISBN: 0-7167-1108-7

Russell B. and Whitehead A, N. (1910). *Principia Mathematica*, Cambridge University Press, Cambridge, UK

Russell B. (1945) H*istory of Western Philosophy*, Routledge, Oxford, ISBN: 0415325056

Verhulst P, F. (1845) Recherches mathematiques sur la loi d'accroissement de la population, *Nouv. mem. De l'Academie Royale des Sci. et Belles-Lettres de Bruxelles* **18**, 1-41

ENHANCED DISTRIBUTED-SIMULATION USING ORiN AND HLA

Toshihiro INUKAI[1], Hironori HIBINO[2], Yoshihiro FUKUDA[3]

[1]FA Engineering Department, DENSO WAVE Inc.,
1-1 Showa-cho, Kariya-shi, Aichi 448-8661, Japan
[2]Technical Research Institute of JSPMI,
1-1-12 Hachiman-cho, Higashikurume-shi, Tokyo 203-0042, Japan
[3]Faculty of Engineering, Hosei University,
3-7-2 Kajino-cho, Koganei-shi, Tokyo 184-8584, Japan

ABSTRACT

Recent manufacturing industries face various problems caused by the shorter product life cycle, the higher demand for cost and quality, the more diversified customer needs and so on. In this situation, it is very important to shorten the lead-time of the manufacturing system construction. Therefore, the manufacturing simulation has been watched with keen interest. However it is used only for the design stage of the construction. It is not useful for the implementation stage because of a proprietary simulation language, a complex modelling, etc. Our goal is to develop a simulation environment which can be used easily throughout the manufacturing system life cycle. Our approach is based on a distributed architecture, ORiN and HLA. This simulation environment is composed of some manufacturing simulators, real FA devices in the factory, its emulators and so on. By distributing them into one simulation environment on the network, a large-scale simulation and a highly accurate simulation are achieved.

KEYWORDS

Manufacturing systems, Distributed simulation, Virtual factory, ORiN, HLA

INTRODUCTION

Manufacturing simulations are very important to shorten the lead-time of the manufacturing system construction. However, the manufacturing system simulator is not useful at an implementation stage. One of the main reasons is that the simulation models which are made at the design stage cannot be reused at the implementation stage. Therefore, a virtual factory at the design stage and a real factory at the implementation stage cannot be combined efficiently in the system development process.

To solve this problem, current simulators are trying to integrate many functions into themselves. For instance, some robot simulators have the function to convert the simulation language to the proprietary robot language. However it is hard to integrate all functions required in a real factory, because the real factory is composed of many kinds of devices. This approach is confronted with a lot of problems. To cope with the problem, we made conceptual change from INTEGRATION to DISTRIBUTION.

In this paper, we propose a simulation environment which is integrating the real devices into the manufacturing simulation systems on the network. This environment is realized as a distributed real simulation system. The system is composed of ORiN system, soft-wiring system, production cell simulator, ORiN-HLA gateway and so on. By using this system, manufacturing system developers are able to use the same simulation model consistently from the design stage to the implementation stage.

BASIC CONCEPT

The procedure for developing a manufacturing system is commonly based on the waterfall model to reduce a waste of loop-back and re-doing. But still there are many loop-backs on each process. It is difficult to shorten the manufacturing system development time without reducing the loop-backs. Therefore, it is necessary for development time reduction to reduce the "loop frequency" and/or to shorten the "loop time".

To reduce the loop frequency, the upper-layer design process should be highly accurate. To achieve this goal, the FA programming task in the simulation environment is indispensable. However, this causes increase in modelling cost and deterioration of cost-effectiveness. The simulation is not usually used at the implementation stage for these reasons.

As a solution of this problem, we propose an architecture that enables diverting the simulation program to the real device in the implementation stage. The point is to use the same model throughout the manufacturing system life cycle. This means that an implementation task is to embody the exactly same model as the real devices. And this leads to the wide-use of the simulator at the implementation stage. As a result, this also leads to shorter average loop time because of the easier loop back in the simulation.

However, it is easy to imagine the difficulty of creating the simulation environment which is usable in all stages of the manufacturing system construction. The difficulty originates from the fact that the production system is composed of quite a lot of FA devices. Moreover the user programs of those devices are described not in a simulation language but in a ladder language or a robot language, etc. Therefore, we propose architecture of using a real FA device in one simulation environment. By using actual ladder programs or robot programs in the simulation, the simulation accuracy can be improved, and those programs can be reused at the implementation stage.

To achieve this simulation environment, it is necessary to realize the following four functions.
1) Function to abstract a wide variety of FA devices.
2) Function to absorb the differences between the abstracted devices and the real devices.
3) Function to connect the abstracted devices logically.
4) Function to simulate the mechanical motion by the signal from the abstracted device.

In addition, to execute a manufacturing cell simulation in the real production environment such as the production order patterns, it is necessary to make an interaction with the upper-layer simulators such as a production line simulator. Therefore the following two functions are also required.
5) Function to exchange data between the cell simulator and the upper-layer production simulators.
6) Function to manage the logical time and the synchronization between simulators.

SYSTEM OVERVIEW

In the above-mentioned six functions, the function 1) and 2) are very important functions. To realize them, we developed "Open Resource interface for the Network, ORiN" [Inukai T. and Sakakibara S. (2004)]. ORiN is the base system of the following two sub-systems, "Soft-wiring system" and "Cell Simulator". These systems are providing function 3) and 4) respectively.

To accomplish the distributed simulation environment such as 5) and 6), we use "High Level Architecture, HLA" [Hibino H. & Fukuda Y. (2002)]. By using HLA, the synchronization and the logical time management between simulators can be achieved. Figure 1 shows our system overview of a distributed real simulation environment [Inukai T., Hibino H. and Fukuda Y. (2004)].

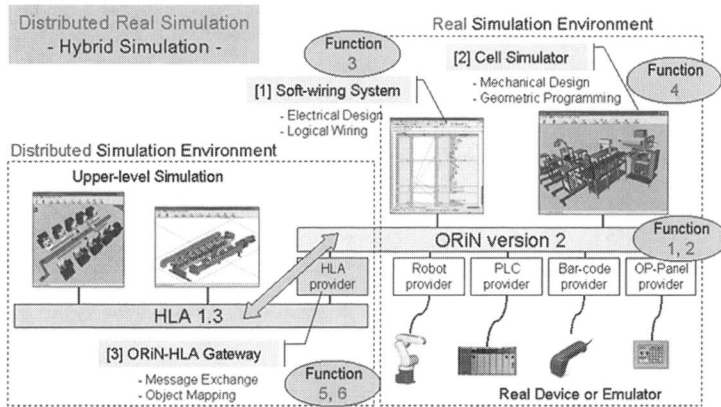

Figure 1: Distributed real simulation environment

ORiN is a software interface for FA devices and the applications. A real FA device is abstracted and is indirectly accessed through the ORiN platform. Therefore the FA applications on ORiN access not a real device but an abstracted device. In short, ORiN can absorb the differences of FA devices. Therefore ORiN applications are executable both in a real factory and a virtual factory.

[1] "Soft-wiring system" provides the function to connect abstracted device logically. By using this system, the information stored in I/O and variable of FA devices can be easily and intelligently transferred to the other FA devices. Moreover different from conventional simulation system, this system can connect not only emulators, but also emulator and real device. In other words, the client program need not distinguish whether the supplied data is from a real device or from its emulator. The difference is completely encapsulated.

[2] "Cell simulator" can provide the function to imitate mechanical motion in accordance with the signals from the soft-wiring system. The mechanical behaviours are represented by two-dimensional tree structure, and its node represents a simple motion. Complex motions are defined as a combination of simple node. By using this simulator, end-user can easily define the motion of production cell.

[3] Synchronization mechanism and logical time management mechanism are very important to achieve the seamless communication between simulators. The functions are provided by HLA and ORiN-HLA gateway. The upper-layer simulators connected to HLA can retrieve the information of a real device through the gateway, and vice versa.

A CASE STUDY

A system shown in Figure 2 consists of a cell simulator, a PLC emulator, a real robot device, and a real bar-code reader, etc. By reading the task instruction from a KANBAN with bar-code reader, a sequence of tasks is performed.

End-users can not only make a program in ladder and/or robot language, but also check a mechanical motion and a cycle time, etc. in a distributed real simulation environment. Therefore, compared with the stressful confirmation task in the real world, the user's load was much reduced.

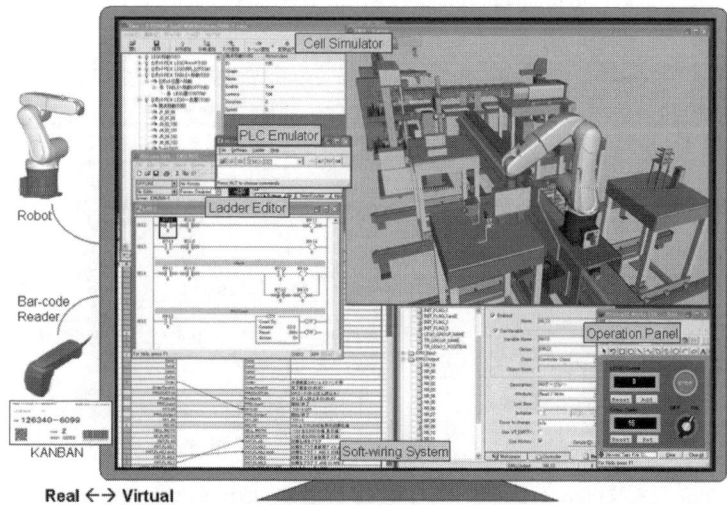

Figure 2: A case study of cell simulation

CONCLUSION

In this paper, we proposed a distributed real simulation environment which composed of ORiN system, soft-wiring system, cell simulation system and ORiN-HLA gateway. By using this simulation environment, manufacturing system developers are able to use the same simulation model consistently from a design stage to an implementation stage. A large-scale simulation and a highly accurate simulation are also achieved. Consequently end-users can perform a lot of tasks in the simulation.

REFERENCES

Hibino H., Fukuda Y. (2002). Manufacturing Adapter of Distributed Simulation Systems Using HLA, Proc. of the 2002 Winter Simulation Conference, pp.1099-1109.

Inukai T., Sakakibara S. (2004). Impact of Open FA System on Automobile Manufacturing, journal of Society of Automotive Engineers of Japan, (in Japanese), vol. 58, No. 5.

Inukai T., Hibino H., Fukuda Y. (2004). The Gateway of Real Factory and Virtual Factory using ORiN and HLA, Proc. of the 5th International Conference on Machine Automation.

OBJECT-ORIENTED EMBEDDED SYSTEM DEVELOPMENT METHOD FOR EASY AND FAST PROTOTYPING

T. Vallius, J. Haverinen and J. Röning

Computer Engineering Laboratory, Department of Electrical Engineering
University of Oulu
P.O. Box 4500
FIN-90014 Oulu, Finland

ABSTRACT

Traditionally, embedded system design requires a considerable amount of expertise, time and money. This complicates the testing of new research results in robotics with real embedded systems, which would be necessary to bring the results into real use. We are studying an easy and fast embedded system development method that enables people without special skills in electronics or embedded systems to create such systems. We hope that this method will ultimately enable utilization of electronics also in research domains where electronics skills are usually not available. In this paper, we present an embedded object based architecture, and the ideology of fitting this architecture into the common object-oriented methods used in software development. We also describe its application to combined software and hardware entities. This paper concentrates on explaining the ideology and architecture of this approach.

KEYWORDS

Object-oriented, embedded systems, embedded object, easy, fast, architecture

INTRODUCTION

Motivation

In academic research involving embedded systems (for example in robotics), research on new ideas usually proceeds as follows: the researcher gets an idea and formulates a hypothesis about it. A model is created to examine the hypothesis. The model is then simulated and modified iteratively several times. Finally, the results are tested in a real embedded system, and depending on the results, the researcher may have to return to one of the previous phases again. Sometimes there exists a significant threshold to real hardware tests. This is because the actual process of building a device to test research results in practice requires a completely new project, another people, lots of expertise, and more time and funding. Tests with real hardware are still often necessary in order to bring research innovations into common use as affordable consumer devices.

We are trying to find out how the creation of an embedded system could be made both affordable and maximally easy, so that the building would not require much expertise in electronics (see the analogy with the PropertyService idea of Mäenpää,Tikanmäki, Riekki & Röning (2004), which enables a non-robotics expert to use robots in research). Thus, different research results in robotics could be tested by the researchers themselves easily in a real embedded system. However, the research is not restricted to robotics, but we try to generalize the results to apply to any research involving embedded systems. We hope that this research will ultimately expand the utilization of electronics also to non-technical areas of science, thus giving totally new possibilities for non-technical research.

Our approach

For software development there are many high level languages available, which enable one to create new software both easily and quickly. One example of them is Microsoft Visual Basic. The ease of use is obtained mostly by an object-oriented approach, visual aids, and a vast amount of ready-made lower-level code. For embedded systems there is not such a high level possibility to create new systems. For hardware development, there are some design methods available, such as Grimpe and Oppenheimer(2001); Kumar et al (1994); Nebel and Schumacher (1996), which in spite of being modular or object-oriented require a lot of expertise in electronics. For embedded software, there also are object-oriented design methods available, some of them even quite innovative, for example Awad, Kuusela and Ziegler (1996) and Object oriented programmable integrated circuits (OOPIC, http://www.oopic.com). Object-oriented embedded system development, which covers both software and hardware design, has been studied by Edwards and Green (2000) in the MOOSE method. Still, the difficulty level of creating an embedded system continues to be very high compared to many high-level software development tools. At the other end, there are the robotics system sets by LEGO, which contain a very easy-to-use user interface and a possibility for fast development of hardware (with Lego bricks), but due to tightly restricted system, they have limited suitability for testing research results or creating anything but simple systems.

We have studied how the high-level software language techniques could be applied to the process of developing an embedded system. We propose an architecture and a development method for embedded systems that is something between LEGO robotics and extended MOOSE, enabling easy building of object oriented embedded systems with minimal limitations.

HARDWARE MODEL FOR AN OBJECT-ORIENTED EMBEDDED SYSTEM DEVELOPMENT METHOD

Introduction

Our method is based on small embedded objects called Atomi-objects. Embedded object means that the Atomi is an object in both software and hardware (embedded system) aspect. Atomis are small electronic boards that contain some sensor circuits, actuator drivers, or other functionality. The software of an Atomi resembles an Automation object (ActiveX Control) by Microsoft. It has properties, methods, and events that correspond to the physical functionality of the Atomi. In other words, one can set different properties of an Atomi (such as intensity of a light), run methods (such as a sequence of positions for a servo), and set an Atomi to respond to events (for example, when heat is below the threshold in a temperature sensor Atomi, the switch property of a heater Atomi is turned on). Atomi boards can be stacked together, and they interconnect through a simple field bus that is extended with a common voltage supply line (see Figure 1). Each board contains a microcontroller unit

(MCU). In other words, the physical architecture is a regular field bus device network, where the Atomis are the nodes of the bus, but which is extended towards object-oriented thinking as combined software and hardware objects, and implemented in a small physical scale.

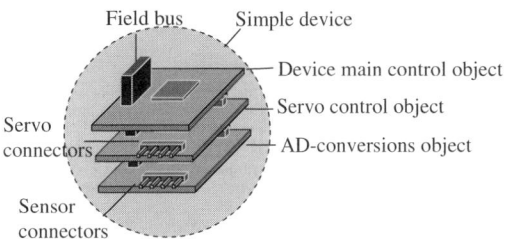

Figure 1 Device built with embedded objects i.e. Atomis

Object-oriented thinking

The advantages of object-oriented techniques (OOT) for software are well known (Booch, 1991, Yourdon, 1994, Martin and Odell, 1992). Many of them apply directly to hardware or an embedded system, such as maintainability, reusability, stability, reliability, faster designing, and extensibility. In our case, one of the main reasons for using object-oriented technology is the goal of making an easy, high-level method to create embedded systems. We consider Atomi as combined software and hardware object: each Atomi contains its own properties, events, and methods, which are related to each Atomi's hardware functionalities. As the benefits of the OOT are achieved via some fundamental elements, such as modularity, encapsulation, abstraction, hierarchy, inheritance, and concurrency, it is relevant to study how they can be realized in Atomis.

Atomis realize modularity, encapsulation and concurrency naturally via their modular architecture. Modularity is important element for the high-level development goal, since it enables the use of library objects i.e. Atomis with ready-made low-level code and hardware. The idea of encapsulation is that the internal data is hidden from the other objects, and only accessible via specific methods. Since the software of an Atomi corresponds directly to the hardware, and Atomis are accessible only through the field bus interface, this realizes naturally. Concurrency means the handling of different events simultaneously. In the Atomi context it is realized through the multiprocessor architecture.

Abstraction and hierarchy are combined through inheritance. Abstraction means presenting the essential characteristics of an object that distinguishes it from other types of objects, and hierarchy means ranking or ordering these abstractions. Inheritance represents a hierarchy of abstractions, in which a subclass (child) inherits from one or more superclasses (parents) (Booch, 1991). In the Atomi architecture, inheritance is realized at two levels. A new object inheriting another object or objects can be realized by just stacking the objects, i.e. Atomis, together and writing new software for the new child object. This we call high-level inheritance (see Figure 2). Here, the child object controls the parent objects, while the interfacing to others takes place via the interface of the new child object. This procedure implements the inheritance of the parents as public class members, since others can access them directly because they are all on the same bus. However, the new child object can also have two interfaces (field bus interfaces and software protocol stacks) and thus implement the inheritance as private class members by attaching objects into the second bus, which is separated from the common bus.

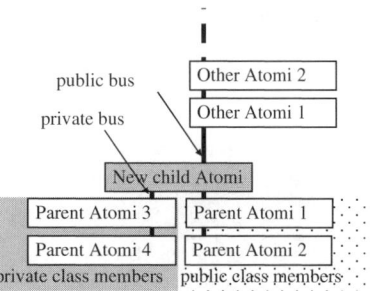

Figure 2 High-level inheritance

Low level inheritance is in question when a completely new Atomi is to be created. In this case, inheritance is realized as a template or base Atomi object. A template Atomi object, which can also be thought of as a generalized abstraction of an Atomi, contains the schematic and layout drawings of an Atomi consisting of the common hardware for the basic Atomi interface. Correspondingly, there is a software template for operating this basic hardware. Thus, inheriting the base Atomi object to create a new object is realized as copying the templates and adding the new components into half-ready schematics and PCB layout and the corresponding control functions for the software. In our test system, the common hardware means an MCU and connectors, and the software means a field bus code with some interface-related code. The low-level inheritance is very important for the flexibility of this architecture, as it enables fully customized Atomis to be created. It also presents a problem for the ease of the Atomi system, as it requires hardware development and hence also some skills and resources in electronics. However, there is still a significant advantage over designing a whole new system, because large parts of the design process for the new Atomi are available as templates.

The characteristics of OOT include the idea of increasing complexity by creating new objects out of a set of other objects. This realizes in the high level inheritance, and it can be realized also by normal aggregation, i.e. by creating a new object that includes other objects. In the physical Atomi architecture however, both inheritance and aggregation realizes similarly, and there is thus not much difference. Since new objects can be created by inheriting/aggregating objects that already have inherited/aggregated objects, the complexity of objects and hence the device can be increased as needed. However, physical restrictions may become a problem at some point. The major restricting factor for object-oriented architecture is the field bus. The field bus capacity defines the real-time capabilities of the system and the maximum number of objects in one bus. However, the architecture can be realized using almost any field bus. Thus, larger systems may use faster buses. Another restricting factor could be the processing power of the MCU, but as the encapsulation suggests, each object can implement its functions hidden from others. Thus, the objects can use any MCU that meets the processing requirements of its functions, as long as it just implements its interface.

DISCUSSION

Towards high-level development

The object-oriented embedded system method has been tested in some devices (Vallius & Röning, 2005a, Vallius & Röning, 2005b, Tikanmäki, Vallius & Röning, 2004) and it is found to be functional. The object-oriented embedded system method brings the development process one step closer to a high-level software programming language: an embedded system can be built by stacking up suitable embedded objects and then just adding a control code to a specific control object. In other words, the lower-level software and hardware are available as objects, and only the high-level control must be developed. This corresponds to high-level software development tools, by which you may create new applications by selecting appropriate software objects from a library and use them with you own control code.

To create a new device, common object-oriented design methods can be used to break up the basic functionality into smaller modules. Usually, the modular diagram of the designed system consists of some functional modules and a control module. In that diagram, the modules correspond directly to Atomi objects, which gives us interesting options. One option is that the control module is a computer. Software creation for a computer is easy because of the great processing power and easy debugging. In the computer one can also use high-level software languages, which makes code generation easy. Another option is to use an Atomi object to control the other objects. If the system is designed properly, the control module could simply contain the main()-function that controls the other Atomi objects. In many software development methods, the control module usually contains a state machine, which controls the other modules. Since a state machine is a very trivial code structure, it is easy to program with almost any programming language. To make it easier, there can also be template state machine code structures, where the user fills in the states and transitions, compiles the code and inserts it into the control module. A graphical state machine editor can be used to make its programming as easy as possible. The control module could even contain a neural network Atomi object. With this approach, some advanced learning methods could be used to teach the device to perform the desired operations. The device could also work without a separate control module. Since all modules can produce events, these events can be set to trigger actions on other modules and thus make the whole device work reactively. These options require still more research and will be considered later.

Cost

If devices designed with this architecture were to be mass-produced, the costs would most likely exceed the cost of an embedded system made with a traditional architecture because of several extra components. On the other hand, if only a few devices were needed of one system, this architecture could be more cost-effective. Depending on the suitability of ready-made modules, the speed of producing a complete device would be faster with the object-oriented embedded system method. The biggest time advantage would most likely come at the debugging phase of the hardware and the software. The hardware and software of a system made with traditional methods or with object-oriented methods, where the objects do not remain independent in the resulting hardware, and the processes do not run in independent processors in the resulting software, the mixing of new hardware and software or existing library objects can cause interference to existing objects. Since the embedded object method is completely encapsulated in software and almost completely in hardware, the risk of malfunctions in library objects is likely to be reduced.

The possibility of high-level system development may also bring cost benefits in the form of a reduced need for expertise. In situations where needed embedded objects are already available, a non-hardware-oriented person is able to create a simple embedded system. provided that the needed objects are available, and no custom objects are needed. Luckily, many sensors and actuators uses common interfaces, such as analog voltage output, serial port, i2c bus, SPI, pulse width modulation, or 8-bit data bus. Thus, with only a few Atomi objects, considerably many kinds of peripherals can be controlled by merely choosing suitable connectors for them.

Success in creating a high-level language embedded system creation method could also inspire IC manufacturers to develop an Atomi-like packaging method for integrated circuits. This kind of a trend can already be seen in the electronics industry, as more integrated and easy-to-use modules come to the market all the time. Only the common interface is missing, and that is what makes Atomi objects feasible.

CONCLUSION AND FUTURE DEVELOPMENT

We have presented an embedded object based architecture and the ideology of fitting this architecture into the common object-oriented methods used in software development. We have applied it with combined software and hardware entities called Atomi-objects and evaluated the pros and cons of such an architecture and design method. The method has also been tested in some devices. This study serves as a basis for further research on high-level development of embedded systems. Some further research will be made towards that goal: to make the Atomi method most feasible, the most common and general Atomi objects should be readily available. The methods for mapping existing design methods to the Atomi architecture will be studied.

ACKNOWLEDGMENTS

This work is supported by the Finnish Academy and InfoTech.

REFERENCES

Awad M., Kuusela J. & Ziegler J. (1996). *Object-Oriented Technology for Real-Time Systems*, Prentice Hall.
Booch G. (1991). *Object oriented design with applications*, Benjamin/Cummings.
Edwards M. & Green P. (2000). An Object Oriented Design Method for Reconfigurable Computing Systems, Design, *Automation and Test in Europe Conference and Exhibition 2000, proceedings.* 27-30 March 2000, Page(s): 692 -696.
Grimpe E. & Oppenheimer F. (2001). Object-oriented high level synthesis based on SystemC, *Electronics, Circuits and Systems, 2001. ICECS 2001. The 8th IEEE International Conference on, Volume: 1,* 2-5 Sept. 2001, Page(s): 529 -534 vol.1.
Kumar S., Aylor J.H., Johnson B.W. & Wulf W.A. (1994). Object-oriented techniques in hardware design, *Computer, Volume 27 Issue 6,* June 1994, Page(s): 64 -70.
Martin J. & Odell J.J. (1992). *Object-Oriented Analysis and Design*, Prentice Hall.
Mäenpää T., Tikanmäki A., Riekki J. & Röning J. (2004). A Distributed Architecture for Executing Complex Tasks with Multiple Robots, *IEEE 2004 ICRA, International Conference on Robotics and Automation, proceedings.* Apr 26- May 1, New Orleans, LA, USA.
Nebel W. & Schumacher G. (1996). Object-oriented Hardware Modelling - Where to Apply and What are the Objects?, *Design Automation Conference, 1996, with EURO-VHDL '96 and Exhibition, Proceedings EURO-DAC '96, European, proceedings.* 16-20 Sept. 1996, Page(s): 428 -433.
Tikanmäki A., Vallius T., and Röning J. (2004). Qutie - Modular methods for building complex mechatronics systems, *International Conference on Machine Automation (ICMA2004), proceedings.* Nov 24 – Nov 26, 2004, Osaka, Japan.
Vallius T. & Röning J. (2005a). Implementation of the "Embedded Object" Concept and an Example of Using it with UML, *The 6th IEEE International Symposium on Computational Intelligence in Robotics and Automation, proceedings*, Jun 27 – Jul 30, 2005, Helsinki University of technology, Finland.
Vallius T. & Röning J. (2005b). Embedded Object Concept with a Telepresence Robot System, *SPIE Optics East 2005, proceedings.* 23-26 October 2005, Boston Marriott Copley Place, Boston, Massachusetts, USA.
Yourdon E. (1994). *Object-Oriented System Design: An Integrated Approach*, Prentice Hall.

INTEGRATED CONSTRUCTION PROCESS MANAGEMENT SYSTEM

Masayuki Takata[1], Eiji Arai[2] and Junichi Yagi[3]

1 Information Processing Center, The Univ. of Electro-Communications
Chofu-shi, Tokyo 182-8585, Japan
2 Department of Manufacturing Science, Osaka University
Suita-shi, Osaka 565-0871, Japan
3 Institute of Technology, Shimizu Corporation
Koto-ku, Tokyo 135-8530, Japan

ABSTRACT

This paper describes an implementation of an Integrated Construction Process Management System, which includes both manufacturing process management features for building parts and also construction progress management features at construction site. To monitor the flow of the building parts, RFIDs are stuck to all of the parts to be managed, and several checkpoints, which we named "gates", are introduced within the coherent process through part-manufacturing plant line, logistic processes and building construction processes. By means of this, building parts can be tracked certainly, and anyone can know the status and the location of building parts at that instant.

KEYWORDS

integrated process management system, part-manufacturing process management, building construction process management, intensive data management system

1 INTRODUCTION

This paper describes an implementation of the integrated construction process management system, which includes manufacturing process management features for building materials, and construction process management features at construction site. Recently, RFIDs are getting popular in logistics and manufacturing industries. The process management system for building construction and building materials manufacturing must cover these two aspects, and the use of RFIDs in construction industries will make the trace-ability of the building materials more accurate. The system we aimed enables efficient project management and diminish the loss in the construction, by providing all information of the both material manufacturing and building construction to all of material designing, material manufacturing, building designing and building construction sites.

2 APPROACH

In this implementation, we aimed to confirm that the system integrity on the whole. In order to make its information processing simple, the process management engine uses only typical durations to process each step in the manufacturing materials or installing them, and the bills of materials.

In order to trace WIPs (Work In Processes), we installed several checkpoints representing boundaries of logical activities, which we named "gates", in the process through material-manufacturing and building-construction. On WIPs passing these gates, RFIDs are read and progress reports are collected. As the WIP's due time for passing the final inspection process gate is deduced from the overall schedule of the construction, the due time for passing each gates can be calculated from the given final due time and the typical durations from one gate to the next. On the other hand, when WIPs pass each gates, the estimated time for passing following gates can be calculated.

In these way, for each building materials types, we can obtain both due time for all demands and for all gates, and actual or estimated time for all WIPs. By associating each demands and each WIPs in the order of time passing a certain gate for each materials type, we carry out the allocation of demands and WIPs. In the case of due time of allocated demands is earlier than estimated time of associated WIP passing by, we assume that tardiness is expected and some action is required.

3 THE IMPLEMENTATION
3.1 Gates

In order to trace WIPs, we have set up nine gates within both material-manufacturing and building-construction processes, as follows; (1) Design approved, (2) Ordering raw material, (3) Start processing, (4) Assembling, (5) Shipping out from the manufacturing plant, (6) Carrying into the construction site, (7) Distributing within the construction site, (8) Installing building material in the building, (9) Final inspection. It is easy for the system to change the number of gates, to change typical durations from gate to the next.

When the WIPs pass gates, following processes are took place. (1) Reading RFID on the WIP, (2) Converting to the WIP identifier, (3) Logging the time passing the gate, (4) Logging the physical position of WIP, (5) Logging the result of the post-process testing. These data are accumulated within an actual achievement database resides in the shared data space, described later.

3.2 Tracking Works with RFIDs

At each gate, the system gathers actual achievement information by means of RFIDs. In this implementation, we used read-only type RFIDs with 128 bit length identifiers. Generally speaking, as each building materials consists of multiple parts which are manufactured independently in the manufacturing line, single building material may contain multiple RFIDs in it. On reading RFIDs of a building material, some RFIDs may respond and some may not, but the tracking engine should handle these information properly. In some cases, an assembled WIP may be dis-assembled and re-assembled find much more matching combination.

So, the tracking engine should have following features.
1. Identifying the WIP from partial RFIDs information.
2. Keeping RFID identifiers of all parts consisting the WIP.
3. Keeping tree-structured information including assembling order and part structure, for the case of dis-assembling and re-assembling.

Furthermore, it is expected that reading some particular RFIDs instead of entering some information manually, such as operator name, physical location of the work-cell, and others. So, the tracking system can judge whether given identifier represents some WIP or process equipment.

3.3 Allocation

In this implementation, we use simple algorithm described in the Section 2 to allocate demand to corresponding WIP. The basic data, which are the typical durations from one gate to the next or previous gate and the bills of materials, are given and stored in the shared data space among multiple processing agents, described later. The typical durations for due date deduction may differ from those for calculating estimated times for following gates. Both durations are defined for all product types, but the users can define other values to override them for respective types.

We named the lists of same type WIPs, arranged in the order of expected time passing a predefined gate, as "preceding list." In other hand, we named the list of same type demand, arranged in the order of due time at a predefined gate, as "priority list." Their priority is defined only by the due time order, and we ignore any other aspects.demands.

All demands and WIPs are processed in building material type by type, and in the case of the material type made from multiple parts, the demands of those parts is newly created at the gate at which those parts are assembled together, according to the bills of materials. The due time of the parts are given from the due time of the assembled WIP at the same gate. The estimated time of a WIP to pass the assembling gate is calculated as the latest one of those times of its parts.

3.4 User Interface

In this subsection, we describe the user interface screens implemented. The user interface of this system was developed as the application of WWW (World Wide Web) system, in order to make them accessible from not only desk-top computer systems but also portable data terminals.

Figure 1 shows the status display for one particular association of the demand and the WIP, including the parts which demands and WIPs consists of. In each table entry, the upper row contains date information in the format of YYYYMMDD, and the lower one contains time in the format of HHMMSS. The table entries show the due time of the demand for that gate with

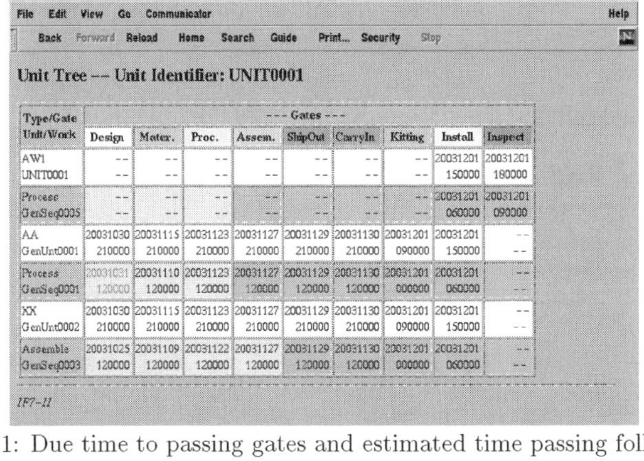

Netscape: Unit Tree:

File Edit View Go Communicator Help

Back Forward Reload Home Search Guide Print... Security Stop

Unit Tree -- Unit Identifier: UNIT0001

Type/Gate Unit/Work	--- Gates ---								
	Design	Mater.	Proc.	Assem.	ShipOut	CarryIn	Kitting	Install	Inspect
AW1 UNIT0001	-- --	-- --	-- --	-- --	-- --	-- --	-- --	20031201 150000	20031201 180000
Process GenSeq0005	-- --	-- --	-- --	-- --	-- --	-- --	-- --	20031201 060000	20031201 090000
AA GenUnt0001	20031030 210000	20031115 210000	20031123 210000	20031127 210000	20031129 210000	20031130 210000	20031201 090000	20031201 150000	-- --
Process GenSeq0001	20031031 120000	20031110 120000	20031123 120000	20031127 120000	20031129 120000	20031130 120000	20031201 000000	20031201 060000	-- --
XX GenUnt0002	20031030 210000	20031115 210000	20031123 210000	20031127 210000	20031129 210000	20031130 210000	20031201 090000	20031201 150000	-- --
Assemble GenSeq0003	20031025 120000	20031109 120000	20031122 120000	20031127 120000	20031129 120000	20031130 120000	20031201 000000	20031201 060000	-- --

IF7-II

Figure 1: Due time to passing gates and estimated time passing followings

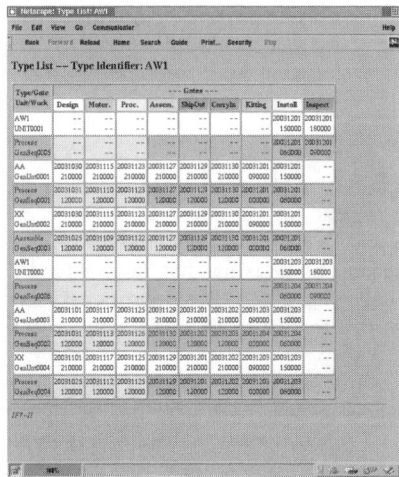 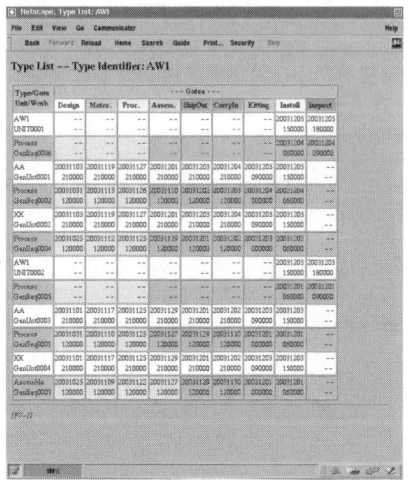

Figure 2: Status display for all works of specified building material type

Figure 3: Status after changing due time

while background color, show the actual achievement time of the WIP to pass the gate with green background, and the estimated time of the WIP to pass the gate with blue background. The red characters show the tardiness expected. The neighboring entries consisting upper demand line and lower WIP line show the allocated pair. As the allocation is subject to change, the display contents change when due dates are changed or WIP passes new gate.

Figure 2 shows the status display for one particular type of building materials, including the parts which demands and WIPs consists of. This example shows that two building materials are currently under processing, and one WIP recovered its tardiness at the first gate, by shorten the duration for processing from the first gate to the second, but another WIP has delayed at the last gate it passed and tardiness at the gate for the final inspection is expected. In order to recover such situations, one may try to shorten the duration from the gate to the next, or may postpone the due time for the final inspection of the building materials installation.

Figure 3 shows the case of postponing the due time of the final inspection of the unit with unit-ID UNIT0001 for four days. As the result, the due time of the unit UNIT0001 and that of the unit UNIT0002 are reversed, and the parts set which are going to be used for those units are exchanged.

4 THE INFRASTRUCTURE SYSTEM

In this study, we used the system named "Glue Logic" [1,2] as the infrastructure to support multiple-agent processing system. The Glue Logic, which is developed by the Univ. of Electro-Communications, includes the active database and the network transparent programming environment, and supports data processing in the event driven programming paradigm. Figure 4 shows the overall structure implemented using the Glue Logic infrastructure system.

The active database is a subclass of the database systems, of which databases have an abilities to behave when it finds some changes of its contents, without waiting for external actions. The change of the contents includes; (1) when data is changed, and (2) when some relations are formed. On the other hand, the behavior executed on these incidents includes; (1) changing contents of the

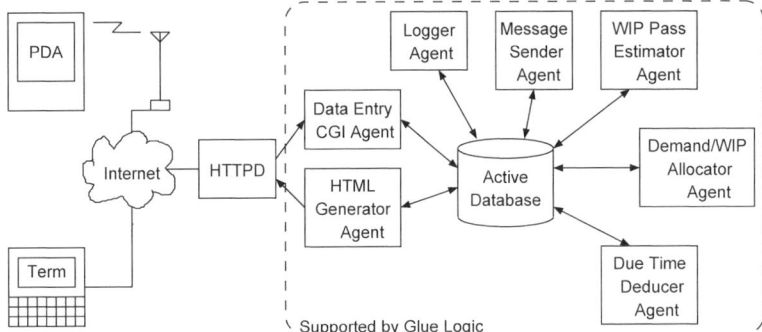

Figure 4: Overall Structure of the Implementation

database, (2) calculating certain expression and store the result inside and (3) sending message to some client agents.

The Glue Logic is designed to make building manufacturing work-cell control systems easy and flexible, and also coordinates agents by means of followings;
- Providing field of coordination
- Implementing shared data space among agents
- Virtualizing agents within the shared data name space
- Controlling message passing among agents
- Implementing mutual execution primitives
- Prompting agents to start processing
- Adapting control systems to real-time and network processing environment

As the Glue Logic supports event notification and condition monitoring features based on active database scheme, users can easily build real-time and event-driven application agents, only waiting for notification messages from the Glue Logic.

Each agents in an application system can be developed concurrently, and can be added, deleted or changed freely without modifying other existing agents. As the result of these, the Glue Logic compliant agents are easy to re-use, and the users can build large libraries of application agents.

In this implementation, the flow of its data processing is as follows.
1. When a WIP reaches a gate, or when a due time of a demand changes, the corresponding agent is activated via a CGI for the WWW user interface. These agents updates the actual achievement records for WIPs or the due time requirement records for demands. These records are kept in the shared data space in the Glue Logic.
2. When shared data is changed, some messages are sent from Glue Logic to agents which is going to handle the data items. This time, the agents keeping preceding lists or priority lists are informed, and update those contents.
3. When the contents of preceding lists or priority lists are changed, the notification message is sent to the allocation agent. The allocation agent reads preceding lists and priority lists, and then makeup associations between demands and WIPs.

5 EVALUATION

We implement whole system on a Sun Netra T1 processor running on Solaris 8 operating system. Through this implementation, we found following performance considerations.

1. Re-allocation of the demand hardly took place unless due time of final inspection changed or some WIPs pass other preceding WIPs.
2. In many cases, as there are less than tens of WIPs concurrently being processed in the whole system, re-allocation of the demand takes only a few seconds. In the case of building materials consisting multiple parts, as the multiple re-allocation processes occur, it may takes more than ten seconds.

We used multiple agent support systems in the implementation, in order to ease future extension of functionality. From the view point of the system extend-ability, we found followings.
1. All information on the actual process achievements are kept within a database. This makes any other agents being able to utilize these data for processing and user interface purpose within a few seconds, such as status display systems and e-mail sender programs.
2. It seems to be appropriate that the conversion from the identifier of RFID to building material identifier should be done by specialized subsystem in the management system. There may be many class of RFIDs other than expressing WIPs.

In this implementation, we introduced some limitations to simplify the system, as follows.
1. There is no problem solving engine to minimize cost.
   To find best solution of re-allocation, it is required to minimize cost to re-distribute building materials, or in some cases it is required to determine which WIP to be scraped. Solving these problems may need massive computational power, because of the combinational explosion.
2. There is no clear decision rules to be embedded within the system.
   Some incidents can be processed automatically without human interventions, but some require human approvals. There is no clear border and the best way depends on its environment.

6 CONCLUSION

Through this implementation described above, we found that the integrated process management system including both part-manufacturing and building construction is feasible enough. In this financial year, we are now going to carry out a field test, applying this system to the actual manufacturing and construction sites.

ACKNOWLEDGMENTS

This research activity has been carried out as a part of the Intelligent Manufacturing Systems (IMS) international research program: "Innovative and Intelligent Parts-oriented Construction (IF7-II)." We appreciate the kind guidance of each members of this project.

REFERENCES

[1] Takata M and Arai E. (2001). Implementation of a Layer Structured Control System on the "Glue Logic". *Global Engineering, Manufacturing and Enterprise Networks* pp.488-496, Kluwer Academic Publishers.
[2] Takata M and Arai E. (2005). Implementation of a Data Gathering System with Scalable Intelligent Control Architecture. *KNOWLEDGE AND SKILL CHAINS IN ENGENEERING AND MANUFACTURING: Information Infrastructure in the Era of Global Communications*, pp.261-268, Springer.

A ROBOTIZED SYSTEM FOR PROTOTYPE MANUFACTURING OF CASTINGS AND BILLETS

Mikko Sallinen[1], Matti Sirviö[2]

[1]VTT Electronics, Kaitovayla 1, 90571 Oulu, Finland
[2]Simtech Systems Inc.oy, Kukkaromäki 6C5, 02770, Espoo, Finland

ABSTRACT

In this paper, we present a new method for manufacturing prototype castings using robot-based system. The contribution of the paper is new methods and tools for managing very different sizes of work pieces. The tools helps and assists the designer for manufacturing pieces which are not a straightforward to program for the robot. The methods are designed for prototype manufacturing which means lotsize is something between one to ten.

KEYWORDS: robot milling, off-line programming, prototype casting

1. INTRODUCTION

The models for prototypes of the cast objects has been traditionally made by hands. They are made by wood and the form has been generated into a sand box. Nowadays, the wooden or plastic models are made using a milling machine where the first step towards automation has been taken. However, the disadvantages of the milling machine is the price and flexibility and therefore we have been taken a new approach to use robot as a milling machine Sirviö et. al. (2002). Robot-based milling stations have been developed in low intensity over the last number of years. The two main restrictive reasons for use of robot in milling are the rigidity of industrial robots and the difficulty of flexible off-line programming. Rigidity of the robot systems is not anymore such a problem like several years ago. However, in the last years, another problem has been high (or high enough) absolute accuracy of the robot manipulators.

The machining of moulds or other prototypes has been made so far using a milling machine Boomenthal (2000). The difficulty of programming 5-axis machining as well as the lack of required accuracy has been forced to use three to five times expensive milling machine instead of a robot. The robotic milling that has been done so far has been concentrated on milling on soft materials like different kind of foams or wood to avoid the problems of rigidity Veergest et. al. (1998). The machining of large work pieces and path planning for material removing has been studied in Jager et. al. (2001).

This paper is organized as follows: in chapter 1 an introduction to the topic is given. Chapter 2 describes the prototype manufacturing process, the use of robot system is described in chapter 3. Results from actual tests are in chapter 4. Finally conclusions are given in chapter 5.

2. DESCRIPTION OF THE PROCESS

The principle of the short series production is explained as follows. The geometry of the casting will be machined into the sand. After machining and coating the mould, the work piece will be cast. Depending on the properties of the casting, this process is repeated until a piece with sufficient qualifications has been achieved. This may require one to ten iteration steps. Numerous castings that are produced in small batches, including prototypes and spare parts, they will be produced by substitution of the pattern making process with advanced robotics and utilization of a digital library created from a CAD model, thus economizing in terms of materials and storage space used and time spent. Moreover, patterns will be manufactured by using the same system by only substituting sand mould with plastic material. This will increase end user's productivity over 50% and decrease costs substantially, especially when prototypes are developed. From a technical point of view, new products, of enhanced shapes will be created in extremely short times and in as many items as desired. Finally, the milling and mould making can be totally isolated from the foundry environment, a fact that will decrease dust production substantially. In figure 1, the prototype production could be made without pattern making process.

3. ROBOTIC SYSTEM FOR PROTOTYPE MANUFACTURING

The robot system consists of CAM phase, where the milling paths are generated, off-line programming phase, where the robot programs are postprocessed and actual running phase where the milling itself is carried out.

3.1 CAD / CAM

In the system, raw manufacturing paths are generated in CAM software. Raw means here that no machine dependent information is needed, only information of tools. The system is designed to be open such that designer can use any commercial CAD/CAM software he is used to and the output is in APT format. Therefore, the threshold of taking use of the proposed technology is as low as possible. In the CAM phase, user selects the general milling parameters such as tool contact angle and deepness of milling. These can be selected similar way as when using milling machine because

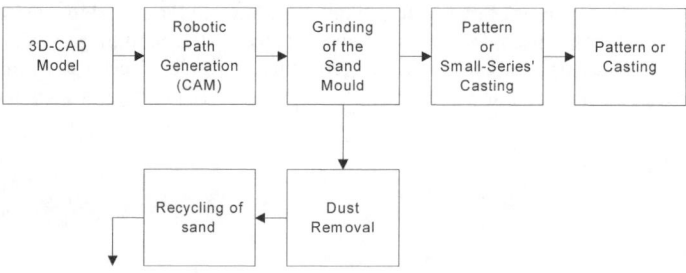

Figure 1. Foundry process with robotic sand grinding system

the material to be milled is relatively soft for the robot. However, the amount of points have to be reduced due to limited performance of the robot controller. This can be seen as an actual velocity of robot arm.

3.2 Off-line programming

3.2.1 Calibrations of the robot work cell

The fundamental problem of robot systems is a requirement for use of one program in several robots. Usually it is not possible even if they are manufactured by the same manufacturer and they should be the same-like, i.e. same model. The reason comes mainly from the manufacturing deviations between the different manipulators. To overcome these inaccuracies the kinematic model and coordinate transformations can be corrected and in that way improve the accuracy of the robot systems. When using robots attached with sensors to different kind of measuring and manipulating task and when they are off-line programmed, following three calibrations should be done: calibration of the sensor internal parameters, hand-eye calibration and calibration of the robot.

3.2.2 Managing with various size of work pieces

Machining of large work pieces is carried out by splitting the paths and CAD models of the work pieces. This is carried out by geometrical information of the robot workcell, i.e. the reachability of the manipulator arm with tools used in the cell. By using this set-up, the full performance of the robot workcell can be used. Using the classical approach, a very large robot would have been chosen. However, the disadvantage of this way is lack of accuracy because the larger robot you choose, the more inaccurate it is. By using medium-size robot, it is a compromise of accuracy but we can still manufacture large work pieces up to several meters. After splitting the paths and CAD information, paths are optimized for proper robot respectively. In the optimization phase, path is converted into form that movements of the robot are minimized between the points in the path. The optimization includes reach and collision check to prevent the unexpected situations.

3.3 Actual machining

3.3.1 Localization of work piece in robot work cell

To be able to manage the split pieces, each piece has to localize very carefully. In the localization, we use methods presented in previous paper Sallinen & Heikkilä (2000). The idea is to fit the measured points to the reference model of the work piece. Method is fast and flexible to use. The method is open for different surface forms including plane surface, cylinder surface and spherical surface. The localization is here carried out using touch sensor which is rather reliable to use in foundry environment.

3.3.2 Robotic Milling

The postprocessor outputs the programs ready-to-run in the robot including commands for running of spindles, tools and tool changers. Depending on the equipments of the robot, the machining parameters such as velocity, rotation speed of the tool and length of the path has to be find out. These depend also from the sand material in a terms of size of the grains and hardening material.

To solve these parameters, a certain test-run has to be go trough to find out the best surface quality which is one of the main objectives of the system. The inaccuracies coming from the flexibility of the robot are compensated in a optimizing phase where paths are planned such that movements from one point to another includes minimum amount of movement. This affects a path where movements of the robot are minimized and relative inaccuracy between the points is reduced. The other motivation is the surface quality required in the prototype manufacturing. In those products, typically allowed tolerance is between 6...10% of the dimensions of the work piece. So in a typical medium size product dimension length of 500 mm, allowed tolerance is between 30...50 mm and that is definitely under performance of the robot system.

4. TESTS IN THE FOUNDRIES

The methods have been tested in simulation and actual production and results are very good. Using the optimization, the paths that was not able to run normally, could be run. The usability of the optimization has been improved based on the comments from the users. Also the splitting of the CAM paths and CAD models was tested with a success. The actual milling process is described in figure 5. The system was built up and tested in two different robot systems: ABB IRB6400 with S4C controller and KUKA KRC 150L110 with KRC2 controller. Also both electric and air pressure spindles was tested. Both of the robot systems with different spindles and tools was working fine.

5. CONCLUSIONS

In this paper, we present methods for manufacturing variety sizes of sand moulds using robot. Using the flexible control of the robot system, a cost-effective production of small series can be achieved. The proposed method consists of three different phases: CAD/CAM, off-line programming and actual milling. The whole system has been tested in actual foundry environment with very promising results.

References

Bloomenthal M., Riesenfeld R., Cohen E., Fish R., (2000), An Approach to Rapid Manufacturing with Custom Fixturing, *IEEE Int. Conf. on Robotics and Automation*, San Francisco, USA, pp. 212-219.

Jager P. J., Broek J. J., Horvath I., Kooijman A., Smit A. (2001). An Effective Geometric and Kinematical Analysis of Ruled Surface Feature Manufacturability for Rapid Prototypind. *Proc. Of DETC'01. ASME 2001 Eng. Technical Conference and Computers and Information in Engineering Conference*. Pittsbourg, PA, Sep. 9-12. 2001.

Sallinen M., Heikkilä T., (2000), Flexible Workobject Localisation for CAD -Based Robotics, *Proceedings of SPIE Intelligent Robots and Computer Vision XIX: Algorithms, Techniques, and Active Vision*. Boston, USA, 7 - 8 Nov. 2000. USA. Vol. 4197 (2000), pp. 130 - 139

Sirviö M., Väinölä J., Vapalahti S., Sallinen M. (2002), Automatic Line for Manufacturing Prototype Castings and Billets in Environmentally Friendly Robotic Cell, *Proceedings of the International Conference on Machine Automation* (ICMA2002), 11-13.9.2002, Tampere, Finland.

Veergeest J., Tangelder J., Horvath I., Kovacs Z., Kuczogi G., (1998), Machining large complex shapes using a 7 DOF tool, *IFIP SSM'98 Symposium*, Chryslr Tech. Center, 9-11 Nov 1998.

TOWARDS HUMAN-PROFILE BASED OPERATIONS IN ADVANCED FACTORY GOVERNANCE SYSTEMS: CONTEMPORARY CHALLENGES FOR SOCIO-TECHNICAL SYSTEMS DESIGN?

F.M. van Eijnatten[1] and J.B.M. Goossenaerts[2]

[1] Human Performance Management Group,
[2] Information Systems Group,
[1,2] Research School for Operations Management and Logistics (BETA)
Department of Technology Management,
Eindhoven University of Technology, 5600 MB Eindhoven, NL

ABSTRACT

This paper is about the requirements for an advanced factory governance system. Five capital assets are distinguished: Natural, artificial, human, social, and financial. A factory's operations involve and affect these five capital assets. To scope workers' activities with respect to objectives that exist for these capital assets, this paper applies insights from Manufacturing System Design Decomposition. It is argued that in an advanced factory governance system the workers do also engage in governance and management activities. We introduce the decision-object hierarchies and an extended generic activity model, and explain how human profiles are derived from these. Sociotechnical systems design can be used in such a system to balance 'control by the system', and 'self-control' by the workers involved.

KEYWORDS

Information systems design methodology, information storage, advanced factory governance, capital assets management, human-centered manufacturing, socio-technical systems, decision-objects hierarchy, manufacturing systems design decomposition, generic activity model, decisional reference model.

INTRODUCTION

Socio-Technical Systems Design (STSD) is a dominant human-factors design approach to develop a human-centered, technology enabled, team-based, manufacturing system (Van Eijnatten, 1993). STSD asserts that the human factor is of vital importance for the successful functioning of intelligent manufacturing systems (Vink et al., 2002). STSD was used in the European Union 5th Framework IST project 'PSIM' (Participative Simulation Environment for Integral Manufacturing Enterprise Renewal) that was part of the global Human-Machine Coexisting Systems project (HUMACS). HUMACS was aimed at developing a 'Human-Factors Centered Manufacturing Enterprise' in which people give full play to their capabilities from each and every perspective with full sense of fulfilment and satisfaction (Yamada, 2002).

In the PSIM project (Van Eijnatten, Ed., 2002) both a procedure and dedicated software were developed and tested in five companies in Europe, Japan and the US. The PSIM participative simulation prototype (Little et al., 2001; Bovenkamp et al., 2002) supports companies in developing new organizational structures in which humans can perform healthier and better. The participating companies evaluated PSIM positively as a 'breakthrough' innovation in the field of business and work.

A virtual plant environment (Goossenaerts et al., 2002; Matsuo & Matsuoka, 2004; Shin et al., 2004) is an advanced information environment that supports operations, decision-making and transformations in the factory. The drivers for the decision-making increasingly stem from the public domain and are characterized by a growing range of concerns such as *quality, safety and health, environment friendliness, security, etc*. In the ideal case, the objectives or functional requirements that govern the factory's evolution are aligned with economic and sustainability objectives for capital assets in the ecosystem. The integral effect of this alignment activity on the factory is mediated by the reflective activities of governance, management, and design & analysis.

Factory governance systems that adhere to the principles of Socio-Technical Systems Design – predominantly autonomy and self-regulation – are better harnessed to align factory operations with the evolving development goals articulated in the ecosystem. It is a challenge to find a proper balance between self-regulation by human agents and control by enacted systems. In this paper the relationship between ecosystem development goals and reflective activities in the factory are explored. On both sides, interesting insights are gained. Factory Governance requires an intensive collaboration among agents at the scales of human, business and the public domain. A comprehensive methodology to derive human profiles that can support human operations with shifting objectives in advanced factory governance systems draws on three bodies of knowledge: (i) Institutional Analysis and Development (IAD) Framework (Ostrom, 1990/1994); (ii) A Decomposition Approach for Manufacturing System Design (Cochran et al., 2001); and (iii) A Decisional Reference Model.

It is assumed that a management activity derives new objectives for governing activity systems (in the factory) that it can influence by choices for certain decision variables.

ECOSYSTEM, CAPITAL, AND INDICATOR SYSTEMS

The firm exists in an ecosystem. The economic behavior of firms and other agents and the distribution of assets among them involve mechanisms that allocate society's resources among their many alternative uses. Whereas the ecosystem is not as designable as the firm, approaches exist to make it intelligible, as regards its links to resources and capital, and to influence the ways in which capital is affected by it. An example of such an approach is the Institutional Analysis and Development (IAD) framework (Ostrom, 1990). IAD is a framework for designing policy experiments, empirically tested theories and models linking institutions and the sustainability of common-pool resource systems (Ostrom et al., 1994). The IAD framework has a systematic theoretical focus on the impact of rules and norms on individual incentives in complex ecological-economic systems and accounts for dynamic system interactions at multiple tiers of analysis. In the variant of this framework that Rudd (2004) developed, five types of *capital assets* are specifically included: Natural Capital, Manufactured Capital, Human Capital, Social Capital, and Financial Capital.

Each capital asset is composed of a 'stock' that provides a 'flow' of *value*, goods and services that people can use to help them meet objectives and achieve their aspirations. The flow characteristics of the different kinds of capital vary strongly. Financial Capital for instance represents obligations, and is liquidated as money for trade, and owned by legal entities. Capital assets are the primary referent for *indi-*

cator systems. An indicator system must consider what comprises the capital stock, what comprises the flow, and what quantity of the flow can be used sustainably (e.g., fishing as addressed by Rudd, 2004) or is targeted (e.g., economic growth). When an indicator system is available, a threat to capital assets or a *pressure state* can be represented in terms of a difference between the (expected) values of an indicator and the desirable ones.

INSTITUTIONS AND ECONOMIC AGENTS

Institutions are 'rules-in-use' that influence actor (economic agent) incentives and behaviors. They include both norms and rules. The IAD framework distinguishes three main situations in which institutions operate: operational (ecosystem operations) in which the focus is on the day-to-day impacts of rules and norms; collective choice and constitutional. The focus in *collective action situations* is on the types of rules that are chosen for the activity. Rules can be input- or output-oriented, production- or conservation-oriented, or regulatory- or market-oriented. At the *constitutional level*, the focus is on whom has the rights and power to set lower level rules regarding access to, and utilization of, resources.

In response to the (emerging) pressure state, the relevant governing entities may perform ecosystem redesign experiments. The effectiveness of various design options is compared and evaluated with respect to the indicator systems. The factory and its workers exist in the ecosystem and rely on its capital assets. In particular, the economic agent has to account for its use of capital assets using indicator systems that are also meaningful to the stakeholders in the ecosystem. For certain capital assets, such as the financial assets, strict control procedures are enacted. For other assets, such as the decision objects related to social and natural capital there is less need for control and more room for self-regulation.

DECISION OBJECTS AND THEIR HIERARCHIES

The main items influencing the decision-making are: *(i)* The *decision objective* or set of objectives a decision has to meet. Objectives indicate which types of performances are targeted. Global objectives refer to the entire production system and, according to the principle of coordination, are consistently detailed to give local objectives to all (produc-tion) units; *(ii)* The *decision variables* are the items upon which a decision-maker can make decisions that al-low him or her to reach the objectives. The decision *criteria* guide the choice of the decision-making; *(iii) Constraints* are the limitations on possible values of variables. Decision constraints limit the freedom of a decision-maker to select any arbitrary value for its decision variables; *(iv)* A *performance indicator* is an aggregated piece of information allowing the comparison of the performance of the system to the system's objectives. Performance indicators should be consistent with objectives because it is necessary to compare performances targeted (objectives) and performances reached (indicators), and with decision variables because those variables will have an effect on the performance monitored (controllability).

David Cochran's Manufacturing System Design Decomposition (Cochran et al., 2001) offers a tool to separate objectives from the means to achieve them, to relate low-level activities and decisions to high-level goals and requirements, to understand the interrelationships among the different elements of a system design, and to effectively communicate this information across the organization. Figure 1 (Part A) illustrates the decision-object hierarchy for the top objective of maximal return on investment. In an advanced factory governance system, also the decision-object hierarchies for other top objectives such as safety, health, security, environment, and disaster reduction are defined. Safety and health objectives are expressed with respect to human capital. Security- and disaster-reduction objectives are expressed with respect to several kinds of capital, with action means in the realm of the social capital. Environment objectives are expressed with respect to natural capital stocks and flow.

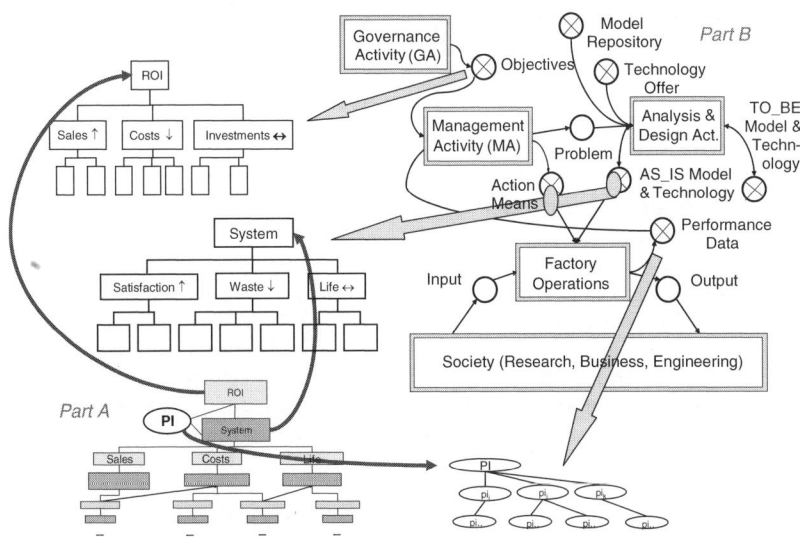

Figure 1: Linking one decision-object hierarchy to the EGAM

FACTORY GOVERNANCE

Figure 1 (part B) depicts any kind of operations (object system) and its relations to decision activities and the environment. In the figure a high-level Petrinet notation is used, crossed circles (stores) denote persistent data sets, and arcs from places to activities (or processes) liberally follow the control/ resource/ input/ output conventions of the generic activity model (GAM). The object system performs a function in the environment, and (performance) objectives are expressed and evaluated for it. The environment is the source of inputs and the sink (market) for the outputs. The model is called an *Extended Generic Activity Model* (EGAM) because it also includes the reflective activities that influence the operations. The governance activity expresses objectives for the object system, taking into consideration relevant constraints (natural, social, etc.) that exist for the capital assets in the factory's environment. The management activity monitors the operations and signals a problem if targets are not met. It will call upon the analysis & design activity to analyse the problem of the object system, to create new designs (TO_BE model & technology), and to compare performance. Governance and management activities decide about the implementation of a new design in the object system.

A Factory is a technical structure (part of the Artifactual Capital) with its operation prescriptions. Within an environment, and using social flows, this technical structure has allocated Natural Capital (space, time, and material artifacts) to productive uses in such a way that the top-level objectives are achieved. Usually this results in a cellular structure on top of which hierarchies are built for the aggregate reflective activities. Within the Factory, the Social Capital has been refined to meet the various top-level objectives that derive from the Factory's mission statement and from the Factory's embedding in society. Each member of the work force (human capital) has a profile which reflects the various tasks the member can perform with a performance that is consistent with the related objectives: production tasks, roles in training, safety and health enhancement, disaster reduction, etc. An extended profile also includes the decision-object hierarchies that are related to the operational situations in which the person may find him or herself and to the responsibilities vested in the person. Depending on the kind of event that occurs – expected (prediction), anticipated (contingency), or unexpected (threats to the capital assets, adaptiveness) – the person will perform prescribed operations or engage in reflective activities with the purpose to bring the operations or situation in line with objectives. In the most general case, the input can be of any kind, ranging from a routine-production order, over a new guideline on toxic materials, to the occurrence of a disaster or attack.

ADVANCED FACTORY GOVERNANCE

Figure 1 shows how the sub-hierarchies of objectives, decision variables and performance indicators (for ROI, Part A) are linked to the EGAM for factory operations (Part B). A similar action must be performed for all relevant hierarchies of decision objects. As also the factory itself will have a structure, for each organizational element some of the decision objects (its scope, a projection of the overall hierarchy) will matter, and all reflective activities must be assumed. The new demands on factories will require us to do additional objective breakdown for non-financial (i.e., natural, artifactual, social capital assets). As eco-system objectives may be subject to change, the question is how to ensure continuous alignment.

For each kind of capital asset, the question is how the reflective activities are best allocated. The more mobile a capital asset is, e.g. financial capital, or the larger share in the time or impact on assets the operations have, e.g. manufacturing activities in JIT production facilities, the more need there is for control of the operations themselves. In the case of emergencies on the other hand, there is need for autonomy and immediate and effective reflection and response.

CONCLUSIONS

Advanced factory governance systems require a mix of controls and autonomy to continuously achieve objectives for all allocated assets. Basic ideas from Socio-Technical Systems Design – predominantly autonomy and self-regulation – might be combined with characteristics of capital assets, in order to arrive at a better balance between the amount of control that is executed by the factory system, and the amount of self-control that is left to the teams of human agents. A (cell) situation-specific mix of governance, management and operational powers with respect to all relevant kinds of assets is expressed in a profile. In relation to natural, human, and social capitals more autonomy is likely. For instance, the amount of environmental protection could be left to the discretion of the human stakeholders. But also aspects of safety and security are open to certain human autonomy over the system. Factory governance systems should leave maximum degrees of freedom for the way (order, pace and method) humans execute their work. What is actually left to the discretion of the human beings will influence positively the motivations and subsequent responsible performances of these agents in an intelligent manufacturing system. In a total asset context, where operations are challenged by frequent adjustment of objectives, or by the occurrence of rare unwanted events, Socio-Technical System Design offers instruments to determine and maintain a proper balance between self-regulation by human agents and automatic control by the factory-governance system.

REFERENCES

Bovenkamp M. van de, Jongkind R., Rhijn G. van, Eijnatten F. van, Grote G., Lehtela J., Leskinen T., Little S., Vink P., and Wäfler T. (2002). *The E/ S tool: IT Support for Ergonomic and Sociotechnical System Design.* In: Yamada S. (Ed.), *Humacs Project: Organizational Aspects of Human-Machine Co-existing Systems* (pp. 67-81). Tokyo, Japan: IMS/ HUMACS Consortium, CD-Rom, March.

Cochran D.S., Arinez J.F., Duda J.W., and Linck J. (2001). A Decomposition Approach for Manufacturing System Design. *Journal of Manufacturing Systems,* **20:6**, 371-389.

Eijnatten F.M. van (1993). *The Paradigm That Changed the Work Place.* Assen/ Stockholm: Van Gorcum/ Arbetslivscentrum, 316 pp. (Anthology: Historical overview of 40 years of STS, with contributions of Hans van Beinum, Fred Emery and Ulbo de Sitter).

Eijnatten F.M. van (Ed.) (2002), *Intelligent Manufacturing Through Participation: A Participative Simulation Environment for Integral Manufacturing Enterprise Renewal.* Hoofddorp, The Netherlands: TNO Arbeid/ PSIM Consortium/ Tokyo, Japan: IMS/ HUMACS Consortium, CD-Rom, March 2002.

Eijnatten F.M. van, and Vink P. (2002). Participative Simulation in the PSIM Project. In: Eijnatten F.M. van (Ed.) (2002).

Goossenaerts J.B.M., Reyneri C., and Berg R. van den (2002). The PSIM Environment Architecture. In: Eijnatten F.M. van (Ed.) (2002).

Little S., Bovenkamp M. van de, Jongkind R., Wäfler T., Eijnatten F. van, and Grote G. (2001). The STSD Tool: IT Support for Socio-Technical System Design. In: Johannsen G. (Ed.), *Proceedings 8th IFAC/ IFIP/ IFORS/ IEA Symposium on Analysis, Design, and Evaluation of Human-Machine Systems* (pp. 409-414). Kassel: IFAC/ HMS.

Matsuo T., and Matsuoka Y.(2004). Integrated Virtual Plant Environment for Analyzing Chemical Plant Behavior. In: Taisch M., Filos E., Garello P., Lewis K., and Montorio M. (Eds.), *International IMS Forum 2004: Global Challenges in Manufacturing, Part 1* (pp. 507-514). Milano: Polytecnico di Milano, Department of Economics, Management, and Industrial Engineering, Print: Grafica Sovico srl, Biassono (Milano).

Ostrom E. (1990). *Governing the Commons: The Evolution of Collective Action.* Cambridge, UK: Cambridge University Press.

Ostrom E., Gardner R., and Walker J. (1994). *Rules, Games, and Common-Pool Resources.* Ann Arbor, MI: University of Michigan Press.

Rudd M.A. (2004). An Institutional Framework for Designing and Monitoring Ecosystem-Based Fisheries Management Policy Experiments. *Ecological Economics,* **48:1**, January, 109-124.

Shin D.P., Han K., Choi S.J., and Yoon E.S. (2004). Integrated Intelligent Management of Process Safety, Health, Environment and Quality in the IMS/ CHEM Framework. In: Taisch M., Filos E., Garello P., Lewis K., & Montorio M. (Eds.), *International IMS Forum 2004: Global Challenges in Manufacturing, Part 1* (pp. 499-506). Milano: Polytecnico di Milano, Department of Economics, Management, and Industrial Engineering, Print: Grafica Sovico srl, Biassono (Milano).

Vink P., Eijnatten F.M. van, and Berg R.van den (2002). Participation: The Key to Intelligent Manufacturing Improvement. In: Yamada S. (Ed.), *Humacs Project: Organizational Aspects of Human-Machine Coexisting Systems* (pp.1-9). Tokyo, Japan: IMS/ HUMACS Consortium, CD-Rom, March, Invited paper (key note speech) to the 20th International Conference on Conceptual Modeling (ER 2001), November 27-30, Yokohama, Japan. International Workshop on Conceptual Modeling of Human/ Organizational/ Social Aspects of Manufacturing Activities, HUMACS 2001.

Yamada S. (2002). Global Perspectives of the PSIM Project. In: Eijnatten F.M. van (Ed.), *Intelligent Manufacturing Through Participation: A Participative Simulation Environment for Integral Manufacturing Enterprise Renewal* (pp. 1-8). Hoofddorp, The Netherlands: TNO Arbeid/ PSIM Consortium/ Tokyo, Japan: IMS/ HUMACS Consortium, CD-Rom, March 2002.

RELATION DIAGRAM BASED PROCESS OPTIMIZATION OF PRODUCTION PREPARATION PROCESS FOR OVERSEA FACTORY

Shuichi Sato[1], Yutaka Inamori[1], Masaru Nakano[1],
Toshiyuki Suzuki[2], Nobuaki Miyajima[2]

[1] Toyota Central R&D Labs., Inc., Nagakute, Aichi, 480-1192, Japan
[2] Toyota Motor Corporation, Toyota, Aichi, 471-8571, Japan

ABSTRACT

This paper proposes the method for the optimization of the production preparation processes for factories in oversea. The method does not use the existing tasks, but the relations between physical designed and measured variables written in a relation diagram. The relation diagram is one of the seven new tools for quality control. The new method can optimize the process based on the physical relations and the essential constraints on the task order with the genetic algorithm. The new technique was evaluated using a hot forging trial process and a 40% improvement of the lead time can be seen in comparison with the sequential trial.

KEYWORDS

production preparation, process optimization, design structure matrix, relation diagram, genetic algorithm, project scheduling

INTRODUCTION

For manufacturing companies today, strategic and timely product development is essential to survive. Value chains including the market, the production, and the supply of parts have to be considered for the world-wide point of view. Subsequently, some manufacturing companies are moving their factories abroad. On the other hand, most companies continuously perform quality control activities (QC), in order to produce high quality goods to satisfy consumers.

We have developed the process optimization technology that can be applied to the production preparation processes for factories outside of Japan. In order to shorten the lead time of the production preparation for oversea factories, manufacturing companies focus on two points: First, measuring process data such as the temperature of a part after being heated for a hot forging process is focused. Second, the design standard is considered. We analyzed the production preparation process for oversea factories and found the following results. Measuring process data helps dividing big and complicated problems into smaller and simpler sub-problems. Furthermore this reduces the influence of uncontrolled elements in the latter stage of the production preparation. We also found that the design standard can change the dependency relationship between the different tasks. The existing study (Sato *et al.*, 2003) proposed the optimization technique for the oversea production preparation process by considering the dependency relations between the different tasks, measuring the process data and the design standard together. That approach uses the physical relations between the designed variables, measured process data, and performance measures. By using the physical relations, the dependency relations between tasks are generated in the Design Structure Matrix (Yassine *et al.*, 1999) and the process is optimized by considering the difference of the verification accuracy among the different trial phases. But that approach has two major problems. One problem is the difficulty to reveal the physical relations in the matrix expression when the number of the design variables, measured process data, and performance measures is large. The other problem is the impossibility to consider the essential constraint of the task order coming from the engineer's experience. The proposed technique has been developed in order to overcome such problems.

With the new technique, the engineer writes the physical relation in the relation diagram expression as shown in Figure 1, which is one of the seven new tools of the QC. The new optimization algorithm considering the strong constraints on the task order is proposed based on the Genetic Algorithm (Holland, 1975) with Partial Matched Crossover (Goldberg, 1989). The new technique is evaluated using a hot forging trial process and the result showing at the end of this paper confirms the efficiency of the proposed approach.

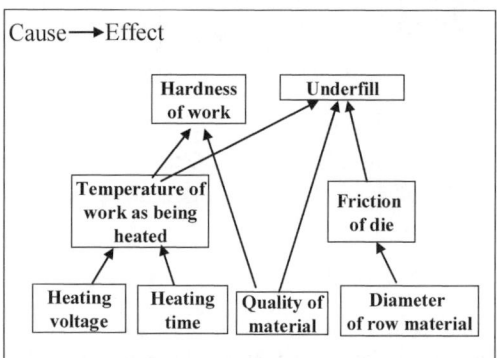

Figure 1: Relation diagram for physical relations between designed variables

APPROACH

We focus on the fact that different engineers perform the production preparation in different ways. We think we should consider the relations behind the process, which are more general and don't depend on the individual engineer. We have concluded that the physical cause-effect relations of the object to be designed are the fundamental factors. Figure 2 shows the proposed hierarchical model of the production preparation process. The proposed optimization technique is based on this model. The lowest level is composed of the physical cause-effect relations of the object to be designed and the company's design standards such as the standard design requirement and the standard design sequence. The company's design standards are important factors to compete with rival companies. In the middle level, the dependency relations between design and/or preparation tasks exist. The structure shows the dependency relations between the tasks are constrained by the physical cause-effect relations and the company's design standards. The product design and the production preparation process exist on the dependency relations between the tasks and the essential constraints on the task order. The essential constraints on the task order will be described in detail, further below. With respect to such a hierarchal model for the production preparation process, we developed an optimization technique as shown below.

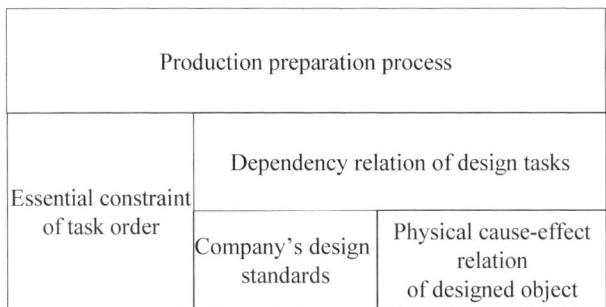

Figure 2: Hierarchical relation model of production preparation process

Description in relation diagram

The current technique (Sato *et al*., 2003) requires the engineer to input the physical relations between the designed variables, measured process data, and performance measures in the matrix. However, adding the relations in the matrix is difficult for most practical cases. We found that the matrix expression is useful to analyze the process, but the engineer is hesitant to use the matrix expression to visualize their knowledge.

Incidentally the engineers usually use the seven fundamental tool of the QC as the numerical method for the quality control activity. Furthermore they have the new seven tool of the QC as the linguistic method. These tools are used as the basic techniques for business reengineering and problem solving in the production area. The relation diagram is one of the seven new tools of the QC. This diagram is the method to describe the cause-effect relations if many causes are interacting with each other. Many engineers are familiar with describing the relation diagram for problem solving.

The proposed method in this paper uses the relation diagram to visualize the physical relation as seen in Figure 1, and subsequently transforms the diagram to the matrix formation.

Optimization algorithm

In the actual process, there are many causes that constrain the task order strongly coming from something except for the dependency relation between the tasks. One example is about the time required to complete each task. The engineers have to do the tasks in the earlier stage, which take long time to be performed. Another example is the situation that some tasks have been completed when the target process starts. The optimization algorithm used in the current technique (Sato *et al*., 2003) cannot consider the essential constraint on the task order except for the dependency relation between tasks to generate the task order.

In this paper, one of the modern heuristic methods in the artificial intelligence research field, Genetic Algorithm (GA) (Holland, 1975), is used. This method can consider various constraints flexibly by modifying the fitness function. The expressions of the essential constraints and the chromosome of the GA are explained in the following sections. The crossover operation method and the fitness function to evaluate each chromosome are also described subsequently.

Expression of essential constraints

In the proposed method, "1" is assigned for the dependency relations between the tasks as illustrated in Figure 3. "10" is used for the essential constraint coming from something except for the dependency relations between the tasks. For example, if the essential constraint on the task order is that the task 2 has to be performed after the task 5, "10" is assigned to the cell as seen in Figure 3. If there are no dependency relations between the tasks and the essential constraints, the cell contains '0'.

Essential constraint on task order

	Task 1	Task 2	Task 3	Task 4	Task 5	Task 6	Task 7	Task 8
Task 1		0	0	1	0	0	0	0
Task 2	0		0	0	10	0	0	0
Task 3	0	0		0	0	0	1	0
Task 4	1	0	0		0	0	0	0
Task 5	0	1	0	0		0	1	0
Task 6	0	0	0	1	0		0	0
Task 7	0	1	0	0	0	0		0
Task 8	0	0	0	0	0	0	0	

Figure 3: Description of essential constraint on task order

Expression of chromosome

The gene is expressed through the task identification. The chromosome shows the task order through the task identification as seen in Figure 4. Therefore the length of the chromosomes corresponds to the number of the target tasks.

Identification number of task

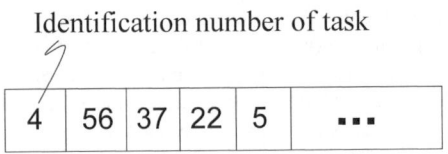

Figure 4: Expression of chromosome

Crossover operation

If the regular crossover operation (Holland, 1975) is applied to the aforementioned chromosome, the chromosome must have the same genes, i.e., the same task identifications in almost all cases. This means that the chromosome generated by the crossover operation does not express the task order. Therefore, if the regular crossover operation is used, the efficiency of searching the best task order degrades significantly. In the proposed method, the special crossover method called Partially Matched Crossover (Goldberg, 1989) is used. This method was initially developed for the traveling salesman problem (TSP), where the order of the places the salesman should visit is resolved. This approach can also be used to resolve the problem which we handle.

Calculation of individual's fitness

Each generated chromosome includes the corresponding DSM. The proposed method uses the following functions to calculate the individual's fitness.

$$F = C_1 \sum_{i=1}^{N} \sum_{j=i+1}^{N} m_{ij} + C_2 \sum_{i=1}^{N} \sum_{j=i+1}^{N} m_{ij} (j-i)^2 \qquad (1)$$

where m_{ij} shows the element of the line i and the column j in the DSM, N is the number of tasks, C_1 and C_2 are two coefficients. The first term of Eqn. 1 has the effect to reduce the number of "10" and "1" in the upper-right field of the matrix. In the DSM expression, the order of the matrix represents the task order, and the "1" in the upper-right field represents the point for the back loop of the process. Therefore, the first term has the effect to reduce the possibility of the back loop coming from the dependency relations between the tasks and satisfy the essential constraints. The second term shortens the distance from the "10" and/or "1" to the diagonals, which is represented as $j-i$ in the equation as seen in Figure 5. This effectively makes the size of the back loop smaller with the satisfaction of the essential constraint.

Figure 5: Expression of chromosome

CASE STUDY

The new technique to improve the oversea production preparation was evaluated using a hot forging trial process. This process can be divided into the following three trial phases.

- Trial phase with an experimental set up
- Domestic trial phase by machines used after starting the production
- Overseas trial phase

A total of 95 physical parameters in this process are extracted as shown in Table 1. The process optimization using the presented method was able to improve the lead time by around 40%, in comparison with the sequential trial. Furthermore, the proposed method realized the optimized process while satisfying all the essential constraints. Figure 6 shows the part of the matrix which includes the essential constraints on the task order. Figure 6(a) shows the result of the method without consideration for the essential constraints on the task order. The task group in Figure 6 shows the tasks which should be performed together. Task A and Task E compose one group. Figure 6(b) shows the result of the method considering the constraints. The task groups in both cases are the same. But the order of the tasks differs between Figure6(a) and (b). Only the process in Figure 6(b) satisfies the essential constraints.

TABLE 1
COMPOSITION OF PHYSICAL PARAMETERS

Category	Number of items
Quality	9
Cost & safety	4
Raw material	3
Cutting	6
Heating	6
Forging	42
Trimming	16
Thermal refining	5
Shot blast	4
Total	95

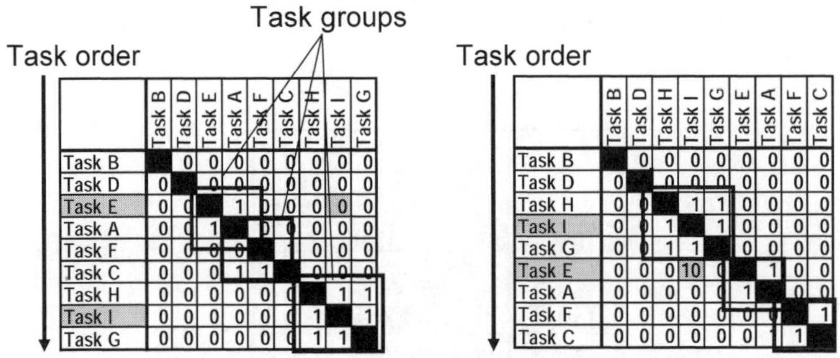

a) Without use of essential constraints   b) With use of essential constraints

Figure 6: Efficiency by considering essential constraints

CONCLUSION

The new technique does not start with the existing tasks, but with the physical cause-effect relations between designed, adjusted, intermediate and goal variables. These physical relations are described in the relation diagram. Required tasks are generated based on these physical cause-effect relations. The proposed technique was evaluated using a hot forging trial process that includes 95 physical variables. An improvement of the lead time of around 40% was realized using a process optimization with the essential constraints, as compared to the sequential trial.

REFERENCES

Goldberg D.E. (1989). *Genetic Algorithms in Search, Optimization, and Machine Learning*, Addison-Wesley.
Holland J.H. (1975). *Adaptation in Natural and Artificial Systems*, University of Michigan Press.
Sato S., Inamori Y., Nakano M., Suzuki T. and Miyajima N. (2005). Analysis Method for Overseas Production Preparation Process. *Journal of Japan Society of Mechanical Engineering* 71:705, 322-329.
Yassine A., Falkenburg D. and Chelst K. (1999). Engineering Design Management: An Information Structure Approach. *Journal of Production Research* 37:13, 2957-297.

CYBER CONCURRENT MANUFACTURING INTEGRATED WITH PROCESS ENGINEERING AND 3D-CG SIMULATION
-PRODUCT DESIGN, PRODUCTION SYSTEM DESIGN, AND WORKSTATION SYSTEM DESIGN AS A CASE STUDY ON "CURTAIN WALL" CONSTRUCTION WORK-

Kinya Tamaki
School of Business Administration, Aoyama Gakuin University, 4-4-25 Shibuya, Shibuya-ku, Tokyo, Japan

ABSTRACT

The research in the last fiscal year (2003), we have indicated a conceptual framework of "Cyber Concurrent Manufacturing (CCM)" system. In order to continue the research in the last fiscal year, a method of modeling in detail by using a Process Engineering tool is proposed. The method is applied to a case study which is to model construction processes of the "curtain wall" installation in a virtual construction site. The feature of this method is to define the total processes with keeping mutual relationship between (1) product design, (2) production system design, and (3) workstation system design. Furthermore, it is able to previously verify the result of the model data of (1) to (3) by various 3D-CG simulators before starting an actual construction.

KEYWORDS

Cyber manufacturing, process engineering, 3D-CG simulation, construction management, intelligent manufacturing system (IMS), product design, production system design, and workstation system design.

INTRODUCTION

This research has been performed as a work package study involved in "Innovative, and Parts-oriented Construction (IF7-II)" project. IF7-II project is one of Intelligent Manufacturing System (IMS) program which was proposed by Japan Ministry of Economic and Trading Industry since 1989. This research member is as follows: Aoyama Gakuin Univ., Waseda Univ., Osaka Univ., Tokyo Institute of Electric and Communication, Shimizu Corp., Tostem Corp., and Hitachi Zosen Information Systems Corp.

As a working research group in this project in the last fiscal year, we have indicated a conceptual framework of "Cyber Concurrent Management (CCM)" system by utilizing both for virtual manufacturing processes and for real-field based manufacturing processes, which are covered with following phases: (1) product design, (2) production system design, and (3) workstation system design. Figure 1 shows the research background of the CCM system in the current fiscal year (2004) from the last fiscal year (2003).

The purpose of this paper is to propose the concept and method for modeling engineering process by using a process modeling tool (it is henceforth called PE tool), and verifying validity of the modeled process results by the various 3-dimensional computer graphic (3D-CG) simulators, based on the CCM system. As a case study for that, we built a test bed system for the verification method of modeling according the construction process of the "curtain wall" which constitutes the exterior wall of a high-rise building to the PE tool, and the model data based on various 3D-CG simulators.

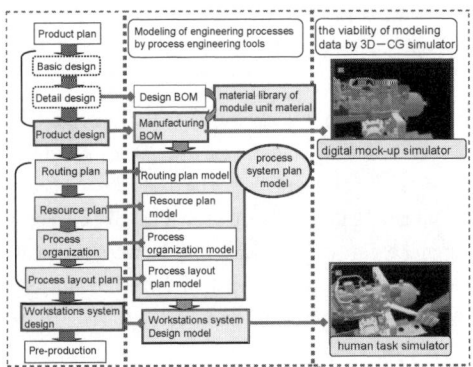

Figure 1: Conceptual Framework of Cyber Concurrent Management System

RESEARCH SUBJECTS OF CCM SYSTEM

Figure 2 illustrates that the CCM system treats with a range of the engineering processes from product design, production system design, to workstation system design. First, using PE tool, as it is in the left side of Figure 2, modeling of each phase of construction processes is performed as follows: (1) product design, (2) production system design, and (3) workstations system design. That is to perform: (1) product design which is considered of assembling sequence and assembling efficiency based on a structure of BOM (bill of materials) consisting of curtain wall materials, (2) production system design based on the assembling sequence of the product design, and (3) workstation system design paying attention to human operation order for assembling materials in a construction site.

Next step is "intermediate deliverables" of the construction process data of each phase in PE tools: (1) product design; storing to the library of materials and conversion from Design BOM to Manufacturing BOM, (2) production system design; integration of assembly process and delivery processes of material handling, (3) workstation system design; work organization of two or more contractor's group work is carried out the resources plan by using the PE tool. Since the construction process of such each phase is modeled, the technique called process engineering is used.

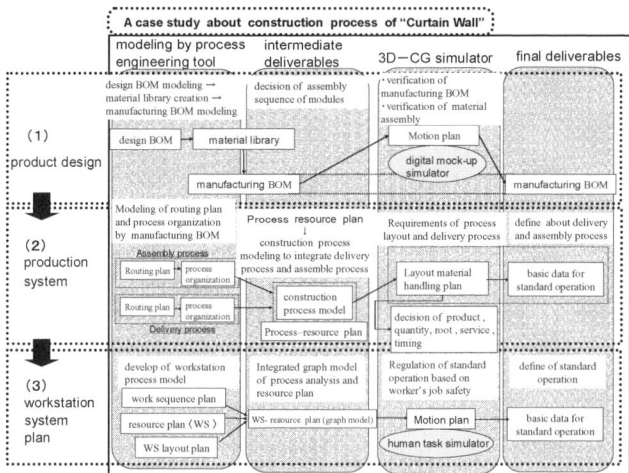

Figure 2: Research Range of Process Engineering Modeling and Modeling Data Verification in the CCM

MODELING AND VERFICATION OF PRODUCT DESIGN PROCESS

As a precondition, in the detailed design stage of a product design, the curtain wall materials are divided per module materials based on "Design BOM", and the "materials library" is created. The "Design BOM" creates the structure which can respond to two or more product kinds and option materials flexibly by dividing to the module materials of curtain wall materials and changing those module materials into a "materials library".

Next, at the stage when considering of assembling sequence, the "Manufacturing BOM" is created. Moreover, in order to create "Manufacturing BOM" from the "materials module" stored in this materials library, it is necessary to determine the assembly method and an assembly order about the materials. Below, the procedure of operating process modeling of a production design stage is describes: 1) modeling "Design BOM" upon the part drawing of the detailed design stage of a product design, 2) modeling "materials library" out of module materials based on the part information on Design BOM, and 3) modeling "Manufacturing BOM" based on the materials library. The validity of the process data of each product design stage is verified by 3D-CG digital mock-up simulation, and the basic data of the verified Manufacturing BOM as a "final deliverables".

MODELING AND VERIFICATION OF DESIGN PROCESS FOR PRODUCTION SYSTEM

According to a construction project, it is necessary to deliver required materials, tools, machines, etc. in the construction process of a curtain wall to the working area or a destination. Moreover, according to a construction schedule, it is necessary to assemble the delivered materials, carrying out suitable work organization. The following procedures perform such a construction process as modeling of the production system design by using the PE tool: 1) the routing planning of materials assembly work, 2) process organization of the materials assembly work by two or more work contractors, 3) the resources planning for a delivery process, and 4) the site resource planning of the construction place corresponding to integration between the assembly work and the resource delivery process just-in-time.

The model is created for the integration production processes of materials delivery and materials assembly with the PE tool. Integration of both these processes realized the modeling of the production process in connection with work organization at large "which worker works in a required work area using the materials and quantity of a required kind at the time of necessity."

MODELING AND VERIFICATION OF DESIGN PROCESS FOR WORKSTATION SYSTEM

Workstation system design is performed focusing on the work routing of the human workers for each site workstation after process organization (workstation unit) in a construction working area. That is, in the workstation system design, the work process "using what resources workers use and what operation they do" is designed in consideration of human workability. Modeling of the workstation system design by the PE tool is performed by the following procedures: 1) the site work routing plan corresponding to the materials assembly, 2) resource planning required for materials assembly work, 3) the site workstation layout planning of a working area, 4) a setup of the prerequisite of task planning of work organization of two or more work contractor of operation, and 5) regulation of the task planning of work organization of two or more work contractor of operation, and analysis of work load

The model notation of the structure of the group work by cooperation, i.e., the work organization, was carried out with the PE tool among the workers who become the standard and the workers of 1 upstairs of a construction story in the case study. In case verifying the validity of modeling of the work system design using a human task simulator, analyze load mitigation and working efficiency paying attention to the workload and the workability of human work. Furthermore, the analysis to the posture and work load of human work which attaches vertical material was shown by applying the simulator. By performing load analysis of modeling of the process of these human task, or human work, data can be used as basic data these results at the time of creation of a "work standard document."

CONCLUSION AND FUTURE RESEARCH SUBJECTS

The conceptual framework of "Cyber Concurrent Manufacturing (CCM)" system was indicated under a series of join research IF7-II project. In order to continue the research in the last fiscal year (2003), a method of modeling in detail by using a Process Engineering tool is proposed in this paper. Based on the proposed modeling and its verification method, this fiscal year focused on "modeling of the design engineering processes and the manufacturing processes by the PE tool", and "verification of the simulation of modeled data by the 3D-CG simulators", in connection with construction of curtain wall materials. The method is applied to a case study which is to model construction processes of the "curtain wall" installation in a virtual construction site.

REFFERENCES

Kinya Tamaki et al. (1999). Development of Virtual and Real-field Construction Management Systems in Innovative, Intelligent Field Factory, *Proceeding of ISARC*

Summary Research Report (2004). *Innovative, and Parts-oriented Construction (IF7- II)* Project, IMS International Joint R&D Support Program, IMS Promotion Center

WIRELESS DATA TRANSFER APPLIED ON HYDRAULIC SERVO

O. I. Karhu[1], T. K. Virvalo[1], and M. A. Kivikoski[2]

[1]Institute of Hydraulics and Automation, Tampere University of
Technology, P. O. Box 589, 33101 Tampere, Finland
[2]Institute of Electronics, Tampere University of Technology,
P. O. Box 692, 33101 Tampere, Finland

ABSTRACT

In hydraulic servo systems, especially in mobile applications, there might be great advantages if there was no need for wiring between actuators and users and/or a main controller. Most of the wires in hydraulic servo systems carry measurement and control signals. Therefore, wireless transfer of feedback signals and output of the controller is studied. Experimental results are shown and the performance and possibilities of wireless data transfer in these kinds of control applications are discussed.

KEYWORDS

wireless, data transfer, latency, closed-loop, control, hydraulic servo

INTRODUCTION

Wireless closed-loop control has been studied by a couple of different groups in recent years. These projects are discussed by Hörjel (2001), Ploplys & Alleyne (2003), Ploplys (2003), and Kawka & Alleyne (2004). The majority of research has been carried out using PC hardware, which is not an optimal solution for most applications. Many standards for wireless communication are designed for fast file transfer instead of low latency. In this research the objective is to find a solution that makes it possible to achieve a minimal sampling and control interval.

EQUIPMENT

The equipment used in the experiments is shown in Figure 1. It consists of two units, the controller unit and the hydraulic unit. Everything in the hydraulic unit is either mounted to the hydraulic test rig or located very near it. The controller unit is located a couple of meters away from the hydraulic unit.

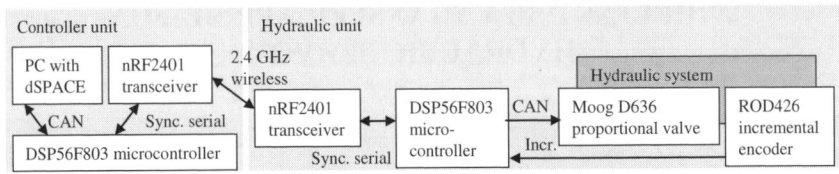

Figure 1: The test equipment

A flexible user interface is needed to develop different controllers and control parameters as well as to record measurements. A desktop PC with a real time controller and a connector board including CAN from dSPACE was used for this purpose. The dSPACE processor can be programmed from Simulink.

802.11b network adapters have been successfully utilised in wireless closed-loop control for example by Ploplys (2003). The problem with 802.11b is rather long latency. Using UDP instead of TCP minimizes the latency to approximately 2 ms. Minimal sampling interval is restricted to 4-5 ms because the round-trip of a small data packet takes about two times the latency. Bluetooth in closed-loop control has also been researched. Range and reliability would suit this project but again the problem is latency. According to Hörjel (2001), minimum latency achieved using Bluetooth is 18 ms which is much too long for the studied case. Other ready-made radio modems are usually designed for sending small, not time-critical packets over long distances. Although some of them have adequate bit rates, the latency is usually not presented in data sheets. There are also different non-standard transceiver circuits. They are available at different bit rates, ranges, modulations and frequency bands. Some circuits perform intelligent functions such as bit error recognition or address field processing.

Experiments were started using the nRF2401 transceiver from Nordic Semiconductor because it had detailed timing information on its data sheet, Nordic Semiconductor (2004), and seemed to have low latency. The nRF is capable of bit rates up to 1 Mbit/s but selecting a slower bit rate gives more range and reliability. The bit rate can be selected quite low because the nRF adds very little overhead to data packets and there is no minimum packet size. The frequency channel is programmable for frequency hopping. The low transmitter power restricts the range to approximately ten meters at open space.

The system needs two microcontrollers: one to connect the valve and the encoder to the transceiver and another to connect the other transceiver to the dSPACE. The microcontrollers should have enough performance to work as the main controller of the system. DSP56F803 hybrid controllers from Freescale were chosen because they suit control applications well. The DSP has a CAN controller which allows easy connections to the dSPACE and CAN valves.

CONTROLLERS AND SOFTWARE

The equipment was used to test a simple proportional controller and a state controller. First the DSP in the controller unit was only used in transferring data between the controller realized in dSPACE and the transceiver. Then the same controller was realized in the DSP and dSPACE was only used as an interface to enter controller parameters and to log controller data. The software on the DSPs is shown as a block diagram in Figure 2. The DSP in the hydraulic unit runs the same program regardless of whether the main controller is the dSPACE or the controller DSP. Compared to a wired arrangement, this wireless setup has a lag of 2 ms in the control loop. If a packet is lost, the previous control signal is held. As Kawka & Alleyne (2004) state, for many hydraulic systems this suits better than to immediately output a zero. If 10 successive packets have been lost, the DSP closes the valve. This 10 packet hold causes an additional delay of only 20 ms if the communication link is completely lost but makes the system very tolerant of small packet losses.

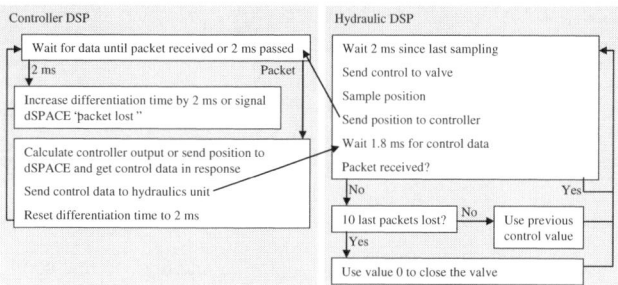

Figure 2: The software on the DSPs

When the dSPACE is the main controller, the DSP in the controller unit sends incoming nRF packets immediately via CAN to the dSPACE. The DSP also sends packet loss information that can be used to prevent speed and acceleration from distortion. The Simulink controller polls the CAN buffer of the dSPACE and when a packet is received, responds with the controller output. The controller DSP then sends the controller output via the transceivers. The model is shown in Figure 3. It also illustrates the velocity calculation block with packet loss compensation. If a packet is lost, the block increases the previous position value by the previous increase. This equals to holding the last velocity.

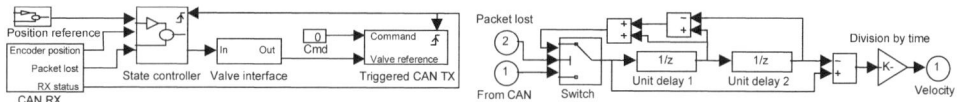

Figure 3: The Simulink model of the state controller and its velocity calculation block

The problem with the controller on the dSPACE is that the position and control data have to be transferred via CAN. These transmissions take relatively long, approximately 15% of the cycle. When the main controller is implemented in the controller DSP, the output is calculated immediately as the position data arrives. Furthermore, the differentiation can easily process packet losses. The CAN may now be used to transfer controller parameters from dSPACE and to receive measurement data. The arrival of these packets is not time-critical as long as the latency is below the human reaction time.

RESULTS

The proportional controller seemed to have the same behaviour regardless of whether the arrangement was wired or wireless. The proportional controller is thus not sensitive to a lag of a few milliseconds in the control loop or moderate packet losses. Selected state controller parameters are shown in Table 1. In the wireless arrangement the lag and lost packets made the system unstable at lower velocity and acceleration gains than in the wired reference setup. Because the KV and KA are used to damp the oscillation caused by high proportional gain, also the KP had to be reduced a little. As can be seen from Figure 4, the velocity stays almost equally stable at both cases. The proportional gain affects the settling time within 0.1 mm of the error signal. The wired reference setup has a settling time of 0.3 s in inward movement whereas the wireless arrangement has a settling time of 0.6 s.

TABLE 1
STATE CONTROLLER PARAMETERS

Parameter	Wired	Wireless
KP (position feedback gain)	270	230
KV (velocity feedback gain)	2.0	1.5
KA (acceleration feedback gain)	0.1	0.07

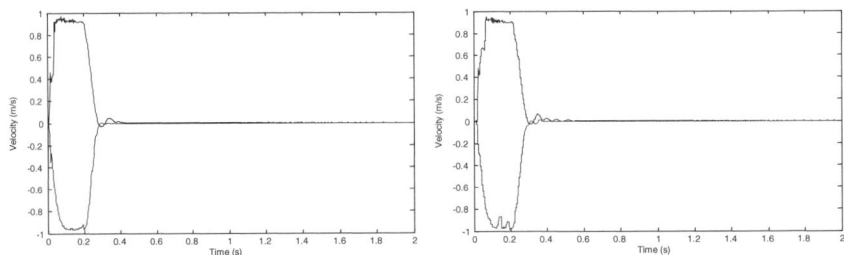

Figure 4: A movement of 0.2 m using the wired controller and the wireless controller on the DSP

The bit rate of the transceiver was configured to 250 kbit/s. During the measurements the packet loss from the hydraulic unit to the controller unit was monitored. The loss altered between 10% and 20%. Most of these losses are single packets. With current compensation methods the losses increase the effective sampling and control interval temporarily from 2 ms to 4 ms. Erroneous packets were seldom received, approximately once in ten minutes. For future work some detection for errors will be added.

The distance between the transceivers was about 2.5 m. The packet loss stayed below 20% as the distance was increased to 5-6 m. Due to the laboratory environment, a longer distance could not be experimented. The reinforced concrete walls of the laboratory attenuate the signal strongly, which makes control applications practically impossible. The RF power of 1 mW is too low for ranges above 10 m in open space. There are also devices that use high transmission power at the same 2.4 GHz frequency band. A lot of wireless network activity might even totally block the nRF communications. On the other hand, the same transceiver circuit is used in wireless PC equipment such as wireless game controllers, so there should be at least some compatibility with wireless LANs.

CONCLUSIONS

The data transfer of a closed-loop control system can be done using wireless transceivers. A state controller can be implemented but it will perform a little worse than a wired controller. With proportional controller there is no difference at all. Some compensation method for lost packets is required especially with the state controller. The nRF2401 transceiver circuit suits well for short range applications at least when neither interfering signals nor obstacles are present. Improvement could be achieved by employing a frequency hopping algorithm or by increasing transmitter power. 802.11b network adapters could be adequate as well if there were no need for sampling intervals below 4 ms.

REFERENCES

Hörjel, A. (2001). *Bluetooth in Control*. M.Sc. thesis, Lund Institute of Technology.

Kawka P. and Alleyne A. (2004). Wireless servo control for electro-hydraulic positioning. *Proceedings of Bath Workshop on Power Transmission and Motion Control*, 159-172.

Nordic Semiconductor. (2004). *Single chip 2.4 GHz transceiver nRF2401*. Product specification.

Ploplys, N. (2003). *Wireless Feedback Control of Mechanical Systems*. M.Sc. thesis, University of Illinois.

Ploplys N. and Alleyne A. (2003). UDP Network Communications for Distributed Wireless Control. *Proceedings of the American Control Conference*, 3335-3340.

THE CHALLENGES ON THE DEVELOPMENT OF MOBILE CONTROLLED RFID SYSTEM

Mikael Soini[1], Lauri Sydanheimo[1] and Markku Kivikoski[2]

[1]Tampere University of Technology, Electronics Institute, Rauma
Research Unit, Kalliokatu 2, 26100 Rauma, FINLAND
[2]Tampere University of Technology, Electronics Institute,
Korkeakoulunkatu 10, 33720 Tampere, FINLAND

ABSTRACT

The Mobile Controlled RFID System (MCRS) has been developed for remote tracking and control of RFID tags. MCRS utilizes the packet switched GPRS (General Packet Radio Service) network in its operation. However, to guarantee the necessary QoS (Quality of Service) level for critical applications can be a problem using the GPRS network. Therefore the main goal of this paper is to study how different applications, mainly MCRS can operate in variable QoS conditions. At the same time the influence of enhanced security features are studied in relative to the operability and usability of MCRS. The security level is increased by user certification and by VPN tunnelling.

KEYWORDS

Automation, Mobile tracking and control, RFID systems, Security, Wireless Communication

INTRODUCTION

The motif for this paper rises from previous work concentrating on the Mobile Controlled RFID System [1]. With this system the tracking and control of RFID tags can be carried out remotely through the GSM (Global System Mobile communications) network's GPRS service. Tag tracking solutions can be exploited in many practical situations such as manufacturing processes, material flow, supply chain and warehouse management, logistics, and security. Research in this paper concentrates on the QoS issue in GPRS networks. The paper shows how the quality of the GPRS network connection affects the operation of MCRS and other applications. GPRS QoS parameters here are: mean delay, jitter, and packet loss proportion. The object is to study how variations in these parameters influence the amount of transferred packets and data transfer times. The simulations are executed with and without Virtual Private Network (VPN) tunnelling. VPN is used to increase data security in remote connections from public to private networks. In this case the VPN connection is based on the Point-to-Point Tunnelling Protocol. The extensive practical GPRS network operability measurements are the reference for the simulations.

MOBILE CONTROLLED RFID SYSTEM

MCRS enables the remote tracking of RFID tags through the GPRS network. The system architecture consists of RFID tags, one or more RFID readers, and a Mobile Gateway Server (MGS) that controls readers and communicates with the user mobile. MCRS operation may be briefly as follows. A reader lists every tag detected in its reading range. This list is updated on tag arrivals in or departures from the reading range; all changes are sent to the MGS. The MGS automatically informs the user about the movements of those tags that are set to follow-up. The user can also check all the tags that are within range of a certain reader at a particular time.

GENERAL PACKET RADIO SYSTEM AND QOS ISSUES

GSM is the world's most widespread digital mobile phone standard for cellular circuit-switched communications. GPRS is a technology that utilizes the upgraded GSM networks radio interface offering packet-switched networks and always-on connections for the user [3]. Modern mobile phones and developed networks make remote solutions possible. Still, high-level performance in GPRS communication systems is not guaranteed. The ETSI standard [4] defines five different QoS attribute classes: precedence, delay, reliability, and mean and peak throughput. Combinations of these attributes can define different GPRS QoS profiles.

However, there are some limits affecting QoS in GPRS such as 1) only one QoS profile can be used for a given PDP (Packet Data Protocol) address, 2) QoS profiles are vaguely specified, and 3) GPRS radio only supports best-effort traffic [5]. The first of these indicates that the above mentioned ETSI standard determines a specific QoS profile to an end-to-end connection for a subscriber, not for an application. Thus if different levels of QoS are needed a new PDP context or logical network connection [6] must be activated for every application using the limited address space of the network. Secondly, loosely specified GPRS QoS standards lead to compatibility problems between different manufacturers' devices. Thirdly, the radio access network is designed for best-effort traffic. Thus handovers, IP-address changes, signal strength weakening, limited bandwidth, and contention for resources are properties that decrease the quality of GPRS.

QOS SIMULATIONS

The paper presents functionality of three different applications in variable GPRS conditions. These applications are Internet Explorer v6.0 (IE), WS_FTP Pro 7.61 (FTP), and MCRS. In simulations the quality of the GPRS connection is varied by different measures (delay, jitter, packet loss) and the goal is to study how the variations in these parameters and the security level (VPN) affect the data transfer time and the amount of transferred packets. The reference for the simulations is the operation of GPRS class 10 (typical for mobile phones) in practical environments [2]. The reference values are: *Capacity Downlink* 26kbps, *Capacity Uplink* 15kbps, *Packet Loss* 3%, *Mean Delay* 690ms, and *Jitter* 350ms. The functionality of the IE was tested by downloading a test page (100kB, 79 jpeg pictures) through the simulated network. The functionality of the FTP was studied by transmitting a test packet (100kB, zip file) through the simulated network. Access to an MGS that controls the RFID system is through a certification page where the user is identified and authenticated as a legitimate user. The size of the certification page used in these simulations is 449 bytes.

Delay effect. Figure 1 shows how delay affects applications. **MCRS** tolerates very long delays. VPN tunnelling increases the amount of transferred data by 30 to 80 %. The access time to the MGS increases linearly as a function of delay. VPN does not affect access times. **IE** tolerates even extremely long delays very well and the connection stays open but data transfer capacity that is successfully received jpeg pictures decreases strongly as a function of increased delay with and without VPN. **FTP** tolerates long delays up to eight times the reference level but after that the connection is lost. It can be seen that data packaging (~25%) is carried out in VPN. Download time increases linearly as a function of delay. If VPN is not used connection tolerates delays twice as long.

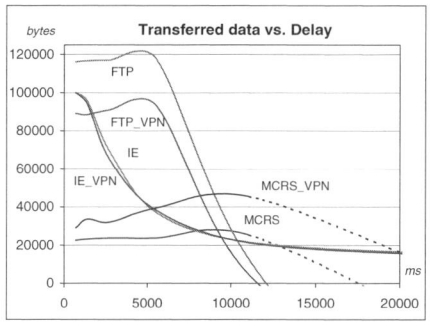 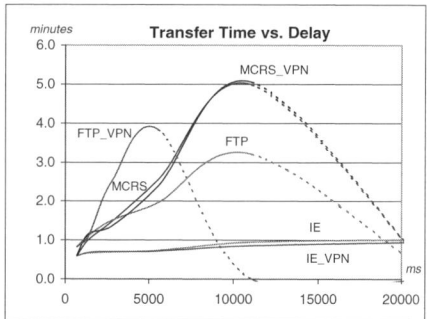

Figure 1: Transferred data and transfer times as a function of delay

Jitter effect. Figure 2 shows how jitter affects applications. **MCRS** tolerates large variations of delay well. VPN increases the amount of transferred data by 30 to 40 %. Access time to the MGS increases again linearly as a function of jitter but nevertheless jitters up to 5 seconds are tolerated. VPN does not significantly affect access time. **IE** tolerates very long jitters and the connection holds up but the received jpeg pictures decreases as a function of increased jitter. The effect of VPN on the application can be seen as a 10 to 50 % decrease in the amount received jpeg pictures when jitter increases. The download time remains fairly constant. **FTP** with and without the VPN handles well even 5-second jitter variations. VPN does not significantly affect the operation.

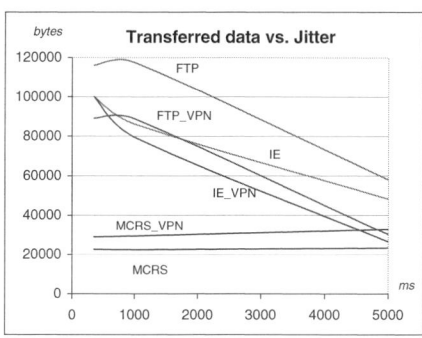 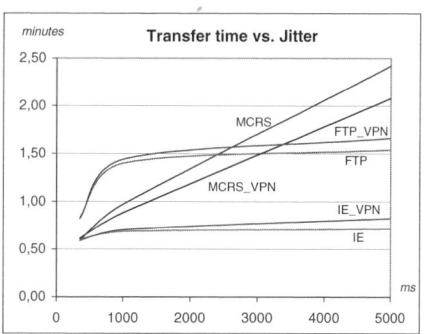

Figure 2: Transferred data transfer times and as a function of jitter

Packet loss effect. Figure 3 shows packet loss effect (3 to 12%). **MCRS** tolerates 6 % packet loss well and the system without VPN behaves well even with packet loss up to 12 % though the access time doubles from the 6 % case. VPN increases the amount of transferred data about 20 % when packet loss is doubled. In the **IE** case doubling the packet loss to 6 % decreases the received jpeg pictures from the full 79 to on average of 59 without VPN. With VPN the decrease is from 79 to 24 jpeg pictures. The downloading time in the 6 % packet loss situation is 30 % longer with VPN and in the 12 % packet loss situation the downloading time is 70 % longer. In **FTP** the 6 % packet loss halves the data transfer rate with and without the VPN, and there is no significant difference in data transfer time either. The packet loss of 12 % makes the operation of FTP impossible in both cases.

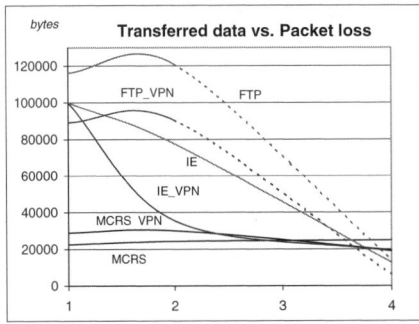 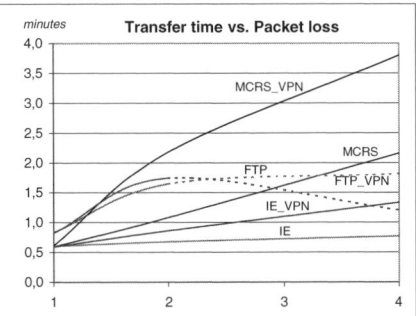

Figure 3: Transferred data and transfer times as a function of packet loss

CONCLUSIONS

Simulation shows how different applications operate in varying GPRS conditions. The QoS in GPRS is difficult to apply in applications that would need a guaranteed level of link quality for their operation. Therefore QoS was studied from the point of view of how these applications operate in variable GPRS conditions. The research studied extreme limits for operation and it is natural that the usability of the applications is very poor close to these limits.

MCRS tolerates delays and jitters very well whether VPN is used or not. The difference is that when using VPN the amount of data transferred is considerably higher than without it. The same is valid in situations where the amount of lost packets increases. As the GPRS link quality decreases drastically the system operates in some fashion but usability is lost. The simulated access time from mobile to MGS varies between 30 and 35 seconds depending on whether VPN tunnelling is used. The time is so long because a large of amount of data (>20kB) is transferred during this process compared to the actual page size (0,5kB). From earlier research it can be noted that without the certification phase the access time to the MGS is less than 5 seconds in practice. Thus there is development work to do to attain a highly secure and usable system.

In the future mobile UMTS networks will come in to evermore practical and wider use. The reference [5] presents the evolved QoS properties of UMTS compared to GPRS. This together with enhanced data transfer capacity should help to implement remote control applications even with modest real-time requirements.

References

[1] Soini M., Eeva T., Sydanheimo L. and Kivikoski M. (2004). The Mobile Controlled RFID System. *11th IFAC Symposium on Information Control Problems in Manufacturing (INCOM2004)*.
[2] Ruohonen T., Ukkonen L., Soini M., Sydanheimo L. and Kivikoski M. (2004). Quality and reliability of GPRS connections. *2004 IEEE Consumer Communications and Networking Conference (CCNC2004)*.
[3] Samjani A. (2002). General Packet Radio Service. *IEEE potentials*. **21:2**, 12-15.
[4] ETSI EN 301 113 v6.3.1 (2000). Digital cellular telecommunications system (Phase 2+), General Packet Radio Service (GPRS), Service description, Stage 1.
[5] Koodli R. and Puuskari M (2001). Supporting Packet-Data QoS in Next Generation Cellular Networks. *IEEE Communications Magazine*. **39:2**, 180–188.
[6] Bettstetter C., Vogel H. J. and Eberspacher J. (1999). GSM Phase 2+ General Packet Radio Service GPRS Architecture, Protocols, and Air Interface. *IEEE Communications Surveys*, **2:3**, 2-14.

WIRELESS COMMUNICATION WITH BLUETOOTH HEARING PROTECTOR

Mika Oinonen, Pasi Myllymäki, Matti Ritamäki, Markku Kivikoski

Tampere University of Technology, Institute of Electronics, P.O. Box 692, FIN-33101, Tampere, Finland

ABSTRACT

In today's mobile world, it is important to be easily accessible via a cellular phone. It is also important to protect the hearing in noisy environments. Often these two requirements must be fulfilled simultaneously. As a solution, a prototype of a Bluetooth hearing protector was constructed. A microstrip antenna was designed using a trial-and-error method. The device was installed inside a high-quality passive hearing protector and a boom microphone was attached to the earcup. The developed Bluetooth hearing protector enables both wireless communication and hearing protection, while also eliminating the need for inconvenient cables, thereby improving safety and accessibility.

KEYWORDS

Wireless communication, Bluetooth, hearing protection, safety, wearable, inverted-F antenna

INTRODUCTION

Hearing protection is vital in noisy environments. Strong impulse noise, especially, is hazardous and can cause permanent hearing loss. Traditional passive hearing protectors attenuate noise efficiently and protect the inner ear from loud noise. However, when wearing a passive hearing protector, it is almost impossible to use a cellular phone, and important calls may be missed if the ringing tone is not heard. It is possible to connect a cellular phone into an electronic hearing protector with a cable. While this provides communication and hearing protection at the same time, the cable can be inconvenient in many work tasks. A Bluetooth radio link between the cellular phone and the hearing protector makes it possible to communicate with others and work in noisy environments without inconvenient cables. The cellular phone can be kept in the pocket and incoming calls can be answered simply by pressing a button on the earcup when the ringing tone is heard inside the earcups.

During the last few years, the wireless Bluetooth radio link has been replacing cables, e.g. in computer products. Bluetooth connection can also be used to replace the cable between the cellular phone and the hearing protector, provided that the cellular phone supports a Bluetooth connection. Bluetooth is a short-range radio link, which operates in the globally available licence-free 2.4 GHz ISM band. The audio data rate is 64 kb/s, as in the GSM system (Bray 2001). Risk of interference with other devices is minimized by the use of a frequency-hopping spread-spectrum transmission scheme (Bravo-Escos 2002). Bluetooth profiles, which are defined in the Bluetooth specification, describe how a particular application can be implemented, and which parts of the core Bluetooth protocol should be used to support the profile (Bray 2001). In a headset application only a fraction of the capabilities of Bluetooth is utilized. The Bluetooth audio connection between a cellular phone and a headset can be realized by using either the headset- or handsfree- profile. To operate with all Bluetooth mobile phones, the headset must support both profiles. Nowadays there are several kinds of Bluetooth chips and modules available from various manufacturers. Many of the off-the-shelf modules adapt themselves directly to headset-type applications.

THE CONSTRUCTED PROTOTYPE

A prototype of a Bluetooth hearing protector was constructed. The prototype PCB includes an Avantwave BTR110B v0.2 class 2 Bluetooth transceiver module, a number of additional electronics and a microstrip inverted F-antenna (IFA).

An inverted-F antenna is a compact antenna with a height of about one tenth of the wavelength (Olmos 2002). The layout of the developed IFA antenna is shown in Figure 1. The IFA antenna consists of two sections: the inverted-L radiating section (*a-b-c*) and the matching section (*b-d-e*). The antenna is fed at point *a* and grounded at point *e* (Olmos 2002). The smaller the spacing between points *a* and *e*, the lower the resonance frequency and the higher the impedance (Ali 2000). The final layout was obtained through trial and error. The antenna is connected to the Bluetooth module by a 50 Ω transmission line. The designed antenna was integrated on the same printed circuit board as the Bluetooth module and the additional electronics. The printed circuit board has a length of 35 mm and a width of 31 mm. Thus, the device is relatively easy to install inside the earcup.

There are loudspeakers in both earcups and a single boom microphone is located in front of the mouth. The microphone is directional and insensitive to background noise, which makes it practical in noisy environments. A lithium-ion rechargeable battery provides the 3.6 V supply voltage. The Avantwave BTR110B module includes all the Bluetooth software, and supports both headset and handsfree profiles. The prototype is compatible with virtually any Bluetooth cellular phone. The module is based on CSR BlueCore2™ Flash –chip.

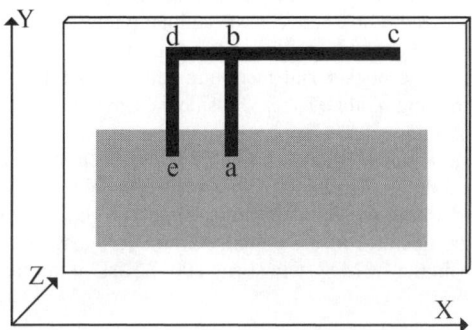

Fig. 1. The layout of the inverted F-antenna. The measurements were d-c = 24,5 mm, d-b = 5.0 mm and a-b = 6.8 mm. The grey area depicts the ground plane on the bottom side of the PCB.

THE MEASUREMENTS

Performance of the developed antenna was measured. The reflection coefficient of the antenna was measured, and the antenna was fine-tuned on the basis of this. The radiation patterns were measured in all three directions. Finally, the antenna was integrated on the same PCB as the Bluetooth module and the device was tested with three different GSM cellular phones.

The reflection coefficient of the antenna (S_{11}) is shown in Figure 2a. The S_{11} was measured with a Hewlett Packard 8722D 50 MHz – 40 GHz network analyser. S_{11} is less than –10 dB in the frequency range of 2320 – 2560 MHz. S_{11} has a minimum of –12.5 dB at 2450 MHz. The quality factor (Q) of the antenna is 10.4, which means that the designed antenna functions well.

The radiation patterns of the microstrip antenna were measured in a radio anechoic chamber at Tampere University of Technology. The microstrip antenna was attached to a wooden rod, which was placed on a turntable. The antenna was placed at a height of 120 cm. A sine signal with a frequency of 2.45 GHz and amplitude of 0 dBm was fed to the microstrip antenna from a Rohde & Schwartz SMR-20 signal generator. The radiation pattern was measured with HP 11966E double-ridged waveguide horn antenna, which was placed at a height of 120 cm and at a distance of 3 m from the microstrip antenna. The signal from the HP horn antenna was analysed using a HP 8539A spectrum analyser with 85712D EMC auto-measurement personality.

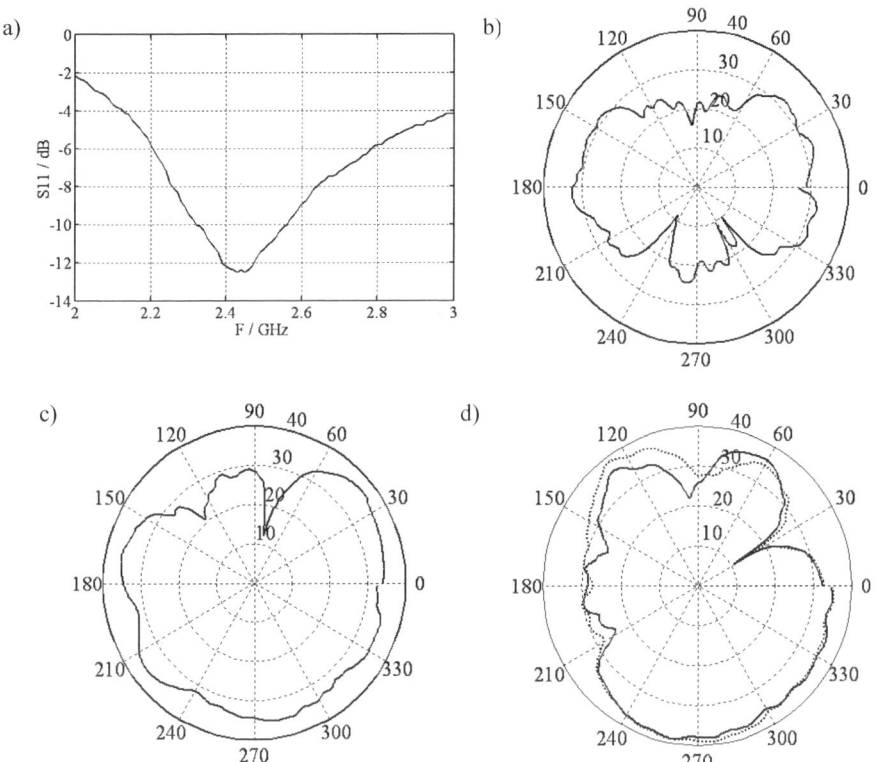

Fig. 2. Reflection coefficient (S_{11}) of the antenna (a), and radiation patterns of the antenna in XZ-plane (b), XY-plane (c), and YZ-plane (d).

First, the radiation pattern was measured in the XZ-plane (refer to Figure 1) and the antenna was fully rotated counter-clockwise around the Y-axis. Next the radiation patterns were measured in XY- and YZ-planes in a similar manner. The measurement results are shown in Figures 2b, 2c, and 2d, respectively. The radiation pattern measurements show that the antenna is quite unidirectional in the XY- and YZ-planes. In the XZ-plane the radiation pattern is smaller and its shape is more elliptical than circular. This must be taken into account when determining the exact placement of the Bluetooth PCB within the earcup. The radiation pattern should be large in horizontal directions, but it may be smaller in vertical directions. For example, the cellular phone may be on charge on a desk, and the user may walk around freely on the same floor. Figure 2d shows that the plastic earcup has no significant effect on the radiation pattern. In fact, the measured radiation pattern is slightly greater with the Bluetooth antenna installed inside the earcup. Thus, the plastic earcup will probably cause no problems.

The prototype was tested in operation using three GSM cellular phones from different manufacturers, and employing both headset and hands-free profiles. The phones were Sony Ericsson T68i, Nokia 3650, and Siemens SX-1. The prototype worked well with every phone and with both profiles. The sound quality was rated quite good, and speech intelligibility also improved because the hearing protector attenuates noise disturbance. When a directional noise-cancelling microphone is used, no background noise is transmitted. The operating range of the device was approximately 10 m, and the sound quality was rated good. The current consumption of the device was measured. The supply current was 4 mA in standby mode and 40 mA in transmitting mode.

CONCLUSIONS

For most people, nowadays, a cellular phone is indispensable. However, in noisy environments its use can be severely restricted, especially if a hearing protector is being worn.

A prototype of a Bluetooth hearing protector was developed to overcome this problem. The developed device provides hearing protection while also permitting communication via a cellular phone without the need for inconvenient cables. A microstrip antenna was also developed for the device. The antenna was measured and performed well. The prototype supports both headset- and hands free profiles. In tests it was shown to be functional with three different cellular phones.

REFERENCES

Bray J. and Sturman C.F. (2001). *Bluetooth Connect Without Cables*, Prentice Hall.

Bravo-Escos M. (2002). Networking gets personal. *IEE Review* **48:1**, 32 – 36.

Olmos M., Hristov H.D., and Feick R. (2002). Inverted-F antennas with wideband match performance. *Electronics Letters* **38:16**, 845 - 847.

Ali M. and Hayes G.J. (2000). Analysis of Integrated Inverted-F Antennas for Bluetooth Applications. *IEEE-APS Conference on Antennas and Propagation for Wireless Communications,* Waltham, MA USA, 21 – 24.

DEVELOPMENT OF LOCAL POSITIONING SYSTEM USING BLUETOOTH

T. Hirota[1], S. Tanaka[1], T. Iwasaki[1], H. Hosaka[1], K. Sasaki[1], M. Enomoto[1] and H. Ando[1]

[1] Graduate School of Frontier Sciences, the University of Tokyo,
Hongo 7-3-1, Bunkyo-ku, Tokyo, 113-8656, JAPAN

ABSTRACT

A local positioning system utilizing propagation loss of Bluetooth wireless technology is presented. The distance between a stationary station and a mobile station (both located indoor) carried by a person is derived from the propagation loss. The position of the person is calculated from distance measurements on three different fixed stations. The positioning accuracy was 1.2m at average fixed station installation density of one station in every 8.3m2. The system is intended for tracking worker's position in a work environment.

KEYWORDS

Indoor positioning, Bluetooth, Propagation loss, Compensation, Attenuation by human body

INTRODUCTION

Location-based services, such as car navigation systems, restaurant and hotel guides for mobile phones are becoming popular. Most of these systems obtain location information from GPS(Global Positioning System), which is limited to outdoor use. But needs for indoor navigation systems in museums, department stores, and for work assistant system in hospitals or factories exist, and an indoor positioning system combined with GPS will provide user's position data seamlessly from outdoor to indoor and broaden the application fields of location-based services. Although several methods for indoor positioning have been proposed, they have some shortcomings.
Typical requirements for tracking worker's position are: a) Accuracy of 1m in order to track worker's position and action in indoor work environment b) Low power consumption to enable operation of 24 hours with batteries c) Bi-directional communication capability for exchanging location-based information between the mobile unit and the system. Current GPS technology is limited to outdoor use. Positioning systems using infrared or ultrasound are sensitive to physical interferences, because the line of sight between the transmitter and the receiver must be kept clear (for example, see Priyantha et al.(2000) or Want et. al. (1992)). This will be a severe limitation for tracking people's positions and

movements. Communication with RFID (Radio frequency identification) is not bi-directional (for example, see Hightower et. al. (2000)), and sensitive range of passive RFID is too short for tracking people. Wireless LAN devices consume higher electrical power (typically about 1[W]) than devices of other methods. Moreover, TDOA (Time difference of arrival) system becomes expensive because it needs high-resolution clock to measure small arrival time difference of radio wave. This paper describes development of a positioning system based on measurement of propagation loss of Bluetooth wireless transmission.

POSTIONING SYSTEM USING BLUETOOTH

In our proposed system, several base stations with Bluetooth wireless module are placed in the measurement area, and a person carries a mobile unit also equipped with a Bluetooth module. Pseudo distances between the mobile unit and the base stations are derived from the propagation losses. Position of the mobile unit is calculated from these pseudo distances using a trilateration algorithm (described in Kitasuka (2003)).

The relationship between the received signal power Pr[W] and the transmission signal power Pt[W] in free space at distance d is calculated from a propagation model. We used the free space model and the two-ray ground propagation model that account for both the direct line-of-sight path and the path reflected from the ground (described in Fall (2001)). Since real propagation is a combination of the two models, we will introduce constants α and β to combine the two models to express a practical propagation model defined by the following equation.

$$P_r = \frac{\alpha P_t}{d^\beta} \tag{1}$$

Constant α depends on the system, such as gains of antennas and the wavelength, and constant β depends on the environment. The experimental data have shown good agreements with the proposed propagation model. The values of the parameters obtained by curve fitting on the experimental result, were $\alpha = 2.01$ and $\beta = 2.59$. Theoretically, the propagation models mentioned above are only applicable to far filed region. In the present case, far field region begins approximately 1.2[m] from the antenna, which is 10-wave length of 2.45[GHz] radio wave.

Influence of the human body and variance of propagation loss

Since the frequency of Bluetooth is 2.45GHz, absorption by human body affects the distance measurement based on propagation loss. Figure 1 (a) shows the relationship between the distance and the propagation loss when a human body is in the propagation path. A person with a mobile unit on the front of the body stood away from a base station, and propagation loss was measured as the person turned 90, 135, and 180 degrees, at distances of 1[m] to 7[m] at 1[m] interval. The angle facing the base station was defined as 0°. When the angle was larger than 90°, the mobile unit lost the line of sight to the base station and the propagation loss increased by more than 10 [dB], which was equivalent to an error of approximately 10% of the pseudo-distance. This result indicates that compensation for the human body influence is necessary.

Figure 1 (b) shows the variance of propagation loss at distance of 3[m] and 7[m]. The result indicates that the variance of propagation loss is smaller at shorter distance.

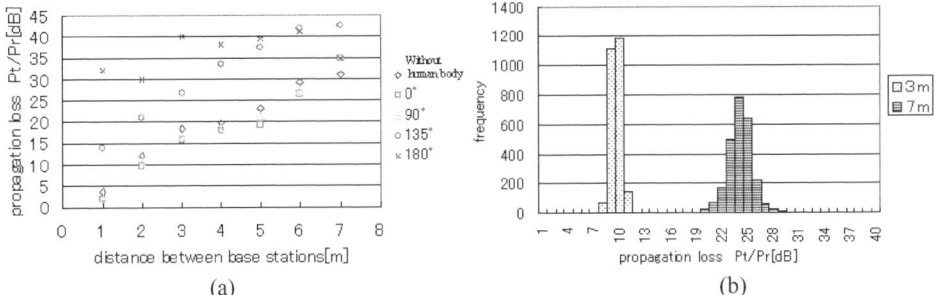

Figure 1: (a) Relationship between the distance and the propagation loss with presence of a human body, (b) Relationship between distance and variance of propagation loss

Hardware system

The present system consists of base stations and a mobile unit. One of the base stations is connected to a laptop PC, which serves as a position calculation server. Other base stations and a mobile unit are composed of a Bluetooth module and a microcomputer. Size of the base station and the mobile unit are both 60×103×28[mm], and their weight is 130[g](including a dry cell battery) as shown in Figure 2 (a). First, the mobile unit establishes a communication link with every base station. Then the mobile unit measures the transmission power to each base station, and the corresponding base station measures the received signal power. The position server collects these data through communication links among the base stations and calculates the mobile unit's position using the method described in the previous section.

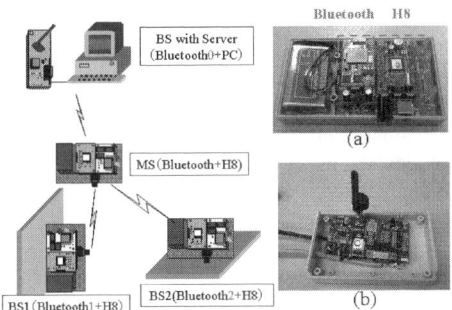

Figure 2: Positioning system hardware
(a) Mobile station and base stations (b) base station connected to the position server

EXPERIMENT

The positioning system described in previous section was applied for tracking workers' positions in a shipyard, because workers' positions and movements in the work area are useful for safety analysis and task management. Since the arrangement of the work area changes frequently in shipyards, the proposed system has merits over other positioning methods because of its simple and easy installation. Six base stations were installed in a measurement area of 15×20[m]. A worker carried the mobile unit. Actual position was recorded by visual observation. In this situation, RMS_error was 3.4[m].
Since the accuracy of the present system decreases as the distance between the mobile unit and the base

station increases, we increased the placement density of the base stations. Installation area for the six base stations was changed from (a)15×20 [m] (0.02 station per m^2), which was the case described above, to (b)7×7[m](0.12 station per m^2). As a result, RMS_ error decreased to 1.3[m] as shown in Figure 3.

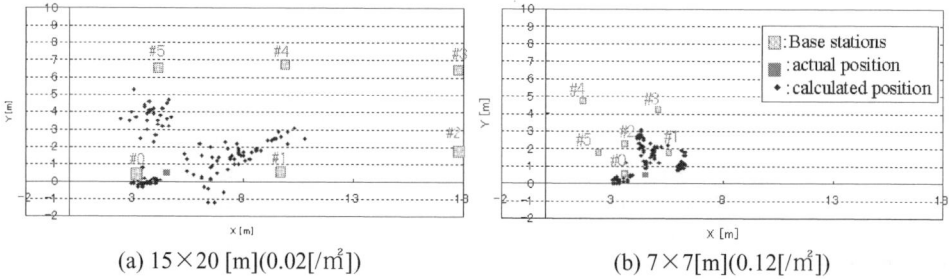

(a) 15×20 [m](0.02[/m^2])   (b) 7×7[m](0.12[/m^2])

Figure 3: Calculated position and density of base station installation

In order to compensate for the influence of human body on propagation loss measurement, a second mobile station was attached on the wearer's back in addition to the one on the front. First, the positions of the two mobile units were calculated independently, and then the middle point of the two positions was adopted as the worker's position. Incorporating this compensation method, in area (a), RMS_error was reduced to 2.3[m]. In area (b), RMS_error was reduced to 1.2[m].

CONCLUSIONS

1) Positioning system based on pseudo distance derived from propagation loss of Bluetooth wireless technology has been proposed and evaluated. 2) Attaching mobile units on the front and the back of the human body compensated for the error caused by absorption by human body. RMS-error wad reduced from 3.4[m] to 1.3[m] by this compensation method. 3) RMS_error was reduced from 3.4[m] to 2.3[m] by increasing the placement density of the base stations. 4) Overall RMS-error was 1.2[m] by incorporating two compensation methods.

REFERENCES

Fall, K. and Varadhan, K. (2001). The ns Manual, The VINT Project, A Collaboration between researchers at UC Berkeley, LBL, USC/ISI, and Xerox PARC, http://www.isi.edu/nsnam/ns/.

Hightower J., Borriello G. and Want R. SpotON (2000): An Indoor 3D Location Sensing Technology Based on RF Signal Strength. *UW CSE Technical Report #2000-02-02*.

Kitasuka T., Nakanishi T. and Fukuda A. (2003) Location Estimation System Using Multi-Hop Wireless Network. *Journal of Information Processing Society of Japan*, **44:SIG10**.

Priyantha N.B., Chakraborty A. and Balakrishnan H. (2000). The cricket location-support system. *Proceedings of MOBICOM 2000*, 32-43.

Want R. et. al.. (1992). Active Badge Location System. *ACM Transactions on Information Systems*, **10:1**, 91-102.

ANALYSIS OF MULTIPLE OBJECT IDENTIFICATION WITH PASSIVE RFID

K. Penttilä, L. Sydänheimo, M. Kivikoski

Tampere University of Technology, Institute of Electronics, Rauma Research Unit, Kalliokatu 2, FI-26100 Rauma, Finland.

ABSTARCT

Ability to identify multiple items simultaneously, without errors is one of the major advantages achievable with RFID (Radio Frequency Identification) technology. Current specifications and standards include a wide range of different protocols that enable above. This paper analyses the identification performance of multiple objects with two commonly used algorithms. Theoretical results are supported by practical measurements done in laboratory conditions. Results show how identification speed and reliability depends on the size of the tag population, the range, and tags' mutual alignment.

KEYWORDS

Automation, Aloha Protocols, EPC Tree Algorithm, Multiple Object Identification, Passive tags, RFID

INTRODUCTION

Multiple object identification is a benefit in several automation and mechatronics application cases. Accurate item identification, high operation efficiency, reliable control and real time monitoring or tracking are achievable. Furthermore, passive RFID (Radio Frequency Identification) technology enables applications far from those that are possible with currently used item identification systems. Application examples include supply chain management, item control and management during manufacturing, warehousing, transportation, etc.

Current RFID standards and specifications include wide variety of different protocols enabling multiple object identification. However, their operational performance within commercial devices varies a lot, and therefore it is important analyse the operation of pure algorithms. In order to verify the value of the theoretical results, we measure simultaneous identification up to 64 passive tags with commercially available reader and tags, using 915 MHz frequency band, 4 W EIRP transmission power, and EPC tree algorithm (Auto-ID Labs 2001). The bit rates and the time slot intervals defined in this specification are used in theoretical analysis in order to maintain consistent examination.

Multiple object identification with RFID technology has been analysed in some publications (Lim & Mok 1998), (Vogt 2002), but focusing only on Aloha protocols and more sophisticated algorithms, such as ID arbitration and code division multiple access. As passive RFID systems are typically designed for extremely low cost applications, sophisticated algorithms and more efficient microprocessors requiring systems are not applicable. Therefore, this paper focuses on simple protocols suitable for identifying low cost tags.

PASSIVE RFID SYSTEM

An RFID system consists of tags, readers, and an application host. The readers communicate wirelessly with the tags to obtain the information stored on them. The data sent by the reader is modulated and backscattered from a number of tags. The cheapest RFID tags with the largest commercial potential are passive, harvesting energy from the reader's communication signal to power up their operation and communication with the reader (Auto-ID Labs 2001), (Vogt 2002). RFID communication consists of a number of communication cycles. Each cycle consists of three sections: first, the reader sends an activation field to the tags. Then, the reader sends a command to the tags, and finally it sends a CW field that the tags modulate and backscatter back to the reader. The reader's command field defines the content of the tags' replies. Communication bit rates are 70.18 kb/s for forward link and 140.35 kb/s for backward link (Auto-ID Labs 2001).

ANTICOLLISION ANALYSIS

This chapter analyses EPC tree algorithm and Aloha protocols. EPC tree algorithm (Auto-ID Labs 2001) is chosen according to its wide popularity. Aloha protocols are included in ISO 18000-6 standard and also used by some RFID manufacturers (Vogt 2002).

EPC tree algorithm

EPC tree algorithm defined by Auto-ID labs (Auto-ID Labs 2001) goes through all possible code combinations as a binary tree. It optimises the number of required time slots by ignoring those leaves that do not respond without any further requests. Moreover, any collisions between replying tags do not interfere with the identification procedure as the reader does not need to know the contents of tags' replies, only whether any replies occur or not. This is because of the well-synchronized reply window: it has eight slots and each tag will modulate the requested 3-bit section of its identification code to one slot. Chose of the slot is based on the content of the reply. The eight slots allow each different set of the tree bits to occupy different slots ($2^3 = 8$). The actual duration of the total identification procedure of a number of tags lies between the maximum and minimum curves, depending on the alignments of the identification codes of the tags in the current binary tree. These maximum and minimum curves are presented in Figure 1. 64 bit tag identifiers were used in calculations. The derivation of these curves is presented in publication (Penttilä et al. 2004).

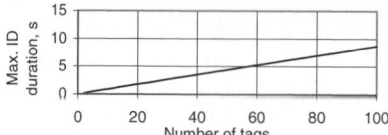 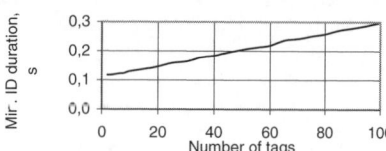

Figure 1 Maximum (left) and minimum (right) identification duration with EPC tree algorithm

Aloha family protocols

With Aloha protocols, messages are sent whenever needed without checking the communication channel (Wieselthier et al. 1989). Collisions lead retransmission with a random delay. According to the protocol, whenever a terminal has a radio packet to transmit, it transmits the packet without checking the channel. Possible collisions lead to retransmission of packets with random delay. With the use of slotted Aloha protocol, time is divided in to slots of one packet duration. Each tag may reply at most once in a slot. Framed Aloha protocol uses time frames that are divided into a number of slots. Now each tag may reply at most once in frame (Wieselthier et al. 1989). Figure 2 illustrates the average duration of identification slots of tag populations varying from 10 to 100. 64 bit tag identifiers were used in calculations. Framed Aloha shows slightly superior performance, being however clearly less efficient that EPC tree (see Figure 2).

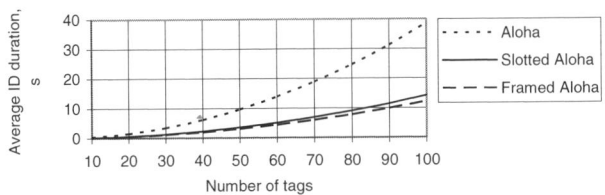

Figure 2 Average identification duration with Aloha, slotted Aloha and framed Aloha protocols

MEASUREMENTS

Measurements were taken in TUT RFID laboratory, Tampere, Finland. The arrangement is shown in Figure 3. Tags were placed on a specific grid that has either 64 or 16 blocks (see Figure 3, right side). The tag populations included 4, 9, 16, 25, 36, 49 and 64 tags. The reader and tags were commercially available, operating under Class I specification (Auto-ID Labs 2001). The centre of the block of tags and the centre of the reader antenna were placed at the same height. The tag grid and the reader antenna face each other at a distance of either 1 or 2 m. Measurement data was collected during 1 and 5 minutes.

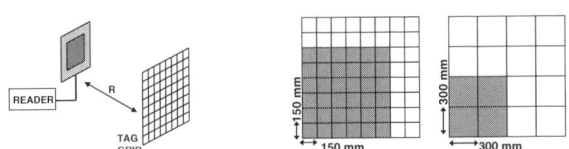

Figure 3 Measurement setup (left) and tag grids (right)

The studied factors were the influence of the identification range, the tag population, tags' mutual alignment, time used for identification, and identification reliability. First, reducing the range clearly increases the number of successful identification cycles. However, the number of tags identified is not increased. Second, decreasing the number of tags increases the identification certainty. Third, the identification performance decreases when tags' are located closer to each other. Fourth, increasing the time does not increase the number of tags identified, but the number of successful identification cycles multiplies by the same factor as the time is multiplied. Finally, Figure 4 shows the percentage of identified tags as a function of the size of tag population for each measurement cases, where lines correspond squares and dashed lines correspond triangles, coloured correspondingly.

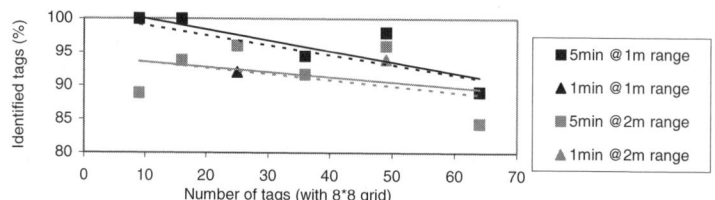

Figure 4 Percentage of successfully identified tags with the 8*8 tag grid.

CONCLUSION

In this paper we analysed several factors influencing the multiple object identification with passive RFID technology. The measured results show that the 100 % identification reliability can be achieved only with small tag populations. The major factors affecting to the certainty besides the number of tags are the distance between the reader and the tags and the beam width of the reader antenna that determines the range. The increase of the time interval used for identification does not have significant impact to the reliability. Furthermore, the measurements showed that the mutual alignment of tags has an impact to the reliability. As the distance between tags increases, the tags will interfere less with each other.

Item specific antennas and tag attachments will become an essential factor when designing fast, passive RFID with the option of multiple object identification. Specific limitations of multiple object identification with passive RFID technology lie within object and tags mutual alignments. Furthermore, as the electromagnetic fields easily reflect from metallic surfaces and attenuate to non-conducting materials the fabrication material of objects to be identified has a great influence to the identification accuracy.

ACKNOVLEDGEMENTS

The authors would like to thank the Finnish National Technology Agency and Nokia Foundation for financing the research done for this paper.

REFERENCES

Auto-ID Labs. *860 MHz – 930 MHz Class I Radio Frequency Identification Tag Radio Frequency & Logical Communication Interface Specification*. Published on 14 Nov. 2001

Lim A., Mok K. A study on the design of large-scale mobile recording and tracking systems. *IEEE Proceedings of the 31st Hawaii International Conference on System Sciences*, 6 – 9 Jan.1998. Kohala Coast, HI USA. Vol.7, pages 701-710.

Penttilä, K., Sydänheimo L., Kivikoski M. Analysis of Multiple Object Identification with Passive RFID *Proceedings of the 5th International Conference on Machine Automation, ICMA.* 24-26 Nov. 2004. Osaka, Japan. pp. 559-564

Vogt H. Efficient Object Identification with Passive RFID Tags. *International Conference on Pervasive Computing*. Zurich, 2002.

Wieselthier J.E., Ephremides A., Michaels L.A. An exact analysis and performance evaluation of framed ALOHA with capture. *IEEE Transactions on Communications*. Feb. 1989. **Vol. 37, issue 2**, pages 125-137.

MODELING ELECTROMAGNETIC WAVE PROPAGATION IN PAPER REEL FOR UHF RFID SYSTEM DEVELOPMENT

M. M. Keskilammi, L. T. Sydänheimo and M. A. Kivikoski

Tampere University of Technology, Institute of Electronics,
Rauma Research Unit, Kalliokatu 2, FI-26100 Rauma, FINLAND

ABSTRACT

In passive radio frequency identification systems (RFID), data and power is transferred between a reader and an identification device wirelessly by means of electromagnetic waves, Finkenzeller (2003). Antenna solutions, in both the identification device and the reader, are crucial to the performance of radio frequency identification systems. To improve the performance of these RFID systems application specific antennas can be used for challenging items including metals, liquids or lossy material. This paper describes the simulation model for radio wave attenuation in paper reel. Simulated values for propagation in different grades of paper are presented. Theoretical background is also discussed.

KEYWORDS

RFID, Antennas, Automation, Communication Systems, Sensors

INTRODUCTION

According to their operation frequency, radio frequency identification systems are dividable into low frequency and high frequency systems. In low frequency systems, a magnetic field is used in the coupling between the identification device and the reader, and various loop solutions are used as antennas. In low frequency systems, the reading distance is short and the reading distance depends on the areas of the antenna coils and their mutual positions. In high frequency systems, an electric field is used in the coupling, and the antennas used are usually dipole, folded dipole or microstrip antennas. Out of these, dipole and folded dipole antennas are omnidirectional, whereas a microstrip antenna is directional. In high frequency systems, the identification device is either active or passive. Active identification devices comprise a radio transmitter and a battery, whereas passive systems use the energy obtained from the reader. In high frequency systems, the reading distance is longer than in low frequency systems. This paper concentrates on passive high frequency RFID systems operating in ultra high frequency (UHF) band.

Roll-like bulk goods, such as paper or cardboard rolls, have to be identified with 100% reliability when the roll is handled at a factory, warehouse, when loading a conveyer chain or at the warehouse of a printing house. A roll is identified in a controlled situation, wherein the position of the roll with respect to its cylinder axis is known, i.e. the roll is either in a vertical or in a horizontal position. As far as the antenna of the identification device is concerned, this means that the polarization plane of the antenna is known. In contrast, the position angle of the roll around the cylinder axis is not known. In other words, when the identification device to be arranged in the roll uses a directional antenna element, the direction of the maximum of the antenna radiation beam is not known. If an identification device arranged on the surface of the roll is used in this kind of a situation, in the worst case the identification device is on the opposite side of the roll and the direction of the radiation beam of the antenna of the identification device is opposite to the direction from which the reader makes the identification. This means that reliable identification is very unlikely in such a situation.

The dipole and folded dipole antennas generally used in radio frequency identification devices are usually omnidirectional, i.e. they emit electromagnetic radiation in all directions. However, these antenna types have low amplification. Furthermore, the frequency bands used by radio frequency identification devices have an officially regulated highest permitted transmission power, i.e. directional antenna structures can be used for improving the transmission of an identification device, if required. The use of directional, i.e. amplifying antenna structures, such as a microstrip antenna or an antenna array, allows the electromagnetic radiation power transmitted by the antenna to be directed more efficiently in the desired direction. This improves the coupling between the identification device and the reader antennas in the direction of the maximum of the radiation beam of the directional antenna compared with omnidirectional antennas, whereas the coupling is weaker outside the radiation beam than with omnidirectional antennas.

DIELECTRIC PROPERTIES OF PAPER

The relative permittivity in copy paper or in other paper qualities consisting mostly of wood fibers is typically from 2 to 4 decreasing with frequency. In coated paper the permittivity increases even up to 8 due to high amount fillers like $CaCO_3$ added. The change in the moisture content of paper doesn't change much the dielectric constant of paper itself, though the dielectric constant in water is 80. This is because in paper, water molecules are associated with polysaccharide chains and cannot rotate freely. Rotation is possible only if the field is parallel to the chain axis. Because of the chain orientations in paper are random, only a small fraction of the paper molecules have perfect alignment with the electric field. This makes the effective dielectric constant much smaller than it would be in liquid water, Niskanen (1998). However, the increase in moisture content increases dielectric losses in paper. In paper with anisotropic fiber orientation, the dielectric constant is largest in the direction of the fiber orientation angle i.e. typically in the planar directions. In z-direction the dielectric constant is smaller. See Figure 1.

The real part of the relative permittivity ε_r of paper increases with increasing density ρ and the behavior follows with reasonable accuracy the Clausius-Mossotti relation, Niskanen (1998)

$$\frac{\varepsilon_r - 1}{\varepsilon_r + 2} \propto \rho \qquad (1)$$

The imaginary part of the relative permittivity, loss tangent tanδ, increases linearly with density. Factors effecting on the electrical properties of paper are given in Matsuda (2002). Dielectric constant is affected by paper density, fiber orientation, crystalline cellulose and pulp components (lignin, hemicellulose, etc). The amount of dielectric loss depends on ionic conduction losses, inclusion of

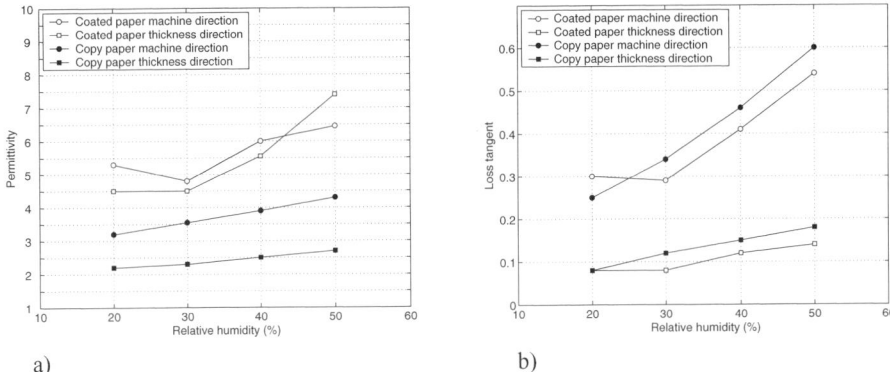

Figure 1: Electrical properties of the coated paper and copy paper as a function of relative humidity at a frequency of 1 MHz. a) Permittivity b) Loss tangent, Simula et al. (1999).

organic and inorganic ions, adsorbed ions, carboxyl groups, fiber morphology, polarization losses, rotation and oscillation of polar material, fine structure of cellulose and pulp components.

SIMULATION RESULTS FOR THE COUPLING BETWEEN TWO DIPOLE ANTENNAS THROUGH THE PAPER REEL

For most paper reel identification applications the transponder should be attached to the core of the paper reel. This is how the identification of the reel can be done over its whole lifecycle. However, the performance of the communication link between the reader unit and the transponder is limited due to losses in paper. In the following the effect of loss tangent of the paper on the coupling between two 915 MHz dipole antennas is studied.

The height of the simulated reel is 1200 mm and the diameter of the reel is 1000 mm. The core diameter inside is 76 mm. In the simulations one dipole was inserted inside the reel in the middle of the reel core while the other dipole was outside the reel. The axial position of both dipoles in relation to the reel was 600 mm. The separation of dipoles was 538 mm.

The free space loss can be evaluated using Friis transmission formula, Balanis (1997)

$$P_r = P_t G_r G_t \left(\frac{\lambda}{4\pi r}\right)^2, \qquad (2)$$

where is P_r, P_t, G_r and G_t are the received and transmitted powers and antenna gains respectively. The term is called the free space loss factor where λ is the wavelength used and r is the separation of the antennas. For two dipoles ($G_r = G_t = 1.63$) with 538 mm antenna separation in 915 MHz frequency ($\lambda = 328$ mm) the coupling in free space is -22.05 dB.

In Figure 2, the effect of loss tangent of the paper on coupling between dipoles is presented. In the simulation the relative permittivity of paper was 2.0. The value of loss tangent varied from 0.05 to 0.5. The coupling between the antennas is decreased from the free space coupling with increasing loss tangent value. The simulation results agree well with previous studies with two dimensional layer model, Keskilammi et al. (2000).

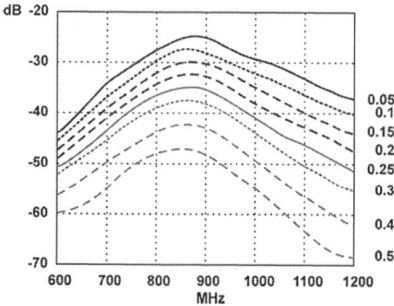

Figure 2: Coupling between two 915 MHz dipole antennas through paper as a function of frequency for eight loss tangent values. Relative permittivity of paper is 2.0.

SIMULATION RESULTS FOR DIPOLE ANTENNA INSIDE THE PAPER REEL

To find out the effect of change in dielectric properties of paper to the properties of dipole antenna the following set-up was simulated. The dipole antenna was inserted between the reel core and bulk paper. The dipole antenna was set in the middle of axial height of the reel. The height and the diameter of the reel were as in the previous simulation. The reel core inner diameter was 76 mm with wall thickness of 16 mm.

The Effect of Permittivity

First relative permittivities of 2.0, 2.5, 3.0, 3.5, 4.0, 5.0 and 6.0 for paper were simulated. In the Figure 3a) the change in return loss due to change in relative permittivity of paper is presented. The loss tangent for the simulations is $\tan\delta = 0.1$.

In the Table 1 the resonance frequency, return loss and bandwidth as a function of relative permittivity of paper are presented.

TABLE 1
SIMULATED RESULTS FOR DIPOLE ANTENNA

ε_r	Resonance frequency (MHz)	S11 (dB)	Bandwidth (MHz)	Bandwidth (%)
2.0	630	15.0	568…703 = 135	21.4
2.5	590	13.7	536…659 = 123	20.8
3.0	560	13.0	512…624 = 112	20.0
3.5	540	12.5	492…595 = 103	19.1
4.0	510	12.0	474…568 = 94	18.4
5.0	480	11.5	446…519 = 73	15.2
6.0	450	10.9	424…478 = 54	12.0

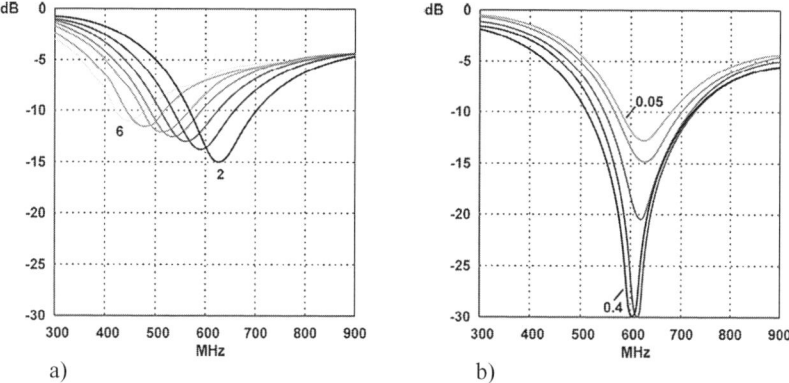

Figure 3: Simulated return loss S11 (dB) of a dipole antenna as a function of frequency.
a) Effect of relative permittivity (2.0, 2.5, 3.0, 3.5, 4.0, 5.0, and 6.0).
b) Effect of loss tangent (tan0.05, 0.1, 0.2, 0.3 and 0.4).

As the relative permittivity of both paper and the reel core increases the resonant frequency of dipole antenna decreases noticeably as expected. Also the -10 dB bandwidth narrows, as the return loss gets worse with the increasing relative permittivity.

The effect of loss tangent

The effect of loss tangent on properties of dipole antenna was studied for values 0.05, 0.1, 0.2, 0.3 and 0.4. The simulations were repeated for three different values of relative permittivity 2.0, 4.0 and 6.0. The relative permittivity of the reel core was 3.0 and the loss tangent 0.1. In the Figure 3b) the change in return loss due to change in loss tangent of paper is presented.

In the Table 2 the resonance frequency, return loss and bandwidth as a function of loss tangent of paper are presented.

TABLE 2
THE EFFECT OF LOSS TANGENT ON RESONANCE FREQUENCY,
RETURN LOSS AND BANDWIDTH OF A DIPOLE ANTENNA.

$\tan\delta$	Resonance frequency (MHz)	S11 (dB)	Bandwidth (MHz)	Bandwidth (%)
0.05	630	12.9	579...685 = 106	16.8
0.1	630	15.0	568...703 = 135	21.4
0.2	620	20.6	548...718 = 170	27.4
0.3	610	31.2	530...725 = 195	32.0
0.4	600	29.7	515...723 = 208	34.7

For the simulated cases the return loss increases with increasing the value of loss tangent. Also the resonant frequency decreased slightly when increasing losses. The -10 dB bandwidth increases as the return loss increases and the matching of the antenna gets better.

CONCLUSIONS AND FUTURE WORK

The electrical properties of material, the identified object is made of, change the characteristics of the transponder antenna fastened to the object. To maintain the performance of the RFID system in the vicinity of challenging materials the antenna element has to be tuned according to application. To test the performance of these application specific antennas for the paper reel application the electromagnetic model for the reel was created. The dielectric properties for the model were taken from the literature. Little information of dielectric properties of paper at higher frequencies is available.

First the field attenuation in paper was studied by simulating the coupling between two dipoles with paper in between. The attenuation increased from 3 dB to 25 dB from the free space attenuation value as the loss tangent value increased from 0.05 to 0.5. In the means of antenna separation in free space this means that the distance between the antennas is increased from 0.53 meters to 0.8-8 meters.

Second the effect of change in dielectric properties of paper to the properties of dipole antenna inserted inside the paper reel was analyzed. Increasing the relative dielectric constant of the paper lowered the resonant frequency of the dipole antenna. The change in loss tangent of the paper did not affect the resonant frequency remarkably, but the change in antenna matching was noticeable.

In the future the research will concentrate on testing new application specific antenna geometries for the paper reel RFID transponders with the proposed model.

REFERENCES

Finkenzeller K. (2003), *RFID Handbook, 2nd Ed.*, John Wiley & Sons Inc., New York, USA

Niskanen K. (1998), *Papermaking Science and Technology, Book 16: Paper Physics*, Fapet, Helsinki, Finland

Simula S., Varpula T., Ikäläinen S., Seppä H., Paukku A., Niskanen K. (1999), Measurement of the Dielectric Properties of Paper, *Journal of Imaging Science and Technology*, **43:5**, 472-477.

Matsuda S. (2002), *Handbook of Physical and Mechanical Testing of Paper and Paperboard, 2nd Ed.*, Dekker, New York, USA

Balanis C.A. (1997), *Antenna Theory, Analysis and Design, 2nd Ed.*, John Wiley & Sons, Inc., USA

Keskilammi M., Salonen P., Sydänheimo L. and Kivikoski M. (2000), Radio Wave Propagation Modeling in Paper Reel for Novel Radio Frequency Identification System, *IEEE, JamCon2000, Technology for Economic Development*, Aug. 11-13, 2000, Ocho Rios, Jamaica

EFFECT OF CONDUCTIVE MATERIAL IN OBJECTS ON IDENTIFICATION WITH PASSIVE RFID TECHNOLOGY: A CASE STUDY OF CIGARETTE CARTONS

Leena Ukkonen[1], Mikael Soini[1], Daniel Engels[2], Lauri Sydänheimo[1] and Markku Kivikoski[1]

[1]Tampere University of Technology, Institute of Electronics, Rauma Research Unit,
Kalliokatu 2, FI-26100 Rauma, Finland
[2]Massachusetts Institute of Technology, Auto-ID Labs,
77 Massachusetts Avenue, Bldg. 35-205, Cambridge, MA 02139, USA

ABSTRACT

This paper presents a comparison of the performances of two different passive tag antenna designs attached to cigarette cartons. The aluminium foil in the cigarette packs makes the identification of cigarette cartons difficult using passive RFID technology. Therefore, a novel microstrip patch-type tag antenna for passive RFID of cigarette cartons was designed. The performance of the novel tag antenna is compared to the performance of a label-fabricated folded dipole-type tag antenna. The maximum read ranges of a single tagged carton and two tagged cartons are measured and compared. The effect of the aluminium foil in the cigarette packs is studied by carrying the measurements out also using cigarette packs without the foils and an empty carton. The novel tag antenna performed superior to the folded dipole tag antenna on full cartons of cigarettes.

KEYWORDS

Automation, Communication system, Information equipment, Information storage, Measurement

1. INTRODUCTION

The increasing use of passive radio frequency identification (RFID) systems at ultra-high frequency (UHF) spectrum requires finding solutions for RFID tags to be attached to different products and packages (Foster & Burberry, 1999). RFID is being adopted for a wide range of applications, such as applications within the supply chain, like tracing pallets, cases and individual products (Raza et al., 1999, Glidden et al., 2004). Other emerging applications of RFID are identification of paper rolls and numerous applications in health care industry (Raza et al., 1999). RFID system consists of a reader unit, reader antenna, host computer and a transponder (i.e. tag). A tag contains a microchip and an antenna. The microchip stores the identification data of the tag. Passive RFID tags have no internal source of energy and thereby they get all the energy for functioning from the electromagnetic field sent by the reader. Communication between the tag and the reader is based on backscattering: reader sends commands to tag which then responds to the reader by backscattering its identification data. The identification data is modulated into the backscattered electromagnetic wave using load impedance modulation (Finkenzeller, 2003). The components of an RFID system are presented in Figure 1.

At present, one of the biggest challenges is tagging objects that are totally made of or contain conductive materials in their structure. Conductive materials next to antennas operating at UHF spectrum affect the performance of the antennas for example by lowering the radiation efficiency and changing the resonance frequency. Conductive materials also reflect the electromagnetic wave radiated by the antenna and therefore they affect the radiation pattern and radiation directions of the antenna (Raumonen et al., 2003). In addition, the conductive materials attenuate the incident electromagnetic wave and therefore the electromagnetic wave does not propagate well or at all through the conductive material (Reitz, Milford & Christy, 1993).

Cigarette cartons, which contain ten individual cigarette packs, have been a difficult object to identify using RFID technology because the individual packs are wrapped with aluminium foil. This foil contains thin layer of paper coated with 0.25 μm thick layer of pure aluminium that is highly conductive. The aluminium layer is coated with very thin polyester layer. The structure of a cigarette carton and an individual cigarette pack are presented in Figure 2.

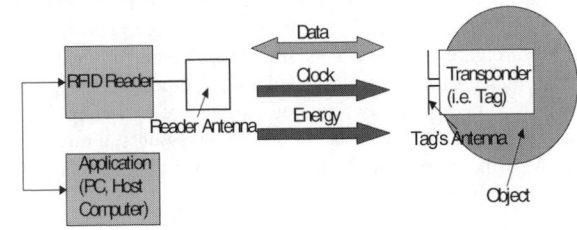

Figure 1: The components of an RFID system

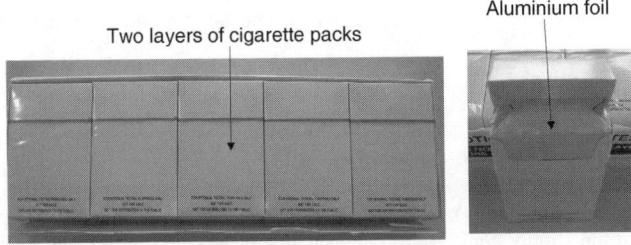

Figure 2: A cigarette carton and a cigarette pack

Conventional label-fabricated RFID tags do not work with enough reliability when attached to a cigarette carton. Therefore, a novel tag design using microstrip patch antenna integrated on a cigarette carton was designed. The cigarette carton with individual packs was used as a substrate material of the microstrip patch antenna and the dimensions of the antenna structure were designed to fit on the carton and at the same time to achieve the 915 MHz resonance frequency. 915 MHz is the UHF center frequency used in RFID in North and South America.

2. THE ANTENNA DESIGN

As previously noted, metallic structures near antennas affect their performance in many ways. Placing a conductive surface near an antenna has advantages and disadvantages. In some cases a metallic plate near an antenna can act as a reflector causing the directivity to increase. Also a number of antenna types need a conductive ground plane to function properly. In these cases a metallic surface can be used to improve the performance of the antenna.

On the other hand, if the antenna does not use a ground plane in its function, the wave radiated by the antenna is almost totally reflected from the metallic surface since metal is highly conductive. When electromagnetic wave reflects from metallic surface a 180 degree phase shift occurs (Cheng, 1993). This reflected wave cancels the incoming wave and therefore the radiation efficiency of the antenna decreases. These negative effects are strongest when the antenna is very near (for example at a distance of a couple of millimetres or less) the metallic surface (Raumonen et al., 2003).

The basic structure of the antenna design is shown in Figure 3. The patch-type tag needs a ground plane to function, and the metallic ground plane makes the antenna more stable and well functioning even though the cigarette carton contains conductive aluminium foil. Also, a folded extension of the ground plane is added to improve the performance of the antenna. The dimensions of the antenna were optimised using a computer simulation tool based on finite element method (FEM). The simulated radiation pattern of the antenna, which is typical for microstrip patch-type antennas, is presented in Figure 4. The simulated bandwidth of the antenna is 236 MHz and the return loss (S11) at the 915 MHz resonance frequency is -17.52 dB. These values indicate a relatively wide bandwidth and sufficient impedance matching.

The simulated input impedance of the antenna at 920 MHz is $Z = (927 - j4.84)$ Ω which is, considering the use of a microstrip patch-type antenna, relatively close to the impedance of the identification microchips (Alien Technology's straps). The matching impedance of the straps is $Z = (1200 - j145)$ Ω. The relatively high matching impedance of the patch antenna leads to sufficient power transfer from the microchip to the antenna and vice versa.

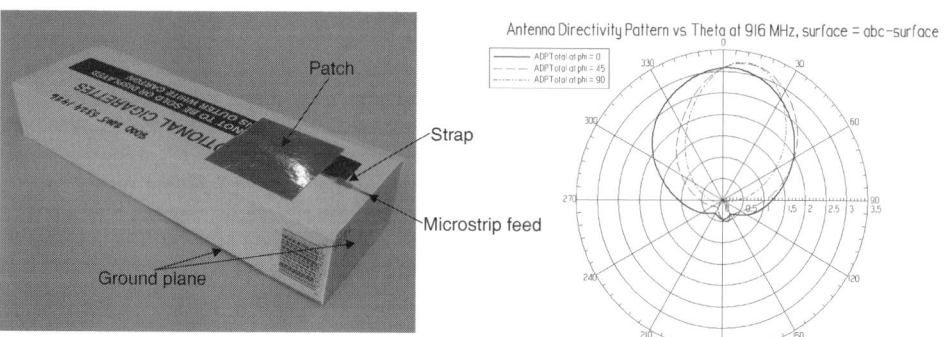

Figure 3: The structure of the patch-type tag antenna

Figure 4: Simulated radiation pattern

3. READ RANGE MEASUREMENTS

Read range measurements were carried out using a ThingMagic reader unit (Mercury 2, version 1.2.9 software) and a linearly polarised reader antenna. The read range measurement set up is shown in Figure 5. When the maximum read range was measured, the criteria for reliable identification was that the reader continuously identified the tag for at least one minute at the maximum reading distance. To study the effect of the aluminium foil on the read range, the measurements were carried out using packs with the foils, packs without foils and an empty carton.

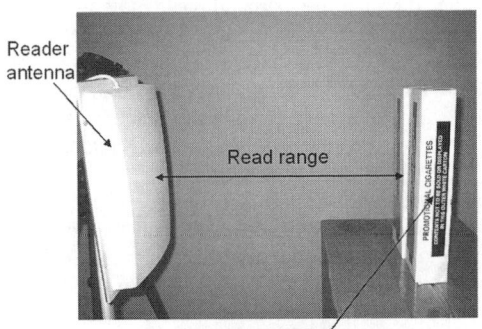

Figure 5: The read range measurement setup

3.1 Read Range Measurements of One Carton and Comparison to the Measurements with a Conventional Tag

Read range measurements were carried out with two individual integrated patch-type tags and two conventional folded dipole-type tags. The read ranges are an average value of the read ranges achieved with the two individual tags. The folded dipole-type tag is presented in Figure 6.

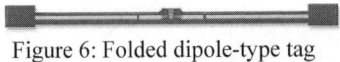

Figure 6: Folded dipole-type tag

Table 1 shows a comparison of the read ranges achieved with both tag types. It can be observed that when the aluminium foil is not removed from the packs the carton cannot be identified when folded dipole tag is used. However, with the integrated patch-type tag a read range of 1.05 m is achieved. When the foils are removed from the packs, the read range of the integrated patch-type tag almost doubles. The read range of an empty carton is only slightly longer than the read range of a carton with packs without the foils. When the foils are removed from the packs or an empty carton is measured, typical read ranges of over 2 m are achieved with the folded dipole tag.

TABLE 1
COMPARISON OF MAXIMUM READ RANGES OF THE TWO TAG TYPES

Tag antenna	Packs with foils	Packs without foils	Empty carton
Integrated patch	1.05 m	1.95 m	2.00 m
Folded dipole	0 m	2.40 m	2.75 m

3.2 Read Range Measurements of Two Cartons

To study the effect of multiple carton identification on read ranges, the tags were identified in pairs next to each other and on top of each other. The reading positions are shown in Figures 7 and 8. Both of the tags had to be read reliably at the maximum reading distance.

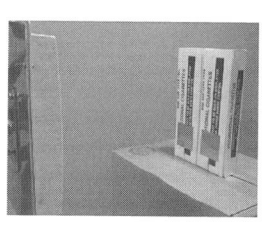

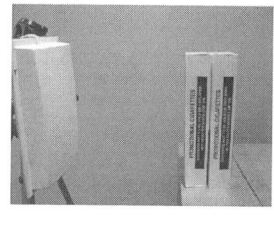

Figure 7: Reading position with the cartons next to each other

Figure 8: Reading position with the cartons on top of each other

Tables 2 and 3 present the results of the read range measurements. Also in the case of two cartons the identification cannot be carried out using folded dipole tag when the foils are in place in the packs. It was also observed that the read range of the integrated patch-type tag dropped when two cartons were read side-by-side or on top of each other. It can also be noted that when the cartons are on top of each other and the foils are removed from the packs the integrated patch has longer read range. The patch-type tag can exploit the packs as a substrate material, and therefore identification of two cartons on top of each other is more reliable than when the folded dipole tag is used. The folded dipole tag does not need a substrate material for functioning, and therefore the cigarette packs between the tags attenuate the incident wave and thereby shorten the read range. In general it can be stated that increasing the number of tags to be read simultaneously shortens the read range.

TABLE 2
MAXIMUM READ RANGE COMPARISON OF TWO CARTONS NEXT TO EACH OTHER

Cartons next to each other	Packs with foils	Packs without foils	Empty carton
Integrated patch	0.50 m	1.20 m	1.20 m
Folded dipole	0 m	1.80 m	1.80 m

TABLE 3
MAXIMUM READ RANGE COMPARISON OF TWO CARTONS ON TOP OF EACH OTHER

Cartons on top of each other	Packs with foils	Packs without foils	Empty carton
Integrated patch	0.35 m	0.65 m	0.50 m
Folded dipole	0 m	0.45 m	1.25 m

4. CONCLUSIONS

This paper presents a case study of identification of cigarette cartons with passive RFID technology. Two types of tags are tested and the achieved read ranges are compared. The aluminium foil in the cigarette packs makes the identification of the cartons difficult. Therefore, a novel microstrip patch-type tag antenna for passive RFID of cigarette cartons was designed.

The performance of the new tag antenna design was compared to that of the conventional, folded dipole-type tag antenna. It was observed that the aluminium foil in the cigarette packs affects the read ranges significantly. With the novel patch-type tag antenna the maximum read range was 1.05 m when the foils were in the cigarette packs. When the folded dipole tag was tested, the read range was 0 m when the foils were in the packs. Removing the foils from the cigarette packs lengthens the read ranges to approximately 2 m. It was also observed that two cartons next to or on top of each other can be read simultaneously. Reading both tagged cartons simultaneously shortens the read ranges.

5. ACKNOWLEDGEMENTS

Authors would like to thank William R. Sweeney from Philip Morris USA for support.

6. REFERENCES

Cheng D. K. (1993). *Fundamentals of Engineering Electromagnetics*, Prentice-Hall, pp. 304-330.
Finkenzeller K. (2003). *RFID Handbook, 2nd Edition*, John Wiley & Sons 2003, pp. 7-9.
Foster P. R. and Burberry R. A. (1999). Antenna Problems in RFID Systems. *IEE Colloquium on RFID Technology (Ref. No. 1999/123)*, pp. 3/1-3/5.
Glidden R. et. al. (2004). Design of Ultra-Low-Cost UHF RFID Tags for Supply Chain Applications. *IEEE Communications Magazine*, **42:8**, pp. 140-151.
Raza N., Bradshaw V. and Hague M. (1999). Applications of RFID Technology. *IEE Colloquium on RFID Technology (Ref. No. 1999/123)*, pp. 1/1-1/5.
Raumonen P., Sydänheimo L., Ukkonen L., Keskilammi M., and Kivikoski M. (2003). Folded Dipole Antenna Near Metal Plate. *Proc. IEEE Antennas and Propagation International Symposium*, **1**, pp. 848-851.
Reitz J. R., Milford F. J. and Christy R. W. (1993). *Foundations of Electromagnetic Theory*, Addison-Wesley, pp. 454-462.

CURRENT LIMITER COMPLICATES THE DYNAMIC CHARACTERISTICS OF SERVO MOTOR

Pakorn Serikitkankul, Hiroaki Seki, Masatoshi Hikizu, and Yoshitsugu Kamiya

Department of Mechanical Systems Engineering, Kanazawa University,
Kanazawa, Ishikawa, 920-1192, Japan

ABSTRACT

In this paper, effects of a current controller on dynamic characteristics of the servo motor system are studied. The current controller regulates motor current to well control motor torque, and prevents overloaded motor current. It makes a servo motor to be easily controllable; however, it complicates some dynamic characteristics of the servo motor system. Then, development of high-speed and high-accuracy positioning system is proposed. The simulation results show that the performance of the modified positioning system is better than that of the previous system.

KEYWORDS

Dynamic characteristics of servo motor, Current limiter, Positioning system.

INTRODUCTION

Presently, servo motors are widely used in many applications such as robotic applications, home appliances and industrial automation. The typical servo motor system that is frequently used in above mentioned applications consists of an electric motor and several cascaded control loops, which are a position control loop, a velocity control loop and a current control loop. The current control loop is used to control motor current and to prevent overloaded motor current. The velocity control loop controls motor velocity, and the position control loop commands the velocity control loop in order to rotate the servo motor to the desired position. This paper studies effects of the current limiter on the dynamic characteristics of servo motor and proposes high-speed and high-accuracy positioning system.

CURRENT CONTROL

Since induced voltage from motor armatures, which complicates velocity and torque control of servo motor, must be eliminated and overloaded current must be prevented, a current controller, consisting of a current limiter and a current amplifier as illustrated in Figure 1(a), is added to the servo motor system. The current limiter prevents motor overloaded current by limiting current command. The current amplifier regulates motor current, and help to minimize the effects of induced voltage from motor armatures. Functionality of the current amplifier is proved below. Since effects of induced voltage, E(s), to the motor current could be approximated a linear function, the transfer function of the current loop in Figure 1(a) would be defined in Eqn. 1, and if the integrating gain, K_A, is large enough, the effects of induced voltage, E(s), could be omitted.

$$I(s) = \left[\frac{K_B S + K_A}{(R_a + \alpha K_B)S + \alpha K_A}\right]V(s) - \left[\frac{S}{(R_a + \alpha K_B)S + \alpha K_A}\right]E(s) = \left[\frac{\frac{K_B S}{K_A}+1}{(\frac{R_a + \alpha K_B}{K_A})S + \alpha}\right]V(s) - \left[\frac{\frac{S}{K_A}}{(\frac{R_a + \alpha K_B}{K_A})S + \alpha}\right]E(s) \approx \frac{V(s)}{\alpha} \quad (1)$$

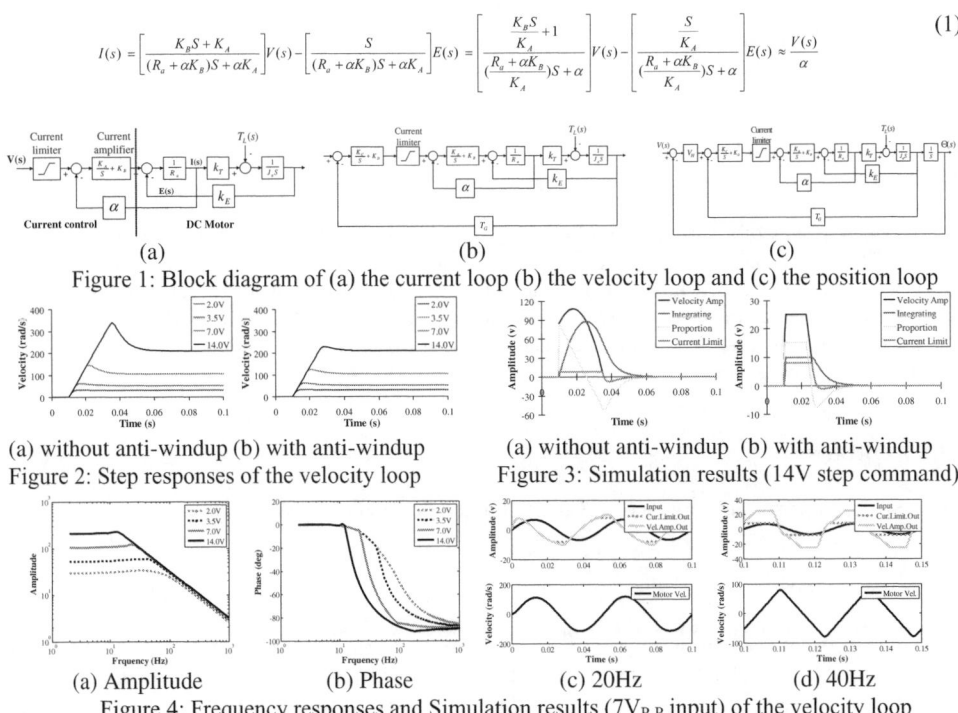

Figure 1: Block diagram of (a) the current loop (b) the velocity loop and (c) the position loop

(a) without anti-windup (b) with anti-windup
Figure 2: Step responses of the velocity loop

(a) without anti-windup (b) with anti-windup
Figure 3: Simulation results (14V step command)

(a) Amplitude (b) Phase (c) 20Hz (d) 40Hz
Figure 4: Frequency responses and Simulation results ($7V_{P-P}$ input) of the velocity loop

VELOCITY CONTROL WITH CURRENT LOOP

A velocity control loop maintains a motor velocity run by the velocity command from a position amplifier. It compares measured motor velocity to the velocity command, and, then, commands a current loop to adjust motor velocity in order to minimize the velocity error. Numerical simulation of the velocity loop shown in Figure 1(b) is used to study the effect of a current loop on a velocity loop. Simulation parameters are set as follows: K_C is 800, K_D is 6, α is 2, K_A is 4000, K_B is 1, R_a is 8.7Ω, k_E is 0.187V/(rad/s), k_T is 0.188Nm/A, J_a is 5.59×10^{-5}Kg.m^2, T_G is 0.0668V/(rad/s), the current limiter is set to ±8V, and step velocity commands are 2.0V, 3.5V, 7.0V and 14.0V. Two systems, which are a velocity loop with anti-windup and without anti-windup, are simulated. For the velocity loop with anti-windup, the integrating limit and proportional limit are set at ±10V and ±15V, respectively. The step responses of both systems are shown in Figure 2 and 3 respectively. The frequency responses and some simulation results of the velocity loop with anti-windup are shown in Figure 4.

Figure 2 shows that the velocity loop without anti-windup has high overshoot response and longer settling time than that of the velocity loop with anti-windup because of current saturation and integral windup. In regard to current saturation, it is the current limiter in the velocity loop; however, the current limiter is one of the important parts of a velocity loop since it prevents a servo motor from overloaded current. Regarding integral windup, velocity error is fast accumulated in the integrating amplifier to the level that the current command reaches saturation. After its saturation, output of the integrating amplifier is still increased, but it no longer matters. Then, when motor velocity reaches the desired velocity, the velocity loop is generally used to maintain motor velocity at the desired velocity. However, in this case, it could not do so since output of the integrating amplifier is so high that it could not be reduced fast enough. Therefore, overshoot occurs as shown in Figure 2(a) and 3(a). In order to prevent integral windup, anti-windup must be added to the integrating amplifier. It can reduce overshoot and settling time of step responses as shown in Figure 2(b) and 3(b). However, the velocity loop with anti-windup still degrades some dynamic characteristics of the servo motor. The frequency responses at high frequency are degraded, as demonstrated in Figure 4, especially, when high-level input is fed into the system due to saturation of the current limiter.

POSITION CONTROL WITH CURRENT LOOP

A position control must drive a servo motor so as to hold the position of the motor at the desired position commanded by an external source. The position amplifier amplifies the position error between the measured position and the desired position, and, then, this amplified position error is used to drive the servo motor to minimize the position error. The typical position loop is shown in Figure 1(c). The simulation results of the position loop are shown in Figure 5 and 6. The simulation parameters are set as follow. The position gain, V_H, is 100. The step position commands are 0.04, 0.16, and 0.63 radian, respectively. Other parameters are the same as that of the previous section. From the simulation results, this positioning system has underdamped responses, large overshoot, ringing and long settling time because of effects of high position gain and saturation.

DEVELOPMENT OF HIGH SPEED AND HIGH ACCURACY POSITIONING SYSTEM

For the purpose of development of the high-speed and high-accuracy positioning system, the position loop controller has been modified to work in either the velocity control mode or the position control mode. Therefore, the position amplifier, V_H in Figure 1(c), has been replaced with a non-linear amplifier in the case of velocity control mode and linear amplifier in the case of position control mode. Functionality of the modified position amplifier is described below. The linear region of position error is set to $[-\varepsilon,+\varepsilon]$, and the initial mode of the position amplifier is position mode, with a linear gain (V_H). Then, the condition for switching between the position mode and velocity mode is described below. For condition 1, if the initial position error is not in the region of $[-\varepsilon,+\varepsilon]$, the amplifier mode is changed to the velocity mode with input-output characteristic functions as demonstrated in Eqn. 2. For condition 2, if the current mode of the amplifier is the velocity mode, and the current motor velocity is approximated zero, the amplifier mode is set to the position mode (linear mode) with a linear gain V_H.

$$V(x)=V_{H(NL)}\sqrt{x-\text{sgn}(x_{err(i)})\times\varepsilon} \quad \text{where } x \geq \text{sgn}(x_{err(i)})\times\varepsilon, \quad V(x)=-V_{H(NL)}\sqrt{\text{sgn}(x_{err(i)})\times\varepsilon-x} \quad \text{where } x < \text{sgn}(x_{err(i)})\times\varepsilon \quad (2)$$

where $V_{H(NL)}$ is non-linear gain, x is the motor position, and $x_{err(i)}$ is the initial position error.

Simulation of a positioning system with a modified position amplifier is studied. the simulation parameters are set as follows. The linear gain, V_H, is 100. The non-linear gain, $V_{H(NL)}$, is 6. The ε is 0.01. The step position commands are 0.04, 0.16, and 0.63 radian, respectively. Other parameters are the same as that of the previous section. By choosing $V_{H(NL)}$ and ε, a trial and error technique has been used as shown in Figure. 10. It shows that 6 of $V_{H(NL)}$ and 0.01 of ε give the best system responses. The simulation results of the modified position loop are shown in Figure 7 and 8. Phase plan diagrams of the modified system and the transfer characteristics of the modified position amplifier are shown in Figure 9. Comparing step responses in Figure 7 with that of the pervious system in Figure 5, it could be understood that the modified positioning system can reduce overshoot, ringing and settling time.

The modified position amplifier can drive velocity loop in a manner that saturation effect is minimized as shown in Figure 8. In conclusion, the simulation results show that the modified position amplifier can improve speed and accuracy of the positioning system. How the modified positioning system switches between velocity mode and position mode is demonstrated in Figure 9. According to that figure, the position amplifier is initially in position mode. After feeding a desired position into the system, large position error will occur, and, then, the amplifier mode will change to velocity mode. Then, the motor will be accelerated and then decelerated following the velocity profile. After motor velocity crosses zero value with its position error within the region of $[-\varepsilon,+\varepsilon]$, the amplifier mode will change to position mode. Finally, the motor will be driven so that its position reaches the desired position.

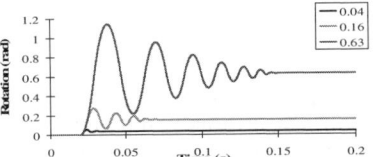

Figure 5: Step responses of the position loop

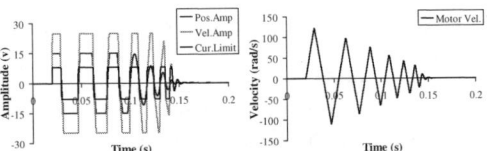

Figure 6: Internal signals of the position loop where step position command is 0.63 radian.

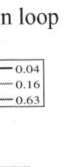

Figure 7: Step responses of the position loop with the modified position amplifier

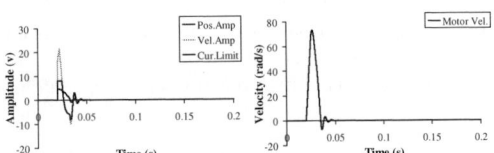

Figure 8: Internal signals of the modified position loop where the step position command is 0.63 radian

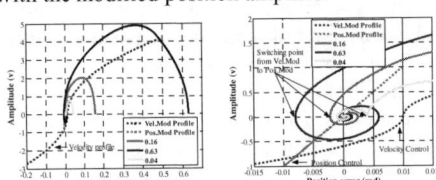

Figure 9: Phase-plan diagram of the modified position loop

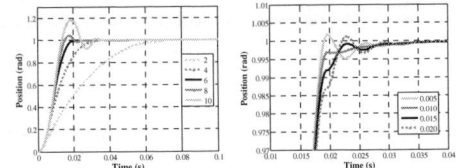

Figure 10: Unit step responses of the positioning system (varying $V_{H(NL)}$ and ε respectively)

CONCLUSION

This paper studies effects of a current loop on dynamic characteristics of the servo motor system. The simulation results show that the saturation effect of a current amplifier and integral wind-up complicates some dynamic characteristics of the servo motor; however, the integral windup effect can be reduced by anti-windup. In terms of development of the positioning system, the modified position amplifier can improve speed and accuracy of the positioning system.

Reference
F. Sakai, Y. Kamiya, H. Seki, and M. Hikizu (2000). Analysis of Non-linear Dynamic Characteristics of Motor Driven by Conventional Servo Amplifier. *Transactions of the Japan Society of Mechanical Engineers* **66:667**, 189-195. (In Japanese)

ACTIVE SUSPENSION SYSTEM WITH HIGH-SPEED ON/OFF VALVE (APPLICATION OF PREVIEW CONTROL WITH ADAPTIVE DIGITAL FILTER)

Hironao YAMADA[1] and Takayoshi MUTO[1]

[1] Department of Human and Information Systems, Gifu University,
1-1 Yanagido, Gifu 501-1193, Japan

ABSTRACT

The aim of this study is to propose an active control hydro-pneumatic suspension system composed of high speed on/off solenoid valves. In order to realize a robust control system, we adopted preview control method utilizing an ADF (Adaptive Digital Filter). The experiment was performed on a bench system which physically simulates induced vibrations from the road. These results confirmed that in the case of the preview control method, the damping efficiency is higher than in the case using the conventional sky-hook control method. Furthermore, the control performance of preview control using an ADF is superior to the conventional preview control system. Therefore, it is expected that the active suspension system, which is developed in this study, could produced at low cost and achieve a high level of reliability.

KEYWORDS

Preview Control, Adaptive Digital Filter, ON/OFF Valve, Active Suspension, PWM Control.

INTRODUCTION

In recent years, an active control suspension system has come into use thanks to progress in the development of micro processors, sensors, and actuators [1]-[4]. By applying these new techniques to the active suspension system, it is expected that the both riding comfort and vehicle control stability will be improved. However, one of the problems associated with these developments is that the introduction of active suspension inevitably makes vehicles complex, heavy, and expensive.

Conventional active suspension systems normally use a pressure control valve or a servo valve as a system control device. These valves, however, have defects in that they are expensive and have a weakness in oil contamination. In contrast to these valves, a fast switching on/off valve can be considered an attractive device for overcoming the above-mentioned defects. The reason for this is that the on/off valve has a simple configuration and, furthermore, can be a logical interface between a computer and hydraulic systems. Hence, a low cost and highly reliable system may be developed, under the recent condition that the systems are equipped with computers.

In the present study, we propose an active control hydro-pneumatic suspension system composed of high speed on/off solenoid valves which are driven in a PWM (Pulse Width Modulation) digital mode. In order to realize a robust control system even when measurement errors exist, we adopted a preview control method equipped with an ADF (adaptive digital filter). The experiment was performed by using a bench system which physically simulates induced vibrations from the road and also simulates the body mass of an automobile. The results obtained from the system using preview control with an ADF were compared with those obtained from a system using a conventional sky-hook control.

CONSTITUTION OF THE ACTIVE SUSPENSION SYSTEM

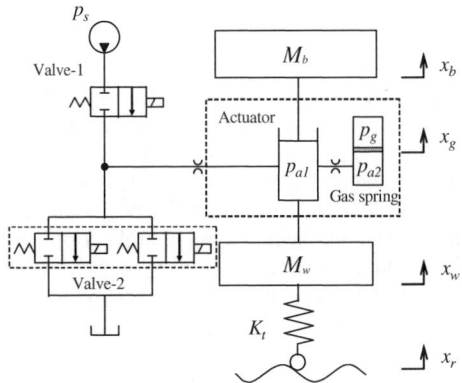

Figure 1: The outline of the system used for the experiment

The configuration of the new system proposed in this study is shown in Figure 1. The pressure control valve, which is used in a conventional active suspension system, is replaced by two on/off valves (valves 1 and 2). The valves are of a high-speed solenoid type. In this system, the supply flow rate to the cylinder is controlled by either valve 1 or valve 2 according to the PWM-signal. The constitution of the on/off valve is illustrated in Figure 2.

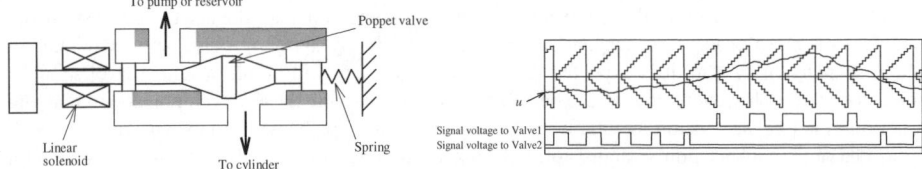

Figure 2: Structure of a high-speed ON/OFF valve   Figure 3: PWM (Pulse-Width-Modulation)

The poppet in the valve is actuated by the on/off input voltage applied to the solenoid and thus the flow rate through the valve is controlled in digital mode in accordance to a duty signal. In the following, the principle of the PWM method adopted in this study is demonstrated based on the experimental results shown in Figure 3.

In Figure 3, the signal u denotes the control input (=Duty) and the triangle waves are carrier waves. The input signal voltage to valves 1 and 2 is generated by comparing the control input u with the triangle carrier waves as shown in the figure.

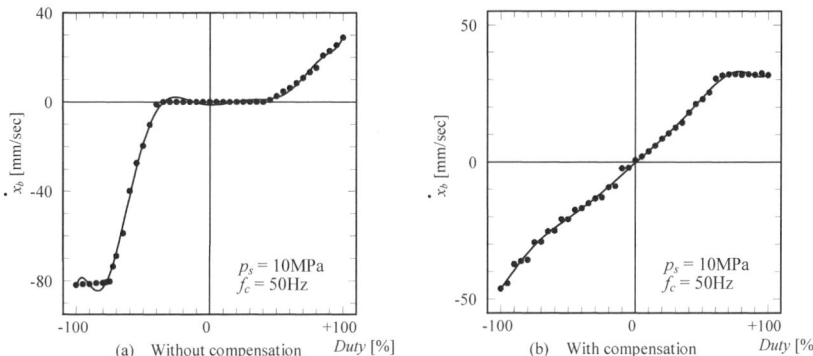

Figure 4: The Duty-velocity characteristics of an ON/OFF valve

Next, Figure 4 shows the valve characteristics between suspension cylinder speed and duty when the valves are driven by the duty signal. It is seen in Figure 4(a) that a dead zone exists in the vicinity of the origin because of the delay time of the valve. This kind of nonlinear characteristic, however, can easily be compensated into a linear one by adopting an appropriate compensation method. That is, in order to cancel the influence of the dead zone, we can impose a wider pulse width than the estimated Duty by the amount of the delay time. The results of compensation are shown in Figure 4 (b).

CONSTRUCTION OF THE BENCH SYSTEM

The dynamic performance of the proposed system was investigated experimentally using a quarter-car test bench system as shown in Figure 5.

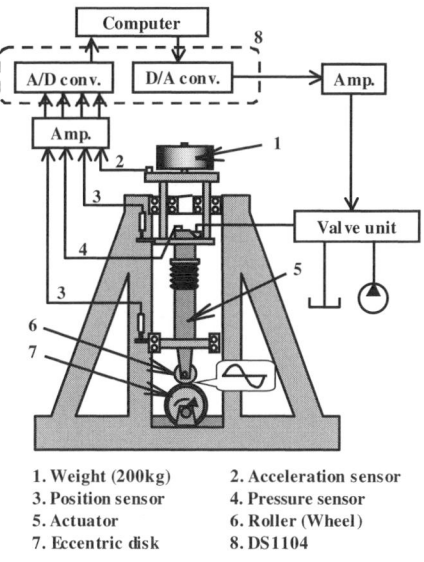

1. Weight (200kg) 2. Acceleration sensor
3. Position sensor 4. Pressure sensor
5. Actuator 6. Roller (Wheel)
7. Eccentric disk 8. DS1104

Figure 5: Quarter-car test bench system

In the bench system, the hydraulic actuator (5) supports the 200 kg mass (1) which corresponds to a quarter of a car's body mass. The eccentric disk (7), which is a physical model of the displacement from the road, is attached at the lower part of the hydraulic actuator and is driven by a hydraulic motor. The amount of disk eccentricity is set to 5 mm. Each displacement of the body mass and wheel is detected by position sensors, while cylinder pressure is detected by a pressure sensor. These signals are transmitted to a computer through an A/D converter and thus the control input u to the valve is calculated so that the vibration of the car body can decrease in the minimum level.

PREVIEW CONTROL METHOD WITH ADAPTIVE DIGITAL FILTER

In our advanced work, we adopted a sky-hook control method to confirm that the active suspension system using the digital valve has adequate control performance almost equivalent to the conventional one composed of a pressure control valve[5]. Thus, the method of preview control using an ADF is adopted in this study as the more effective controller for robustly decreasing vibration. Figure 6 shows the schematic diagram of the preview control using an ADF.

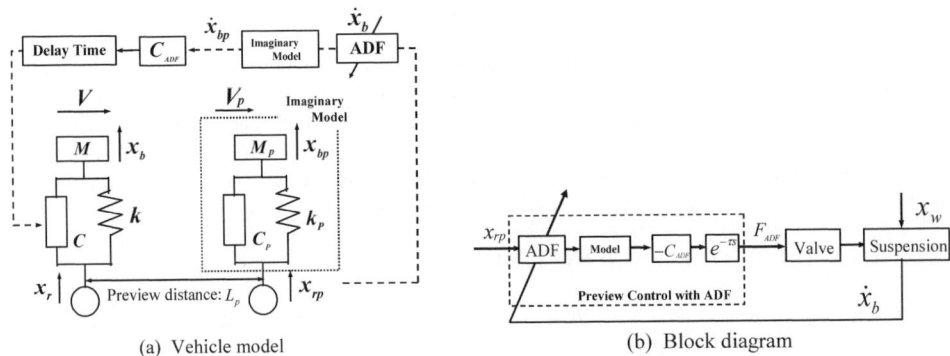

(a) Vehicle model   (b) Block diagram

Figure 6: Preview control using an ADF

We assumed that the displacement of the road surface in the forward of an automobile can be detected before T_d seconds by using some sort of sensor. (In the experiment, we actually used the displacement of the eccentric disk (in Fig.5) which is detected one period before. Here, road displacement x_{rp} and body displacement $\dot{x}_b$ are considered as an input signal and a error signal, respectively. The road displacement x_{rp} is compensated by the ADF using error signal $\dot{x}_b$. In this method, the LMS (Least Mean Square) algorithm is used for renewal of the ADF's coefficients. The formula for the renewal of the filter coefficients is

$$W_{k+1} = W_k + 2\mu \cdot \varepsilon_k \cdot X_k \quad (1)$$

where W_k is the filter coefficient, μ is a step size, ε_k is the error signal ($=\dot{x}_b$), and X_k is the input signal ($=x_{rp}$).

Equation of motion for the active suspension:

$$M_p \ddot{x}_{bp} + c_p(\dot{x}_{bp} - \dot{x}_{rp}) + k_p(x_{bp} - x_{rp}) = 0. \quad (2)$$

In the preview control part, $\dot{x}_{bp}$ is calculated using the above equation (2). Thus, the control input F_{ADF} is obtained as $c_{pre} \cdot \dot{x}_{bp}$. The valves are controlled after the system delay time τ is compensated for. Here,

$$\tau = T_r - T_d, \quad (3)$$

where T_d is the compensation time for the system delay and the value is determined by trial and error. (In the experiment, $T_d = 60\text{ms}$.)

EXPERIMENTAL INVESTIGATION OF THE DYNAMIC PERFORMANCE OF THE SYSTEM

As the experimental condition, the supply pressure $p_s = 10\text{MPa}$, the road amplitude $x_r = \pm 5\text{mm}$, and the control gain $C_{ADF} = 4 \times 10^3 \,\%/(\text{m/s})$. In order to evaluate the robustness of the controller, time error ΔT_d is added as shown in Figure 7. Point A is the present position on the road and point B is the preview position forward of point A. The error time ΔT_d (=+30%, +50%, +100% of T_d) is added to the compensation time T_d.

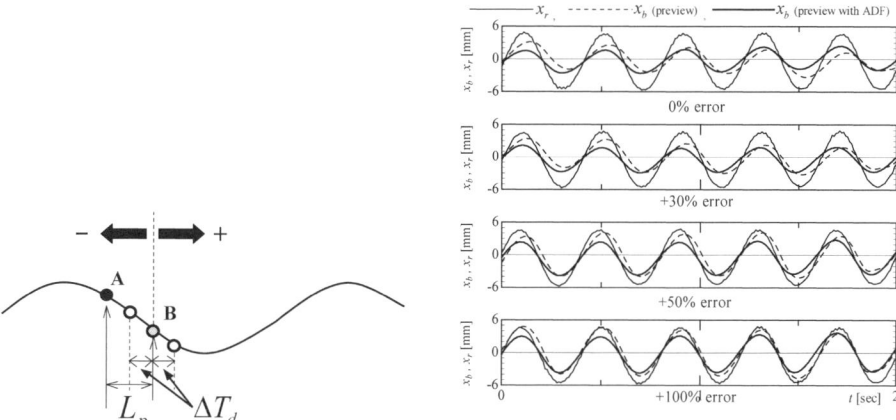

Figure 7: Amplitude of the road surface

Figure 8: Experimental results of preview control and preview control using an ADF

Figure 8 shows the experimental results obtained from the bench system along with various errors in road measurement. In the figures, x_r and x_b denote the displacements of wheel and body mass, respectively. Here, curves of x_b with preview control are indicated by the broken line in the diagram and curves with preview with ADF are drawn by the thick line. In this experiment, the rotation frequency of the eccentric disk f_d is set to be 2.5 Hz and the carrier wave frequency of PWM f_c=20 Hz. As seen in the figures, at the larger error the damping efficiency becomes relatively lower when the conventional preview control system is adopted. On the other hand, the preview control using the ADF has a higher damping efficiency even when a large error exits.

Additionally, the damping efficiency of the individual controller was also investigated using a frequency response test in a frequency range of 0.75 – 2.5 Hz. As shown in figure 9. At first, in the case of preview control (both with ADF and without ADF), the damping efficiency is better than the case using the sky-hook control mainly around the low frequency area.

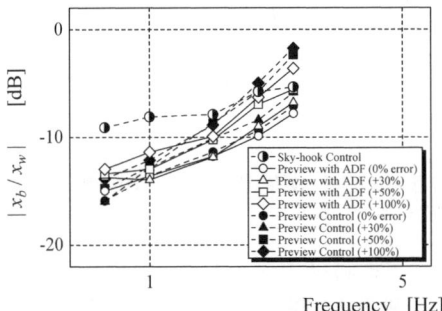

Figure 9: Frequency characteristics

Next, the control efficiency both of the simple preview control and the review control using the ADF are reduced when the error ΔT_d increases. Even under such a condition, however, the preview control using the ADF shows a higher damping efficiency compared to that of the simple preview control. However, the control efficiency both of the simple preview control and the preview control using the ADF are less than that of the sky-hook control in the high frequency area when error ΔT_d is large. The control performance will be expected to improve by combining both the preview control using the ADF and the sky-hook control.

CONCLUSION

In this study, we proposed an active control hydro-pneumatic suspension system composed of high speed on/off solenoid valves, which are driven in a PWM (Pulse Width Modulation) digital mode. In order to realize a robust control system even when measurement errors exist, we adopted preview control method utilizing an ADF. The experiment was performed on a bench system which physically simulates induced vibrations from the road and also simulates the body mass of an automobile.
These results confirmed that in the case of the preview control method (both using ADF and without ADF), the damping efficiency is higher than in the case using the conventional sky-hook control method. Furthermore, the control performance of preview control using an ADF is superior to the conventional preview control system under conditions that contain measurement errors. Therefore, it is expected that the active suspension system utilizing high speed on/off valves, and which is equipped with preview control using an ADF, could me produced at low cost and achieve a high level of reliability.

References
1) Foag, W., (1989), A practical control concept for passenger car active suspensions with preview, Proc. Instn. Mech. Engrs. **203**, 221-230.
2) Kawakami, H., Urababa, S., Inoue, H., and Ichimaru, H., (1991), Development of Soarer Active Control Suspension, Toyota Technical Review, **41:1**, 64-76.
3) Yoshimura, T., Nakaminami, K., Kurimoto, M. and Hino, J., (1999),Active suspension of passenger cars using linear and fuzzy-logic controls, Control Engineering Practice, **7:1**, 41-47.
4) Fukao, T., Yamawaki, A. and Adachi, N., (2002), Adaptive Control of Partially Known, Systems and Application to Active Suspensions, Asian Journal of Control, **4:2**, 199-205
5) Yamada, H., Suematsu and Y., Muto, T. (1995), Sky-hook Control of Active Suspension System Composed of High Speed ON/OFF Valves, 10.Fachtagung Hydraulik und Pneumatik, Germany, 551-564.

EMBEDDED DISTRIBUTED SUB CONTROL SYSTEM BASED ON HYBRID CONTROLLER

M. Lammila, T. Virvalo and E. Lehto

Institute of Hydraulics and Automation, Tampere University of Technology
Tampere, Finland

ABSTRACT

Modern forest machines, like forward loaders and harvesters, are still manually controlled hydraulically driven, but strongly computerized machines. There are some fully or semi automatic functions, but most often a driver acts as a controller. Goals of improving energy utilization of forward loader boom and at the same time intensify the control of the boom require a lot of computation power. Modern hybrid controllers seem to be good solution to these kinds of requirements and applications

KEYWORDS

Embedded controller, hybrid controller, hydraulic boom

INTRODUCTION

Natural development during last decade has brought more integrated electronics also into mobile machines and their hydraulic components like valves, pumps, and actuators. Quite a flexible method is to use components that can be connected to the control system via a data bus. However, very few valves with field bus interfaces have been available in the past, BoschRexroth (2005), Sauer-Danfoss (2005), Ultronics (2004). On the other hand machine builders have accepted field buses very well. The CAN bus is one of the most common data buses used in real time fluid power applications, especially mobile hydraulic control systems. Lately the manufacturers of hydraulic components have also developed valves and accessories which are available with a CAN bus interface, Moog (2005), Parker (2005). There are suitable sensors for fluid power applications like position, pressure, and temperature sensors with the CAN bus interface, Balluff (2005), Dynisco (2005), Pepperl-Fuchs (2005).

One important topic in R&D of mobile machines is the improving of energy utilization. Total power levels, for instance, in forest machines are 150-200kW. Low efficiency leads to over-sizing of components and heat problems as well as to poor economic results.

APPLICATION

The studied application is shown in Figure 1. Timbers are loaded from ground to the palette of a loader or unloaded from the palette to ground. When a load is lowered the lift cylinder has an overrunning load. When a load is raised the jib cylinder has overrunning load, respectively. A question is can these overrunning loads be utilized in energy recovering. The whole working range of the boom is shown in Figure 1. The end points of the studied movements are marked to the Figure 1.

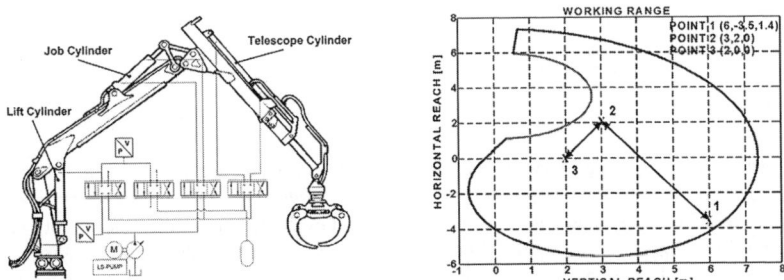

Figure 1. Example boom and working range

The studied movements in the unloading case are back-and-forth motion between the points (3,2,0) and (6,-3.5,1.4), Figure 1. When the load is lowered oil from the lift cylinder is partly used in the telescope cylinder and extra oil is stored into the accumulator. The required oil during the return movement of the telescope cylinder is taken from the accumulator. According to these requirements the size and pre-charging pressure of the accumulator is fixed.

PROPOSED CONTROL METHOD

A driver gives the required velocity of each DOF with his joy-sticks. The lift cylinder chamber and accumulator line pressures are measured, Figure 1. If the chamber pressure is a certain amount higher than the accumulator line pressure, the flow from the lift cylinder is directed to the accumulator line via the valve V2 otherwise to the tank line via the valve V1. The volume flow to the accumulator line can be stored into the accumulator or utilized directly in the telescope cylinder. With the measured pressure difference over the control notch and the estimated valve spool position an adaptive and fairly accurate flow measuring device is achieved, Kannisto 1 (2002), Kannisto 2 (2002), and Kannisto 3 (2002).

The main tasks of the controller are: Realize the Velocity State-Controllers (VSC) of each DOF. Measure the cylinder chamber pressures and supply line pressure. Compute volume flows of valves. Use the computed piston velocities and accelerations as feedback signals of the VSC. Compute all VSC-algorithms. Decide which valves are active. Realize the current controls of the valve solenoids. Communicate with the main controller.

MAIN ALGORITHMS IN COMPUTATION

At the beginning of every control cycle all pressures and outputs of joy-sticks are measured as well as on/off-type position reference points. Square roots of absolute pressure differences are computed. The actual velocities are estimated taking into account the estimated square root, the nominal volume flow of the control valve and the slightly filtered control signal from the joy-stick.

The velocity errors between the reference signal (from joy-stick) and the velocity feedbacks (estimated velocities) are computed and set inputs to the VSCs. The gains of the VSCs (integral gains, velocity and acceleration feedback gains) are tuned for smooth dynamic behavior. The output of the lift cylinder VSC is then handled by the valve switching logic between valves V1 and V2. When the pressure difference is lower than the lower limit of the pressure hysteresis the flow is directed to the tank line via the valve V1. When the pressure difference is higher than the lower limit but lower than the higher limit of the pressure hysteresis the valve V1 is still operated and the flow is directed to the tank line. Mostly the flow through the valve V2, in the cases of overrunning loads, is used in the telescope cylinder, but also partly to load the accumulator. In order to get roughly approximation for the orientation of the boom the positions of the pistons are computed by integrating the estimated velocities of the pistons. Three on/off switches are used in each DOF as reference points to improve the position computation accuracy. The forward kinematics of the boom is then used to compute finally the orientation of the boom.

SIMULATION RESULTS

The verified model of the studied hydraulic boom is used. The behavior of the pressures in the lift cylinder and accumulator are presented in Figure 2, when the load of 400kg is lowered. Oil from the lift cylinder is partly used to drive the full stroke of the telescope cylinder and partly used to raise the pressure of the accumulator. The responses of valve 1 and 2 during this movement are shown in Figure 2. When pressure difference between the lift cylinder and the accumulator is lower than the certain limit the flow is directed via the valve 1 to the tank line. Otherwise the flow is directed via the valve 2 to the accumulator.

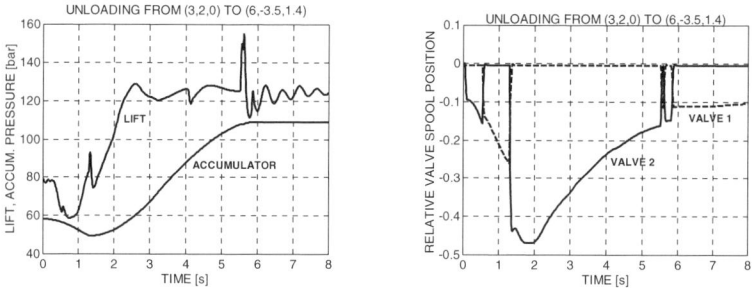

Figure 2. Simulated responses of pressures and valves

The return movement takes place without the load. Oil required in the telescope cylinder is taken from the accumulator. According to simulation the energy consumption in the back-and-forth motion with the traditional control system is 166kNm. The energy consumption with the proposed control system is 122kNm

EVALUATION CAPABILITY OF HYBRID CONTROLLER

The following I/O-capabilities are needed: Nine pressure sensors, four outputs of joy-stick, ten analogue valve control outputs, one CAN node, and twelve on/off inputs. It might be possible to use pressure sensors and/or a joy-stick with CAN bus interfaces.

Digital-Signal-Processors have previously had a serious limitation in control applications; lack of suitable I/O. Motorola's DSP56F803 processor and its newer version are designed to fulfil a gap between traditional Micro Controllers and high speed DSPs [Freescale 2005]. These kinds of DSPs are often called a hybrid controller. Main features which make this processor suitable for this application are; a CAN 2.0B module to connect controller to other modules in the system (joystick, other controller modules, etc.), a PWM-module to control proportional valves and ADC-channels to measure pressures. One nice feature is also a quadrature decoder interface which gives possibility to connect incremental encoder directly to the controller without any extra electronic. To achieve a reliable current control of proportional valves it is necessary to close the current control loop via a hybrid controller. This control solution gives more options to improve long term stability and diagnostic of valves. Because this hybrid controller has not enough I/O-capacity it is necessary to distribute the control of this sub-system. Best and most flexible solution is to use one hybrid controller per DOF. The preliminary tests show the speed of this hybrid controller is enough to realize needed control algorithms, especially when applied only on one DOF. Based on tests the estimated computation time is approximately 1ms. This guarantees that 2ms sampling time of the control loop can be used, which is short enough for these kinds of applications and needed algorithms.

CONCLUSIONS

According to preliminary simulations and estimations the following conclusion can be presented: It is possible to reduce energy consumption of forward booms in unloading work with proposed control system about 25% in certain work cycles. Only pressure sensors are needed. Hybrid controller has suitable I/O and computation capacity for computation of one DOF.

FURTHER STUDIES

In the next stage the study is extended to loading process. Then the jib cylinder is in the same role as the lift cylinder in unloading. Experimental studies will be also carried out. If results are still promising the study is extended to deal with all four DOFs of the boom.

REFERENCES

Balluff, www.balluff.com, 2005
BoschRexroth www.boschrexroth.com, 2005
Dynisco, www.dynisco.com, 2005
Freescale, www.freescale.com, 2005
Kannisto, S. & Virvalo, T. 1, 2002. Approximate velocity feedback using estimated volume flow rate of servo valve. 8th Mechatronics Forum, Netherlands, 24-26 June, 2002. s. 764-774.
Kannisto, S. & Virvalo, T. 2, 2002. Approximate Velocity Feedback Using Hydraulic Valve Flow Modelling. 5th JFPS, Nara 2002, November 14, 2002. 2. Japan. s. 501-506.
Kannisto, S. & Virvalo, T. 3, 2002. Valve Manufacturer Specifivation Based Hydraulic Velocity Control. IMECE2002, November 17-22, 2002, New Orleans, Louisiana, USA. 2. 7 s
Moog, www.moog.com, 2005
Parker, www.parker.com/iqan/, 2005
Sauer-Danfoss, www.sauerdanfoss.com, 2005
Ultronics, www.ultronics.com, 2004

ACTIVE NOISE CANCELLATION HEARING PROTECTOR WITH IMPROVED USABILITY

Mika Oinonen, Harri Raittinen, Markku Kivikoski

Tampere University of Technology, Institute of Electronics, P.O. Box 692, FIN-33101, Tampere, Finland

ABSTRACT

Low frequency performance of a passive hearing protector can be improved significantly by active noise control methods. In a laboratory environment, the low frequency attenuation of a passive hearing protector can be improved by as much as 25 dB. However, when the device is used in real-life situations, several usability aspects must also be considered. A prototype of an active noise cancellation hearing protector was constructed. The main goal was to achieve comfortable and unnoticeable operation. Stability issues were also taken into account. The developed prototype is comfortable to use because of its improved stability and reduced nonlinear distortion.

KEYWORDS

Hearing protector, stability, active control of sound, safety, wearable, human friendly

INTRODUCTION

With a passive hearing protector, noise attenuation decreases at frequencies lower than 500 Hz. However, it is relatively easy to attenuate noise having a frequency of 500 Hz or less by implementing active noise control (ANC) methods. Thus, it is possible to construct a very efficient hearing protector by combining an active noise cancellation system with a high-quality passive hearing protector.

Active control of sound is based on destructive interference. When two sound waves of equal magnitude but opposite in phase interfere, they cancel each other out. A personal hearing protector is well suited to the application of active noise control, because the space is confined and the characteristics of the plant are known, or at least can be measured. A relatively simple system can be used. For example, a feedback system implemented by means of analog electronics can be used for generating a quiet zone near the ear. When designing an active noise cancellation hearing protector, it is relatively easy to achieve good performance in a laboratory environment. For practical purposes, there are several design aspects that must be taken into account. First, there is a trade-off between active attenuation, usability, and the operating frequency range. A large amount of active attenuation requires a large amount of gain, which may lead to stability problems and increased distortion.

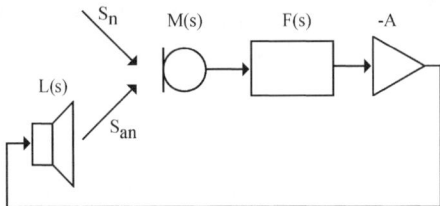

Fig. 1. The block diagram of the feedback-type active noise cancellation system.

The transfer function of the acoustic environment (the plant) must also be taken into account when designing the filters that define the operating frequency range. In the present case the plant consists of the earcup, the mechanical construction of the hearing protector, the microphone, the loudspeaker, and the head and ear of the user. As with any system with negative feedback and high gain, the active noise control system may become unstable under certain circumstances. A block diagram of an active noise cancellation system is shown in Figure 1. S_n is the noise signal, M(s), F(s), -A and L(s) are the transfer functions of the error microphone, the filter, the amplifier, and the loudspeaker (the secondary source), respectively.

A loud low-frequency signal can saturate the amplifier. When this occurs, no signal can pass through it without becoming distorted. For example, when a low frequency tone saturates the amplifier, a higher frequency tone also becomes distorted. For example, head movement and walking cause changes in the pressure of the air inside the earcup. These infrasound pressure variations can be extremely large in magnitude when compared with audible sounds. The microphone also converts these strong infrasound signals into electric signals, which may get distorted because of the supply voltage limitations. The movement of the earcup may also cause instability. For example, an adaptive ANC headset developed by Rafaely maintained stability during minor changes in the fit, but became unstable when the headset was suddenly moved or subjected to an impact (Rafaely 1997).

In addition, the sensors of an active noise control system may be saturated if the noise level exceeds the dynamic range of the sensors. The saturation generates harmonic distortion (Kuo 2004). However, in the present case the only sensor is located in the quiet zone. It is, therefore, unlikely that the sensor will be saturated. Instead, the loudspeaker and amplifiers are more likely to be saturated because of the higher signal level.

THE IMPLEMENTED PROTOTYPE

A prototype of an active noise cancellation hearing protector was implemented according to the block diagram shown in Figure 1. The prototype was implemented using analog feedback-type system. The operating frequency range is defined by high-pass and low-pass filters. The low-pass filter was designed in order to ensure stable operation at the upper end of the frequency range, whether the earcup is tightly fit, partially open, or fully open.

An electronic solution to the saturation problem described in the previous chapter was developed. An automatic gain control (AGC) circuit, which adjusts the amount of active attenuation, was developed (Oinonen 2004). When a loud low frequency signal is present, the amount of active attenuation is reduced in order to avoid saturation.

Stability at higher frequencies was ensured by careful design of the low-pass filter and acoustic plant. The filter was adjusted so that the device will be stable with tight or loose fit. The goal in designing the acoustic plant was to avoid sharp resonance peaks in the transfer function. Any cavities, which could store acoustic energy were avoided.

The developed prototype was installed inside a high-quality passive hearing protector. The secondary source was installed in a small enclosure, and the error sensor was placed near the secondary source. The complete secondary source and error sensor –assembly was installed so that it is located near the ear. The assembly described above was installed in both earcups of the hearing protector. A 3.6 V lithium-ion –battery was installed inside one earcup and the ANC controller was installed inside the other earcup.

IN-EAR MEASUREMENTS

The performance of the implemented prototype was measured using in-ear method. The sound pressure level at the entrance of the ear canal was measured without hearing protection, with passive hearing protection and with active hearing protection.

A Sennheiser KE 4-211-1 electret microphone capsule was placed at the entrance of the ear canal. The microphone was connected to a self-made battery-powered microphone pre-amplifier. The amplified signal was recorded by a Sony DCD-D8 portable DAT recorder and analysed using a Brüel & Kjaer 2260 Investigator sound level meter. Fast time constant and A-weighting were used. The measurements were made in a room equipped with sound-absorptive material covering the walls and ceiling. Two different kinds of noise stimuli were used. The first stimulus was a 30-second sample of pink noise generated by a self-made digital pink noise generator. The other stimulus was a 30-second sound recorded inside a moving tank, and it was played by a Sony ZA5ES DAT player. The signal sources were connected to an active speaker system via an AMC Stereo Preamplifier 1100. The active speaker system consists of two Genelec 1030A two-way monitors and one Genelec 1092A subwoofer unit.

The first stimulus was pink noise. The sound pressure levels without hearing protection (solid line), with passive hearing protection (dotted line) and with active hearing protection (dash-dot –line) are presented in Figure 2a. A sound sample of a moving tank served as the second stimulus. The sound pressure levels without hearing protection (solid line), with passive hearing protection (dotted line) and with active hearing protection (dash-dot –line) are shown in Figure 2b.

The dynamic range of the device was also tested. Pink noise was used as a stimulus and the sound pressure level was increased gradually from 80 dB SPL to 100 dB SPL. It was clear that the effect of

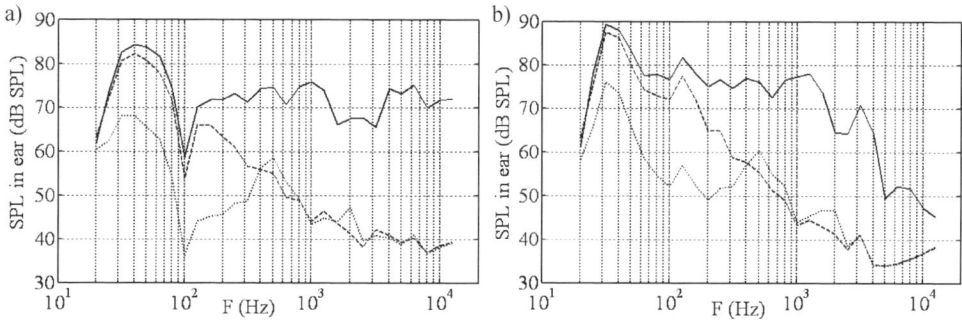

Fig. 2. Third octave band levels of the noise with pink noise (a) and tank noise (b).

the active attenuation gradually disappeared when the AGC circuit reduced the gain of the controller. Audible distortion was not detected until the sound pressure level reached 97 dB SPL inside the earcup. However, with loose fit between the hearing protector and the head, distortion could be heard at lower sound pressure levels.

The in-ear measurement results show that the developed device is able to actively attenuate low frequency noise up to a maximum of 20 dB. This is a significant improvement in the low frequency performance of a hearing protector. The measured active attenuation is almost same for both stimuli. The automatic gain control circuit reduces the active attenuation when needed, which makes the device more usable in high noise level environments. The drawback of the gain reduction is that active attenuation performance is reduced at the same time. The device also improves comfort and speech intelligibility because it reduces significantly the low frequency boom, which is typical of passive hearing protectors due to their poor low frequency attenuation. Because the prototype improves low frequency noise attenuation, it reduces the risk of hearing loss and thus improves safety. Although the AGC circuit reduces distortion and extends the dynamic range of the device, further research is still needed for very high SPL environments.

CONCLUSIONS

One problem associated with active hearing protectors is that a loud low frequency sound can saturate the system, which is heard as distortion. A prototype of an active noise cancellation hearing protector had been developed earlier, and now special attention was paid to improving the comfort and stability of the device. As a solution, an automatic gain control circuit was incorporated, and both acoustical and electrical designs were improved in order to ensure stability. In-ear measurements were made. The measurement results show that the developed prototype significantly improves the low frequency attenuation of a passive hearing protector. The listening tests demonstrated that the AGC circuit makes the device more comfortable to use. Further, there was no sign of instability.

ACKNOWLEDGEMENTS

This work was supported by Oy Silenta Electronics Ltd, a Finnish hearing protector manufacturer and TEKES, The National technology agency of Finland.

REFERENCES

1. Rafaely B. (1997). Feedback Control of Sound. *Ph. D. Thesis, University of Southampton, UK,*

2. Kuo S.M., Wu H. Chen F., and Gunnala M.R. (2004). Saturation Effects in Active Noise Control Systems. *IEEE Transactions on Circuits and Systems-I: Regular Papers* **51:6**, 1163 – 1171.

3. Oinonen M.K., Raittinen H.J., and Kivikoski M.A. (2004). An Automatic Gain Control for an Active Noise Cancellation Hearing Protector. *Active 2004 – The 2004 International Symposium on Active Control of Sound and Vibration*, Williamsburg, VA USA.

SUPPRESSING MECHANICAL VIBRATIONS IN A PMLSM USING FEEDFORWARD COMPENSATION AND STATE ESTIMATES

M. J. Hirvonen and H. Handroos

Institute of Mechatronics and Virtual Engineering, Department of
Mechanical Engineering, Lappeenranta University of Technology
P.O.Box 20, FIN-53851 Lappeenranta, FINLAND

ABSTRACT

The load control method for suppressing mechanical vibrations in a Permanent Magnet Linear Synchronous Motor (PMLSM) application is postulated in this study. The control method is based on the load acceleration feedback, which is estimated from the velocity signal of a linear motor using the Kalman Filter. The linear motor itself is controlled by a conventional PI-velocity controller, and the vibration of the mass is suppressed from an outer control loop using feed forward acceleration compensation. The proposed method is robust in all conditions, and is suitable for contact less applications e.g. laser cutters. The algorithm is first designed in the simulation program, and then implemented in the physical linear motor using a DSP application. The results of the responses are presented.

KEYWORDS

Acceleration Compensation, Kalman Filter, Linear Motor, Velocity Control, Vibration Suppression

INTRODUCTION

Nowadays fast dynamic servomotors are becoming quite common in several machine automation areas. This sets new demands on mechanisms connected to motors, because it can easily lead to vibration problems due to fast dynamics. On the other hand the non-linear effects caused by motor and machine mechanism frequently reduce servo stability, which diminishes the controller's ability to predict and maintain speed. As a result, the examination of vibrations that are formed in a motor as well as of the mechanism's natural frequencies, has become important.

The traditional approach to the dynamic analysis of mechanisms and machines is based on the assumption that systems are composed of rigid bodies. However, when a mechanism operates in high-speed conditions, the rigid-body assumption is no longer valid and the load should be considered flexible. The flexibility of a mechanism causes a disturbing velocity difference between reference- and load velocity, especially in the fast transient state.

Conventionally the motor control is assumed to be a velocity controller of a motor. In that case the vibrations of the tool mechanism, reel, gripper or any apparatus connected to the motor are not taken into account. This might reduce the capability of the machine system to carry out its assignment and impair the lifetime of the equipment. Nonetheless, it is usually more important to know how the load of the motor behaves.

There are two complementary methods to improve the dynamic behaviour of the machine system. The first is to make the mechanism more rigid, but this method usually makes the response slower. The second is to take the dynamic behaviour of the mechanism into account in the control strategy. The latter method is of interest to us. Motion control technologies have been widely used in industrial applications. Due to the fact that good technologies allow for high productivity and products of high quality, the study of motion control is a significant topic.

The aim of the proposed controller is to drive the load to a reference in such a way that the load follows the desired value as rapidly and as accurately as possible, but without awkward vibration. One of the most traditional methods to suppress resonance in the electromechanical system is to allow only small and slow changes in the reference command. For example different kinds of filters are used in a reference signal to suppress mechanical vibrations. *Dumetz et al.* (2001) have studied bi-quad and low pass filters in a control loop but also as a reference filter. The closed loop filter makes possible to compensate poles and zeros of the transfer function from the motor side, and the reference filter compensates poles of the transfer function in the load side. Another widely used filter for vibration suppression is the Notch filter (Ellis et al., 2000). The drawback of the filtering is the low sensitivity to parameter variations and also this method reduces the dynamical properties of a servo system.

A more promising method is to use acceleration compensation to suppress load vibration. In this method the motor is controlled by a simple PI –controller and load acceleration can be measured or estimated and used as a compensation feedback. *Kang et al.* (2000) and *Lee et al.* (1999) have used this kind of a method successfully in the vibration control of elevators. The weakness of using acceleration feedback is that the signal is usually very noisy. If the system is observable, it is possible to estimate the state variables that are not directly accessible to measurement using the measurement data from the state variables that are accessible. By using these state-variable estimates rather than their measured values one can usually achieve an acceptable performance. State-variable estimates may in some circumstances even be preferable to direct measurements, because the errors of the instruments that provide these measurements may be larger than the errors in estimating these variables.

CONTROLLER DESIGN

In control system design, the mechanism can usually be simplified for a 2-dof system, when only the first fundamental natural mode is taken into account. The two-mass-spring model of the linear motor system is introduced in Figure 1.

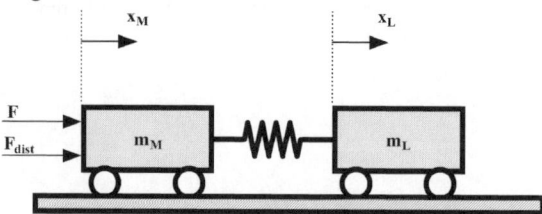

Figure 1: Two-mass-spring model of the PMLSM.

The transfer function from the controller force F to the load velocity $\dot{x}_2$ is the following:

$$\frac{\dot{x}_2}{F} = \frac{bs + k}{m_1 m_2 s^3 + b(m_1 + m_2)s^2 + k(m_1 + m_2)s} \quad (1)$$

where b is the damping constant, k is the spring constant, and m_1 and m_2 are motor and load mass, respectively. In theory, the conventional linear controller (PI/PID) can suppress the vibration of the load in the linear system. There are small gain margins in the root locus where the system is stable. However, when controlling the load by a simple PI controller, the velocity becomes unstable very quickly when the gains are increased. The physical linear motor application is also highly non-linear, and therefore conventional controllers fail in the suppression.

Due to the instability problems it is therefore necessary to have other control strategies than those based on a PI corrector. In the proposed controller the load acceleration compensator is added to a conventional velocity PI controller in order to reduce mechanical vibration, which can be assumed to be a disturbance force added to a flexible load. The advantage of the proposed method is that it suppresses vibrations without degrading the overall velocity control performance. In Figure 2, there is the structure of the proposed controller. K_m and K_a in the figure are the motor constant and the compensation gain, respectively.

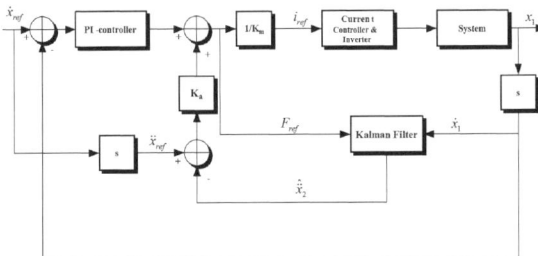

Figure 2: Control system diagram.

The force reference of the controller is the following

$$F_{ref} = \left(v_{ref} - v_1\right)K_p + \int\left(v_{ref} - v_1\right)dt \cdot K_i + \left(\frac{dv_{ref}}{dt} - \hat{a}_2\right)K_a \quad (2)$$

where v_1 is the motor velocity, $\hat{a}_2$ is the load acceleration estimation, K_p and K_i are the proportional- and integral gains of the velocity controller and K_a is the compensation gain. The values are introduced in Table 1 in the appendix.

The classical control system theory assumes that all state variables are available for feedback. In practice, however, not all state variables are available for feedback. Therefore, we need to estimate the unavailable state variables. There are several methods to estimate unmeasurable state variables without a differentiation process. The acceleration of the load in the controller is estimated using the Kalman filter (Kalman, 1960). The use of the estimated acceleration is based on the fact that the estimated acceleration is preferable (delayless and noiseless) to the measured and filtered signal. The Kalman filter is an optimum observer, meaning that the observer gain, here called the Kalman gain, is optimally chosen, whereas with a linear observer the gains are positioned arbitrarily.

EXPERIMENTAL RESULTS

For the Kalman filter a linear state-space model of the mechanical system is derived. The friction and other nonlinearities are assumed to be system noise, which the Kalman filter handles as a random process. The estimated states of the system are velocity of the motor v_1, velocity of the load v_2, and spring force F_s, i.e. the state vector is:

$$\mathbf{x} = \begin{bmatrix} x_1 \\ x_2 \\ x_3 \end{bmatrix} = \begin{bmatrix} v_1 \\ v_2 \\ F_s \end{bmatrix}. \qquad (3)$$

The state matrix **A**, input matrix **B** and output matrix **C** are described as:

$$\mathbf{A} = \begin{bmatrix} -\dfrac{b}{m_1} & \dfrac{b}{m_1} & -\dfrac{1}{m_1} \\ \dfrac{b}{m_2} & -\dfrac{b}{m_2} & \dfrac{1}{m_2} \\ k & -k & 0 \end{bmatrix}, \mathbf{B} = \begin{bmatrix} \dfrac{1}{m_1} \\ 0 \\ 0 \end{bmatrix}, \qquad (4)$$

$$\mathbf{C} = \begin{bmatrix} 1 & 0 & 0 \end{bmatrix}$$

where b is the damping constant, and k is the spring constant. The control input u is in this application motor thrust F_e. These matrices are discretized for the real-time Kalman filter. The process noise covariance **Q** in this application is:

$$\mathbf{Q} = \begin{bmatrix} 100 & 0 & 0 \\ 0 & 10 & 0 \\ 0 & 0 & 1 \end{bmatrix} \qquad (5)$$

and the measurement covariance is scalar due to one input for the Kalman filter, and it is $R=0.01$. The acceleration estimation $\hat{\ddot{x}}_2$ used in the compensation loop is measured from the estimated spring force $\hat{F}_s$ by dividing it by load mass m_2, i.e. acceleration estimation is:

$$\hat{\ddot{x}}_2 = \frac{\hat{F}_s}{m_2}. \qquad (6)$$

The derived acceleration compensation is first tested and implemented in the control of the simulation model, which is introduced in (Hirvonen et al., 2004). The whole simulation environment was carried out in Simulink due to simple mechanics. In the modelling of a linear motor, a space vector theory is used, and main non-linearities are taken into consideration. After testing the control in the simulation model, it was implemented in the physical linear motor application.

The motor studied in this paper is a commercial three-phase linear synchronous motor application with a rated force of 675 N. The moving part (the mover) consists of a slotted armature, while the surface permanent magnets (the SPMs) are mounted along the whole length of the path (the stator). The permanent magnets are slightly skewed (1.7°) in relation to the normal. Skewing the PMs reduces the detent force (Gieras, 2001). The moving part is set up on an aluminum base with four recirculating roller bearing blocks on steel rails. The position of the linear motor was measured using an optical linear encoder with a resolution of approximately one micrometer.

A spring-mass mechanism was built on a tool base in order to act as a flexible tool (for example, a picker that increases the level of excitation). The mechanism consists of a moving mass, which can be altered in order to change the natural frequency of the mechanism and a break spring, which is connected to the moving mass on the guide. The mechanism's natural frequency was calculated at being 9.1 Hz for a mass of 4 kg.

The physical linear motor application was driven in such a way that the proposed velocity controller was implemented in Simulink to gain the desired force reference. The derived algorithm was transferred to C code for dSPACE's digital signal processor (DSP) to use in real-time. The force command, F^*, was fed into the drive of the linear motor using a DS1103 I/O card. The computational time step for the velocity controller was 1 ms, while the current controller cycle was 31.25 µs.

Figure 3 shows a comparison of the velocity responses in non-compensated and compensated systems. The light line is the velocity response, when a conventional PI – velocity control of the motor is used. The load of the system vibrates highly reducing the efficiency of the system. The thicker line is the velocity response of the load when the acceleration compensation is used. The velocity follows the reference signal accurately; even the system stiffness is relatively loose. The small ripple in the compensated response is due to a small inaccuracy of the acceleration estimation. Also PI –velocity control affects the ripple for the system response because it is unable to compensate for all non-idealities in the motor.

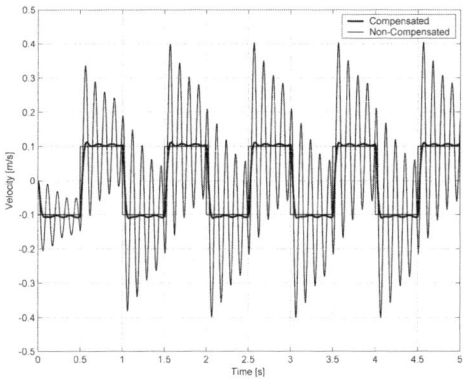

Figure 3: The comparison of the non-compensated and compensated velocities.

CONCLUSIONS

In the study, a load control method for a PMLSM is introduced and successfully implemented in the physical linear motor application. The motor is controlled by the conventional PI –controller, while the acceleration of the load is compensated from the outer control loop. The acceleration of the load for a compensation feedback is estimated using the Kalman Filter. The vibration of the load is considerably reduced and the proposed controller perceived to be stable in all conditions.

APPENDIX

TABLE 1
System Parameters

Parameter	Value
Motor Mass [m_1]	20kg
Load Mass [m_2]	4kg
Proportional Gain [K_p]	10000
Integral Gain [K_i]	0.1
Compensation Gain [K_A]	220
Spring Constant [k]	13700N/m
Damping [b]	6 Ns/m

References

Dumetz E., Vanden Hende F. and Barre P.J. (2001). Resonant load Control Method Application to High-Speed Machine tool with Linear Motor. *Conf. Rec. Emerging Technologies and Factory Automation* **2**, 23-31.

Ellis G. and Lorenz R. D. (2000). Resonant Load Control Methods for Industrial Servo Drives. *IEEE Industry Application Society Annual Meeting* **3**, 1438-1445.

Gieras J. F. and Piech Z. J. (2001). *Linear Synchronous Motors: Transportation and Automation Systems*, CRC Press, Boca Raton, USA.

Hirvonen M., Pyrhönen, O. and Handroos, H. (2004). Force Ripple Compensator for a Vector Controlled PMLSM. *In Conf. Rec. ICINCO 2004* **2**, 177-184.

Kalman R. E. (1960). A New Approach to Linear Filtering and Prediction Problems. *Transaction of the ASME – Journal of Basic Engineering*, 35-45.

Kang J.-K. and Sul S.-K. (2000). Vertical-Vibration Control of Elevator Using Estimated Car Acceleration Feedback Compensation. *Trans. on Industrial Electronics* **47:1**, 91-99.

Lee Y.-M., Kang J.-K. and Sul S.-K. (1999). Acceleration Feedback Control Strategy for Improving Riding Quality of Elevator System. *Conf. Rec. IAS* **2**, 1375-1379.

CHARACTERIZATION, MODELING AND SIMULATION OF MAGNETORHEOLOGICAL DAMPER BEHAVIOR UNDER TRIANGULAR EXCITATION

Jorge A. Cortés-Ramírez.[1], Leopoldo S. Villarreal-González [1] and Manuel Martínez-Martínez.[2]

[1] Centro de Innovación en Diseño y Tecnología, CIDyT, del Instituto Tecnológico y de Estudios Superiores de Monterrey, ITESM. Monterrey Campus. Monterrey 64849, Nuevo León, México. jcortes@itesm.mx
[2] Recinto Saltillo Aulas 1, ITESM Saltillo Campus. Saltillo, Coahuila, México.

ABSTRACT

Vibration control of vehicle suspensions systems has been a very active subject of research, since it can provide a very good performance for drivers and passengers. Recently, many researchers have investigated the application of magnetorheological (MR) fluids in the controllable dampers for semi-active suspensions. This paper shows that; the characterization of a damper can be made through of the physical characteristics of the MR fluids, current and damper design characteristics. A constitutive model can be determined by simple power equation in function of the electrical current. In addition it is shown that the use of ADAMS software is an excellent computational tool to simulate dynamic mechatronics systems. In other hand, a reconfigurable system is designed to be adjusted according to the circumstances and is able to respond by a position change or by itself just as the MR suspension do it.

KEYWORDS

Magnetorheological Fluids, Damper, Mechatronics, Vibration, Computer Simulation.

INTRODUCTION

Magnetorheological (MR) fluids belong to the general class of smart materials whose rheological properties can be modified by applying an electric field, [El Wahed Ali, K. (2002)]. MR fluids are mainly dispersion of particles made of a soft magnetic material in carrier oil. The most important advantage of these fluids over conventional mechanical interfaces is their ability to achieve a wide range of viscosity (several orders of magnitude) in a fraction of millisecond [Bossis, G. (2002)]. This provides an efficient way to control vibrations, and applications dealing with actuation, damping, robotics and mechatronics have been developed [Bossis, G. (2002), Yao, G.Z. (2002) and Nakamura, Taro (2004)]. In the other hand, by the use of dynamic simulations software is possible to analyze the behavior and performance of systems consisting of rigid or flexible parts undergoing large displacement motions [Ozdalyan, B. and Blundell M.V. (1998)].

PURPOSE

Vibration control of vehicle suspensions systems has been a very active subject of research, since it can provide a very good performance for drivers and passengers [Yao, G.Z. (2002)]. Recently, many researchers have investigated the application of magnetorheological (MR) fluids in the controllable dampers for semi-active suspensions. This work has the purpose of characterize, identify the mathematical model and simulate the behavior of a magnetorheological fluid in car suspension systems.

METHODOLOGY

To reach the purpose previously pointed out, firstly, the characterization is made by means of experimentation and by using a prototype damper. The displacement of the damper is measured by stages meanwhile known compression forces are applied under the influence of different magnetic fields. Subsequently, the constitutive model is developed throughout the mathematical identification of the relationships Force-Displacement, and Equivalent Damping Coefficient-Displacement. Polynomial expressions are derived in function of electrical current as independent variable and displacement, force and velocity as dependent variables. Finally, the simulation is carried out in two parts. Part one; uses a program in which the constitutive model is used in order to adjust the damper resistance based on the necessary current and according to different modes of behavior that can simulate several kinds of road. And part two; the damper resistance is read by the module ADAMSVIEW of MSC ADAMS software in which a suspension system has been modeled for describing the damper displacements at different virtual road conditions.

SYSTEM DESCRIPTION

The MR fluid used for this analysis, shown in Figure 1, is mainly a dispersion of iron powder 99.9%, as the soft magnetic material, in a carrier oil, and it was developed at ITESM, Campus Monterrey. The iron particles size distribution has a mean value of 15.53µm with standard deviation of 2.624µm. The particles are irregularly shaped and the mass fraction of the solid phase is 60%. The kind of oil used is commercial engine oil. The total period of precipitation exceeds 40 days, without movement. The viscosity of the MR fluid varies from 800 cP to 150,000 cP according to magnetic field applied. And, under the influence of a magnetic field the liquid phase separates from particles after more than 24 hours. The system used for the experiments is composed by the following components and presented in Figure 2.The damper is a prototype made of aluminum with 0.112 m of length, 0.014 m of diameter and 3.6 x10^{-6} m^3 of capacity. The common oil used inside the damper has been replaced with the magnetorheological fluid, which under no current presents a similar behavior as the original fluid.

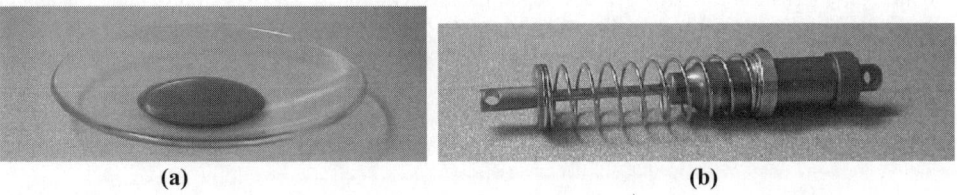

Figure 1: **(a)** Magnetorheological fluid and **(b)** prototype damper.

A coil has been designed to be capable of produce a magnetic field of 70.8kAm^{-1} at a current of 3A$_{DC}$, and it was designed to be located around the damper, identified as A, in Figure 2a. A special fasten extremity was designed to fix the upper damper part to the universal test machine, identified as B in Figure 2a. The universal test machine used for this work is the SHIMADZU AG-1 250KN, which allows force measurements accuracy of ± 1% of indicated test force. The module ADAMSVIEW of MSC Software is used to create a virtual prototype of a suspension system and to view key physical measures that emulate the data normally produced physically. An equivalent damper coefficient (EDC) concept has been used. If the piston rod is translated at a velocity $\dot{x}$, this will require that the fluid trapped on one side of the piston squeeze through the spaces between the piston and the cylinder. The fluid action opposes the motion with a magnitude given by Eqn. (1), where c is the equivalent damping coefficient. It is equivalent because the force exerted by the damper on the mass must not deviate from this expression no matter how fast or slow we move the mass [Cochin Ira and H.J. Plass. (1990)].

$$F = -c\dot{x} \tag{1}$$

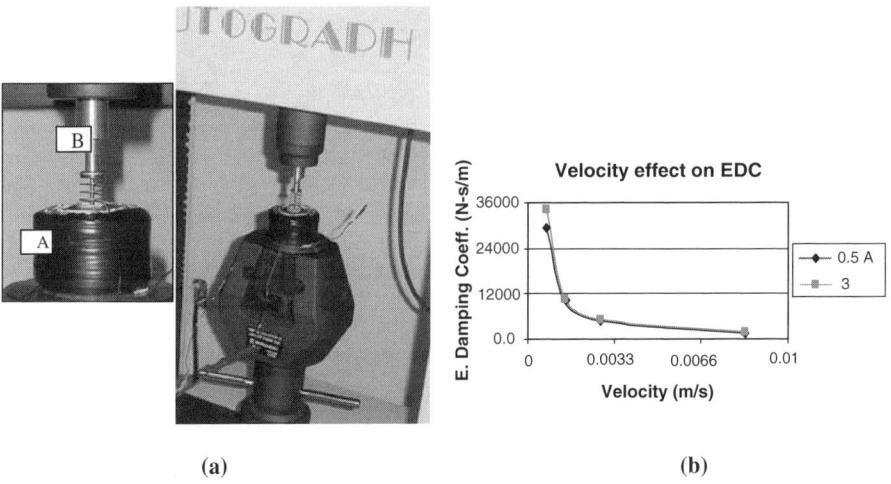

Figure 2: (a)Experimental set up. A; Coil and B; Fastener. And, (b) EDC behavior at different velocities.

RESULTS

Characterization of MR Damper

Experimental Work. The characterization of the magnetorheological damper has been done to obtain an expression, which represents its performance capabilities under different magnetic fields. Such expression lets establish the way in which a controllable damping system can be fully used. Firstly, it is necessary to get the set of data for the determination of force-displacement and EDC-displacement relationship. The damper is fixed on the branches of the universal test machine; meanwhile a coil is located around the damper body, as shown in Figure 2a. The test were done both under triangular excitation at a constant velocity of 0.0007 m/s and at different electric current intensities through the coil, that vary from 0.5 to 3 A. The velocity of 0.0007 m/s is selected because it represents low velocity, high equivalent damping coefficient in addition to have a clear influence of the electrical current, such as is shown in Figure 2b. A similar behavior has been found in reference [Yao, G.Z. (2002)]. The relationship obtained by experiments is shown in Figure 3.

Constitutive Model

Mathematical Identification.

The force-displacement relationship is obtained directly from the tests done, and the EDC-displacement relationship is obtained through the use of the EDC concept using the constant velocity of the tests and the force obtained from the following mathematical model. The constitutive model is obtained by mathematical identification of relationships force-displacements gotten from the test. Power equations, Eqn. (2), have been found in function of the displacement, δ, and electrical current, i. Figure 3 and Eqn. (3), shown results for an applied current of 3 Amperes.

$$f = a\delta^b \qquad (2)$$

Where f is the force required to overcome the resistance to compress the damper. And, δ is the displacement given by compression in the damper.

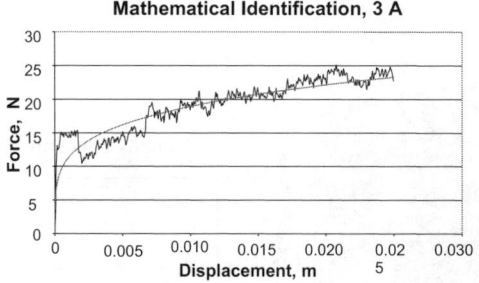

Figure 3: Mathematical identification of relationship force-displacement.

$$f = 11.263\delta^{0.2257} \qquad (3)$$

Once, all equations have been established, the constants a and b were plotted, as shown in Figure 4, to obtain general polynomial expressions, Eqn. (4) and Eqn. (5), in function of the current.

$$a = -0.0079\,i^2 + 1.0958\,i + 8.1546 \qquad (4)$$

$$b = 0.011\,i^2 - 0.0209\,i + 0.1869 \qquad (5)$$

Finally, a general power equation, Eqn. (6), constituted by two polynomial expressions has been obtained:

$$f = (-0.0079 i^2 + 1.0958 i + 8.1546)\delta^{0.011 i^2 - 0.0209 i + 0.1869} \qquad (6)$$

EDC is obtained and plotted, as Figure 5a shown, based on the constant velocity used in tests and the force obtained from equation (6) at 0.005, 0.01, 0.015, 0.02 and 0.025 m displacements. Similar to the previous analysis a general power equation, Eqn. (7), has been obtained:

$$EDC = (1.5321 i + 11.669)\delta^{0.011 i^2 - 0.0209 i + 0.1868} \qquad (7)$$

The connection between the mathematical model and the software can be given by introducing the equivalent damping coefficient expression in function of the displacement.

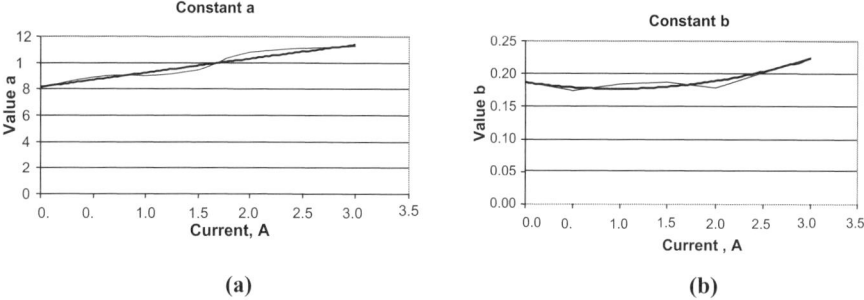

Figure 4: Analysis of (a) Constant **a** and (b) constant **b**.

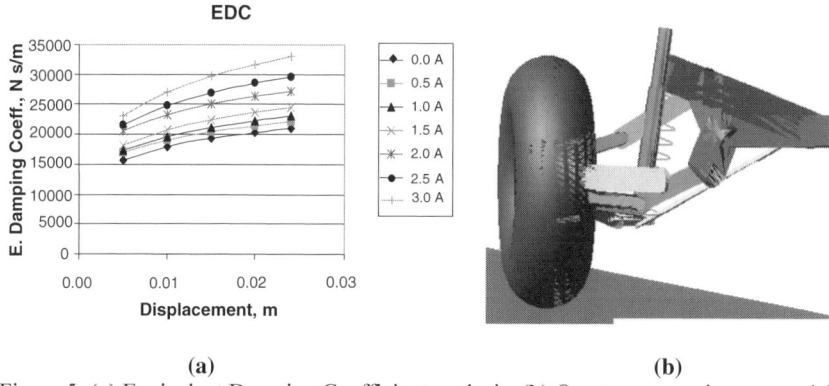

Figure 5: (a) Equivalent Damping Coefficient analysis. (b) Quarter suspension car model.

Simulation of MR Suspension System

The use of computational software has played an important role in design. Computational techniques are being used to complement, reinforce and specially to reduce time and money spent on experiments and practical applications. **Part one.** Adjustment of the damper resistance according to constitutive model. A quarter suspension car has been designed in ADAMSVIEW software, as shown in Figure 5b, based on a commercial car. The analysis of the suspension was done by simulating a collision between the car and an object at a velocity of 16.6 m/s. Once the design is completed, the damper coefficient value was modified by introducing a set of data points, which permits the software, based on an internal function, interpolate the discrete data. Such interpolation represents the EDC equation. **Part two.** Damper displacements at different virtual road conditions. According with the results obtained from the comparative analysis, a strong difference behavior between passive and semi-active suspension systems exist. The passive system shows a drastic change in the damper deformation and chassis displacement, meanwhile the semi-active system shows an adaptive behavior according with the respective damper displacement. When the MR damper is under a low magnetic field the suspension system presents a smoother reaction compared with that of the passive suspension and a higher magnetic field. According with the results obtained from the analysis, it has been demonstrated that the equation obtained for the ECD made possible an appropriate response of the suspension system based on the magnetic field induced. Once the behavior of the MR suspension system has been demonstrated, a control algorithm is necessary to be developed and implemented, so that, the system responds according to the road conditions and the comfort required by the human being.

CONCLUSIONS

A magnetorheological fluid has been specially developed and incorporated into a damper prototype also specially used for this purpose. A set up with a designed load cell was used independent and also was mounted in an Autograph Shimadzu system in order to determine the force, velocity and displacements at different forces. The constitutive model is given by a mathematical power expression constituted by two polynomial expressions, which are in function of the electrical current. The suspension system is taken from a real model actually in use for a commercial automobile characterized by its design and excellent performance. The simulated system shown the movements and quantify the forces and displacements. The results obtained from a comparative analysis shown strong differences between passive and semi-active suspension system. From the experiments and simulations done, it has been shown that; the characterization of a damper can be made through of the physical characteristics of the MR fluids, current, damper design and spring characteristics. In addition it has been shown that the use of ADAMS software is an excellent computational tool to simulate dynamic mechatronics systems. Finally a reconfigurable suspension system has been analyzed. Its ability to change its rheological properties in addition to its quickly response to the circumstances makes the MR technology a feasible way to develop other reconfigurable systems. Future work involves the introduction of a couple systems in the simulator in order to reproduce real events for driving, to determine the details of mechatronics control and to improve the coil's design for its implementation in a complete prototype. A control algorithm is necessary to be developed and implemented, so that, the system responds according to the road conditions and the comfort required by the human being.

NOMENCLATURE

A	Ampere	f	Force required to overcome damper resistance
a	Power equation constant		
b	Power equation exponential constant	i	Current through the coil
c	Equivalent damping coefficient	MR	Magnetorheological
cP	Centipoises	m	Meter
DC	Direct Current	s	Seconds
EDC	MR Equivalent Damping Coefficient	N	Newton
δ	Damper displacement or deformation	$\dot{x}$	Piston rod velocity
F	Force exerted by the damper		

REFERENCES

Bossis, G. (2002). Magnetorheological Fluid. *Journal of Magnetism and Magnetic Materials.* **252**. 224-228.

Cochin Ira and H.J. Plass. (1990). *Analysis and design of dynamic systems*, Harper Collins, New York, NY.

El Wahed Ali, K. (2002). Electrorheological an Magnetorheological Fluids in Blast Resist Design Applications. **Materials & Design, 23.** 391-404.

Nakamura, Taro. 2004. Variable Viscous Control Of A Homogeneous ER Fluid Device Considering Its Dynamic Characteristics *Mechatronics* **14.** 55-68.

Ozdalyan, B., Blundell M.V. (1998). Anti-Lock Braking System Simulation and Modeling in ADAMS. *International Conference on Simulation.* 140-144.

Yao, G.Z. (2002) MR Damper and its Application for Semi-Active Control of Vehicle Suspension System. *Mechatronics* **12**. 963-973.

SOFT-SENSOR BASED TREE DIAMETER MEASURING

Vesa Hölttä

Control Engineering Laboratory, Helsinki University of Technology
P. O. Box 5500, FI-02015 TKK, Finland

ABSTRACT

The forest harvester used for felling timber is a complex machine with a high degree of automation. To work properly, automatic functions need accurate measurements. In this paper tree diameter measurement is improved using different filtering and smoothing algorithms. The cases where smoothing is done during stem processing and after the stem has been processed are treated separately. Validation using manually measured data indicates that the methods that are presented improve the performance of the diameter measurement considerably.

KEYWORDS

Measurement, filtering, smoothing, Kalman filtering, Kalman smoothing

INTRODUCTION

Currently the vast majority of wood felled in Finland is felled with a forest harvester. In spite of the considerable amount of automation that helps the harvester operator in his work, felling timber can still be seen as handicraft. The operator must be a trained professional who is able to navigate the harvester in the woods without damaging the environment, and to choose the trees to cut so that future growth of the forest is guaranteed. Achieving these goals is compromised if the operator must concentrate on too many secondary tasks. Consequently, the need of operator interventions in less important tasks should be minimized. This is not possible unless the operator can rely on the automatic functions.

One field where a computer can outperform a human operator is bucking, i.e. selecting the points where the stem should be cut to logs. The price of a log is determined by its volume, but also by its grade (stock, paper-wood, etc.). The grade can often be changed by choosing the cutting points differently. Optimizing the bucking such that the value of the stem is maximized is a well-suited task for a computer. In order for the optimization to be successful, the length and diameter measurements that are fed to the optimization algorithm must be reliable. However, several factors can degrade the quality of the measurement signals during stem processing, thus resulting in non-optimal bucking. One criterion of the grade of a log is the top diameter. If the top diameter is measured incorrectly, the log may be classified to the wrong grade. Usually this means that a stock becomes a less valuable paper-wood log.

In addition to the optimization of bucking, accurate length and diameter measurements are needed also for computing the volume of the logs. The volume determines the price of the log, and consequently inaccurate measurements cause financial loss either to the seller or to the buyer of the wood. The problem is highly relevant, since measurements done by the harvester were used as the delivery measurement for 65 % of the timber harvested in Finland in 2002. In standing sales of privately owned forests the amount is even larger, 87 %. (Metsäteho, 2003) In other countries the volume of the wood is measured on the roadside before transportation, or at the saw mill, because the harvester measurement is not considered to be as reliable and objective as other methods. If using the harvester measurement became widely accepted, the cost of this second measurement could be saved.

In this work a soft-sensor approach is used to improve the accuracy of the stem diameter measurement. Using a soft-sensor means that instead of measuring a process variable directly with one physical sensor, measurements form several sensors and other knowledge of the process are incorporated using software to obtain an even more accurate measurement.

MECHANICAL TIMBER HARVESTING

There are different methods for mechanised timber harvesting differing by their philosophy and the machines needed. In North America the full-tree and tree-length methods are common whereas in Scandinavia the cut-to-length method is dominant. In the cut-to-length method the trees are felled, delimbed and bucked (i.e. cut to logs) with a forest harvester. The harvester is equipped with measuring devices that measure the length and diameter of the stem. Optimization algorithms choose the bucking such that the value of the stem is maximized. Once the stems are processed, a forwarder carries the logs to the roadside for further transportation. A cut-to-length forest harvester can be divided into four main parts: engine and power transmission, cabin and controls, crane and harvester head. The diesel engine is used for rotating the supply pumps of work hydraulics and hydrostatic transmission. The supply pump of work hydraulics delivers hydraulic power to the crane, to the harvester head and to all the auxiliary functions of the machine. The hydrostatic transmission consists of a variable displacement pump, of a variable displacement hydraulic motor and of mechanical transmission to the wheels. The cabin is equipped with the controls that are needed for operating the functions of the harvester and with a display module, which gives the operator information on the harvesting process and on the state of the harvester.

The most complex part of the harvester is the harvester head, which has a large-scale effect on the overall timber harvesting efficiency, and on the quality of the harvested timber. Its main functions are sawing, feeding, delimbing of branches, and measuring log length and diameter profile. Trees are felled and stems are cut to logs with a hydraulically actuated chain saw. Once sawing is complete, the stem is fed to a new cutting point with the hydraulic feeding rollers. To prevent the feeding rollers from slipping, the rollers are pressed hard against the stem with a hydraulic cylinder. In front and behind of the feeding rollers there are delimbing knives, which wrap around the stem. As the stem is fed to the next cutting point, branches are cut when they meet the delimbing knives. Delimbing knives also prevent the stem from falling out of harvester head grasp during the feeding operation. The diameter of the stem is measured using the delimbing knives. The setup is depicted on the left in Figure 1 where the delimbing knives can be seen holding the stem against the frame of the harvester head. Both delimbing knives are fitted with a potentiometer that gives a voltage that is proportional to the position of the knife. This measuring arrangement assumes that the stem stays in contact with the harvester head frame and that the delimbing knives touch the surface of the stem. If these assumptions do not hold, a measurement error will be introduced.

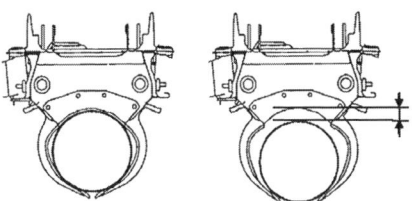

Figure 1: Correct stem position (left) and a "hanging" stem (right).

DIAMETER MEASUREMENT PROCESSING

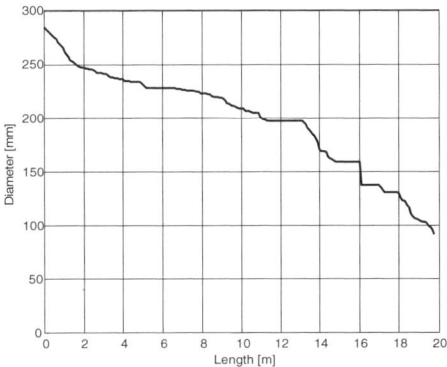

Figure 2: Typical diameter profile of a stem. Note the level sections and the drop in diameter.

A typical diameter profile of a stem is presented in Figure 2. The diameter in millimeters is on the vertical axis and the length in meters is on the horizontal axis, so that the bottom of the tree is on the left and the top is on the right. There are two characteristic unnatural features in the diameter profiles measured by a harvester. First, there are long level sections in the diameter profile, i.e. the diameter of the stem seems to remain constant. Second, there are abrupt drops in the profile. Both can be seen in Figure 2. Since the delimbing knives measure the diameter of the stem, they must follow the surface of the stem as closely as possible. Moreover, the stem should stay at all times against the frame of the harvester head. Loss of contact between the stem and the frame of the harvester head causes the distance between the stem and the frame to be added to the diameter. If the stem loses contact with the harvester head (on the right in Figure 1), the delimbing knives open, giving a diameter measurement that increases towards the top of the stem. The measuring system of the harvester requires diameters to be monotonically decreasing, and outputs a constant diameter value until the diameter decreases again. The result is a level section in the diameter profile. When feeding ends and the stem stops, the delimbing knives grasp the stem with maximum force, forcing the delimbing knives against the stem and the stem against the harvester head. This can be seen as the sharp narrowing in the diameter profile if the delimbing knives were open or if the stem was not against the harvester head. Thus the measurement following a sharp narrowing can be considered to be more accurate than the ones before the narrowing. Other sources of error with less significance are e.g. branches between the harvester head frame and the stem, hysteresis in the potentiometers that measure the position of the delimbing knives, incorrect calibration of the measuring device, and the functioning of the diameter measurement processing algorithm in some special cases.

Different aspects must be taken into account when designing algorithms for processing the diameter measurements to get more accurate results. Due to the large amount of disturbances that is present in the working environment of the harvester, any algorithm that attempts to correct the functioning of some system in the harvester must be robust. Currently the harvester measures the diameter online while the stem is processed without using future measurements to estimate the current diameter (filtering case). The estimate is used whenever a diameter measurement is needed, for instance to predict the tapering of the stem and to compute the volume of a log. In this paper three approaches for processing the measurements are discussed: online filtering, online smoothing and offline smoothing.

A dataset consisting of 479 logs was used to evaluate the performance of the methods. The diameter profile of each log in the dataset was measured manually. The measurements were taken with a measuring interval of 0.3 meters at an accuracy of 1 mm. The diameter was measured along two perpendicular lines, and the mean of the measurements was taken to be the final value. The manual measurements were regarded to be correct and were compared to the diameter profile measured by the harvester. The sum of square errors at each measurement point and at the cutting points divided with the number of the measurement points were used as metrics for the performance of the methods.

ONLINE FILTERING

First a simple linear approximation was used. The diameter measurements on the level sections of the stem profile are rejected and estimated by fitting a linear function in least squares sense to the valid measurements. Estimated diameters are obtained by evaluating the function at the measurement points.

A second approach was to use a Kalman filter. Kalman filters are estimators that are used for deducing the true value of a variable in a dynamical system. If the measurements given to a Kalman filter contain normally distributed uncorrelated noise, then the estimates are optimal with respect to all quadratic functions of the estimation error. (Grewal et al., 1993) The Kalman filter needs a model of the system to work. Tree tapering curves have been studied previously to some extent, for example polynomial models were used by Laasasenaho (1982) and in mixed linear regression models by Lappi (1986). These approaches were not used in this study because of the relatively high complexity of the models. It can be concluded from the abovementioned studies that the parameters of the models vary by geographic region and by tree species, so an adaptive model is needed. The current solution is to use a tapering matrix which is updated during harvesting, and this is the approach that was used also in this study. In this application a first order time-variant filter was used. The tapering matrix is a model of the change of the stem diameter between two consecutive measurements at each relative height. If the measured diameter stays constant, the magnitude of the noise covariance is increased to account for the increased uncertainty in the measurements. This results in reducing the weight that is given to the difference between the measured and estimated diameter values when the next diameter estimate is computed. The measurement noise in this application is neither normally distributed nor uncorrelated, which degrades the optimality of the estimates. However, a Kalman filter is still worth studying because of its ability to handle measurements with varying uncertainty. Kalman filtering was also applied such that the output of the filter is used at the level sections of the stem profile. When a level section is found, the tapering rate of the filtered profile is set to be the same as the current tapering rate obtained from the Kalman filter.

Finally, a backward approximation method was used. The method tries to fix previous measurements by recalculating them. Each time the algorithm detects a large drop in the diameter profile it checks if there is a level section before the drop. If this is the case, the algorithm connects the beginning of the level section with the measurement at the bottom of the drop.

The two first filters improve the accuracy of the diameter measurement for the whole stem as well as in the cutting points. Since the Kalman filter is based on a model of the stem profile, the results are dependant of the quality of the model. Even if the model was good for the majority of the trees on a stand, most likely some of the trees would have a very different profile due to environmental conditions, and their filtering result would be poor. Using Kalman filtering only at level sections of the stem profile and backward approximation are poor solutions since they follow the original measurement too closely, improving the diameter estimate only locally. Although the filtering methods give good results, because of their lack of robustness they should not be used if smoothing methods are applicable.

ONLINE SMOOTHING

The simplest solution is to smooth the diameter profile by computing the average of the measurements inside a window, i.e. a portion of the diameter profile containing an equal amount of measurements on each side of the current measurement. The window is centered at the measurement that is being estimated. A second approach is to fit a polynomial to the values inside the window. Like in the averaging method, some measurements are considered before and after the current measurement. Polynomial fitting may give hugely erroneous results if the number of fitting points is small compared with the order of the polynomial. This may occur if many fitting points are removed because of level sections in the diameter profile. The basic polynomial fitting method can be improved by assigning different weights to different measurements. A large weight is assigned to points that are situated after large drops in the diameter profile, which is justified by the characteristics of the measurement process. A Kalman filter can be also used to smooth measurements. The model for the tapering of the stem and changing of the noise were implemented like in the Kalman filter online filtering method.

Online smoothing methods are better and more robust than the online filtering methods discussed earlier, because in smoothing it is possible to use measurements before and after the point to be estimated. Thus smoothing should be used instead of filtering whenever possible. The simple average yields good results both when the whole stem and when the cutting points are considered. The result is most of all due to the fact that the averaged profiles are smooth and resemble the true stem profiles. Windowed weighted polynomial fitting is good in the sense that it gives profiles that pass near the bottoms of large drops in the diameter profile. Polynomial fitting suffers from the unpredictable behavior of polynomials when there are not enough fitting points. The algorithm discards measurements that are on level sections of the stem profile. This leads essentially to extrapolating the diameter when long level sections are encountered. In the methods where the amount of measurements to be taken into account can be changed, increasing the window size is advantageous. Increasing the number of measurements improves the robustness of the algorithms and makes it possible for them to take forthcoming changes in the diameter earlier into account. Like in filtering case, also in smoothing the quality of the results given by the Kalman filter depend on the quality of the stem tapering model. The better robustness of the smoothing approach when compared with the filtering approach may compensate some flaws of the model.

OFFLINE SMOOTHING

The first algorithm that was used was linear approximation of incorrect measurements. The method searches the first and last point of each level section. The first point is connected with a line to the point that follows the last point of the level section and the line is used as the diameter estimate instead of the original measurement. This method can be generalized by fitting a polynomial to all the measurements in the least squares sense. The measurements that are situated on the level sections of the profile are not taken into account since they can be considered erroneous.

The next method added to the previous one weighting of the points that follow large drops in the diameter profile. As stated before, the points following large drops are closer to the true diameter of the stem, because the delimbing knives have most likely been in contact with the stem at these points. In the algorithm the weighting factor is decreased in order to take into account the increasing uncertainty as measuring continues after the drop. If a level section is detected, the weight is returned to the normal level. This method can be changed by using weighting coefficients that depend of the size of the drop in the diameter profile. Large drops are given a larger weight in order to get the fitted polynomial closer to the measurements following large drops. Another adjustment to the weighted polynomial fitting method is to change weighting according to the smoothness of the stem. This is motivated by the experimental result that weighted polynomial fitting is good for stems containing many drops in the diameter profile, whereas polynomial fitting is more suited for smooth stem profiles. The two improvements to the ordinary weighted polynomial fitting can be used also together so that the overall weighting is determined by the smoothness of the stem and local weighting by the size of the drops in the diameter profile.

The most significant difference is between the methods based on linear approximation and on polynomial fitting. The former works locally and improves diameter accuracy only little or not at all. The polynomial fitting methods that can utilize better the global knowledge of the diameter measurements have a notable effect on the accuracy of diameter measuring. Offline smoothing could be advantageous particularly considering volume determination. After the logs have been bucked, their volume is saved to the harvester database. If the stem profile was smoothed offline before the volume is computed, accuracy could be better due to a more accurate diameter measurement. The methods that are based on polynomial fitting or weighted polynomial fitting produce very smooth stem profiles. Smoothness could be an advantage if the profiles are used to update a stem narrowing matrix or some other prediction tool.

CONCLUSIONS

Several methods were presented to improve the quality of stem diameter measurements. More reliable automatic operations in forest harvesters help the operator to concentrate more on planning and other tasks that cannot be done automatically. In addition to this, more accurate diameter measuring has several advantages, like maximizing the value of the timber, making the work more efficient, and eliminating the need of measuring the timber twice. Diameter measurement error along the whole stem and in the cutting points can be reduced by using the measurement processing algorithms presented. When the sum of square errors in all measurement points is considered, the improvement is with online filtering up to 10.2 %, with online smoothing up to 15.3 %, and with offline smoothing up to 18.9 %. When the sum of square errors in the cutting points is considered, the improvement is with online filtering up to 9.8 %, with online smoothing up to 10.7 %, and with offline smoothing up to 14.1 %.

The results obtained in this paper should be applicable to all harvesters that have a measurement system similar to the one used in this study. However, all manufacturers use their own algorithms and implementations in their machines, which may cause the methods presented in this paper to perform differently. The compatibility of the algorithms must be verified case-by-case. Many of the methods presented in this paper decrease the error in the diameter measurement. However, when considering the value of the results, it must be taken into account that the dataset used to validate and compare the methods is relatively small, which decreases the reliability of the results. The methods contain also parameters that have been adjusted according to the properties of this dataset. Further studies will be needed to show if these parameters need to be changed for different felling sites.

REFERENCES

Grewal, Mohinder S. and Angus P. Andrews (1993). *Kalman Filtering.* Prentice-Hall, Inc., New Jersey.
Laasasenaho, Jouko (1982). *Taper curve and volume functions for pine, spruce and birch.* Finnish Forest Research Institute, Helsinki.
Lappi, Juha (1986). *Mixed linear models for analyzing and predicting stem form variation of scots pine.* The Finnish Forest Research Institute, Helsinki.
Metsäteho (2003). *Metsäteho website.* URL: http://www.metsateho.fi/.

STUDY ON-MACHINE WORK PIECE MEASUREMENT ON 5-AXIS CONTROLLED MACHINING CENTER

Shunsuke Nakamura[1]   Yukitoshi Ihara[2]

[1]Major in Mechanical Engineering, Osaka Institute of Technology Graduate School,
5-16-1 Omiya, Asahi-ku 535-8585 Osaka, JAPAN
[2]Department of Mechanical Engineering, Osaka Institute of Technology,
5-16-1 Omiya, Asahi-ku 535-8585 Osaka, JAPAN

ABSTRACT

The study presents an application of machined work piece measurement system with the laser displacement sensor and C_S axis control on five-axis controlled machining center. Generally, post-process measurement of multi-face machined product is carried on a coordinate measuring machine (**CMM**), however, this way cannot achieve either high productivity because of loading and unloading of work piece nor low cost with expensive CMM. To solve this problem, On-machine work piece measurement with the laser displacement sensor is proposed in this paper. The main objective of this research is to establish work piece measurement system on five-axis controlled machining center and to develop measurement software for On-machine work piece measurement, which collaborate commercial CAD/CAD software.

KEY WORDS

On-machine measurement, five-axis controlled machining center, work piece measurement, mold machining

INTRODUCTION

Nowadays the style design becomes increasingly important for consumer electronics and consumer goods industries. Thus die/mold shape becomes too complicate to be machined for conventional 3-axis machining center with high speed and high accuracy. In addition, even for industrial parts such as automotive parts or aircraft parts, not only complex shape which 5-face machining is needed but also high dimensional accuracy for expanded function and capacity are required. On this ground, demand of 5-axis machining center increases rapidly, which enables to machine complex shape parts by obtaining cutting tools' multiple degree of freedom. Generally, one axis measurement by using caliper or micrometer is not suitable for products that are machined by 5-axis machining center because it is too complicate to be measured all form and dimensions. Thus, in real manufacturing process, machined products are loaded to expensive measurement-only machine such as CMM. What is more, a large amount of time is needed for complex shape measurement even by measurement-only machine. In this reason, total productivity is not so high although 5-axis machining center is introduced for efficient machining. Thus, manufacturer who use 5-axis machining center is eager to get quicker and higher productive manufacturing system, which enables maximum capacity of 5-axis machining center's productivity.

Post measurement systems using some sensor and C_S control on machine tools have been developed for confirming shape and dimension. In these systems, measurements are carried on the machine tool's table, so work piece loading for the measurement is not required. There are two proposed measuring sensors for On-machine work piece measurement. One is a contact triggering touch probe with a simple mechanism called 3-D touch probe. The other is non-contact laser displacement sensor with sophisticated electronic optical device. 3-D touch probe is commonly used in On-machine measurement, and On-machine measurement with correcting process is already proposed [1]. However, the proposed system is only used for dimensional measurement and not suitable for free form shape measurement such as die/mold shape. On the other hand, laser displacement sensor has a possibility for a free form shape because it has an ability that acquire large amount of points at high speed. Nakagawa showed the advantage of laser displacement sensor on free curved surface measurement [2]. In this report, On-machine measurement of a mold which had free curved surfaces was carried on the 3-axis controlled machining center using laser displacement sensor. When normal vector of measurement surface is inclined to laser, error arises. Moreover, the angle between laser and normal vector is larger, the measurement is impossible. In this case, the measurement can be done by giving two additional degrees of freedom to the laser displacement sensor. The additional degree of freedom is realized by adding rotary axis of 5-axis machining center. On the 5-axis machining center, laser always radiate perpendicularly to the measurement surface.

In this research, we try to use orientation control of the spindle in order that a laser sensor can always direct perpendicularly to the measurement surface.

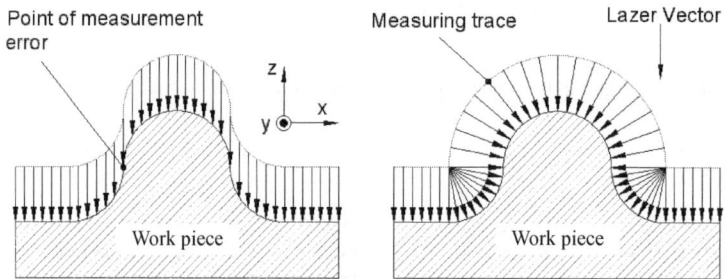

(a) Conventional method under 3Axis control (b) New method under 5Axis control

Figure 1: Measuring by conventional method and proposed method

CONCEPT OF MACHINED MEASUREMENT ON 5-AXIS CONTROL

With laser displacement sensor on 3-axis controlled machine tool, laser vector is traced as shown in Figure 1 during measuring operations. Conventionally, On-machine work piece measurement is carried by tracing surface in simultaneous 2-axis control on 3-axis controlled machining center. The principle of the laser displacement sensor using this research is shown in Figure 2. Triangulation method is adopted for the laser displacement sensor. As compared with other measuring devices, measurement resolution of dimension is good, and measurement can apply even if measuring surface is glossy metal. Although the sensor has the excellent feature for shape measurement, measurement error arises if laser vector is inclined against measurement surface. And it is impossible to measure surface if laser vector is almost parallel against measurement surface. Normally, die and mold has a steep inclination which causes measurement error. This is a critical problem for precision shape measurement which has 3-dimentional free form such as die/mold. So, it is necessary to direct laser to the normal direction of the surface as correctly as possible for executing surface measurement with high accuracy. Then, as mentioned above, we try to use additional two rotary axes of 5-axis machining center for surface measurement. Laser vector is traced by using orientation control of the spindle on five axis machining center as shown in Figure 1(b). With this method, not only measurement error is reduced but also all surface measurement in particular die of convex is enabled. For these reason, On-machine work piece measurement on 5-axis controlled machining center has an advantage for work piece measurement after machining.

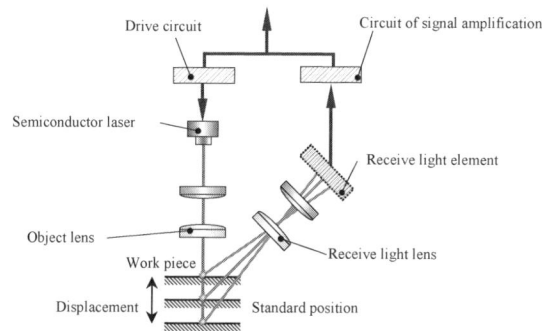

Figure 2: Theoretical mechanism of laser displacement sensor

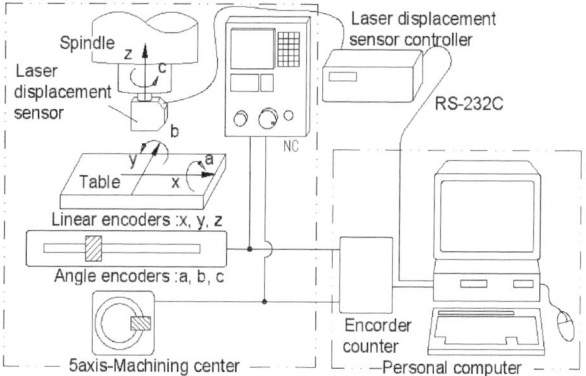

Figure 3: On-machine measurement system

ON-MACHINE WORK PIECE MEASUREMENT SYSTEM

The machined work piece measurement system consists of 5-axis controlled machining center, commercial CAD/CAM software, a personal computer, a laser displacement sensor (includes controller), and encoder counter for PC as shown in Figure 3. Signal lines of linear scales and angle encoders which are equipped in the machine tool for position feedback control are divided and connected to the encoder counter boards which are installed in the personal computer, in order to acquire the exact position information of 5-axis machining center. Digital data can be outputted and inputted by Ethernet system between PC and 5-axis machining center. RS-232C is used to obtain measured data by the laser displacement sensor between the counter of the laser displacement sensor and PC. The 6-axis control machine tool is used in the study as a multi-axis machine tool made by *MORI SEIKI CO., LTD*. It provides machining capabilities beyond the standard 3-axis control machine because of its flexibility. Additionally, the machine has C_s axis which controls the rotational angle of the main spindle of the 5-axis control machine, which has two rotational axes, A and C (as B) axis as shown in Figure 4. The rotary motions around X axis as well as Y axis and the Z axis are designated as A, C (as B) and C_s respectively. The inclination and rotation of the work piece is executed by A and C axis respectively, while the C_s axis controls laser device direction for realizing high measurement accuracy.

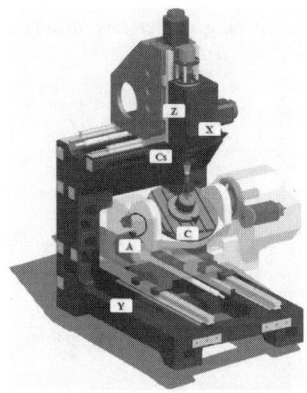

Figure 4: 5-axis machining center

MEASUREMENT SOFTWARE

In this system, commercial **CAD/CAM** system: **Pro/ENGINEER** is used for modeling as well as machining and measurement. Additionally we use versatile post processor setup application: **GPOST** which is collaborated with **Pro/NC**. The post processor setup application can change parameters and add a routine macro easily. Thus, we setup two post processors for simultaneous 5-axis control machining and measurement with laser displacement sensor. On CAM section (cutter location generation support system) generates CL data both for machining and for measurement. CL for machining has an inclination to normal vector of a free form surface for enhanced machining efficiency as shown in Figure 5. As opposed to it, CL for measurement may always suit normal of a free form surface as shown in Figure 5. CAM software can easily set up parameter of the inclination angle to the normal direction of curved surface. NC data for machining center is created by post processor. NC data is composed of the coordinate value of cutting point and angle expression of the tool orientation. Post processor for the measurement generates NC data for the measurement. NC data for the measurement has some special routine macro (called '*switch motion*') in order to arrange measurement data with a personal computer easily. Carrying out pick feed, measurement is performed so that the whole surface may be traced. Therefore, plunging operation which does not trace the measurement surface, retraction operation, and rapid traverse positioning operation follows as an additional operation. It is necessary to notify PC about these non-measurement times in order to hold down consumption of a useless memory. Then, operation which notifies the switch of measurement (: *switch motion*) was introduced. The *switch motion* is the specific motion for notifying software about the joint of measurement by changing coordinate value. The motion consists of Z-axis movement and dwell. Secondary, we have developed PC software for the measurement on 5-axis machining center because both coordinate values and laser data are required in measurement with high accuracy. Furthermore, it is necessary to convert the 3-dimensional form data of a measured model from the acquired coordinates which include angular position of rotary table and displacement from the laser sensor. Then we developed software with these function. The main functions of the software is as follows; inputting CL and NC data which were generated from commercial CAM for measurement, reading machine coordinates and laser displacement sensor value. And the software monitors start and end coordinates of switch motions, which are estimated based on the CL&NC information previously acquired. Moreover, it has the function to store the machine coordinates in the measuring point based on CL data to the specific file promptly.

EXPERIMENT

We executed experiment to verify the feasibility of On-machine work piece measurement with laser displacement sensor on 5-axis machining center in our research. Its major objectives were to:
☞ Confirm that the measurement system perform the intended task smoothly.
☞ Verify that the laser displacement sensor can measure glossy surface of die, which is finish-machined by ball end mill.

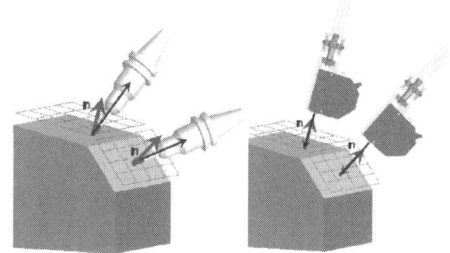

Figure 5: Machining and measuring

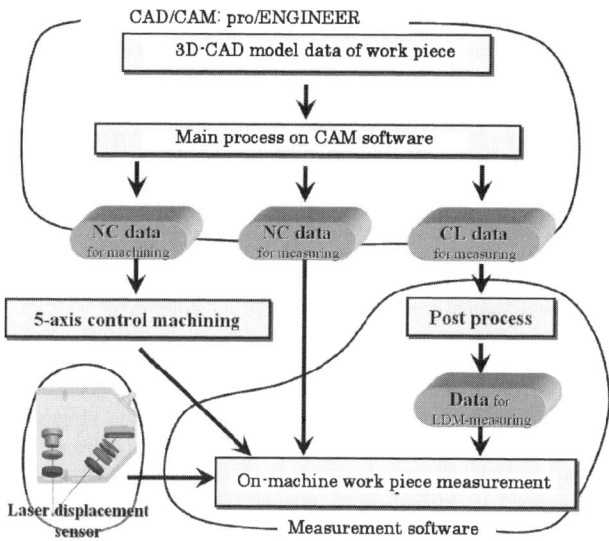

Figure 6: Procedure of on-machine work piece measurement

Simple convex shape was used as the object model. The work piece material of the object was aluminum alloy. Convex is formed into width of 5 mm and height of 5 mm. In the measurement process, we used only 3 linear axes for easy to calculate coordinate system. Measurement time is approximately 9 min with 70 mm/min feed rate. Measurement interval is 70 ms and amount of measurement point is 5000. The surface roughness after machining was 4 μm in Ra.

RESULT AND DISCUSSION

As a result of the experiment, measured shape is shown in Figure 7. The variation of the data obtained by the measurement was 8 μm. The formula of *Ra* (ISO 1302: 2002) was used as the evaluation method of variation. It has confirmed that smooth execution of the system is seen from measurement result. Additionally, it was shown that metal surface measurement is possible. The result shows the feasibility of measurement system with laser displacement sensor and machine's coordinates as a reference.

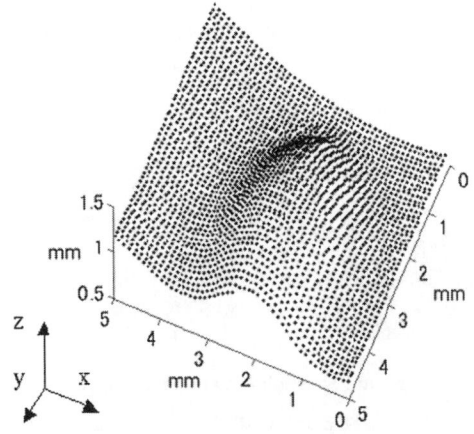

Figure 7: Shape of measured surface

CONCLUSIONS

In order to achieve On-machine measurement with laser displacement sensor on 5-axis machining center, a system for On-machine measurement based on using additional two rotary axes on a 5-axis machining center is proposed in this paper. The conclusions are as follows:
(1) We explained the concept of a measurement method with laser displacement sensor on a 5-axis machining center.
(2) We confirmed feasibility of developed system by validation experiment.

REFERENCES

[1] Nakagawa H., Hirogaki T., Kaji Y., Kita Y., Kakino Y. (2003). In-situ Suitable Controlled Scan of Laser Stylus for Point Measuring of Free Surface, Journal of the Japan Society for Precision Engineering, **69:10,** 1423-1428.
[2] Kakino Y., Ihara Y., Iwasaki Y., Matsubara A., Otsubo H.(1993). Study on Amendable Machining System by Using Machining and Measuring Center --Amendable Grinding of 2-dimensinal Parts with High Accuracy--, Journal of the Japan Society for Precision Engineering, **59:10,** 1689-1694.
[3] Japitana F. H., Morishige K., Takeuchi Y. (2004). 3-Dimensional Machining of Groove with Edges on Oblique or Curved Surfaces by means of 6-axis Control Ultrasonic Vibration Cutting, 2004 Japan-USA Symposium on Flexible Automation, JL004.
[4] Tanaka F., Yamada M., Kondo T., Kishinami T., Kohmura A. (2004). Software System for Sculptured Surface Machining Based on 3+2-axis High Speed Machining on a 5 axis Machining Center, 2004 Japan-USA Symposium on Flexible Automation, JL006.

A NEW METHODOLOGY TO EVALUATE ERROR SPACE IN CMM BY SEQUENTIAL TWO POINTS METHOD

W. M. He[1], H. Sato[2], K. Umeda[1], T. Sone[3], Y. Tani[2], M. Sagara[4] and H. Matsuzaki[4]

[1] Department of Precision Mechanics, Chuo University,
Tokyo 112-8551, Japan
[2] Institute of Industrial Science, University of Tokyo,
Tokyo 153-8505, Japan
[3] Makino R&D Center, Kanagawa 243-0308, Japan
[4] Toshiba Machine, Shizuoka 410-8510, Japan

ABSTRACT

The sequential two points method (STPM), which was developed to identify the straightness motion of machine tools and cut surface of machined parts, is applied to evaluate the straightness error motion of the probe in a Coordinate Measuring Machine (CMM). Repeating the measurement on the objective plane, a method to make a planelike error form is proposed by combining the measured straightness error motions. If the procedures are made for the respective coordinates, error space can be evaluated. This could be more easily done by any other conventional method. A compensation method is proposed, which is possible to improve measurement accuracy without changing the hardware configuration of the CMM.

KEYWORDS

Coordinate Measuring Machine, Sequential Two Points Method, Straightness, Flatness, Error Space, Error Compensation, Manufacturing System, Measurement

INTRODUCTION

A Coordinate Measuring Machine (CMM) is indispensable in a manufacturing system as an apparatus of accurate measurement. It has come to be used in production lines together with machine tools as well as in specific rooms for measurement. Much effort has been made to develop methods of calibration and compensation of a CMM (Zhang et al. (1985), Kunzmann & Wäldele (1988), Evans et al. (1996)) and to evaluate accuracy performance of machines (Kunzmann et al. (1995), Balsamo et al. (1997)). In spite of the effort, it has been difficult to obtain definite and reliable results. One reason would be that the ballplate which was used as the standard for the calibration was difficult to handle and to obtain sufficient accuracy.

The sequential two points method (STPM) was developed to evaluate the straight form error of machine tools. By this method it is possible to identify both error forms for the tool post and the machined part by a single feed traverse of the tool post in terms of simple algorithm for the data (Tozawa et al. (1982), Tanaka & Sato (1986)). The method has advantages in that any standard of specific straightness is not required, the algorithm is simple, repetiive measurement accuracy is high, and the result is stable even under vibration environment.

In this investigation, the following targets are set to evaluate the straightness error movement of the probe of the CMM, and consequently its error space, which makes it possible to compensate the error for the work space and to improve performance in accuracy:
1. Identify the straightness error motion of the probe in the CMM,
2. Identify the planelike error form which is composed of a bundle of the straightness error motions for the objective plate due to term 1,
3. Identify the 3D error space and develop a compensation method for 3 axes.

EXPERIMENTAL FACILITIES

Principle of STPM

In Figure 1 schematic view of the STPM is illustrated. When an engine lathe is supposed, two sensors which can measure the relative distance to whatever objective or machined part are located on the tool post with the distance L in-between. When the tool post is fed and the data are acquired stepwise at positions of L multiplied by integers, two data series for the individual sensors are obtained. The data are processed by the following equations,

$$X_k = X_{k-1} + D_{k,A} - D_{(k-1),B} \tag{1}$$

$$Y_k = X_k + D_{k,A} - D_{0,B} \tag{2}$$

$$X_0 = 0, X_1 = 0, Y_0 = 0 \tag{3}$$

where $D_{k,A}$ and $D_{k,B}$ are the measured data by the sensors A and B, X_k and Y_k are the straightness error motion for the tool post and form for the objective respectively. It is shown that error forms could be simultaneously and independently derived.

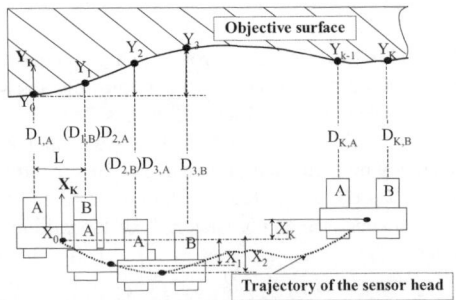

Figure 1: Principle of sequential two points method

Application of the STPM to the CMM is attempted to calibrate and compensate the movement accuracy of the probe in the following, by which machining error in the production system could be improved.

Measurement System

Figure 2 shows the whole view of the CMM measurement system. The machine is a RVA 600, which is a bridge move type and is made by Tokyo Seimitsu Co. Ltd. Specification of the machine is as follows: available space of the measurement; 650×500×300mm³ for X×Y×Z, resolution of the measurement; 1μm, straightness accuracy E_s whose unit is μm;

$$E_s = 4.9 + 4.5(L_t / 1000) \qquad (4)$$

where L_t is span of the length in which the probe traverses. A plate of Al with the size of 400×650×20 and machined by a conventional milling machine is adopted as the objective. Capacitance type of displacement sensor made by Ono Sokki Co. Ltd., is utilized, whose specification is 0-0.5 mm as measurement range, 0.1μm as resolution, 8 mm as the diameter of the sensor, 20 mm as the distance of the two sensors.

The control computer of the CMM is an IBM 6587-JC3, whose operation software is OS2, which is out of the mainstream, Windows, consequently making it difficult to process the acquired data by the STPM. Then a laptop computer, Gateway Solo 2500, is placed for the data acquisition through a Keyence NR-110 AD converter, which was done by a trigger signal sent out from the IBM machine synchronizing with the CNC command. In the Figure 2(b) this flow is depicted. The acquired data were processed in Excel on the laptop computer.

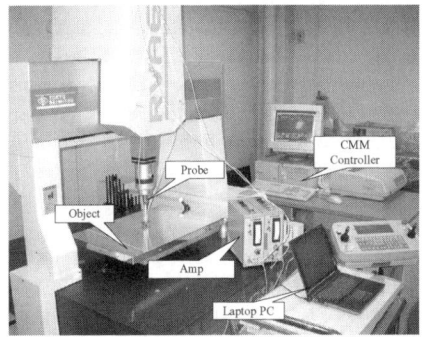
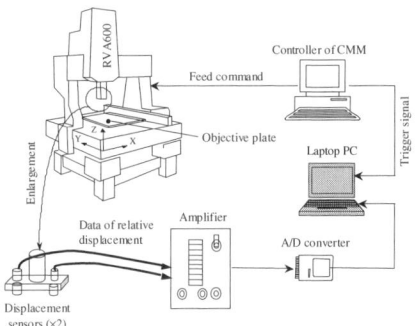

(a) Picture of CMM　　　　　　　　　(b) Conceptual schematic view
Figure 2: Whole view of CMM measurement system

EXAMPLE OF THE MEASUREMENT

Basic characteristics of the straightness error forms of the probe parallel to the X axis are depicted in Figure 3. The first error form was evaluated at Y=0 for a range of 380 mm with an interval of 20 mm and the same evaluations were repeated at the coordinates of the Y axis in the range of 320 mm with an interval of 20 mm, which means that 17 error forms of the probe movement were obtained. In Figure 3 the error forms of the average and the maximum and the minimum fluctuation are arranged. When the evaluation at the left side of the forms can be made zero, those at the right side are supposed to fluctuate to around zero. In the figure these are adjusted to zero for the comparison. The straightness error that can be evaluated by Eqn. 4 is, E_s=6.61μm for L_t=380mm.

If the same attempts were done by a system of the laser interferometer, arrangement for the measurement would be much more time consuming, and the data would easily fluctuate due to the environment. In case of the STPM the operation to evaluate the bundle of the straightness error forms is easily realized by the CNC command of the machine.

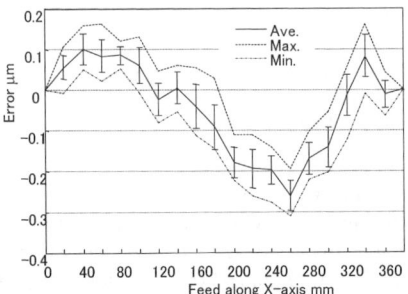

Figure 3: Straightness error motion of the probe along X axis

COMPOSITION OF PLANELIKE ERROR FORM

The error forms in Figure 3 are composed of a bundle of the straightness error forms which are measured parallel to the X axis and are repeated stepwise to the direction of the Y axis.

In coordinate system (X, Y, Z) among the work space of the CMM, a plane which is prescribed by (X_i, Y_j, Z_k) is considered, where i = 0, 1, $\cdots$ l, j = 0, 1, $\cdots$ m, k = 0, 1, $\cdots$ n. A bundle of the straightness error forms parallel to the X axis are measured on a plane of Z_0, that is, (X_i, Y_j, Z_0). In practical application of the STPM, slight discrepancy Δ is inevitably left between the tip of two sensors (Tanaka & Sato (1986)). It has been demonstrated that $i\Delta$ is linearly accumulated at i-th position of the sensors on the measured straightness error form. Then corrected error form can be obtained by subtracting the following accumulation from the raw measured results,

$$Z = i\Delta \quad (i = 1, 2, \cdots l) \tag{5}$$

The evaluation at the right-hand edge of the forms in Figure 3 shows the results with the subtraction. However, it is presumed that residue $\delta_Z(X_1, Y_j, Z_0)$ would remain for the forms obtained for Y_j (j = 1, 2 $\cdots$ m), even if this correction is conducted. At the starting and the finishing conditions at the both ends of the error forms, (X_0, Y_j, Z_0) and (X_1, Y_j, Z_0), the straightness error forms for the direction of the Y axis are measured, which give boundary at the both ends for the bundle of the forms, so that the planelike error form is constructed.

Figure 4 illustrates the flow of composing a planelike error form mentioned above. Z_x and Z_y indicate the straightness errors of the probe for Z axis that are measured by applying the STPM along the X and Y axes respectively.

In Figure 5(a) a planelike error form for the probe movement is constructed from the straightness error forms in Figure 4. The planelike error form for the objective Al plate can be simultaneously evaluated in the procedure, if necessary, as shown in Figure 5(b).

ERROR SPACE AND ITS COMPENSATION

By the method proposed above, a planelike error form for arbitrary Z_k can be constructed, and, it is possible to estimate the error space by constructing enough number of planelike error forms. Figure 6 illustrates an experimental example of detecting an error space for the Z axis. The same procedure can be applied to the directions X and Y which are vertical to Z, and planelike error forms for the X_i and Y_j are constructed. Consequently, error space for 3 axes (X, Y, Z) is constructed. Then it is possible to estimate error components for (X, Y, Z) at arbitrary points among the measurement range of a CMM.

The points on which errors are directly measured are discrete. The error at an arbitrary point can be obtained by applying interpolation or least squares method (LSM). When the error $e_z(X, Y, Z)$ is obtained by interpolation or LSM, the compensation of the error is easily performed by subtracting the e_z from Z. By applying the method, the measurement accuracy is improved without any change of hardware configuration.

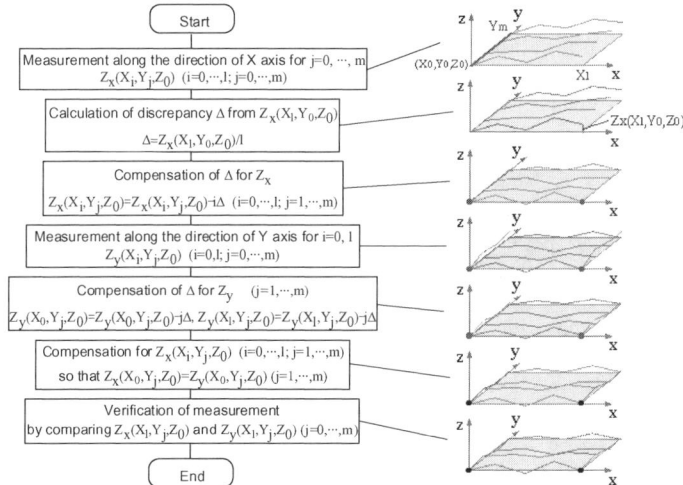

Figure 4: Flow of composing planelike error form for $Z=Z_0$

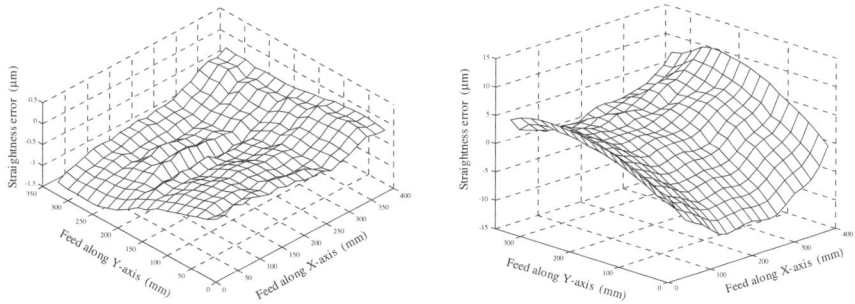

Figure 5 (a) (left): Error curved surface of the straightness motion of the probe
(b) (right): Surface profile of the objective plane

CONCLUSION

It was demonstrated that the sequential two points method could be well applied to evaluate the straightness error motion of the probe of the CMM. The conclusions could be summarized as follows.
1. It was confirmed that the straightness error form can be accurately identified by the STPM. The

measurement operation can be easily extended to planelike error form.
2. An algorithm which constructs planelike error form from a bundle of the straightness error forms was demonstrated, in which a method that gives conditions at the both ends of the error forms is proposed.
3. A procedure which derives error space for measurable range of the machine on the basis of planelike error forms is proposed. This makes it possible to compensate error among the total measurement range, which certainly improves the accuracy performance of the machine.

ACKNOWLEDGEMENTS

The authors express their sincere thanks for the following support by Project "Advancement of Measurement of Macro Error Form (97S21-005)" from 1997 to 1999 by New Energy Development Organization (NEDO), and Project "Advancement of Straightness Error Measurement by Sequential Two Points Method" by Foundation, Promotion of Machine Tool Engineering in 1996 and 1997.

REFERENCES

Balsamo A., Franke M., Trapet E., Wäldele F., Jonge L.D. and Vanherck, P. (1997). Results of the CIRP-Euromet Intercomparison of Ball Plate-Based Techniques for Determining CMM Parametric Errors, *Annals CIRP* **46:1,** 463-466.
Evans C., Hocken R. and Estler W.T. (1996). Self-Calibration: Reversal, Redundancy, Error Separation, and Absolute Testing, *Annals CIRP* **45:2,** 617-634.
Kunzmann H. and Wäldele F. (1988). Performance of CMMs, *Annals CIRP* **37:2,** 633-640.
Kunzmann H., Trapet E. and Wäldele F. (1995). Results of the International Comparison of Ball Plate Measurements in CIRP and WECC, *Annals CIRP* **44:1,** 479-482.
Tanaka H. and Sato H. (1986). Extensive Analysis and Development of Straightness Measurement by Sequential Two-Points Method, *Trans. ASME, J. Eng'g Ind.* **108:3,** 176-182.
Tozawa K., Sato H. and O-hori M. (1982). A New Method for the Measurement of the Straightness of Machine Tools and Machined Work, *Trans. ASME, J. Mach. Des.* **104:3,** 587-592.
Zhang G., Veale R., Charlton T., Borchardt B. and Hocken R. (1985). Error Compensation of Coordinate Measuring Machines, *Annals CIRP* **34:1,** 445-448.

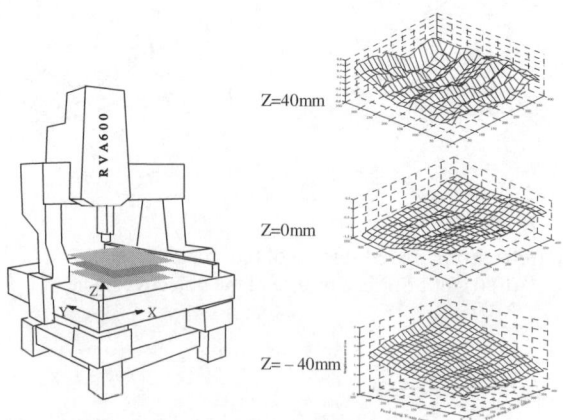

Figure 6: Example of obtaining error space

PRESSURE MONITORING SYSTEM OF GLAND PACKING FOR A CONTROL VALVE

Masanori WADA[1], Masahiro NAITO[1], Hiroshi HOSAKA[1],
Terunao HIROTA[1], Ryouji OKUTSU[2], Kouji IZUMI[2]

[1]/ Inst. of Environmental Studies, Graduate School of Frontier Science, The University of Tokyo,7-3-1 Hongo, Bunkyo-ku, Tokyo, 113-8656, Japan,
[2]/Yamatake Corporation, 4-1-1 Samukawa-Machi, Kohza-Gun, Kanagawa, 253-0113, Japan

ABSTRACT

In control valves, the sealing characteristics of gland packing has a close relation to the pressure of the radial direction of the packing, and tightening the packing in the direction of the axis controls the magnitude of the pressure in radial direction. So far, measurement of this radius direction pressure has been difficult, involving problems regarding the changes of the radial pressure over time or the distribution of the pressure.

Thus, in this paper, a method of measuring radial pressure through grease was proposed using a sensor for measuring the pressure of a fluid, and measurement equipment based on this idea was produced. First, the validity of this measuring method was evaluated. Next, the action of the piled-up packings in a bonnet was clarified by applying this measuring method to the system.

KEYWORDS

Sensing systems, Control valve, Gland Packing, Pressure, Grease, Dynamic seal

INTRODUCTION

Currently, the state of packing is managed by experienced and intuitive of safety members, and this state is distinguished by decomposing control valves. However, inspection by workers is often inaccurate, and too much inspection time reduces the operating ratio of the plant. Thus, a new technology dedicated to failure-diagnosis is demanded.

In order to prevent failure, it is necessary to clarify the behavior of the gland packing. The sealing characteristics of gland packing are strongly affected by the magnitude of the contact pressure between packing and the stem or the bonnet (=Radial Pressure), and this magnitude is controlled by tightening the pressure in the direction of the axis (=Axial Pressure) (Fig. 1). Therefore, in order to prevent failure, it is necessary to clarify the relation between the axial pressure and radial pressure.

So far, the behavior of gland packing has been considered theoretically, and the measurement of its radial pressure has been performed using pressure measuring film, strain gages, etc. [1]-[3]. However, these methods have had problems involving their inability to measure the pressure changes over time and the distribution of radial pressure.

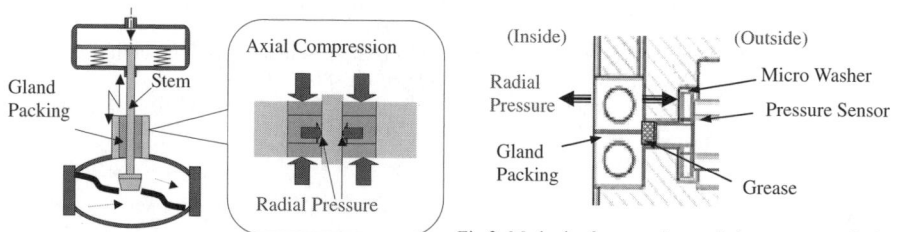

Fig.1 Composition of a control valve

Fig.2 Method of measuring radial pressure of gland packing through grease

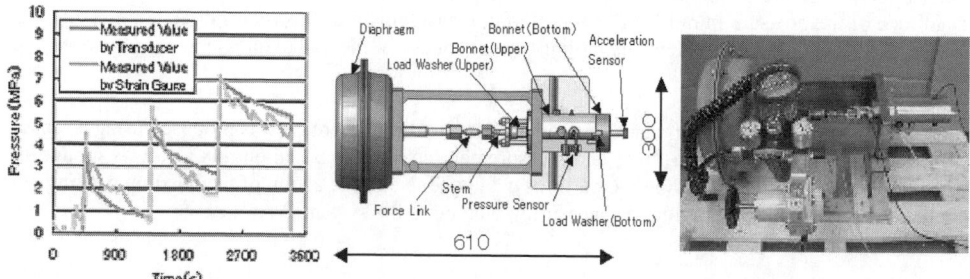

Fig.3 Comparison of measured value

Fig.4 Structure of Measuring Device

Fig.5 Photo of Measuring Device

METHOD OF MEASURING RADIAL PRESSURE THROUGH GREASE AND EVALUATION OF THIS METHOD

The proposed measuring method is shown in Figure 2. A hole for sensors is made on the bonnet and the radial pressure of the packing is measured through grease. The sensor is a Miniature Threaded Pressure Sensor made by ENTORAN.

The packing's exact radial pressure value may not be measured if the packing is deformed by the hole for the grease. Thus the accuracy of the measurement of the radial pressure through grease was confirmed experimentally. A stainless ring of 14.5mm in height and 13mm in thickness that can be stuffed with two pieces of packing was produced. Using this ring, the packing's radial pressures as measured by the pressure sensor through grease and by a strain gauge were compared.
The pressure inside the ring, i.e., the average value of the packing's radial pressure P_r is expressed using the circumferential strain outside the ring ε_θ [1].

The change of radial pressure when pressed by 2, 3, and 4 N·m using a torque wrench every 15 minutes is shown in Figure 3. From the graph, it is seen that the value of radial pressures measured through grease and calculated from the strain outside the ring are mostly in agreement. Therefore, the radial pressure value measured through grease can be stated accurately not only qualitatively but quantitatively.

MEASUREMENT OF THE RADIAL PRESSURE DISTRIBUTION IN THE SYSTEM

The distribution of the radial pressure of the piled-up packings in the system was measured under the conditions described in previous section. The experimental system is shown in Figures 4 and 5. The packings are placed firmly in a stuffing box and the stem can be operated by a control device. The hole for the placement of the sensors was made at the side portion of the stuffing box, and radial pressure was measured by the measuring method described earlier. Axial pressures on both sides were measured using the load cell.

The result is shown in Figure 6 and 7. Ten graphite packings whose inside radius, outside radius, and height are 5mm, 11.5mm, and 6.5mm, respectively, were piled up. The radial pressures are referred A, B, C, D, and E sequentially from the undersurface side. Axial pressure on the side for tightly binding the packings is referred to as the upper pressure and the pressure on the fixed end is referred to as the bottom pressure. The stem was rotated at 0.2Hz.

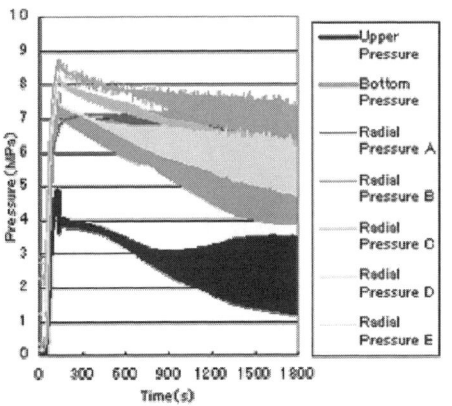

Fig.6 Measurement of Distribution of Radial Pressure

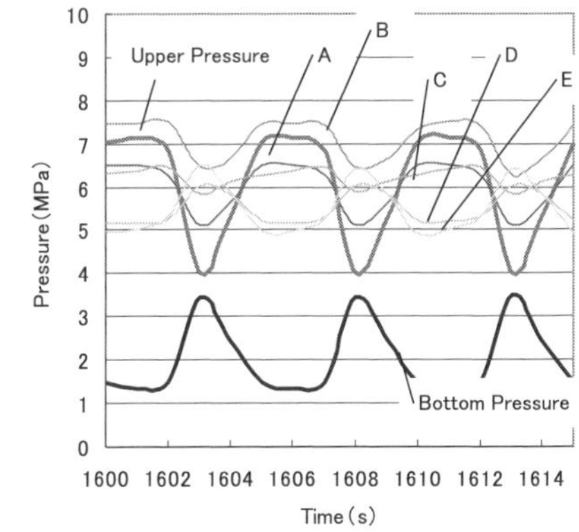

Fig.7 Measurement of Distribution of Radial Pressure (on large scale)

The upper and bottom pressures and the radial pressures change cyclically in accordance with the cycle of the stem operation. The values decrease and the fluctuation ranges increase with time. It is considered that these changes are caused by the influence of stress relaxation and wear of the packings. Moreover, it was confirmed that the phase of the pressure change is reversed in the upper and bottom sides (Fig 7).

CONCLUSIONS

This study evaluated a method for measuring the radial pressure of gland packing of an automatic control valve for the purpose of diagnosing failure, and the distribution of the radial pressure in the system was measured using the newly developed method.

The following results were obtained:
(1) Compressibility of grease does not influence the measurement of radial pressure through grease.
(2) The radial pressure measured through grease is in general agreement with the value calculated based on the circumferential strain outside the stainless ring stuffed with two pieces of packing
(3) The hydrostatic thrust bearing model can be applied to the outflow of grease, and this outflow's influence on the measurement can be disregarded by providing a margin of about one piece of packing on both sides of the sensor hole.

Furthermore, the behavior of the piled-up packings when the stem is in operation can be clarified by applying this measuring method to the system.

REFERENCES

[1] Hisao, T., Fusahito, Y., "Sealing Characterisics of Gland Pacing (1st Report)" Journal of The Japan Society of Mechanical Engineers (C), Vol. 52, No. 477, pp. 1637-1642, 1986. (in Japanese)
[2] Hisao, T., Fusahito, Y., "Sealing Characterisics of Gland Pacing (2nd Report)" Journal of The Japan Society of Mechanical Engineers (C), Vol. 52, No. 477, pp. 1643-1648, 1986. (in Japanese)
[3] Kazuhiro, H., Keiji, H., "Invesigation of Sealing Characterisics of Gland Pacing (1st Report)" Tribologist, Vol. 36, No. 1, pp. 71-78, 1991. (in Japanese)
[4] "JSME Mechanical Engineers' Handbook" B1-38. (in Japanese)

FABRICATION OF A MICRO NEEDLE MADE OF BIODEGRADABLE POLYMER MATERIAL

T. Aoki, H. Izumi and S. Aoyagi

Kansai University
3-3-35, Yamate-cho, Suita, Osaka 564-8680, Japan

ABSTRACT

The aim of this paper is the development of a micro needle made of biodegradable polymer (Poly Lactic Acid, called as PLA). This device is applicable to a blood test system for diabetics. The needle having the size of 1000 μm length, 64 μm height and 127 μm width has been developed. Biodegradable polymer naturally degrades itself in tissues; therefore this material is safe for human body and environment. To achieve the purpose of this study, we have focused on the establishment of wet chemical etching process and micro-molding processes for PLA. The resistively of PLA against common chemical etching solutions and dry etching gases is also investigated.

KEYWORDS

Micro needle, PLA(Poly Lactic Acid), Wet chemical etching, Micro-molding, Chemical resistively

INTRODUCTION

Recently, the number of diabetics is increasing in advanced countries, especially in Japan, according to the change of life-style and the increase of aged people. It is reported by some reference that the ratio of diabetics (including the suspect of it) is about 10% in Japan. To make matters worse, the generation of diabetics is becoming younger. They have to collect their blood for the glucose level measurement at least twice a day, which is indispensable for health monitoring. In this blood collection, they feel pain and fear when needle is pricked. A commercial needle is usually straight and made of metal at present state. The new type needle decreasing pain and fear is strongly desired now. On the contrary, a human being has almost no pain and no fear when he is bitten by a mosquito. The diameter of mosquito's needle is about 30 μm, which is small enough to circumvent the pain spots of a human skin. And also, mosquito's needle has a jagged shape and it can easily cut out skins with vibration. A jagged shaped needle has a merit of decreasing the contact area between the surface of the needle and the dermis of a skin during insertion [1]. Therefore it is considered that the needle can be easily inserted into a skin with vibration. Several micro needles have been reported for biomedical applications, such as drug delivery [2], measurement of cortical biopotentials, etc. However, it is difficult to form a three-dimensional jagged needle shape by micromachining technique. The aim of this paper is to develop such shaped micro needle made of biodegradable polymer, which is assumed

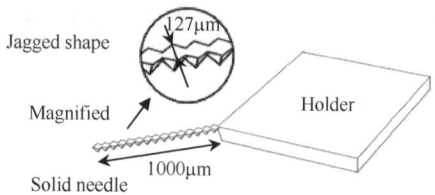

Figure.1: Schematic view of the device

to be used by the diabetics in their blood tests. **Figure 1** shows the schematic view of this needle. In micromachining technology, silicon is usually used as a structural material. Silicon is inert material to a human organism; however, silicon is not safe for use in medical treatment. If a needle is broken and pieces of it are left in organism, they may cause a fatal problem. Hence, biodegradable polymer is used in this study. This material is safe for human, since it naturally degrades itself in organism according to hydrolysis process, and is finally dissolved to harmless materials of water and carbon dioxide [3, 4]. There are small numbers of report about micro-molding process for biodegradable polymer [5]. In this research, a jagged complicated needle is fabricated by this micro-molding method. As a biodegradable polymer, Poly Lactic Acid, called as PLA, is adopted.

CHEMICAL COMPATIBILITY AND MECHANICAL PROPERTIES

In advance of microfabrication, resistance of PLA against several wet etching solutions and dry etching gases, which are commonly used in micromachining process, are investigated. PLA is able to be wet-etched by TMAH solution, and it has resistance against acids such as H_3PO_4, HF. PLA is able to be dry-etched by O_2, SF_6 and CF_4 plasma gases, and it has resistance against CHF_3 plasma gas. **Table 1** shows the summary of chemical compatibility of PLA. **Table 2** shows the comparison of the mechanical properties of representative polymer, Polyimide and PLA. Mechanical properties of PLA are not so inferior to Polyimide, and melting point of the PLA is lower than that of Polyimide.

Table1 Compatibility of chemicals with PLA

Wet etching	TMAH	×
	HF	△
	H_3PO_4	○
Dry etching	O_2	×
	SF_6	×
	CF_4	×
	CHF_3	○

○ : No reaction with PLA (compatible)
× : Reacts with PLA (not compatible)

Table2 Comparison of Polyimide and PLA

	Polyimide	Poly Lactic Acid
Young's Modulus [GPa]	3	3.4
Tensile Strength [MPa]	120	64
Elongation [%]	10	4.1
Glass Transition Temp.	310	61
Melting Point [°C]	450	173
Corporation	DuPont MicroSystems, Ltd.	Shimadzu Corp.
Trade name	PIX-3476-4L	Lacty5000
Manufacturing method	Spin coat	Injection Molding

FABRICATION AND RESULTS

Fabrication Process of Micro Needle

The micro needle is fabricated as shown in **Fig. 2**. A (100)-oriented silicon wafer, of which both front and back surfaces are covered with thermally grown SiO_2, is prepared. SiO_2 mask for wet etching which has jagged shape of needle is patterned by using photolithography and dry etching. Then silicon is anisotropically wet-etched to prepare pyramidal holes. Consequently these holes are connected together and a jagged groove is formed. PLA is molded into this groove and released.

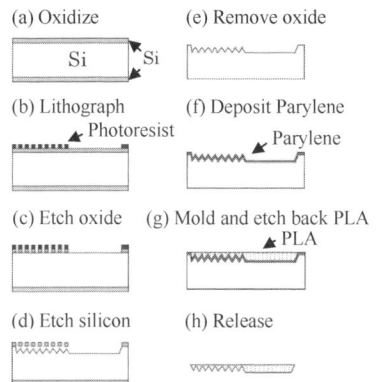
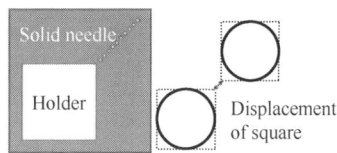

(a) Oxidize (e) Remove oxide
(b) Lithograph (f) Deposit Parylene
   Photoresist   Parylene
(c) Etch oxide (g) Mold and etch back PLA
                  PLA
(d) Etch silicon (h) Release

Figure 2: Fabrication process of micro-needle

Table3 Parameters for evaluating jagged groove

No.	Diameter of circle[μm]	Displacement of square[μm]
1	70	1.5
2	70	2.0
3	70	2.5
4	70	3.0
5	90	2.5
6	90	3.0
7	90	3.5
8	90	4.0
9	90	4.5

Anisotropic Wet Etching

The jagged groove of the micro needle is fabricated by anisotropic wet etching of silicon (100) surface. In order to produce desired jagged shape, optimal value of radius of circles and their distance are investigated. **Figure 3** shows the schematic view of the mask pattern. **Table 3** shows the parameters for evaluating jagged groove. If the displacement of circles in Fig.3 is too close, before pyramidal holes would be made, they are connected to each other and desired jagged shape cannot be formed as shown in **Fig. 4**. On the other hand, the displacement of circles is too apart, the pyramidal holes are not connected easily, which consumes large process time. According to these phenomena, uniform connection of pyramidal holes is not obtained by using only KOH solution as shown in **Fig. 5**. Therefore, a new process of etching silicon groove is developed. First KOH solution is used for pyramidal part, since etching is stopped accurately on (111) surface, second TMAH solution is used for connecting part, since TMAH etching forms more isotropic shape compared with KOH etching. The result of etched groove using both KOH and TMAH solutions is shown in **Fig. 6**.

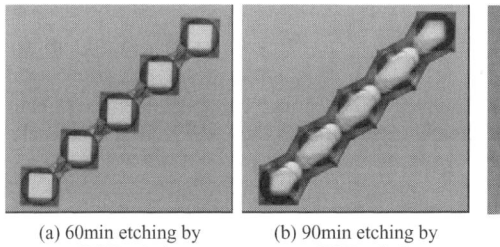

(a) 60min etching by KOH solution
(b) 90min etching by KOH solution

Figure 4: Etched groove when squares are too close

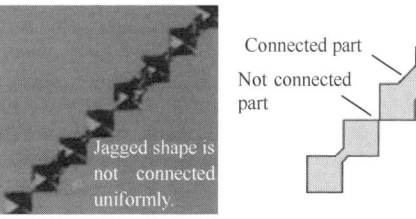

Figure 5: Jagged groove when etched only by KOH solution

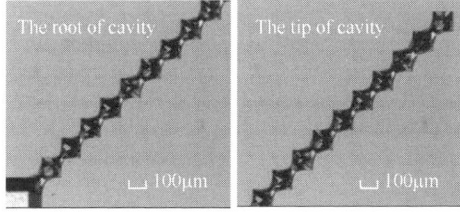

Figure 6: Jagged groove of silicon cavity for micro needle which is etched by both KOH and TMAH solutions

Micro-molding

Next, micro-molding of PLA is investigated. The schematic diagram of compression molding to produce the 3D microstructures of PLA micro needle is shown in **Fig. 7**. First, PLA is re-flowed at above its melting temperature, and is put onto a heated mold (200°C, 3 min.). Second, pressure is applied to the silicon cavity and PLA (14 MPa, 1 min.). Third, PLA is etched back to the level of the silicon surface by oxygen plasma gas. Finally, the micro needle is taken away from the mold by using an adhesive tape. **Figure 8** shows the fabricated PLA micro needle. The size of needle is as follows: length is 1000 μm, height is 64 μm and width is 127 μm.

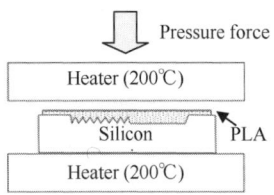

Figure 7: Mechanism of compression molding   Figure 8: SEM image of fabricated micro needle

CONCLUSIONS

A micro needle made by biodegradable polymer is fabricated, which has jagged shape like mosquito's needle. The summary is as follows:
1) The compatibility characteristics of PLA to wet etching solutions and dry etching gases are investigated. PLA has resistively against H_3PO_4, HF solutions and CHF_3 plasma gas.
2) The uniform jagged groove of cavity for the micro needle is fabricated by new method of 2 step anisotropic wet etching by KOH and TMAH solutions. Compression molding method is applied and a jagged needle is surely released.

ACKNOWLEDGEMENT

This work was mainly supported by JSPS (Japan Society for the Promotion of Science).KAKENHI (16310103). This work was also partially supported by MEXT (Ministry of Education, Culture, Sports, Science and Technology). KAKENHI (17656090), "High-Tech Research Center" Project for Private Universities: Matching Fund Subsidy from MEXT, 2000-2004 and 2005-2009, the Kansai University Special Research Fund, 2004 and 2005.

REFERENCES

[1] Oka K., Aoyagi S., Arai Y., Isono Y., Hashiguchi G. and Fujita H. (2002). Fabrication of a micro needle for a trace blood test. *Sensors and Actuators*. **97-98C,** 478-485.
[2] Henry S., McAllister D. V., Allen M. G. and Prausnitz R. (1998). Micromachined needles for the transdermal delivery of drugs. *Proc. of MEMS'98*, 494-498.
[3] Tuji S. and Ikada Y. (1997). Poly Lactic Acid –for medical care • medication • environment–. *KOBUNSHI KANKOUKAI, LTD.*, 1-76.
[4] Urakawa H. and Ohara H. (2000). Crystallinity and molding properties of Polylactic Acid. *Technology Research Laboratory, Shimadzu Corp.*, **56:3・4,** 163-168.
[5] Armani D. K. and Liu C. (2000). Microfabrication Technology for Polycaprolactone, a biodegradable polymer. *J.Micromech.Microeng*. **10,** 80-84.

AN EFFECTIVE STATE-SPACE CONSTRUCTION METHOD FOR REINFORCEMENT LEARNING OF MULTI-LINK MOBILE ROBOTS

M. Nunobiki [1] K. Okuda [1] and S. Maeda [2]

[1] Department of Mechanical and System Engineering, University of Hyogo,
2167 Shosha, Himeji, Hyogo, JAPAN
[2] Department of Quality Assurance, Shin Caterpillar Mitsubishi LTD.
1106-4 Shimizu Uozumi, Akashi, Hyogo, JAPAN

ABSTRACT

One of the problems in reinforcement learning with real robots is to need a large number of trials. This paper proposes a reinforcement learning that uses fuzzy ART for segmentation of state-space. Whenever fuzzy ART encounters a new situation, it generates a new category node to the state-space. We proposed generating methods of new category nodes that inherit the state-value and the policy from a similar node. Proposed methods were estimated from simulations of a two-link manipulator and a multi-link mobile robot. It was confirmed that the proposed method was able to increase the learning speed and reduce the size of state-space.

KEYWORDS

Reinforcement learning, Actor-critic, State-space construction, Fuzzy ART, Inheritance of state-value, Manipulator, Multi-link mobile robot, Action acquisition

INTRODUCTION

We have developed inchworm-type mobile robots to search for life in collapsed buildings. While these robots had neither legs nor wheels, they were able to advance by using vertically undulatory motion of whole body (Takita et al.). These robots demonstrated high mobility (Nunobiki et al.). However, it was difficult to generate suitable walking motions because it was difficult for human to understand the motions of the multi-link robot intuitively. Therefore, reinforcement learning (Suttun and Barto.) is expected for trajectory generation of these robots. This paper deals with actor-critic learning method (Barto and Suttun.). Although the actor-critic methods require minimal computation in order to select an action from a continuous-valued action, the performances are insufficient to apply to the real robots yet (Morimoto and Doya.). A grid-like representation of the state-space was insufficient to the applications (Morimoto and Doya.). This paper used fuzzy ART for sate-space segmentation. It is an incremental category-space construction method (Carpenter et al.). Whenever the fuzzy ART encounters a new situation, it generates a new category node in the category space. Furthermore, we proposed generating methods of a new node that inherit the state-value and the policy from a similar node to increase the learning speed more.

REINFORCEMENT LEARNING WITH FUZZY-ART STATE-SPACE CONSTRUCTION

Figure 1 shows the architecture of proposed system. This system based on actor-critic methods. Using the sensed state from environment, the actor selects an action. The critic evaluates the new state to determine whether the state has gone better or worse than expected. The TD-error represents this evaluation. The TD-error is used for the improvement of the policy and the update of the state-value. If TD-error is positive, it suggests the tendency to select the action should be strengthened for the future. If TD-error is negative, it suggests the tendency should be weakened. We applied the fuzzy ART to generalization of sensed states. In many tasks, most states will never have been experienced exactly before. Fuzzy ART is a kind of self-organized clustering method and it classifies unknown state into the group of approximate states. One state code used for each group.

FUZZY-ART

The fuzzy ART consists of two fully connected layers. The category layer contains category nodes to categorize a given vector. A weight vector Wj shows the representative pattern of category node j. The sensed data are normalized with complement cording. The sensed data vector and its complement vector are input to the input layer. The choice function Ti is defined as equation (1). Where c is a choice parameter, the operator is defined by $(p \wedge q) = \min(p, q)$ and the norm $|x|$ is the sum of its components. The category node i that has the maximal Ti is called a winner node. The winner node is judged to cause resonance or mismatch reset according to the equation (2). If the match function Mi is bigger than the vigilance parameter ρ, resonance is carried out and the weight vector Wi is updated according to the equation (3). Where β is a learning rate parameter. When mismatch reset procedure should be done, other category node that has the next maximal T is chosen again. When any category node is not selected, a new node is generated. In normally, default values are given to the policy and the state-value. In proposed methods, inherited value from a similar node was used. The efficiency was estimated in the simulations of a two-link manipulator and a multi-link mobile robot.

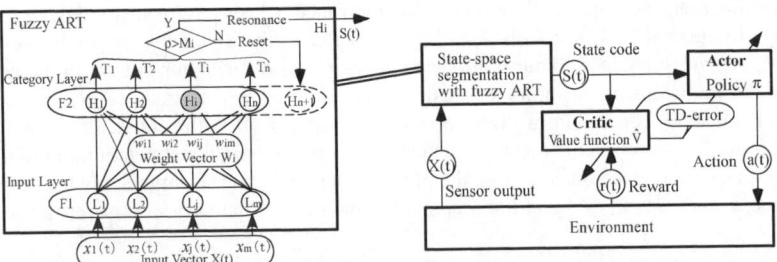

Figure1: Architecture of actor-critic learning method with fuzzy ART

$$T_i = \frac{|x \wedge w_i|}{c + |w_i|} \quad (1)$$

$$M_i = \frac{|x \wedge w_i|}{|x|} \geq \rho \quad (2)$$

$$w_i = \beta(x \wedge w_i) + (1 - \beta)w_i \quad (3)$$

LEARNING FOR HAND REACHING PROBLEM OF TWO-LINK MANIPULATOR

Figure 2 shows a hand reaching problem and the results of learning. A reward of 1 was given when the end effecter reached the goal in 50 steps or less. Each angle of joint was limited to $0<\theta<2\pi$. Each angular velocity was limited under 0.1 rad/step. We addressed a state with the x and y coordinates and the velocity components of the end effecter. The step-size parameter and the credit rate were set to 0.02 and 0.9 respectively. The learning methods were estimated by an attainment rate of task and the number of category nodes. The attainment rate of task was defined as the ratio between number of the succeeded trials and the total trails. We carried out the simulation on each method five times and used the average value. The vigilance parameter was set to 0.98. This result shows that the methods with fuzzy ART were superior to the normal method. Table 1 shows the attainment rates and the number of category nodes at 10,000th trial at various methods and various ρ. The attainment rate increased so that ρ became large. In case that ρ was 0.98, all methods are superior to the normal method. The learning method of inheriting the state-value was superior to other methods. In a normal actor-critic method, we tiled the sate-space with a uniform grid and set the total number of states to 10,000. In learning with fuzzy ART, the number of category units increased so that ρ became large, but the total number of states at 10,000th trials was 202.2 or less.

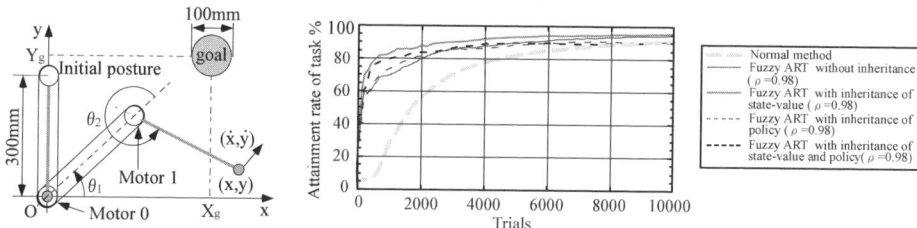

Figure2: Performance of each learning methods for hand reaching problem

TABLE 1
ATTAINMENT RATE OF TASK AND NUMBER OF CATEGRY NODES AT $10,000^{TH}$ TRIAL

State-space segmentation method / Vigilance criterion ρ	Attainment rate of task				Number of category nodes			
	0.95	0.96	0.97	0.98	0.95	0.96	0.97	0.98
Fuzzy ART without inheritance	55.7%	66.9%	83.0%	94.4%	60.4	85.4	121.2	191.0
Fuzzy ART with inheritance of state value	66.5%	76.8%	88.0%	95.4%	56.4	79.6	115.0	182.8
Fuzzy ART with inheritance of policy	57.5%	75.5%	86.0%	90.6%	59.2	78.4	108.8	191.6
Fuzzy ART with inheritance of policy and state value	62.3%	68.2%	80.1%	90.1%	59.8	85.4	124.6	202.2
Normal method	91.1%				10000			

LEARNING FOR MOVEMENT EXPERIMENTS OF A MULTI-LINKED MOBILE ROBOT

We applied proposed method to multi-liked mobile robots. The simple model was used. It consisted of five servomotor modules that were connected serially. The cyclometer was installed to measure the moved distance from the initial position. The moved distance by 20 steps was given as the reward. But a negative reward was given when the robot had moved backward. Each trial began with same posture in which all joints are stretched. 100 trials were carried out. We addressed a state with angles of four servomotors. The step–size parameter and the credit rate were set to 0.02 and 0.9 respectively. The vigilance parameter was set to 0.98. The learning methods were estimated by an attainment rate of task and the total moved distance by 100 trials. Table 2 shows the attainment rate of task at 100th trial and the total moved distance. The attainment rates were low because the number of trials was a little. The learning methods with fuzzy ART were superior to the normal actor-critic learning method. Furthermore, the learning speed had quickened most and the total moved distance was largest in using inheritance of the state-value. The efficiency of the proposed method was confirmed in the experiment of multi-link mobile robot though the number of trials was a little.

TABLE 2
PERFORMANCE OF EACH LEARNING METHOD FOR A MOBILE ROBOT

	Normal method	Fuzzy ART without inheritance	Fuzzy ART with inheritance
Total moved distance	429 mm	485 mm	578 mm
Attainment rate of task at 100th trials	29 %	34 %	36 %

CONCLUSION

We proposed a reinforcement learning method that used fuzzy ART for segmentation of state-space. And we proposed a generating method of a new category node that inherited the state-value of the similar node. The efficiency of proposed method was estimated in the simulations of hand reaching problems and the movement experiments. The learning efficiency was improved more by inheriting the sate-value in the fuzzy ART. The learning speed of proposed method is about 20 times the speed of normal actor-critic method in the hand reaching problems. The size of state-space was decreased very much in proposed method. The efficiency of proposed method was confirmed in the experiments of multi-link mobile robot. Thus, it was confirmed that the proposed method was able to apply to the learning with the real robots.

References

Y. Takita, M. Nunobiki, et al. (1999). An investigation of climbing up stairs for inchworm robot, Proceedings of TITech COE/Super Mechano-Systems Workshop'99, 133-138
M. Nunobiki, T. Takita, et al. (1999). Study on the gaits of an inchworm robot through a narrow path, Advanced Robotics, **13: 3**, 329-330
M. Nunobiki, et al. (2001). An investigation of mobility of inchworm-type mobile robot, Proceedings of the France-Japan Congress of Mechatronics, 113-118.
M. Nunobiki, et al. (2004). Returning motion from the state of falling sideways for articulated mobile robot, JSME International Journal, Series C, **47:1**, 225-232.
R.S. Suttun and A.G. Barto. (1998). Reinforcement Learning; An Introduction , The MIT Press.
A.G. Barto, R.S. Suttun and C.W. Anderson. (1983). Neuronlike adaptive elements that can solve difficult learning control problem, IEEE Trans. Syst., Man, Cyber., SMC-**13:5**, 834-847.
J. Morimoto and K. Doya. (2001). Acquisition of stand-up behavior by a real robot using hierarchical reinforcement learning. Robotics and Autonomous Systems, 36, 37-51.
J. Morimoto and K. Doya. (1998). Reinforcement learning of dynamic motor sequence: Learning to stand up, Proc. of IEEE/RSJ Int. Conf. On Intelligent Robots and System, 3, 1721-1726.
G.A. Carpenter, S. Grossberg and D.B. Rosen. (1991) Fuzzy ART: Fast stable learning and categorization of analog patterns by an adaptive resonance system. Neural networks, 4, 759-711.

CIRCULARLY POLARISED RECTENNA FOR ENHANCED DUAL-BAND SHORT-RANGE WIRELESS POWER TRANSMISSION

J. Heikkinen and M. Kivikoski

Tampere University of Technology
Institute of Electronics
P.O. Box 692, FI-33101 Tampere, FINLAND

ABSTRACT

Due to conversion losses, free-space attenuation, and limited transmitted power only a small amount of DC power can be received from a *rectifying antenna* (*rectenna*). In addition to optimisation of the rectenna efficiency, other methods for increasing the applicability of *wireless power transmission* (WPT) are also needed. The utilisation of *circular polarisation* and operation in two frequency bands in order to enhance the performance of WPT are demonstrated in this paper. The design and performance of a CP (circularly polarised) annular ring-slot antenna and microwave rectifier circuit operating at 2.45 GHz and 5.8 GHz ISM (Industrial, Scientific and Medical) bands are represented. The effect of a simple *electromagnetic band-gap* (EBG) structure on the antenna performance is also demonstrated and discussed.

KEYWORDS

Wireless power transmission, Rectifying antenna, Circular polarisation, Dual-band, Electromagnetic band-gap.

INTRODUCTION

WPT comprises both data transmission and transfer of mere electrical power using magnetic or electromagnetic fields. In the case of data transmission the purpose of transmitted power is solely to carry a piece of information from one place to another. The goal of electrical power transfer, however, is transmission and reception of power high enough to enable the operation of other electrical devices. In short-range WPT, the distance between the transmitter and receiver can be anything from a few millimetres to about ten metres. The main differences between data and power transmission are in the amount of transmitted and received power and in the structure of the transmitter and receiver. If no information needs to be included in power transfer, simple continuous wave (CW) transmission can be utilized. Basically the transmitter then consists of a DC power source, a circuit that converts DC energy to AC energy, and an antenna. Correspondingly, the receiver, i.e. rectenna is composed of an antenna element, an AC-to-DC conversion circuit, and a load. Data transmitter/receiver circuitry can be included into the WPT system if a data link needs to be established.

A common problem for all applications utilising WPT is that due to the limitations in the transmitted power allowed according to regulations, the received power is relatively low. As a result, the conversion efficiency of the rectenna is low, because rectifier diodes operate more efficiently at higher input power levels. In addition, the received power varies if the distance or relative orientation in respect to the transmitter changes. A CP rectenna operating in two frequency bands is demonstrated here in order to diminish some of these disadvantages. Circular polarisation enables nearly constant output independent of the rotational angle of the rectenna, whereas *dual-band* operation increases the operational diversity of the rectenna. The structure of the proposed rectenna is first explained and the measurement results for dual-band operation are then represented. Some of the performance issues are also discussed and the effect of a simple *electromagnetic band-gap* (EBG) structure on the antenna performance is finally demonstrated.

DUAL-BAND CP RECTENNA

Layout and cross-section of the rectenna is shown in Figure 1. The three sections, i.e. antenna, high band rectifier, and low band rectifier, were first designed and measured separately and then combined to form the complete rectenna. Long dotted divider lines in Figure 1 indicate the interface between the antenna and rectifier circuits.

Structure

The dual-band CP antenna was formed as a combination of two shorted annular ring-slot structures designed for operation at 2.45 GHz (low band) and 5.8 GHz (high band). The short section of the low band structure (angle $\alpha 1$) provides a continuous ground to the feed of the high band antenna, whereas the short section of the high band structure (angle $\alpha 2$) provides a continuous ground to the feed of the low band antenna. The feed for the high band antenna was composed of a transmission line $TL1$ and quarter-wave transformers $TL2$ and $TL3$, which were also utilized to match the input impedance of the antenna into 50Ω. Correspondingly the feed for the low-band antenna was composed of a transmission line $TL6$ and a quarter-wave transformer $TL7$.

Rectifiers were composed of HSMS-2862 microwave Si Schottky detector diode pair D, a bypass/storage capacitor C, a load resistor R and a choke inductor L. The diode pair was connected as a voltage-doubler circuit in order to maximise the output voltage. The choke inductor reduces the effect of output DC voltage measurement wires on the performance of the rectifier. The impedance matching circuit of the high band rectifier was composed of transmission lines $TL4$ and $TL5$ and two sections of an open stub ($Stub1$ in Figure 1) between them. Equally, the impedance matching circuit of the low band rectifier was composed of transmission lines $TL8$ and $TL9$ and two sections of an open stub ($Stub2$) between them.

Performance

Measured return loss of both high band and low band antenna and rectifier is represented in Figure 2. All measurements were performed using –5 dBm input power. Good impedance match between antenna and rectifier can be observed in both frequency bands. A second resonant frequency between 6 GHz and 7 GHz for both low band antenna and rectifier can also be noticed.

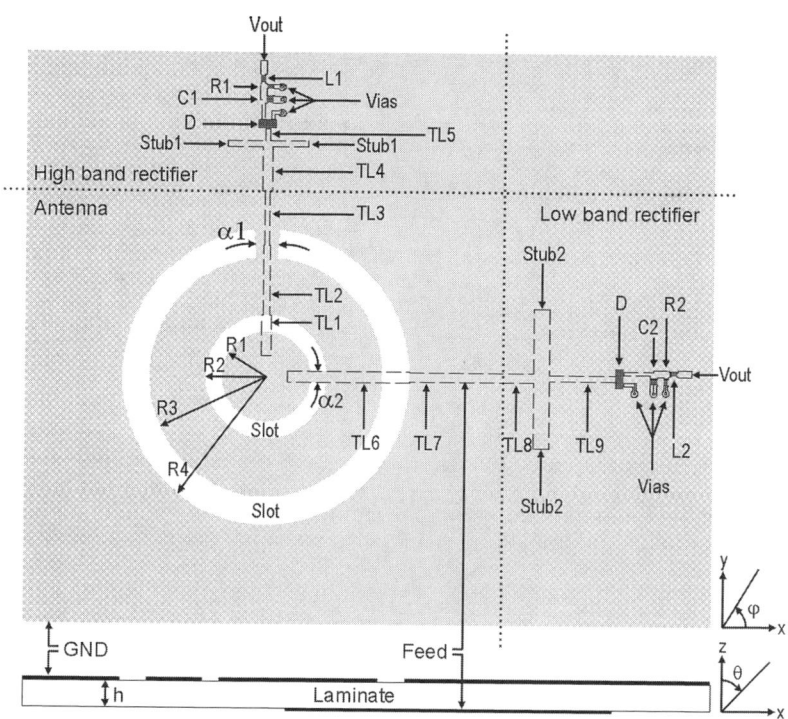

Figure 1. Layout and cross section of the dual-band circularly polarised rectenna. Antenna dimensions: $R1 = 8.1$ mm, $R2 = 11.1$ mm, $\alpha 2 = 11.5°$ (high band), $R3 = 21.5$ mm, $R4 = 26.5$ mm, $\alpha 1 = 10.6°$ (low band), ground plane 89 mm x 78 mm. Antenna feed: $TL1 = 1.9$ mm x 7.3 mm, $TL2 = 1.2$ mm x 12.6 mm, $TL3 = 1.0$ mm x 9.8 mm (high band), $TL6 = 1.9$ mm x 22.7 mm, $TL7 = 1.7$ mm x 18.0 mm (low band). High band rectifier (ground plane 89 mm x 25 mm) consists of single-package diode pair D (HSMS-2862), bypass capacitor $C1 = 68$ pF, load resistor $R1 = 8.2$kΩ, choke inductor $L1 = 6.8$ nH, and impedance matching circuit composed of $TL4 = 1.8$ mm x 8.0 mm, $stub1 = 1.0$ mm x 6.4 mm, and $TL5 = 1.0$ mm x 2.2 mm. Low band rectifier (ground plane 38 mm x 78 mm) consists of single-package diode pair D (HSMS-2862), bypass capacitor $C2 = 68$ pF, load resistor $R2 = 8.2$ kΩ, choke inductor $L2 = 15$ nH, and impedance matching circuit composed of $TL8 = 1.7$ mm x 5.0 mm, $stub2 = 2.9$ mm x 11.7 mm, and $TL9 = 1.0$ mm x 12.0 mm. The laminate (RO4232) has a thickness $h=1.52$ mm, relative dielectric constant of 3.2, and loss tangent of 0.0018.

Measured output DC voltage and efficiency of the high band and low band rectifier circuit at two input power levels (0 dBm and -5 dBm) are also shown in Figure 2. Output voltage was measured with a voltmeter as the rectifier circuit was fed with a microwave signal generator. Figure 2 shows that the efficiency of the rectifier decreases significantly with decreasing input power and increasing frequency.

The performance of the complete rectenna was verified in an anechoic chamber using a linear standard 20 dB gain horn antenna as a transmitter. The output DC voltage was measured while the rectenna was rotated 360° in 45° steps in the XY-plane (Figure 1). Measurements were done at transmission distance of 2 metres. Transmitted power (Pt) was 32 dBm in high band rectenna, and 24 dBm in low band rectenna measurements. Results are represented in Figure 3. Minimum axial ratio of 1.5 dB was measured at 5.6 GHz for the high band rectenna. Similarly for the low band rectenna, minimum axial ratio of 2.3 dB was measured at 2.45 GHz.

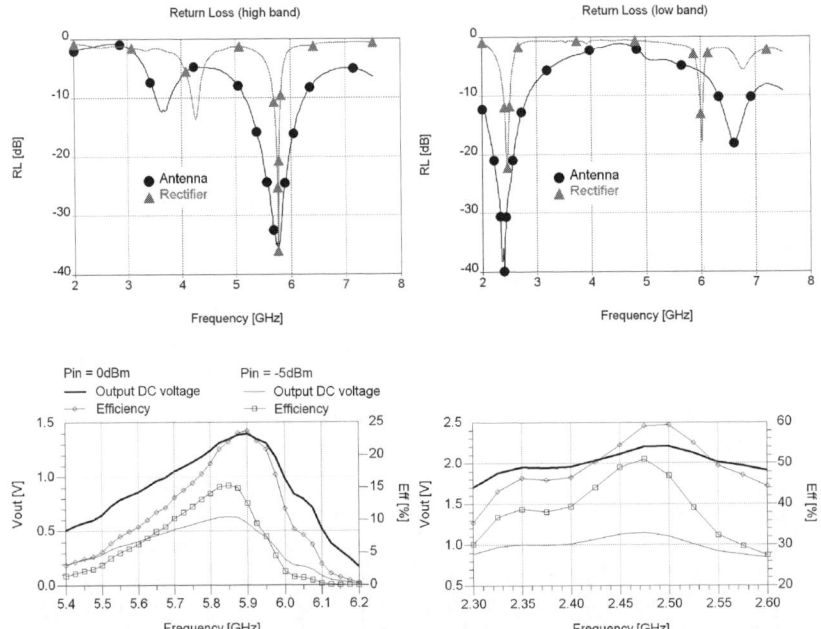

Figure 2. Measured return loss (*RL*), output DC voltage (*Vout*) and efficiency (*Eff*) of the high band antenna and rectifier (left), and low band antenna and rectifier (right).

The measured output voltage of the rectenna was found to be in a good agreement with the rectifier measurement results shown in Figure 3. Maximum efficiency of 19% and 62% was calculated for the high band rectenna at 5.8 GHz, and low band rectenna at 2.45 GHz, respectively.

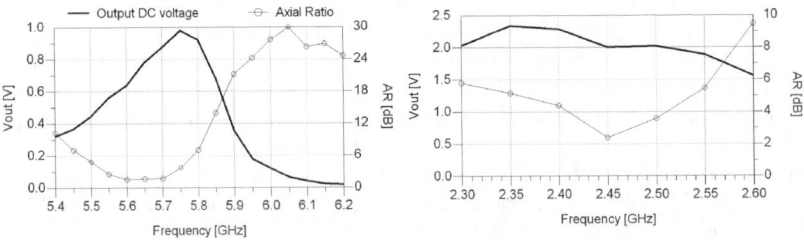

Figure 3. Measured output DC voltage (*Vout*) and axial ratio (*AR*) of the high band rectenna (left), and low band rectenna (right). Transmission distance was 2 metres. Transmitted power was 32 dBm for high band rectenna and 24 dBm for low band rectenna.

Although the rectifier input matching circuits also act as band-pass filters rejecting unwanted harmonics generated by the rectifier diode, some excess resonant frequencies still exist between 6 and 7 GHz (low band antenna and rectifier in Figure 2). The use of additional filter structures in reducing the propagation of energy at these frequencies would require extra space and therefore is not practical. In order to avoid increasing the size of the rectenna the following approach using an electromagnetic band-gap structure was utilised.

DUAL-BAND CP ANTENNA WITH EBG STRUCTURE

In order to reduce electromagnetic propagation from the proposed rectenna at certain frequencies, the effect of introducing a 2D EBG structure to the ground plane of the antenna was studied. The layout of the antenna is shown in Figure 4.

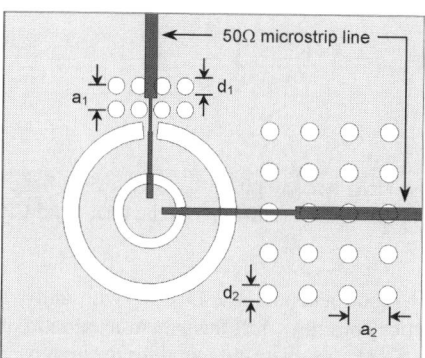

Figure 4. Dual-band CP antenna with a 2D EBG structure. High band antenna: a_1=7 mm, d_1=2.5 mm. Low band antenna: a_2=12 mm, d_2=2.7 mm.

Antenna

The dimensions of the slot and feed structures of the antenna are the same as given in Figure 1. Because the size of the ground plane was also preserved in order to include the space required by the rectifier circuits, both antenna feeds were extended using additional 50Ω microstrip lines.

EBG structure

The studied EBG structure was formed of circles etched to the ground plane of the antenna. Circles having diameter d_1 and d_2 were arranged in a square lattice having period a_1 and a_2 for high band and low band antenna, respectively. The two lattices were located so that they would have minimal effect on the rectifier input matching circuits at the operating frequencies of the rectenna. In general, the guided wavelength corresponding to the stopband centre frequency f_0 is twice the lattice period a. As an example, assuming desired centre frequency f_0 = 6.5 GHz approximately results to a = 14 mm for the used laminate RO4232. The chosen value a_2 = 12 mm for the low band antenna (as well as the starting value a_1 = 7 mm for the design of the high band antenna) was a compromise between desired performance and accommodation of the rectifier circuits. The dimension of the lattice element, in this case the diameter of the circle (d), affects the bandwidth and depth of the stopband. Again, the values d_1=2.5 mm and d_2=2.7 mm were chosen to optimise stopband performance and minimise the effect on the rectifier circuits at the operating frequencies of the rectenna.

Results

The antenna shown in Figure 4 was simulated with and without the EBG lattice using a time-domain 3D EM simulator (CST Microwave Studio). The results for the low band antenna are represented in Figure 5. Clearly, EBG antenna has better harmonic rejection at stopband frequencies, whereas no remarkable effect on the performance can be detected in the operating frequency band. Also, due to EBG, the coupling between high band and low band antenna is decreased in the stopband region.

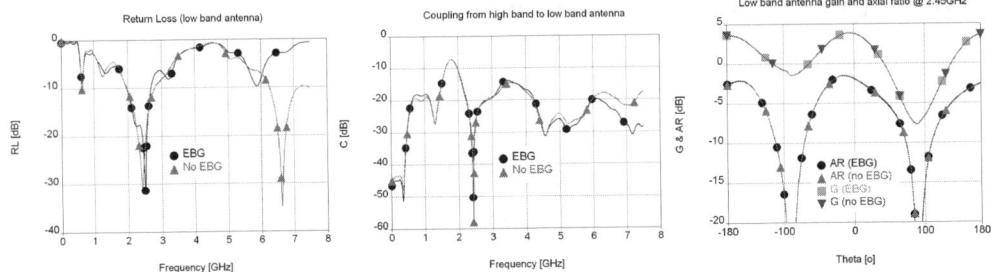

Figure 5. Simulated return loss (*RL*) and coupling (*C*) of the dual-band CP antenna with and without EBG structure. Simulated gain (*G*) and axial ratio (*AR*) of the dual-band CP antenna at 2.45 GHz with and without EBG structure.

Radiation properties of the low band antenna at 2.45 GHz with and without EBG are also shown in Figure 5. As expected, EBG lattice does not affect the gain or axial ratio of the studied antenna in the desired operating frequency band. Similar results were found for the high band antenna.

CONCLUSIONS

A dual-band CP rectenna operating at 2.45 GHz and 5.8 GHz has been designed and evaluated. Successful operation at both frequency bands was verified with measurements while rotating the rectenna relative to the transmitter. A method for improving the harmonic rejection of the antenna was also studied. The proposed EBG structure was found to be simple and effective solution that also resulted in decreased coupling between the high band and low band antenna without affecting their radiation properties in the operating frequency band.

REFERENCES

Brown, W. C. (1984). The history of power transmission by radio waves. *IEEE Trans. Microwave Theory Tech.* **32:9**, 1230-1242.

Chen, W-S., Huang, C-C., Wong, K-L. (2001). Microstrip-line-fed Printed Shorted Ring-slot Antennas for Circular Polarization. *Microwave and Optical Technology Letters* **31:2**, 137-140.

Gonzalo, R., De Maagt, P., Sorolla, M. (1999). Enhanced Patch-Antenna Performance by Suppressing Surface Waves Using Photonic-Bandgap Substrates. *IEEE Trans. Microwave Theory Tech.* **47:11**, 2131-2138.

Heikkinen, J., Kivikoski, M. (2002). Short-range Wireless Power Transfer in Mechatronics, in *Proc. Fourth International Conference on Machine Automation (ICMA'02), Human-Friendly Reliable Mechatronics*, Tampere, Finland, 501-508.

Morishita, H., Hirasawa, K., Fujimoto, K. (1991). Analysis of a Cavity-backed Annular Slot Antenna with One Point Shorted. *IEEE Trans. Antennas Propagat.* **39:10**, 1472-1478.

Radisic, V., Qian, Y., Coccioli, R., Itoh, T. (1998). Novel 2-D Photonic Bandgap Structure for Microstrip Lines. *IEEE Microwave Guided Wave Lett.* **8:2**, 69-71.

Rahmat-Samii, Y., Mosallaei, H. (2001). Electromagnetic Band-Gap Structures: Classification, Characterization, and Applications, in *Proc. 11th International Conference on Antennas and Propagation (ICAP)*, 560-564.

Yang, F-R., Ma, K-P., Qian, Y., Itoh, T. (1999). A Uniplanar Compact Photonic-Bandgap (UC-PBG) Structure and its Applications for Microwave Circuits. *IEEE Trans. Microwave Theory Tech.* **47:8**, 1509-1514.

CONDUCTIVE FIBRES IN SMART CLOTHING APPLICATIONS

Jaana Hännikäinen, Tiina Järvinen, Timo Vuorela, Katja Vähäkuopus,
and Jukka Vanhala

Institute of Electronics, Tampere University of Technology
Korkeakoulunkatu 3, FIN-33720 Tampere, Finland

ABSTRACT

The use of electrically conductive fibre yarns in smart clothing applications as a communication or power transfer medium and electrode materials have been researched. Challenges in usage have been culminated in positive and durable contacts. Therefore, accelerated reliability tests and tensile strength tests have been made to find out proper materials and connection methods for reliable joints. Results show that reliable connections can be made by using appropriate conductive yarns and solder. However, the fibre usage in clothing is very demanding and field tests have to be developed to find out yarns' braking mechanisms and long-term durability in different usage situations.

KEYWORDS

Smart clothing, electrically conductive fibre, connection, accelerated reliability test, wearing comfort

INTRODUCTION

The amount of different electrical devices that are carried also during daily routines increases continuously. Mobile phones and personal digital assistants have got competitors e.g. as forms of heart rate monitors, pace counters, wrist-worn computers, and mobile positioning devices. One of the empowering factors of this trend has been rapid development of electronics, especially in the areas of miniaturization techniques, low-power electronics, telecommunication techniques, and battery technology. These key technologies have also provided possibilities to construct application specific wearable systems i.e. smart clothes. This term encompasses combinations of intelligent textile materials, and electronic and non-electrical devices integrated in ordinary clothing (Rantanen *at al.* 2002). Systems become truly smart when they can sense their environment and act according to the measured stimuli.

Although, we can already implement very compact and good looking smart clothes, we still need to make many compromises. It is difficult to construct small, clothing-like, and lightweight devices, which are also cheap and robust. As users wear these clothes, specific attention has to be paid to wearing comfort, usability, and safety aspects of electrical devices. In smart clothing design these soft values are as important criteria as technical functions. Consequently, we need to develop solutions that are easy to use and maintain their clothing-like properties as well as technical reliability.

In smart clothing applications we usually utilise a distributed electronics architecture, which ensures that any single electronic module is not straining too much textiles and the clothing itself feels good to wear. This distribution of electronics to several pieces of clothing or to several locations in a piece of clothing create needs for communication between different electronic modules. Ordinary wires could be used, since they provide straightforward, inexpensive, and reliable communication medium. However, long wires inside the clothing may cause rigidity and separate connectors are needed between the different pieces of clothing. Wireless solutions are the most practical for that, since extra connections of wires could impair dressing and undressing. On the other hand, wireless solutions can be too complex and e.g. in demanding industry environment the simplest wireless solutions such as inductive coupling may suffer from environment interference. To overcome these difficulties we have studied the use of conductive fibres as one of the key building elements of smart clothing. As a communication medium conductive fibre yarns are as straightforward solutions as ordinary cables. Furthermore, they are lighter and softer to wear than plastic shielded cables, which make them more clothing-like and comfortable. In addition to communication, versatile electrically conductive fibres can also be used as sensing elements.

This paper introduces usage of electrically conductive fibres (ECFs) in smart clothing applications concentrating on the reliability of the connection mechanisms between the fibre and the printed wiring boards (PWBs).

CONDUCTIVE FIBRES

In general, ECFs are used in the prevention of electrostatic shock. Same fibres can also be adapted for wire replacement usage in clothes. To be able to take care of also power transfer between different electronic modules, low resistance and wire shielding is required. The first requirement reduces power consumption and the latter requirement protects users against electrical shock and ensures proper functioning of the system. These were also our main starting points while choosing suitable ECF materials. Conductive fibres become useful for us when several thin filaments form a twisted yarn, which then resembles one core metal cable. In our smart clothing applications, so far, we have mostly used metal clad aramid fibres (DuPont data sheet), which are intended for braided electromagnetic interference shielding in cables and harnesses and for electrical conductions in specialised applications. These fibre yarns have better specific gravity and tensile strength properties than copper offering also flexibility more typical for clothing than metal. Fibres are coated with silver, copper, or nickel metals. The latter may cause allergy reactions while being attached directly to skin. Therefore, only silver or copper coating materials are acceptable. Due to better availability we have chosen to use silver clad fibre yarns.

ECF Usage Experiments

We have employed metal clad aramid fibre yarns to replace plastic shielded cables in lightweight underclothes. One of them is a sensor shirt, which measures user's body surface temperature, respiration rate, and skin's electrodermal activity (Rantanen *at al.* 2001). Altogether nine temperature sensors are attached to the shirt so that they are in close contact to skin allowing skin surface temperature measurements. Long ECF yarns connecting these sensors are unnoticeable for users while plastic shielded cables so close to skin could cause rigidity and feel uncomfortable (Rantanen *at al.* 2001). The most challenging has been the implementation of reliable joints. When connecting yarns to temperature sensors by knotting yarns around component's pins and covering the joints with shrinking plastic after a few usage times the joints will get broken. Therefore, spiral connections were tested instead of knotting. However, the problem still partly remained and we concluded that the edges of heat-shrinkable plastic could be too hard for fibres, which gradually abraded off and disconnected the sensor from the system. We tested also small PWBs as combiners along the sensor connection bus. Yarns were soldered directly to these PWBs. In this case the same problem existed and the yarn got broken near the PWB's edge. The reason could be too hard PWB's edges or the interface between solder and yarns.

We have also sewn metal clad aramid fibre yarns by hand to cloth to form electrodes (Rantanen *at al.* 2001). These kinds of electrodes were used to replace commercial electrodes in the measurement of skin resistance in the sensor shirt. The main function of this measurement was to indicate the sweating level of the user. The simple measurement was not very demanding for the electrodes and they were found to be sensitive enough for our purposes. However, it would be beneficial to be able to sew the yarn by machine to ensure larger electrical conductivity and to ease and accelerate the implementation of electrodes. In the bioimpedance measurement suit we have also tested fabric electrodes (Vuorela *et al.* 2003). Suit's purpose is to measure the impedance of the user and based on that the suit can indicate the total body water level. In this application our goal was to replace commercial silver chloride electrodes with fibre electrodes, since commercial electrodes were found unsuitable for moving people. At this time we have utilised polyamide based silver plated material (Finnesd data sheet). New electrodes were found to be better than commercial ones however, long-term durability, e.g. coping with continuous washings and abrasion has to be found out.

TEST METHODS AND RESULTS

Due to problems that have been occurring in ECF's connections we decided to find out whether we could improve the reliability of these connections by proper connection mechanisms and materials. We have studied the use of two kinds of silver clad aramid fibre yarns. The first yarn is thinner and more twisted than the other one. For a reference material we chose another easily available material made of stainless steel, which is also very strength (Bekaert data sheet). Unfortunately, it has poor conductivity and its joining to PWBs is difficult, because its melting point is much higher than in joining materials. For metal clad aramid fibre yarn connections we used SnPb (tin-lead) solder and for stainless steel yarn connections we used Sn (tin) solder. Lead will be forbidden in consumer electronics and therefore, we also made joints by using electrically conductive adhesives. At first we used Loctite's conductive adhesive (Loctite data sheet) and later we used also Electron Microscopy Sciences' (EMS) adhesive (EMS data sheet). Both adhesives were isotropic and silver filled and they had to be cured by heat. Their conductivity properties are almost similar, but thermal properties of EMS's adhesive are better.

For test methods we chose accelerated environment reliability tests and a tensile strength test. Accelerated environmental tests produce failures in joints by the same damage mechanisms than in real use but in noticeably shorter time. Therefore, these tests can be utilised to estimate long-term behavior of the electronics products. The most probable environment hazards causing damages are temperature and humidity (Yoshinori & Yasuko 1996) and therefore, we decided to include these parameters to our tests. We performed tests in specific test chambers in our institute and during the tests temperature and humidity inside the chambers varied.

Ten pieces of one-type connections have been tested to ensure the reliability of the results. We have tested SnPb and Sn solders and two kinds of isotropic adhesives as joining materials, surface mount and leading through techniques, and also rubber sealant on some joints to prevent abrasion of the yarn against hard solder connection. Each channel of the climate chamber measures continually the voltage over a joint. The increase of this voltage level means higher resistances of the joints. A sample board is shown in Figure 1.

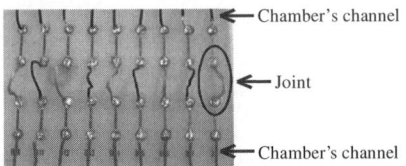

Figure 1. Test board.

Test methods

The first test we made was the temperature and humidity cycling test, which was done according to standard MIL-STD-202F. The whole test lasted ten days. In this test the lowest degree of temperature is minus 10 °C, which is important since in Finland the outdoor temperature can go to degrees below zero. The highest degree of temperature is 65 °C. During this test we used Loctite's conductive adhesive, which glass transition temperature is only 64 °C (Loctite data sheet). Therefore, we were not able to increase the test's temperature any higher than 65 °C. The relative humidity inside the test chambers changed according to the temperature from zero to 90 %.

Since fast changes in temperature strain products much more than slow changes, we decided to perform also a thermal shock test. The distinction between the temperature cycling test and the thermal shock test is the temperature's changing rate. In the thermal shock test the change rate in temperature should be at least 30°C/minute while in ordinary temperature cycling test the rate is not greater than 20 °C/minute. The test was done according to standard Jedec-104A. The test lasted 10 days. In this test standard temperature varied between minus 40 °C and 120 °C. Loctite's adhesive glass transition temperature was under this. So we assumed that it would act differently in this test than in the humidity and temperature cycling test. EMS' adhesive should resist temperature as high as 120 °C according to its data sheet (EMS data sheet).

We also performed tensile strength tests to estimate mechanical strain durability of joints. These were done at the institute of Fibre Materials Science at Tampere university of Technology. In these tests we can find out joints' breaking strengths in Newtons (N). For ordinary textile yarns about 50 samples are tested to get reliable results. However, even 10 samples can give approximate results and we decided to start with that. ECF yarns were joined to PWBs with solder and conductive adhesives introduced in Section Problem statement and test materials. We also studied the use of silicon to soften the contact between yarns and PWBs. The PWB sample's area was about 64 mm^2 and yarn's length about 10 cm. We connected yarns by using through hole as well as surface mount technique. We also studied whether the size of drill holes and the thickness of PWBs have any influence to joint's breaking strength.

Results

In the temperature and humidity cycling test and in the thermal shock test the voltages of the joints were measured at specified intervals. To get reliable results, we calculated the average voltage from ten samples at each interval. The voltages of all the joints were then drawn into same figure during the test time to be able to compare the behaviour of the joints. In the temperature and humidity cycling test voltages of stainless steel yarn joints changed clearer with the temperature than the voltages of metal clad aramid fibre joints. Furthermore, the voltages of adhesive joints are greater than the voltages of solder joints for both metal clad aramid fibre yarns and stainless steel yarns. Connections' behaviour during the test is illustrated in Figure 2. After these tests all the connections were functional and we decided to perform a more demanding test.

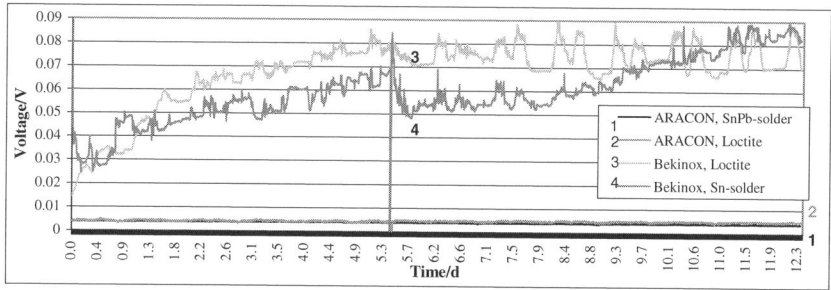

Figure 2. Test results of the temperature and humidity cycling test.

In the thermal shock test stainless steel yarn joints, which were done with Loctite's adhesive varied the most. This is an obvious consequence of adhesive's low glass transition temperature. The second highest changes in voltages happened in stainless steel yarn joints, which were done with EMS' adhesive. Generally voltages over connections varied more in stainless steel yarn connections than in metal clad aramid fibre yarn connections. This is a consequence of two things. First, it was quite hard to joint stainless steel material and therefore its joints might be poor. Secondly, its electrical conductivity was worse than in metal clad aramid fibre yarns. In these both tests solders survived better than electrically conductive adhesives. Closer examination of joints with a microscope showed that during the tests adhesive starts to move away from the connection pad increasing the joint's resistance while solder stays tight in a soldering spot. More specific results from the thermal shock test can be read from (Hännikäinen et al. 2004)

The results from the tensile strength tests are shown in Table 1. Stainless steel material was not suitable for testing and therefore, only the results of metal clad aramid fibres are shown. The joints of stainless steel yarns were easily bad and the tensile strength could not be measured. However, if the joint is reliable, the tensile strength is superior. In the table 1 Avg means the samples' average and Sd standard deviation. The breaking load for metal clad aramid fibres according to its data sheet is 66 N (DuPont data sheet). Results show that through hole joined yarns' breaking loads are higher than in surface mount joints. Breaking loads are also smaller when a PWB is located horizontally. PWBs' thickness and drill holes' sizes has not remarkable influenced to joints' breaking strengths. The strongest joints were made with Loctite's conductive adhesive. The breaking strength of 56.34 N is actually quite close to yarn's breaking load.

CONCLUSIONS AND FUTURE WORK

We have used ECFs as wire replacement as well as electrode materials. Based on the test results made for connections we can conclude that metal clad aramid fibre yarns are more suitable for cables than stainless steel yarns. Since the resistance of stainless steel material is higher than in metal clad aramid fibres and fluctuations according to temperature are larger. In addition, stainless steel yarns are difficult to solder, which makes their usage impractical. Tensile strength tests showed that with proper connection materials and mechanisms we can almost achieve the breaking strengths of the fibre. Solder connections managed better than adhesive connections in long-term tests and adhesive connections survived well in tensile strength tests. Different applications have different demands for joints and therefore, we also need to consider this while choosing the connection mechanisms.

We can also conclude that the truly breaking mechanisms of ECFs joints were not found yet. Instead, temperature and humidity tests proved that we could make reliable connections with ECFs. Temperature and humidity are not as big threats for joints than mechanical chafing. Next steps in the research will be the development of suitable field abrasion tests that can measure fibres' long-term abrasion against hard edges. The research will also be extended to study of ECF shielding, which could widen the usage possibilities of conductive yarns.

TABLE 1. TEST RESULTS IN THE TENSILE STRENGTH TESTS MADE VERTICALLY AND HORIZONTALLY ALIGNMENTS.

Joining materials	Avg/ N	Sd
Through hole, SnPb solder	43,34	3,26
Through hole, SnPb solder, silicone	40,83	6,09
Surface mount, SnPb solder	32,19	13,03
Surface mount, SnPb solder, silicone	41,58	10,64
Through hole, thicker FR4	44,49	6,04
Through hole, smaller drill hole	41,74	5,21
Through hole, Loctite	56,34	1,4
Through hole, EMS	42,74	4,87
FR4 horizontally, through hole, SnPbsolder	30,46	7,78

References

Bekaert Group, Belgium, Data sheet of Bekinox VN-Continuous stainless steel filament yarn.

DuPont, Richmond, Virginia, USA, Product bulletin of Brand metal clad fibre type XS0400E-018.

Electron Microscopy Sciences, Data sheet of Two part conductive silver paint.

Finnesd Inc., Pirkkala, Finland, Data sheet of silver plated nylon thread.

Yoshinori, K. & Yasuko, S. (1996). What is Environmental Testing? *Espec Technology Report* no. 1, Osaka, Japan. Availabale at URL http://www.espec.co.jp/image/english/pdf/technoe1.pdf, 3.6.2004.

Loctite, Data sheet of electrically conductive adhesive, brand code 3880.

Rantanen, J., Impiö, J., Karinsalo, T., Malmivaara, M., Reho, A., Tasanen, M. & Vanhala, J. (2002). Smart Clothing Prototype for the Arctic Environment. *Personal and Ubiquitous Computing.* **6:1**, 3-16.

Rantanen, J., Ryynänen, O., Kukkonen, K., Vuorela, T., Siili, A. & Vanhala, J. (2001). Electrically Heated Clothing. *Proceedings of the World Multiconference on Systemics, Cybernetics and Informatics*, July 22-25, Orlando, Florida, USA, 490-495.

Hännikäinen, J., Järvinen, T., Vuorela, T., Vähäkuopus, K. & Vanhala, J. (2004). Conductive Fibres in Smart Clothing Applications. *Proceedings of the Fifth International Conference on Machine Automation*, November 24-26, Osaka University, Osaka, Japan, 227-232.

Vuorela, T., Kukkonen, K., Rantanen, J., Järvinen, T. & Vanhala, J. (2003). Bioimpedance Measurement System for Smart Clothing. *Proceedings of the 7th IEEE International Symposium on Wearable Computers*, October 21-23, White Plains, NY, USA, 98-107.

DESIGN OF LOW-CLEARANCE MOTION PLATFORM FOR DRIVING SIMULATORS

H. Mohellebi[1], S. Espié[1], A. Kheddar[2], H. Arioui[2] and A. Amouri[2]

[1] Insitut National de Recherche des Transports et leur Sécurité
Arcueil, France
[2] Laboratoire Systèmes complexes – CNRS – University of Evry
Evry, France

ABSTRACT

Nowadays, driving simulators are interactive virtual reality tools that are useful in the human factors studies. The difficulty to reproduce in reality some road situations, mainly for risk and reproducibility reasons, increases the interest for such a tool. Nevertheless, the validation of the experiments carried out on driving simulator is closely related to embedding realism of the driver in the simulated world. Increasing the validity field of such a tool requires integrating the haptic modality and peculiarly inertia effects feedback. We present the design of a driving simulator enhanced with haptic feedback and motion platform that allows 3DOF movement restitution (two DOF are exclusive).

KEYWORDS

Driving simulator, motion restitution, virtual reality, washout, dynamic modelling, lane queuing.

INTRODUCTION

Some tasks that are easily achieved in actual driving situations (e.g. lane shift or queuing), become tedious when they are accomplished using a driving simulator. The lack of coherent sensory stimuli prevents the driver from an adequate control of the virtual vehicle. In order to drive a virtual vehicle, the driver needs to be provided with sufficient information which allows him to control the car as easily as it is the case in most of real situations. Depending on the hardware architecture of each simulator, the feedback strategies might be different, due to the very fact that the control is based on sensory-motor activity. Some studies, aiming to highlight the relevance of kinaesthetic perception in simulator controllability, clearly showed that longitudinal and lateral accelerations significantly reduce the simulator control variability. Consequently, illusion of inertial effects has to be provided to the driver. Such illusion rests on acquired knowledge of the human perceptive system. In the case of continuous accelerations the illusion is generally produced by tilting the driver forward or backward. Such tilt can be interpreted by his/her vestibule system, as either a positive or a negative acceleration, depending on the direction of the tilt (Reid & Grant (1991)). To reproduce transient accelerations, the platform is linearly moved in the same acceleration direction and come back when the acceleration is continuous or vanishes (Seigler & Kemeny (2001)). The implementation of this technique depends on the mechatronics and the architecture of the motion platform (Kheddar & Garrec (2002)). However, designers proposed architectures by seeking as often as possible to supply the driver with stimuli that are as close to those existing in actual situations. The most sophisticated ones bring into play hybrid architectures (X-Y, 6 axis + yaw) and their costs reach up to 100M$ (e.g. the NADS, USA). These simulators seek to simulate all possible driving situations. They are, however, not always able to accurately reproduce braking manoeuvres.

Another approach is possible; it is founded on the design of part-tasks simulators, intended for certain studies or applications (e.g. a particular driving task study) (Seigler & Kemeny (2001)). For these simulators, and concerning movement restitution, the goal is to produce a "sufficient illusion" in order to make possible the achievement of the task. By "sufficient illusion", we mean an illusion that allows the driver to carry out the task by using the same strategies as those he/she would have employed in an actual situation. This is essential to guarantee transferability of results acquired on a simulator to real situations. We designed a driving simulator whose objective is the study of "normal" driving situations (e.g. outside of sliding or harsh braking situations). We will focus on the most common driving situation: car following or queuing driving. Our objective is not to render acceleration in a realistic physical way, but rather to study the minimal inertial effect from which the subject extracts the necessary information to carry out the driving task in a manner comparable to a real driving situation. To do this, we have designed a motion platform equipped with two degrees of freedom. This makes it possible to animate the simulator's cab with a longitudinal movement, on the one hand, and with a weak pitch movement from the driver's seat or a weak tilt of the back of this seat, on the other hand.

PLATFORM DESCRIPTION

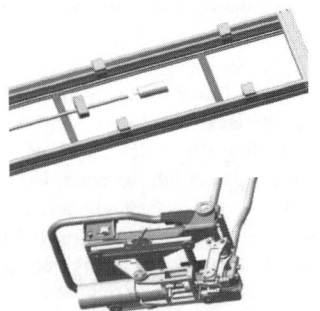 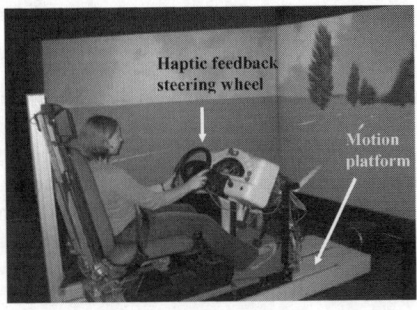

Figure 1: The driving simulator: CAD model of the motion platform and seat (left), overview (right)

The overall system is considered as two independent mechanically linked systems: the rotating driving seat and the longitudinal motion platform (fig 1). Each of them is driven by a single actuator. The motion platform undergoes translation motions according to one direction (front and back) which correspond to driver's acceleration and deceleration. The overall system's design allows having a simple linear model of the motion. The motion base supports the cabin consisting of the seat, the vehicle board and the driver. Because the rotations of the seat are low in amplitude, its induced inertia is negligible comparing to the total mass of the cabin's set. The linear motion of the cabin's set is made thanks to a ball screw/nut transmission mechanism driven by a DC actuator. The technological design was made in order to reduce, even eliminate, mechanical flaws such as backlash, mechanical play, static and dynamic friction, and to be able to design good quality acceleration and jerk based controllers.

MOVEMENT RESTITUTION ALGORITHM

In order to give to the driver the illusion of feeling the inertial effects of the simulated vehicle, the platform is powered by a washout algorithm. A washout algorithm aims to feedback transient accelerations, considering the kinematics and the mechanical/dynamic limitations of the platform (workspace, robustness, bandwidth, etc.). Transient acceleration is obtained by filtering the simulated acceleration signal through a high-pass filter in order to isolate the high frequency component. In this way, the signal collected has non-zero acceleration in the acceleration variation phase and zero acceleration in the continuous acceleration phase. After having filtered the acceleration, the signal produced is integrated twice in order to obtain the desired position profile. Another high-pass filter is integrated for bringing the platform back to its neutral position (fig 2). This is commonly referred to as "washout". The time constant of this filter must be chosen as to allow the platform returning to its initial position under vestibular system's movement perception threshold constraint.

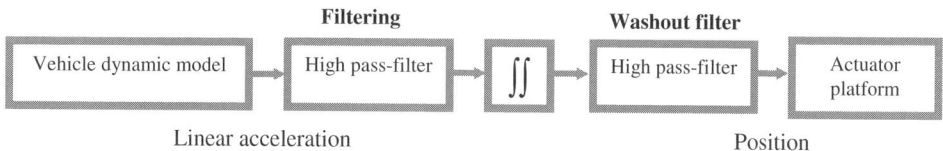

Figure 2: Washout restitution movement algorithm

The mini-simulator mounted on the mobile platform is derived from work carried out jointly between INRETS, LSC and Faros Company. The steering wheel is equipped with haptic feedback (Mohellebi et al. (2005)). Virtual scene rendering is carried out on screens or monitors (up to 150° according to the configuration). The simulator uses INRETS SIM^2 software. Traffic simulation, 3D sound rendering, and scenarios administrator are computed by INRETS ARCHISIM software. The vehicle model used comes from the CNRS CEPA research laboratory.

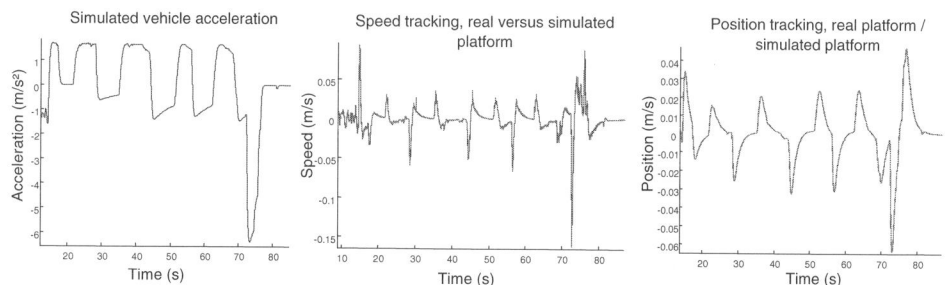

Figure 3: Movement of the platform according to the washout algorithm

A driver placed on the steering wheel of this simulator can cover a virtual route by interacting with a simulated environment. Scenarios make it possible to place the driver in pre-defined and reproducible situations. Data, which can be recorded during driving, concern particularly the driver's actions, the movement of the virtual vehicle and the position of the other vehicles).

In order to carry out a first evaluation of the mobile platform's performance for movement restitution (linear translation), we had a subject driving the simulator. The actuator intended for longitudinal movement restitution was powered by the above-described classic washout algorithm. This algorithm was computed on a control PC which received the acceleration of the simulated vehicle at 1.5kHz.

The acceleration signal obtained during the subject's driving contains acceleration phases, deceleration and continuous accelerations phases. Following the processing of this acceleration by the washout algorithm, this acceleration is transformed into a desired position profile with a tendency to return to the neutral position during the continuous acceleration phase (fig 3). We noted in this one that with a PID corrector, the platform position exactly superposed the desired position. Washout algorithm has been implemented in a two-factor (Back of the seat × Motion base) repeated measure where the seat variable consists of two levels (Back of the seat tilt 'on' or 'off'), whereas the magnitude of the platform motion consists of three levels (Without, Short, or Long longitudinal movement). All of these 2×3 experimental conditions, requires that the subjects drive the simulator for five minutes on average. In (Neimer et al. (2005)) we show that best performance is obtained by having a controlled combination of the back of the seat inclination 'On' and Short longitudinal platform displacement. Rotating the entire seat is not considered since it induces false cues.

CONCLUSION

The proposed driving simulator and the motion cueing algorithm have been used in various psychophysics experiments. Experiments consisted in exploring minimal displacements and subsequent inertial effect restitution to perform file queuing driving situations. Obtained results are presented in (Neimer et al. (2005)) and show the validity of the proposed concepts. Our future work will focus on the development of new control strategies for the platform, which will aim to favour driver control over the virtual vehicle's acceleration. Optimal coupling of visual, haptic and inertia effects restitution will be also investigated.

REFERENCES

Kheddar A. and Garrec Ph. (2002). Architectures de plates–formes mobiles pour simulateurs de conduite automobile. *Appraisal Report, CRIIF.*

Reid L.D. and Grant P.R. (1991). Motion-base development package for NADS. *Transportation Research Center of OHIO.*

Reymond G. and Kemeny A. (2000). Motion cueing in the Renault Driving Simulator. *Vehicle System Dynamics* 34:4, 249-259.

Reymond G. and Kemeny A. and Droulez J. and Berthoz A. (2000). Contribution of motion platform to kinesthetic restitution in a driving simulator. *Driving Simulation Conference.* 33-55.

Seigler I. and Kemeny A. (2001). Etude sur la pertinence de la restitution physique du mouvement en simulation de conduite en fonction des caractéristiques physiologiques et psycho-physiques de la perception du mouvement propre. *Appraisal Report, LPPA.*

Mohellebi H. and Espié S. and Kheddar A. (2004). Adaptive haptic steering wheel for driving simulators. *International Conference on Intelligent Robots and Systems.*

Neimer J. and Mohellebi H. and Espié S. and Kheddar A. (2005). Optimization of Linear Motion Base Dedicated to Normal Driving Conditions. *Driving Simulation Conference.*

HIGH PERFORMANCE LOW COST STEREO PROJECTOR SYSTEM FOR VIRTUAL REALITY

Heikki Kosola, Karri Palovuori

Tampere university of Technology
Institute of Electronics
P.O.Box 692, 33101 Tampere,
Finland

ABSTRACT

This paper describes a novel method for producing projected stereo images with very high quality and reduced cost. Using two standard LCD-projectors with an electro-mechanical shutter it is possible to produce a time division multiplexed stereo image pair with a very high refresh rate. All beneficial characteristics of the LCD-projectors like the spatial stability, good geometry and image sharpness are preserved.

KEYWORDS

Stereo projection, Virtual reality, LCD, Shutter

INTRODUCTION

Motion and position compensated stereographic visualization is an almost essential requirement for any virtual environment setup today. To produce projected stereographic image there are three main techniques available: autostereoscopic, passive stereo and active stereo. In the autostereoscopic display each pixel is divided to subpixels. Each subpixel has its own spatial sector where it can be seen. Spectator's eyes see different set of subpixels and stereoscopic information can be shown. The demand for unpractical number of subpixels and the required optics makes autostereoscopy an unthinkable solution for a large virtual reality display.

The high cost of a stereo enabled CRT or DLP projector could be avoided by using passive methods to separate the images between the viewer's two eyes. Word 'passive' means that user is not wearing active shutter glasses to blank out the unwanted picture. Passive filters are used instead. Both images are projected simultaneously on the screen. Separation is done typically by different polarization of the two images. Both linear and circular polarization are used. Wavelength multiplexing is also used [1]. All of the projection techniques (CRT, DLP, LCD) could be used with passive stereo. Advantages of passive methods are the completely flickerfree image and lighter eyewear. All of the passive methods suffer from less than perfect channel separation.

The most common active stereo approach is to use one CRT-projector per one display surface. Images for user's both eyes are shown sequentially. Shutter glasses are used to select the correct image for each eye. The speed of the phosphor material on the CRT is a compromise between a flickerfree image and an adequate response time. Problems arise especially in stereographic presentations where the CRT is forced to display two (or more) different images sequentially at a high rate. The previous image leaks to the next one so that the channel separation cannot be perfect. A flickerfree visual experience requires at least 75Hz refresh rate per eye at ordinary range of image brightnesses. As the brightness increases, the required image refresh rate rises up to and possibly beyond 500 Hz [2]!

LCD-projectors and displays use a completely different image reproduction method. A constant light source is modulated by a liquid crystal panel. The typical response time of an lcd-panel is 20 - 40 ms. This is just enough to show animated graphics at 30 - 50 frames per second but it is way too slow for time division multiplexing required for shutter glasses. One major advantage of the LCD-projectors is the light output, which is significantly higher than in their CRT counterparts. Also the prices of LCD-projectors are very competitive compared to CRT-projectors with adequate light output. As mentioned above, the LCD-panel can not switch the image fast enough so the switching must be performed externally. This paper introduces a method where the light output of each individual projector is controlled by an external shutter disc. Shutter glasses, worn by the user, select which eye is allowed to see the image of the currently active projector.

THEORY OF OPERATION

The operation principle is simple and elegant. Both eyes have their own projector. The projector modulates its ligh source with the appropriate image and the external shutter switches the picture on and off to the screen. The inherent slowness of the LCD-panel poses no obstacle to a stereographic projection. In our construction the shutter is common for both projectors. Nevertheless, it is possible to use two separate shutters which give more flexibility to projector mounting.

The LCD-projector

The nature of the LCD-projector makes it very attractive to be used with an external shutter. Individual pixels maintain their states over the whole frame period. The external shutter could - but it does not have to - be synchronized with the projector or with the graphics generating computer. In a fast moving scene the unsynchronized image might bring out a 'tearing effect' where the image splits to horizontally unaligned upper and lower portions. If the projector updates its LCD-panels directly with the incoming RGB-signal it would be useful to use the vertical synchronization pulse to control the shutter motor. It would remove the tearing problem and give a constant response time from a graphics drawing to the screen. Both of these are desired aspects in a simulation environment.

For comparison, the DLP projectors offer a slightly better contrast and light output as their LCD counterparts. The DLP technology also enables sequential stereo by itself it still have its limitations. Currently commercial DLP-stereo enabled projectors have frame rates limited to 60Hz per eye which is quite low for high bright images. [3][4]

The electro-mechanical shutter

In our system, a mechanical shutter does the switching between the two images forming the stereo image pair. The shutter is disc shaped and positioned in very front of the projectors. The disc rotation axis is fixed and in the middle of the line crossing both projectors output lenses. One half of the shutter needs to be transparent and another half opaque. To prevent simultaneous illumination of the screen, the opaque portion is extended by the width of the projected image at the shutter disc's plane. The extended opaque section decreases the optical efficiency somewhat from the theoretical maximum of 50%. The width of the extension has to be at least the width of the projected image at the shutter plane. Therefore it is wise to place the shutter as close to the projectors as possible. Increasing the radius of the disc decreases the angular width of the projector's image and therefore increases the optical efficiency. On the other hand, as the diameter of the shutter increases so does the induced audible noise, size of the installation and the alignment problems of the projected image. Figure 1 illustrates the geometry of the shutter disc. If the disc rotates in clockwise direction, the projector 2 would just start to show its image on the screen.

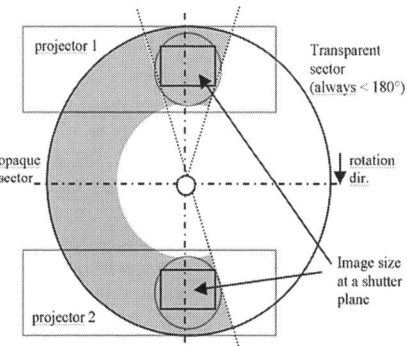

Figure 1: The geometry of the shutter disc

The shutter glasses

The glasses we have been using in our tests and also in the constructed simulation environment are of the CrystalEYES II type. These glasses use an infrared link to get the synchronization signal from a controller. The custom infrared transmitter is controlled with a signal directly from the rotating disc. The phase locking between the electro-mechanical shutter and the shutter glasses is therefore very robust. Even if there is slight fluctuation in the rotational velocity of the disc, the glasses are switching at the precise moment.

IMPLEMENTATION

The construction of the prototype installation is based on two identical Canon LV-7105 LCD-dataprojector. These projectors are of a very common type with XGA resolution (1024 x 768) and about 1100 Ansi lumens of light output. The projectors are mounted inside a wooden enclosure. Without the possibility of the vertical lens shift on the projectors, there are a slight misalignment on the screen. As the position of the projectors obviously has to be slightly offset from each other (in our system they are vertically offset by 12 cm), if the projected images are adjusted for maximal overlap, the slight misalignment results from image keystoning. On a three-meter wide picture, the alignment error was about one to two pixels measured at the upper or lower edges of the image. In many applications, particularly in virtual reality, it is often not mandatory to project the images to overlap exactly as the graphics generating computer has to update the images constantly anyway and can easily incorporate the information of the factual, different image positions. The net result would be a change in the shape and size of the visualizable spatial space of the system. This might be beneficial or detrimental in theory, depending on the case, but in practical applications the difference would be neglectably small. Finally, by sacrificing the non-overlapping parts of the images and presenting them as black, even this small effect can be removed.

The shutter disc is made of acrylic plastic sheet. The opaque areas are covered with metallic paint. The disc is balanced after coating to eliminate vibrations and noise. The diameter of the disc is 160 mm. The projected image covers a sector of 38 degrees from the disc. Therefore, the opaque sector is 218 degrees wide. So about 40 % of the projectors light output is transferred to the screen. The attainable light output is comparable to a single DLP-projector stereo setup. [5]

EXPERIMENTAL RESULTS

The implemented prototype performs as expected. The use of the 100 Hz refresh rate per eye resulted in a completely flickerfree image. The disc type shutter is rotated by the brushless DC-motor that has offered reliable and maintenance free operation for thousands of hours. The projectors used in the prototype have an XGA resolution and 1100 ANSI lumens of light output. With a 3m width of projected image the overall brightness could be better but still is entirely comparable to a projected CRT image. All other characteristics like the sharpness, the spatial stability and the lack of flicker are superior when compared to the CRT counterpart. The constructed prototype has been applied to a person lift platform training simulator system. The simulator has been in everyday use for more than two years now, and the reliability of this new method of stereo projection has been field proven.

CONCLUSION

The described method for producing stereo projection was proven very functional. The major advantage of this method is the high stereo multiplexing rate. It is now possible to use sufficient rates without sacrificing any of the important properties of the projected image.

Possible enhancements in the light output and/or resolution are simply achieved by selecting a proper type of LCD projectors. The stereo projector adjustments can be made even easier when both LCD projectors are located on the same optical axis. This could be done with full or semi-transparent mirrors.

REFERENCES

[1] Helmut Jorke, Markus Fritz, "Infitec – A new stereoscopic visualization tool by wavelength multiplex imaging", Infitec GmbH

[2] Bridgeman, B. and Montegut, M. 1993. Faster flicker rate increases reading speed on CRTs. SPIE Vol.1913 Human Vision, Visual Processing, and Digital Display IV. PP. 134-145.

[3] Christie Digital Systems, Inc. Web site: http://www.christiedigital.com/

[4] Ian McDowall, Mark Bolas, Dan Corr, Terry Schmidt, "Single and Multiple Viewer Stereo with DLP Projectors", Fakespace Labs, Christie Digital Systems

[5] Barco nv. Web site: http://www.barco.com

ANALYTICAL AND EXPERIMENTAL MODELING OF INTRA-BODY COMMUNICATION CIRCUIT

Y. Terauchi[1], K. Hachisuka[1], K. Sasaki[1], Y. Kishi[1], T. Hirota[1],
H. Hosaka[1], K. Fujii[2] and K. Ito[3]

[1] Graduate School of Frontier Sciences, The University of Tokyo,
Bunkyo-ku, Tokyo 113-8656, JAPAN
[2] Graduate School of Science & Technology, Chiba University,
Chiba-city, Chiba 263-8522, JAPAN
[3] Research Center for Frontier Medical Engineering, Chiba University
Chiba-city, Chiba 263-8522, JAPAN

ABSTRACT

Intra-body communication uses human body as the propagation medium. This may become a new wireless communication method for Personal Area Network (PAN) with less power consumption and higher communication security compared to conventional RF methods. A common analytical model of intra-body communication is a combination of capacitive coupling among the human body, electronic devices, and the environment. Experimental results suggest that there are optimal parameters for transmission. We assume that it is a combination of not only the capacitive couplings, but also of a radio wave transmission and of imbalances in the electrical impedances among the transmitter/receiver electrodes attached to the human body.

KEYWORDS

Personal Area Network, intra-body communication, analytical model, human body equivalent phantom, phase measurement

INTRODUCTION

Advancement of information technology has accelerated the spread of ever smaller and lighter information and communication devices such as mobile phones, PCs, and PDAs. It is common to see people carrying more than one of these devices. In the near future, the devices will most likely become wearable. Data and system resources can be shared by connecting multiple devices carried by a single person, similar to computers in offices connected by a LAN. This new type of network was proposed in Zimmerman (1995) as a Personal Area Network (PAN).

Intra-body communication, which uses human body as the propagation medium, is an alternative to conventional radio transmission for short-range wireless communication and proposed as the optimal method for PAN. Since the human body is an electrical conductor, intra-body communication may become a novel wireless communication method with less power consumption and higher communication security compared to conventional RF methods. It also allows a new communication mode of human friendly man-machine interface, because information is transmitted only when body contact is made.

Three types of intra-body communication are shown in Figure 1. The circuit type and capacitive coupling type make a circuit with the surrounding environment, and are suited for communicating with devices that are not attached to the human body (e.g. electronic money, keyless entry system). Several research groups have already demonstrated intra-body communication devices of these types such as Handa, et al. (1997) and Matsushita, et al. (2000). The propagation type does not require external circuits and has more tolerance to external noise compared to the other two. It is suited for communication between devices attached to the body (e.g. PAN).

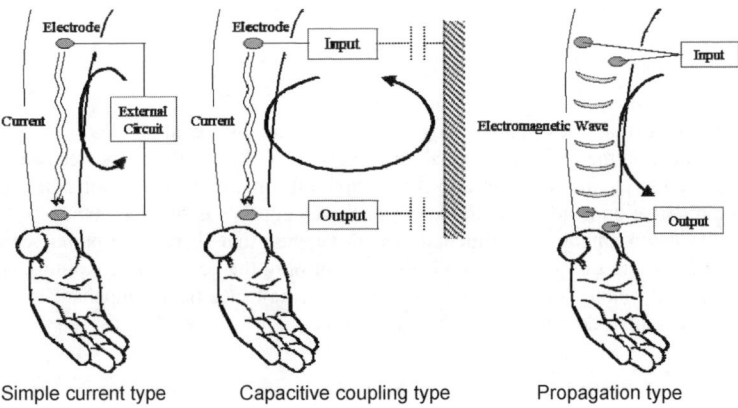

Figure 1: Types of intra-body communication

The authors aim to propose a new wireless communication method intended for PAN. This research focuses on clarifying an analytical model and the mechanism of intra-body communication of the propagation type.

MODELING OF INTRA-BODY COMMUNICATION

Electrical properties of the human body

Electrical properties of the human body were measured to determine the most efficient frequency to send signals. Sine waves of 1 V_{p-p} generated by a function generator (transmitter) were applied to two electrodes attached above the elbow. Two electrodes were also attached to the wrist of the same arm and an oscilloscope (receiver) was used to read the received signal strength. Figure 2 shows that maximum transmission gain is obtained around 10 MHz.

For experiments, special care is taken for electrical isolation between the transmitter and the receiver, including the measurement devices. First, the signal ground is separated because a common ground between the transmitter and the receiver may act as a wired return path for the signal. Such paths would not exist in actual applications for intra-body communication. Using a common AC power line also exhibits a much stronger coupling than the capacitive coupling between the human body and surrounding environment. In order to minimize these undesired couplings, signal generator, amplifier, and oscilloscope are all battery-powered. Separate power supplies are used for the transmitter and the receiver.

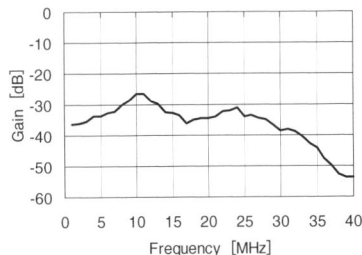

Figure 2: Comparison of transmission gain with frequency

In the experiments, a human equivalent phantom was also used. The conductivity and relative permittivity of the phantom was adjusted to the same value as the human body. Its size was 500 x 500 x 4500 mm^3, the average size of the human arm. Phantoms improve reproducibility of experiments. Its simple shape makes computer based calculations easier.

Four-terminal circuit model

In the kHz range, the effect of electromagnetic waves is considered negligible. A common analysis model of intra-body communication in such frequencies is a combination of capacitive coupling among the electrodes attached to the human body, the devices, and the environment. Figure 3 shows a four-terminal circuit model based on Hachisuka, et al. (2003) and Terauchi, et al. (2003).

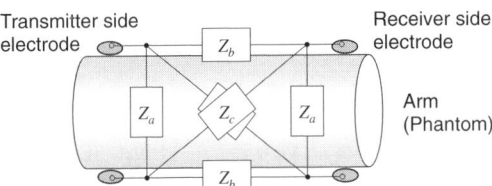

Figure 3: Four-electrode model

Transmission gains calculated from this four-electrode model match the experimental results in the kHz range. It is calculated by the following equation.

$$gain = 20\log_{10}\left(\frac{\dfrac{Z_a Z_c}{Z_a + Z_c} - \dfrac{Z_a Z_b}{Z_a + Z_b}}{\dfrac{Z_a Z_c}{Z_a + Z_c} + \dfrac{Z_a Z_b}{Z_a + Z_b}}\right) \text{ [dB]} \tag{1}$$

It can be seen from Eqn.1 that the difference between the horizontal impedance element (Z_b) and the diagonal impedance element (Z_c) has to be large to obtain a large gain. However, it can also be understood that the difference between the two elements becomes small as the distance between the transmitter and the receiver increases. Also, since the relative permittivity of the human muscle is more than 20,000 at 10 kHz, it is difficult to enlarge the difference between Z_b and Z_c.

In this research, a similar four-terminal circuit model was investigated for a new form of transmission.

In the new model shown in Figure 4, only one electrode each from the transmitter and the receiver makes contact to the human body. This model is conventionally called the two-electrode model.

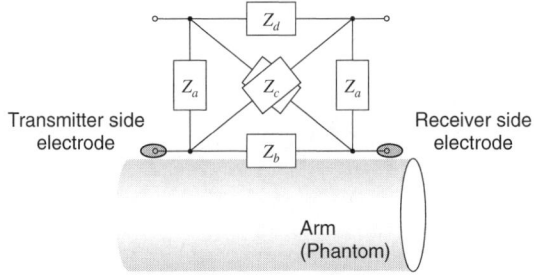

Figure 4: Two-electrode model

The transmission gain of this model is calculated by the following equation.

$$gain = 20\log_{10}\left(Z_a \cdot \frac{\dfrac{Z_c}{Z_c+Z_d} - \dfrac{Z_b}{Z_b+Z_c}}{Z_a + \dfrac{Z_c Z_d}{Z_c+Z_d} + \dfrac{Z_b Z_c}{Z_b+Z_c}} \right) \quad [dB] \qquad (2)$$

The impedance of each element is calculated by FEMLAB (simulator software using the finite element method). Transmission gains are then calculated using Eqn. 2 and compared with experimental results. In Figure 5a, it can be seen that the calculated values and experimental values match well. Figure 5b shows that the gain of the new model is relatively higher than the previous four-electrode model and does not drop as the distance between the transmitter and receiver increases.

If the human body is considered as a ground plane, the two-electrode circuit model may be similar to the behavior of rod antennas. However, this assumption requires further study.

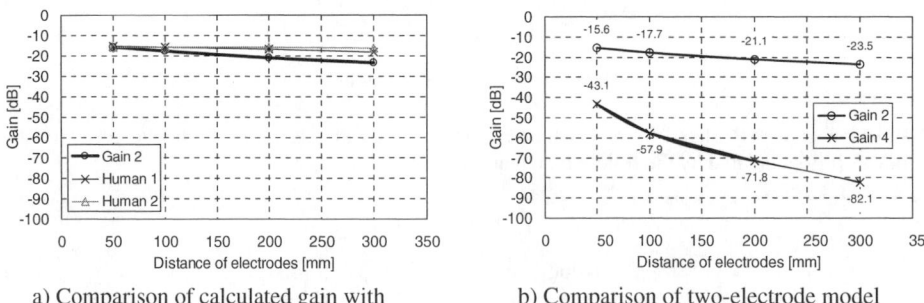

a) Comparison of calculated gain with experimental results

b) Comparison of two-electrode model with four-electrode model

Figure 5: Comparison of transmission gain

MEASUREMENT OF PHASE SHIFT

The authors made a small transmitter with a size of 70 x 100 x 50 mm^3 for phase characteristics measurement of the four-terminal circuit. The signal from the transmitter is sent to the receiver to compare the phase difference between the original signal and the signal transmitted through the phantom. To avoid unnecessary electrical coupling, optical fiber is used to send the synchronization signals. The distance between the transmitter and the receiver is 300 mm.

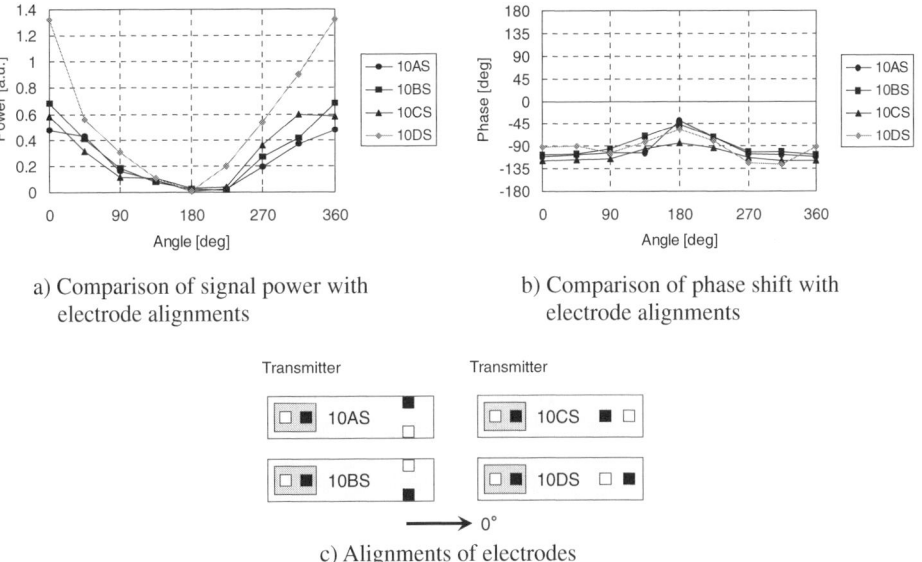

Figure 6: Measurement of phase shifts

Figure 6 shows the result of the experiment at 10 MHz. As the transmitter rotates clockwise, signals were measured every 45 degrees for a total of 8 directions. Maximum transmission gain is attained when the two electrodes of the transmitter are aligned to the direction of the receiver (Figure 6a). Results show that the phase variation between the source signal and the received signal depends on the direction and arrangements of the electrodes (Figure 6b). If we assume that the four-terminal circuit model is correct, there should be a phase reversal when the electrodes of the transmitter are reversed. However, the measurements at 10 MHz show that the phase shift is only plus or minus 45 degrees. The results suggest that there are elements other than capacitive coupling. In the MHz range, the possibility of airborne radio wave transmission also remains (Fujii, et al. (2004)).

TRANSMISSION PATH

In the previous section, the path of transmission still remains unclear. There are three possible paths: (i) Inside the body (through muscles, blood etc.), (ii) Surface of the body (along the skin), and (iii) Airborne (radio wave transmission). An experiment in an electrical anechoic chamber was conducted in order to determine the path. A large conductor plate was placed between the transmitter and the receiver (Figure 7a). The walls of the chamber absorb all electromagnetic waves and there is no reflection so all airborne radio wave transmission is cutoff. The gap between the plate and phantom can be changed. The frequency of 10 MHz was chosen for the experiment.

Figure 7b shows the experimental results with the signal strength calculated by FDTD method. There is only a slight difference between the signal strength measured when the gap is 10 mm and when there is no conductor plate. This suggests that the signal does not travel through the open space. As the gap closes, the received signal strength gradually decreases. When the gap is 0 mm, no signal is received. This suggests that the signal is not propagated inside the human body. This is may be explained by the fact that the relative permittivity at 10 MHz is about 150 for the muscle and over 250 for the blood. High relative permittivity results in the electromagnetic waves to decay in a short distance. The results of this experiment indicate that the transmission path is most likely the surface and proximity of the body rather than the interior.

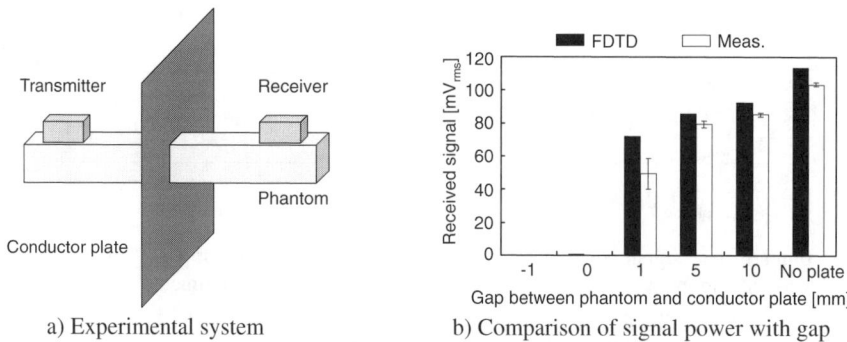

a) Experimental system

b) Comparison of signal power with gap

Figure 7: Investigation of signal transmission path

CONCLUSION

The four-terminal circuit model is effective in calculating the transmission gain in kHz range. The new two-electrode model shows a higher gain compared to the previous four-electrode model. The experimental results match well to confirm the calculation. From the measurement of the phase shifts using the improved transmitter, it was suggested that intra-body communication is a combination of not only the capacitive coupling but also of a radio wave transmission and of imbalances in the electrical impedances among the electrodes of the transmitter and receiver. Additional experiments in the electrical anechoic chamber suggest that the signal is propagated on the surface and proximity of the body. In practical use, intra-body communication devices are to be wearable devices. Further downsizing will be done in following research.

REFERENCES

Fujii K., Ito K., Hachisuka K., Terauchi Y., Sasaki K. and Itao K. (2004). Study on the optimal direction of electrodes of a wearable device using the human body as a transmission channel. *Proceedings of the 2004 International Symposium on Antennas and Propagation* **vol 2,** 1005-1008

Hachisuka K., Nakata A., Takeda T., Shiba K., Sasaki K., Hosaka H. and Itao K. (2003). Development of wearable intra-body communication devices. *Sensors and Actuators A: Physical* **105:1,** 109-115

Handa T., Shoji S., Ike S., Takeda S. and Sekiguchi T. (1997). A Very Low-Power Consumption Wireless ECG Monitoring System Using Body as a Signal Transmission Medium. *Proceedings of the 1997 International Conference on Solid-State Sensors and Actuators*, 1003-1006

Matsushita N., Tajima S., Ayatsuka Y. and Rekimoto J (2000). Wearable Key: Device for Personalizing nearby environment. *Proceedings of the 4th International Symposium on Wearable Computers*, 119-126

Terauchi Y., Hachisuka K., Sasaki K., Hosaka H. and Itao K. (2003). Study on electromagnetic propagation within the human body. *Proceedings of 2003 JSPE Autumn Meeting*, 509, (in Japanese)

Zimmerman T. G. (1995). Personal Area Networks (PAN): Near-Field Intra-Body Communication *MIT Media Laboratory M.S. thesis*

DESIGN OF MULTI SENSOR UNITS FOR SEARCHING INSIDE OF RUBBLE

Kenji Inoue[1], Masato Yamamoto[1], Tomohito Takubo[1],
Yasushi Mae[2] and Tatsuo Arai[1]

[1] Department of Systems Innovation, Graduate School of Engineering Science,
Osaka University,
Toyonaka, Osaka, 560-8531, Japan
[2] Department of Human and Artificial Intelligence Systems, Faculty of Engineering,
University of Fukui,
Fukui, Fukui, 910-8507, Japan

ABSTRACT

"Search balls" are small sensor units for searching inside of rubble. Each ball is not equipped with locomotion mechanism but contains some sensors for searching for disaster victims and a radio transceiver in an impact-resistant outer shell. Many balls are thrown into rubble and fall down while repeatedly colliding; they are scattered inside the rubble. The sensor information from the balls is transmitted on radio out of the rubble and monitored at a safe area. Thus rescuers can search a wide area inside the rubble rapidly. The developed ball has two wireless cameras for search, infrared LEDs for illumination, a radio receiver for communication with outside monitoring computers and a battery; these are packed into an impact-resistant sphere outer shell. This ball can provide the view of its entire circumstance by rotating the cameras using a motor. Just like a brim of a hat, a ring is attached to the shell for suppressing rolling of the ball; it is effective for distribution of balls inside rubble.

KEYWORDS

Rescue, Search, Sensor, Camera, Infrared LED, Wireless Communication, Rubble

INTRODUCTION

At disaster areas created by earthquakes, it is important to find victims buried under rubble as rapidly as possible. In the current rescue activities, because rescuers cannot enter narrow gaps among rubble, they are forced to find victims using a little information such as voice and sound from the victims. Hence rapid search is difficult. Furthermore, for fear of secondary disasters by fire, gas leak and collapse of buildings, disaster areas are also dangerous for rescuers. For these reasons, practical rescue devices, machines or robots for searching are strongly expected. These devices and machines are required to be small, lightweight, cheap, non-flammable, low energy consuming, easy-to-operate and well-adapted to irregular terrain. Recently, many search robots have been studied and developed (Kamegawa (2004), Kimura (2002), Osuka (2003), Perrin (2004), Stoeter (2002), Stormont (2003),

Figure 1: Concept of search ball

Takamori (2003), Tsukagoshi (2002, 2004), Wolf (2003)). But they are large, heavy, expensive and high energy consuming because they have locomotion mechanisms. In addition, they require operation skill. Accordingly, rapid and wide-area search may be difficult using only these robots.

For these problems, we have proposed a concept of ``search ball'' for searching inside of rubble (Inoue (2005)). A search ball is a small sensor unit which is not equipped with locomotion mechanism but contains some sensors for searching for disaster victims, such as cameras, and a radio transceiver in an impact-resistant outer shell. Many balls are thrown into rubble and fall down while repeatedly colliding; they are scattered inside the rubble. Each ball searches the surrounding area with its own sensors. The sensor information from the balls is transmitted on radio out of the rubble and monitored at a safe area. In this way, rescuers can search a wide area inside the rubble rapidly. When the rescuers are removing the rubble to rescue the found victims, the balls are collected for reuse. Search balls can be made small so as to enter narrow space among rubble and have the merits of lightweight, low energy consuming and easy operation. The problem of search balls is that they cannot move actively. In order to cover this weak point, a large number of balls are scattered into rubble.

In the present paper, a new type of search ball is developed: it contains two wireless cameras for search, infrared LEDs for illumination, a radio receiver for communication with outside monitoring computers and a battery. This ball can provide the view of its entire circumstance by rotating the cameras using a motor. Its sphere outer shell is made of impact-resistant and transparent plastic, thus protecting these internal parts from drop impact and collision with rubble. Just like a brim of a hat, a ring is attached to the shell for suppressing rolling of the ball; it is effective for distribution of balls inside rubble. The outside computer identifies the balls inside rubble and acquires the sensor information from them by one-to-one communication.

CONCEPT OF SEARCH BALL

Search balls are rescue devices to search for disaster victims buried under rubble. A search ball is not equipped with locomotion mechanism but contains some sensors and a radio transceiver in an impact-resistant outer shell. **Fig.1** shows the process of searching inside of rubble using search balls.

1) Rescuers throw many balls into rubble. The balls fall down while repeatedly colliding with the rubble, and they are scattered inside the rubble.
2) Each ball searches the surrounding area with its own sensors, and the sensor information is transmitted on radio out of the rubble.
3) The rescuers outside the rubble check the sensor information from all balls and find victims.
4) With the aid of the signals from the balls which detect the victims, the rescuers get gradually close to the victims while removing the rubble.
5) In process of removing the rubble, the rescuers collect the balls for reuse.

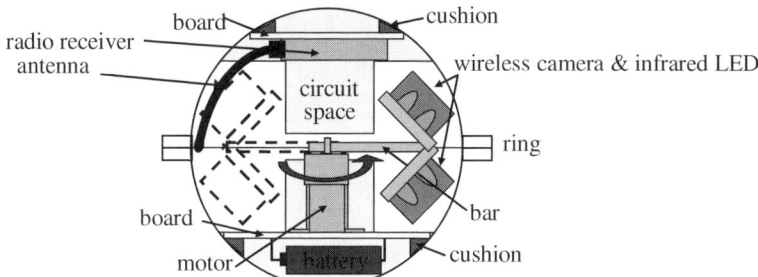

Figure 2: Structure of search ball with rotating cameras

Because of no locomotion mechanism, search balls can be made small so as to enter narrow space among rubble and have the merits of lightweight, low energy consuming and easy operation. The problem of search balls is that they cannot move actively and cannot always be located as desired. In order to cover this weak point, a large number of balls are scattered into rubble; that allows rapid and wide-area search. It is difficult to search the entire area using only search balls. Hence rescuers or rescue robots search the remaining area. Such cooperation of search balls and rescuers/robots would be able to reduce the amount of time required to search inside of rubble.

We suppose to search the inside of a collapsed Japanese-style wooden house; the area to be searched by search balls at once is less than $100[m^2]$ ($10[m] \times 10[m]$). Each ball searches $1[m^2]$ area around itself. The balls are scattered into the area which rescuers cannot see: for example, beneath or behind collapsed beams and inclined furniture.

DESIGN OF SEARCH BALL WITH RORATING CAMERAS

Fig.2 illustrates the conceptual design of a search ball with rotating cameras.

Cameras Providing View of Entire Circumstance

Generally, cameras, microphones, infrared sensors and CO_2 sensors are said to be effective for searching inside of rubble for victims. If rescuers check sensor information and judge whether victims exist, cameras will be most useful sensors. Hence we adopt cameras as the sensors of search balls. Because balls might be scattered around a victim, they will be able to provide some images of the victim from different points of view; it is effective for the judgment by the rescuers. In the future, we will pack other sensors into balls. For example, a microphone permits searching for conscious victims. A microphone and a speaker enable victims to communicate with rescuers.

Search balls are required to find victims around themselves with their sensors. The locations where the balls drop cannot be controlled or the balls cannot move after drop, because they have no locomotion mechanisms: some balls drop into narrow gaps, some balls stop on slopes, and other balls drop to the bottom. Hence it is desirable that the sensors can look all around the ball. For this requirement, the proposed search ball rotates cameras using a motor inside for providing the view of its entire circumstance. As shown in Fig.2, two wireless cameras with $90[deg]$ angle of view are attached to a bar with $45[deg]$ tilted, and this bar is connected to a small motor. Rotating this motor $360[deg]$ around, the ball obtains the view of its entire circumstance. Search balls enter inside of rubble, where it is dark. Hence we attach infrared LEDs around the cameras for illuminating dark environment and rotate them together with the cameras.

Impact-resistant Ball Structure

Because search balls drop into and repeat collision with rubble, impact-resistant ball structure for protecting internal parts is required. For this requirement, the motor connected to the cameras and

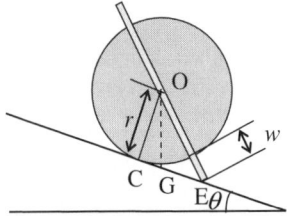 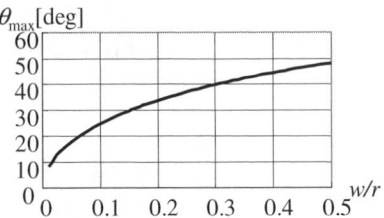

Figure 3: Relationship between ball shape and maximum slope angle

infrared LEDs, a radio receiver and a battery are attached to electronic circuit boards, and the boards are fixed to the internal surface of a small sphere outer shell with cushion material. This shell is made of impact-resistant transparent plastic.

Ball Shape for Suppression of Rolling
In order to cover the weak point of no locomotion mechanisms, some kinds of technique to distribute search balls widely inside rubble are required. For this requirement, we discuss the effect of ball shape on ``rolling''. First a ball drops on top rubble and rolls to its edge, then it drops again on next rubble. If the ball is an entire sphere, it repeats these actions until it reaches the lowest points or horizontal planes. But some balls must stop halfway down to the bottom in order for balls to be scattered widely and evenly inside rubble. Hence it is important to endue balls with different ability to roll. For this purpose, we adopt the ball shape shown in **Fig.3**. Just like a brim of a hat, a ring is attached to the sphere outer shell; r is the radius of the sphere, and w is the width of the ring. As an index of ability to roll, we use the maximum slope angle θ_{max} for the ball not to roll. When this ball remains stationary on the slope of angle θ, the projection of its center of mass on the slope, G, is between the contact point of the sphere on the slope, C, and the contact point of the ring edge on the slope, E. Thus the angle θ is maximum when G coincides with E. Letting O be the center of the sphere,

$$\theta_{max} = \angle \text{EOC} = \cos^{-1}(r/(r+w)) \tag{1}$$

The graph in Fig.3 shows the relationship between θ_{max} and the ratio w/r. By changing w/r, we can endue the ball with different ability to roll: the greater w/r suppresses the ball's rolling. This ball shape also has the effects of irregular rolling and bouncing of the ball. Preparing balls of different w/r and scattering them will bring wide distribution of the balls inside rubble.

Identification of Balls and Acquisition of Sensor Information
A large number of search balls are distributed inside rubble. Thus the wireless communication between the balls inside the rubble and monitoring computers outside requires identification of the balls and acquisition of the sensor information from the balls. Here we suppose to search the inside of a collapsed Japanese-style wooden house; its area is less than 100[m^2]. Because balls are thrown into the house after it is collapsed, there exists the route of entry for each ball, through which the ball enters the inside of the rubble. This route can be a path of communication between the ball and the outside. In this situation, one-to-one wireless communication between each ball and the outside monitoring computer is possible.
As shown in Fig.2 and **Fig.4**, the designed search ball has two wireless cameras, infrared LEDs, a motor and a radio receiver. Before balls are thrown into rubble, a unique ID number is assigned to each ball. The radio frequency of all balls and the computer is matched. After drop, the computer broadcasts the ID of the target ball to all balls. Each ball receives the ID and checks whether it agrees with the assigned ID. If they agree, the target ball turns on its cameras, LEDs and motor. The computer sends motor commands: 1 bit for switching rotation/stop and 1 bit for changing direction of

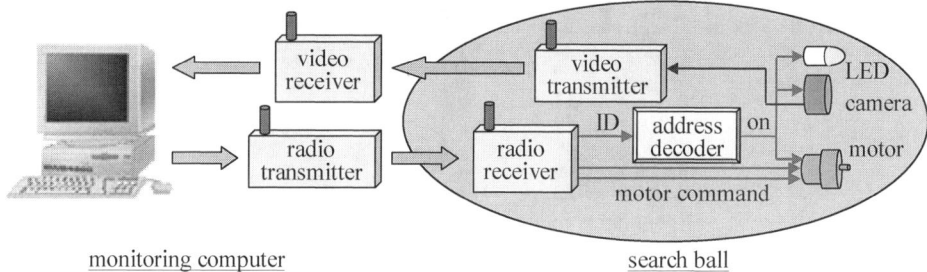

Figure 4: Wireless communication between search ball and monitoring computer

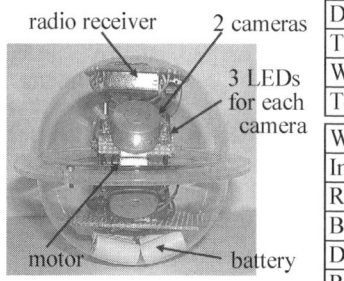

Diameter of sphere	100[mm]
Thickness of shell	2[mm]
Width of ring	10[mm]
Total weight	220[g]

Wireless camera	The ME (RF SYSTEM), 1.2[GHz]
Infrared LED	SLR931A
Radio receiver	TT-01 (CIRCUIT DESIGN), 429[MHz]
Battery	NiMH, 5.0[V]
DC motor	Rated power 0.17[W]
Reduction gear	Reduction ratio 1/196

Figure 5: Developed search ball with rotating cameras

rotation (CW/CCW). Then the target ball transmits the video signals of its cameras while rotating them as commanded, and the computer receives the signals.

DEVELOPMENT OF SEARCH BALL WITH RORATING CAMERAS

Fig.5 shows the developed search ball with rotating cameras. One DC motor with reduction gear rotates two small wireless cameras, and three infrared LEDs are attached to each camera. Two hemispherical outer shells made of transparent plastic are screwed on with each other. The size, weight and components of this ball are also summarized in Fig.5. This ball is made by the combination of commercial products, and the electronic circuit is not fully integrated. Considering the technology of current cellphones, it will be able to be much more miniaturized. The monitoring computer has a radio transceiver (TT-01) and a video receiver (BS-10 by RF SYSTEM).

The developed ball has infrared LEDs for illuminating dark environment inside rubble. We check if humans can be found in the camera images sent from the ball in darkness. The ball and some objects were placed in a dark room (3.5[m](W)x5.5[m](D)x3.2[m](H)) without windows. The distance between them is about 1.0[m]. The LEDs have 450[mW/sr] radiant intensity and 945[nm] peak luminescence wavelength. **Fig.6** shows the camera images when the room lights (6 fluorescent lights of 36[W]) are on and off. As you see, the objects are visible and detectable in dark environment. Especially the cans in narrow space surrounded by obstacles can be seen clearly because of reflection. This situation will be similar to the inside of the rubble.

CONCLUSION

Search balls for searching inside of rubble are explained; scattering many balls with sensors allows rapid and wide-area search. A search ball with rotating cameras is developed. Two wireless cameras

  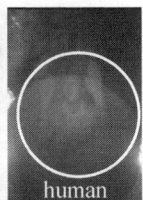

Room lights on   Room lights off

Figure 6: Camera images from ball placed in dark room

for search, infrared LEDs for illumination, a radio receiver for communication with monitoring computers and a battery are packed into a sphere impact-resistant outer shell. This ball can provide the view of its entire circumstance by rotating the cameras using a motor. Just like a brim of a hat, a ring is attached to the shell for suppressing rolling of the ball; it is effective for distribution of balls inside rubble. The monitoring computer identifies the balls inside rubble and acquires the sensor information from them by one-to-one communication.

In the future works, we will pack microphones and speakers into search balls and try further miniaturization. How to get close to the balls which detect victims inside rubble must be solved. Experiments on searching inside of realistic rubble using many balls will be an important issue.

ACKNOWLEDGEMENT

This research was performed as a part of Special Project for Earthquake Disaster Mitigation in Urban Areas (in cooperation with International Rescue System Institute (IRS) and National Research Institute for Earth Science and Disaster Prevention (NIED)).

RERERENCES

Inoue K., et al. (2005). 'Search Balls': Sensor Units for Searching Inside Rubble. *Advanced Robotics* **19:8**, 861-878.

Kamegawa T., et al. (2004). Development of The Snake-like Rescue Robot ``KOHGA''. *Proc. 2004 IEEE ICRA*, 5081-5086.

Kimura H. and Hirose S. (2002). Development of Genbu: Active wheel passive joint articulated mobile robot. *Proc. 2002 IEEE/RSJ IROS*, 823-828.

Osuka K. and Kitajima H. (2003). Development of Mobile Inspection Robot for Rescue Activities: MOIRA. *Proc. 2003 IEEE/RSJ IROS*, 3373-3377.

Perrin D. P., et al. (2004). A Novel Actuated Tether Design for Rescue Robots Using Hydraulic Transients. *Proc. 2004 IEEE ICRA*, 3482-3487.

Stoeter S. A., et al. (2002). Autonomous Stair-Hopping with Scout Robots. *Proc. 2002 IEEE/RSJ IROS*, 721-726.

Stormont D. P., et al. (2003). Building Better Swarms Through Competition: Lessons Learned from the AAAI/RoboCup Rescue Robot Competition. *Proc. 2003 IEEE/RSJ IROS*, 2870-2875.

Takamori T., et al. (2003). Development of UMRS (Utility Mobile Robot for Search) and Searching System for Sufferers with Cellphone. *Proc. First Int. Symp. on Systems & Human Science*, 47-52.

Tsukagoshi H., et al. (2002). Mobile Method of Active Hose Passing through the Narrow Space. *Proc. 2002 IEEE/RSJ IROS*, 841-846.

Tsukagoshi H., et al. (2004). Leg-in-rotor-II: a Jumping Inspector with High Traverse-ability on Debris. *Proc. 2004 IEEE ICRA*, 1732-1739.

Wolf A., et al. (2003). A Mobile Hyper Redundant Mechanism for Search and Rescue Tasks. *Proc. 2003 IEEE/RSJ IROS*, 2889-2895.

MECHATRONICS DESIGN AND DEVELOPMENT TOWARDS A HEAVY-DUTY WATERHDRAULIC WELDING/CUTTING ROBOT

Huapeng Wu, Heikki Handroos and Pekka Pessi

Institute of Mechatronics and Virtual Engineering, Department of
Mechanical Engineering, Lappeenranta University of Technology
P.O.Box 20, FIN-53851 Lappeenranta, FINLAND

ABSTRACT

This paper presents a special robot, able to carry out welding and machining processes from inside the ITER vacuum vessel, consisting of a five-degree-of-freedom parallel mechanism mounted on a carriage driven by two electric motors on a rack. The kinematic design of the robot has been optimised for ITER access and a hydraulically actuated pre-prototype built. A hybrid controller is designed for the robot, including position, speed and pressure feedback loops to achieve high accuracy and high dynamic performances. Finally, the experimental tests are given and discussed.

KEYWORDS

Parallel robot, ITER vacuum vessel, Machining/welding, water hydraulic.

INTRODUCTION

ITER sectors require more stringent tolerances than normally expected for the size of structure involved. The outer walls of ITER sectors are made of 60mm thick stainless steel and are joined together by high efficiency structural and leak tight welds. In addition to the initial vacuum vessel assembly, sectors may have to be replaced for repair. Since commercially available machines are too heavy for the required machining operations and the lifting of a possible e-beam gun column system, a new flexible, lightweight and mobile robotic machine is being considered.
Traditional industrial robots that have been used as general-purpose positioning devices are open chain mechanisms that generally have the links actuated in series. These kinds of manipulators are more suitable for long reach and large workspace, but are inherently not very rigid and have poor dynamic performance at high speed and high dynamic loading under operating conditions. Compared with open chain manipulators, parallel mechanisms have high stiffness, high accuracy and high force /torque capacity in a reduced workspace and have found many applications in manufacturing systems [1][2][3]. Since there are no commercial solutions applicable to the ITER environment, a new robot system, using water hydraulic drives to achieve the required force density, has been developed by the authors in IMVE in Lappeenranta University of Technology and a prototype was built for testing.

STRUCTURE OF VV AND MACHINING PROCESS

The inner and outer walls of the ITER–Vacuum Vessel (VV) are made of 60mm thick stainless steel 316L and are welded together not directly, but with an intermediate so-called "splice plate" inserted between the sectors to be joined. This splice plate has two important functions; to allow access to bolt together the thermal shield between the VV and coils, and to compensate for mismatch between adjacent sectors to give a good fit-up of the sector-sector butt weld. The robot end-effector will have to pass through the inner wall splice plate opening to reach the outer wall. As shown in Fig.1, the assembly processes has to be carried out from inside the vacuum vessel [4].

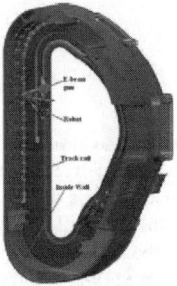

Figure 1: VV Sector to be welded and Path of Robot

The assembly or repair will be performed according to four phases: cutting, edge machining and smoothing, welding and NDT control. The robot acts as a transport device for welding, machining and inspection end-effectors. The welding forces are always small so the forces only come from the weight of the welding device, which may be up to 200 Kg for an e-beam welder. The maximum robot force arises from cutting, when the dynamic force can be up to 3KN.

KINEMATIC MODEL OF PENTA-WH AND DESIGN

The new parallel robot Penta-WH has six degrees of freedom (shown in Fig.2), consisting of three relatively independent sub-structures. One is 3-UPS (Universal-Prismatic-Spherical) parallel mechanism, which contributes the position (x, y, z) of the reference point on the end-effector, the second is a mechanism with 2-UPS legs, which provides two orientations about x- and y-axis, respectively, and the third is a carriage driven by servo motors to drive it on the track rails supported by beams fixed on the both sides of seam of inside wall. A double steel plate construction of carriage keeps the Penta-WH light and stiff. Water hydraulic cylinders have been used as linear drives to offer high force density and easy control.

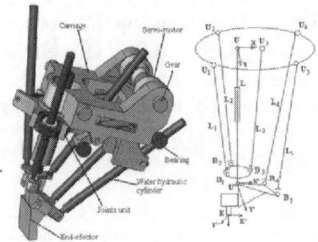

Figure 2: Penta-WH parallel robot and coordinate system

Inverse Kinematics

The inverse kinematics are required to define the parameters of the actuators if the position and orientation of the tool tip are given and the inverse kinematic model is used in position control of the robot, the inverse kinematics model has been given in the reference [5]. According to the frames defined in Fig.2 the models are

$$\vec{L}_j = \vec{E} + R \cdot \vec{EB}_j - \vec{UU}_j \quad \text{for cylinders } L_4, L_5$$

$$\vec{L}_i = \vec{UB}_i - \vec{UU}_i = \vec{UU'} + R' \cdot \vec{U'B}_i - \vec{UU}_i \quad \text{for cylinders } L_1, L_2, L_3 \quad (1)$$

Denoted

$$L_i = f_i(x, y, z, \alpha, \beta) \qquad (i=1,2,3,4,5) \quad (2)$$

Where   R, R´ are rotational transformation matrixes

Forward Kinematic and Jacobian Matrix

The forward kinematics is required to find the position and orientation of the tool tip once the parameters of the actuators are given. The forward kinematics can also be solved from Eqn.2.when the lengths of the linear drivers are given. Since Eqn.2.ontains non-linear items, the forward kinematics is difficult to solve directly. The numeric iterative method has been usually used to solve forward kinematics problems. From Eqn.2, we can obtain the differential motion vector, thus

$$\delta \vec{L} = J[\delta X, \delta Y, \delta Z, \delta \alpha, \delta \beta]^t \quad (3)$$

Where.   J is the Jacobian matrix

$$J = \begin{bmatrix} \partial f_1/\partial x & \partial f_1/\partial y & \partial f_1/\partial z & \partial f_1/\partial \alpha & \partial f_1/\partial \beta \\ \partial f_2/\partial x & \partial f_2/\partial y & \partial f_2/\partial z & \partial f_2/\partial \alpha & \partial f_2/\partial \beta \\ \partial f_3/\partial x & \partial f_3/\partial y & \partial f_3/\partial z & \partial f_3/\partial \alpha & \partial f_3/\partial \beta \\ \partial f_4/\partial x & \partial f_4/\partial y & \partial f_4/\partial z & \partial f_4/\partial \alpha & \partial f_4/\partial \beta \\ \partial f_5/\partial x & \partial f_5/\partial y & \partial f_5/\partial z & \partial g_5/\partial \alpha & \partial f_5/\partial \beta \end{bmatrix}$$

Then the inverse speed equation can be obtained

$$[\dot{L}_1, \dot{L}_2, \dot{L}_3, \dot{L}_4, \dot{L}_5]^t = J[\dot{X}, \dot{Y}, \dot{Z}, \dot{\alpha}, \dot{\beta}]^t \quad (4)$$

When the Jacobian matrix **J** is singular, that is det|**J**|=0, the robot is then in a singular position and cannot put out any Cartesian force.

Static Force

The force capacity of the robot should be investigated, which is for a certain payload inside the workspace to calculate the static forces of cylinders. If we denote Eqn.2.as

$$\delta \mathbf{L} = \mathbf{J} \delta \mathbf{\Theta} \quad (5)$$

Where $\delta \mathbf{L} = (\delta l_1, \delta l_2, \delta l_3, \delta l_4, \delta l_5)$, $\delta \mathbf{\Theta} = (\delta X, \delta Y, \delta Z, \delta \alpha, \delta \beta)$

According to the principle of virtual work, we have

$$P^T \delta L = F^T \delta \Theta \quad (6)$$

Substitute Eqn.5 into Eqn.6 we have

$$P = (J^{-1})^T F \quad (7)$$

The force in the cylinder can be obtained from Eqn.7 once a certain payload is given at the end tip. Considering the friction force is small, this force can also be regarded as the main force in the bearings.

The above models help to investigate the workspace, force capacity, singularity and stiffness to achieve an optimised structure [6]. A multi-body simulation model of robot has also been built to check the deflections, workspace and collisions as well. With optimisation design the robot can reach a larger singularity free workspace than the required $200 \times 200 \times 300 mm^3$, achieve high stiffness up to 400N/μm for the universal joints unit and 315N/μm for the carriage, and has high force capacity able to carry the heavy welding gun and take high machining forces.

CONTROL SYSTEM

The control system includes software and hardware. The hardware as shown in Fig.3 consists of servo water hydraulic drive system and computer control system. Water hydraulic cylinders used as linear actuators have been employed in the robot to offer high force density and easy control. The low frequency vibrations caused by the variable cutting force are neutralised by using pressure feedback control with a high pass filter in the control loop. The water hydraulic drive system includes water hydraulic cylinders, position sensors, pressure sensors and high performance servo-valves.

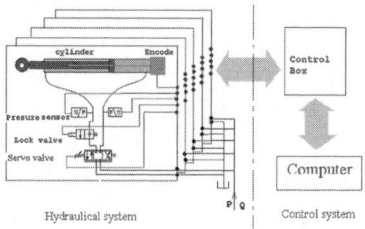

Figure 3: Water hydraulic drive system

Process Programming

For the machining and the welding, the robot controller needs more complex functions to manage two different processes. In the machining, the robot takes the machine tool cut through the VV wall and feeds it a 200mm distance along the seam. In this case the robot works like a 5 axis CNC machine. In the welding, the robot Penta-WH takes the welding gun moving continuously along the joint. To compensate for errors caused by the structure, a camera based seam tracker is used. The error information from the seam tracker is input to the robot controller, where the kinematic program compensates for the errors by sending position instructions to the controller. Both machining and welding functions are integrated and the computer controller carries out these two processes automatically. The main functions of the controller are as follows: Trajectory planning, Kinematics and dynamics, interpolations, position feedback controller, input/output single processors, and teaching function.

Accurate Position Control Algorithm

Fig.4 shows the control scheme. The output commands of the upper level include position and speed references for the servo cylinder controllers.

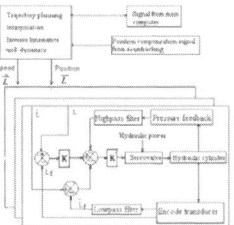

Figure 4: Position control scheme

The servo control loops consist of position loops and speed loops that provide accurate and fast trajectory tracking. The load pressure feedback loops are used for damping the self-excited oscillations normally occurring in natural frequency. The speed loop can eliminate the speed error, while the pressure feedback damps the vibration of the hydraulic actuator. The hydraulic cylinders normally lack damping that make their control difficult by using conventional PID-controllers. The damping can effectively be increased by means of load pressure feedback. The major drawback in using pressure feedback is its negative effect on the static stiffness of the actuator. To overcome this high pass filters are used in the load pressure feedback loops. The high pass filter removes the negative effect of pressure feedback at low frequencies.

PROTOTYPE AND EXPERIMENTS

A prototype was built in IMVE (shown in Fig.5), the robot is fixed on a frame, and to simulate the machining and welding process a moveable table driven by a servo motor is used. With this experimental device several experiments, such as calibrating, workspace investigating, position accuracy and repeat accuracy testing, have been carried out, as well as a cutting test with stainless steel.

Figure 5: Prototype of Penta-WH robot

The end-effector of the robot can reach the required workspace: 300mm in z-axis, ±100 mm in x and y-axis and the orientations about x and y-axis are more than ±20°. The position accuracy is less than ± 0.05mm and repeatability is less than ±0.01mm after calibrating. Fig.6 shows the trajectory of the end tip and the corresponding motion of the hydraulic cylinders. In this experiment the robot tracks the trajectory with a 50kg payload and 500mm/min speed.

CONCLUSION

A parallel robot, driven by five hydraulic cylinders, has been developed to assemble and to repair the Vacuum Vessel of ITER and can accurately and stably hold all necessary machining and welding end-effectors in all positions.
A control system has been designed. A prototype has been built, some tests have been carried out and its position accuracy and repeatability have been investigated and found to be in good agreement with the theory.

REFERENCES

[1] http://www.parallemic.org/Patents.html.
[2] M. Honegger, A. Codourey and E.Burdet .Adaptive Control of the Hexaglide , A 6 dof Parallel Manipulator. In proceedings of the IEEE International Conference on Robotics and Automation. Albuquerque, New Mexico. April 1997.
[3] K.H. Häfele , H.Haffner and P. Spencer , Automatic Fettling Cell- An Example For Applying Computer- Aided Robotics . Industrial Robot. Industrial Robot , Vol.19 No. 5.1992, pp.31-34
[4] L. Jones, Study to Optimise Intersected Welding Robot Design And Machining Characteristics. Final report. January 2002
[5] H. Wu, H. Handroos. Parallel Mechanisms Based On Telescopic Structure and Applications. In Proceedings of the 32nd International Symposium on Robotics, Seoul, April 2001.
[6] H. Wu, H. Handroos, …, L. Jones. Design of Parallel Intersector Weld/Cut Robot for Machining Processes in ITER Vacuum Vessel. International Journal of Fusion Engineering and Design, Vol. 69(2003) pp327-331 .

QUTIE - MODULAR METHODS FOR BUILDING COMPLEX MECHATRONIC SYSTEMS

Antti Tikanmäki, Tero Vallius, Juha Röning

Intelligent Systems Group, University of Oulu
P.O. Box 4500, Fin-90014 Oulu, Finland

ABSTRACT

This article introduces a modular mechatronic device construction method, based on the Atomi concept and Property Service Architecture. The Atomi concept is based on an "embedded- object" based architecture, which applies the common object oriented methods used in the software of combined software and hardware entities called Atomis. The Property Service is software architecture for fast and ease intersystem control and communication. It provides a generic interface for easy, dynamic interfacing to any device over a network, enabling easy modular control of a system consisting of systems. These are generalised methods, suitable for creating any kind of robots or systems through high modularity in mechanics, electronics, and control software architecture. As a test case, the development of Qutie, an interactive mobile robot, has been described.

KEYWORDS

Modular mechanics, mobile robot, embedded systems, Atomi, Property Service

INTRODUCTION

Building a robot is a demanding task. To create a sophisticated combination of mechanics, electronics and software requires a lot of engineering work and expertise. A mobile robot containing a set of sensors and actuators as well as onboard computer and power systems is a complex system that sets great requirement for the electronics. Traditionally, a robot consists of customized electronic boards and specifically designed hardware i.e. sensors and actuators. This kind of architecture has the disadvantages of limited or no expandability and modifications requiring new control boards. Modularity is therefore important for several reasons. As implied before, it makes expanding easy and modifications possible by changing the configuration of modules instead of the complete system. Additionally, there are advantages in maintainability (only the malfunctioning module need to be replaced), reusability (the same modules can be used in different robots), stability and reliability (reused and thus tested modules tend to be stable and thus more reliable), and faster design (system development can be easily divided between development teams). Several modular hardware solutions exist and have been used for building mechatronic systems. For example OOPic (2004) has a board that allows attachment of different kind of sensors and actuators on the board to build a controller for a robot. The number of sensors and actuators per board is very limited, however.

In our approach, called the Atomi concept, small- size embedded objects have been developed to a reach the high level of modularity in the embedded system. Atomi boards are mainly intended for fast prototyping and the creation of easy, high-level embedded systems. The purpose of Atomis is to lower the threshold of creating embedded systems by making the creation of embedded systems both affordable and as easy as possible, so that the building does not require much time or expertise in electronics. The Atomi boards are described in more detail by Vallius et al. (2004). Qutie robot provides a challenging and complex test platform for Atomi boards.

THE ROBOT

Qutie is a mobile robot designed to perform a variety tasks in common environments, such as homes or public places, in interaction with humans and other robots. Further, the robot must have versatile capabilities for interacting with people. The main features of Qutie are a round shape, a belly screen, and furry skin. As the robot is used for research of human-robot interaction, it should be easily modifiable, so that the features of the robot could be changed based on the tests. An overall view of system components is shown in Figure 1. The robot has two computers, a main computer located on the base and a small PC104 computer on the neck and head unit. The parts of the robot are independent and can be separated. The minimal connections between the body of the robot and the neck head mechanism contain only the power and ground lines, and the Ethernet connection between the computers.

The operation of the robot's head and neck mechanism of the robot is shown on Figure 2. The neck has four degrees of freedom, three rotations, and one translation of elevation along the z axis. The linear resolution of each motor's step movement along a moving linear axis is 0.024 mm, and range of operation is limited to 100 mm. The linear motors A, B, and C have flexible joints at the top and the bottom as rotation requires the motors body to rotate. Through the capabilities of its neck, robot can change the orientation of the head camera. In addition, the robot can show "emotions" when interacting with humans. The neck mechanism control Atomi boards are connected to the head computer through a USB connection. The head computer operates all the functions of the head and neck unit, and it is a stand-alone unit requiring only 5V power from the robots base. To reduce the power requirements, the hard disc of the computer has been replaced by a Compact Flash card drive.

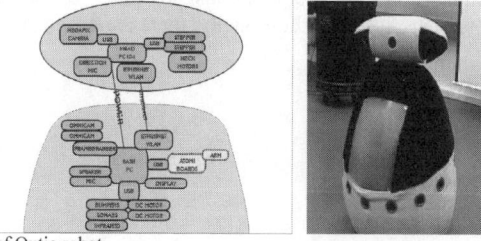

Figure 1. Hardware modules of Qutie robot

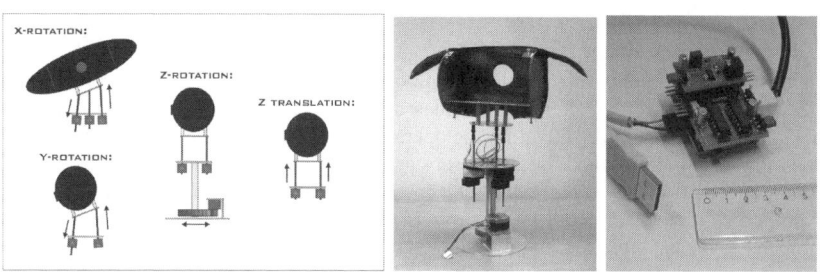

Figure 2, Neck movements (left), Head and Neck unit (center), and Atomi boards (right).

ATOMI MODULES

As the robot contains many actuators and sensors, a very complex electronic system is needed to control the actuators and to transmit information from the sensors to the onboard computer. For purposes of fast system development and prototyping, we have developed an object-oriented embedded system development method, which is based on small embedded objects called Atomis. Atomis are small electronic boards that contain some sensor circuits, actuator drivers, or other functionality.

In Qutie we have tested the suitability of this development method in a real system. The available 'library' objects (i.e. ready-made Atomis) used in building Qutie were a stepper motor controller, a servo controller, switch inputs, LED drivers, an LCD-display, AD converters, USB connection, a DC motor controller, a powering module, and brain Atomis (brains are used for the control software in an Atomi system). The design process for the neck device went as follows. As the neck is actuated by stepper motors, we chose stepper motor Atomis to drive them. We used an USB Atomi to connect the Atomi stack to main computer. Selected Atomis were then stacked together and placed inside the Qutie's head.

To control the two DC-motors driving the wheels of the robot, we used a DC-motor. For connecting Atomis together there are several options. As the communication between Atomis is based on a field bus, they do not necessarily need to be stacked on top of each other, but can also be interconnected with a cable. Thus, each module can be located inside the mechanical module that it is controlling, which provides easy expandability of the robot's. Given this option, we could have interconnected the DC motor Atomi to the other Atomis with a cable. Another option is to make a separate device for driving the DC motors by attaching another Power and USB-Atomi to the DC motor Atomi. Thus there needs to be another available USB port on the controlling computer for the new device. In Qutie's case, there is a separate computer controlling head and the main computer controls the rest of the robot. Thus, the most reasonable option was to make another control device out of Atomis for the robot's base. This option is illustrated in Figure 2.

As the configuration of the robot can be varied by changing the set of sensors and actuators, the robot's software architecture needs to be dynamic and modular. In our previous work, *Mäenpää, Tikanmäki* (2004), we described network-distributed software architecture for operating various resources in the distributed system. In this Property Service, each device like a robot or a stand-alone sensor provides a service containing a set of properties related to the device. In this work, the internal communication of robot's resource has been done with Property Service architecture.

PROPERTY SERVICE ARCHITECTURE

Property Service is a general representation of a network-transparent server that is controlled using property name value pairs. Property Service provides a simple interface for setting and getting different values of the device and for activating the operation of the device. Each device contains a set of properties that can be "get" and "set". Sensors and actuators have standardized properties, related to the data information they receive and their location. There is also a higher level standard for the set of properties. For example all mobile robots provide "mobile robot service" properties that contain properties, for example moving the robot precisely across distances or other functions. This provides a possibility to use different mobile robots with the same higher level control software. For example, commanding the robot to "move one meter forward" has the same effect regardless of robot's size or whether it uses wheels or legs for moving. Each Atomi provides a set of functionalities, and Property Service combines these into a Property Service. As Atomi modules can be attached during power up, new devices and their properties show up on the list of properties of the service immediately. Communication between the head and neck unit and the base computer has been accomplished with these Property Services, through Ethernet. In current implementation, ICE is used as middleware software, but Property Service can also be accomplished over other protocols and physical layer, such as RS232, direct TCP/IP.

DISCUSSION

A robotic platform for human interaction research has been implemented successfully. The modular parts of the robot can be used separately, which allows easy modifiability. The Atomi concept provides the great advance of enabling quickly changes of the setup of the robot's parts, as new Atomi boards can be attached online. The use of Atomi boards also reduced radically the time needed for the development time of the robot. Power optimization of the mobile robot is essential. Atomi boards consume much less power in compared to traditional PC cards that are widely used on robots. Atomi boards can be set to stand-by mode during operation, which reduces the power requirements.

ACKNOWLEDGE

The outlook of the robot has been designed in cooperation with the Industrial Design Laboratory of the University of Lapland. This work has been partly funded by the Academy of Finland.

REFERENCES

OOPic (visited 1.5. 2004) *web pages*, http://www.oopic.com
Tero Vallius, Janne Haverinen, Juha Röning, (2004) *Object-Oriented Embedded System Development Method for Easy and Fast Prototyping*, ICMA Osaka, Japan.
Antti Tikanmäki, Jukka Riekki and Juha Röning (2003), *Qutie - an interactive mobile robot*, 2003 ICAR International Conference on Advanced Robotics, Jun 30 - Jul 4, Coimbra, Portugal.
Microsoft ActiveX Controls, (visited11.3.2004) http://www.microsoft.com/com/tech/ActiveX.asp
Topi Mäenpää, Antti Tikanmäki, Jukka Riekki and Juha Röning (2004) *A Distributed Architecture for Executing Complex Tasks with Multiple Robots*, IEEE 2004 ICRA, International Conference on Robotics and Automation, Apr 26 - May 1, New Orleans, LA, USA.

LINK LENGTH CONTROL USING DYNAMICS FOR PARALLEL MECHANISM WITH ADJUSTABLE LINK PARAMETERS

W. Tanaka[1], T. Arai[1], K. Inoue[1], T. Takubo[1], Y. Mae[2] and Y. Koseki[3]

[1]Department of System Innovation, Division of System Science and Applied Informatics,
Graduate School of Engineering Science, Osaka University
1-3 Machikaneyama, Toyonaka, Osaka, 560-8531, Japan
[2]Department of Human and Artificial Intelligence Systems,
Faculty of Engineering, University of Fukui
[3]National Institute of Advanced Industrial Science and Technology

ABSTRACT

There has always been a workspace problem with parallel mechanisms. Previously we have proposed a parallel mechanism with linear passive joints to adjust link length. This parallel mechanism can achieve different workspaces by adjusting the link length. We have tried to control the link lengths of the parallel mechanism not active but passive using dynamics. This paper proposes the control algorithm of these passive linear joints, to adjust the link lengths using dynamics in the case of a 2-DOF planer prototype.

KEYWORD

Parallel mechanism, Dynamics, Adjustable mechanical parameter, Passive joint control.

INTRODUCTION

Parallel mechanisms have some good advantages compared with conventional articulated arms. The only drawback is its small workspace due to its in-parallel configuration.

Whole of the large workspace may not always be required in one task, and the workspace may be divided into smaller sub-workspaces corresponding to individual tasks as shown in Figure 1. This is our basic idea in the paper [1]. If each sub-workspace can be covered by a corresponding parallel mechanism that is not individuality different but a single mechanism with differently adjusted mechanical parameters, the whole workspace can be achieved by just one machine. Our idea is to cover each sub-workspace by a machine that has adjustable mechanical parameters.

There are some mechanical parameters which can be adjusted, such as base plate parameters, end-effector parameters and link parameters. This paper discussed the parallel mechanism with adjustable

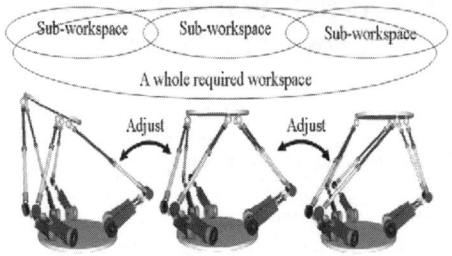

Figure 1 : The combinational workspace

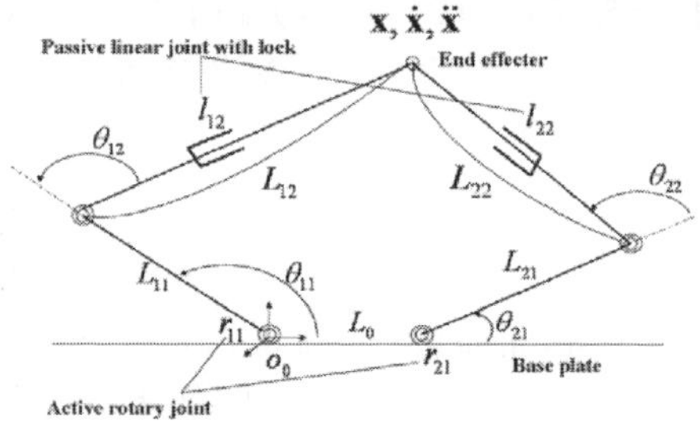

Figure 2 : The 2-DOF planer rotary actuated parallel mechanism with adjustable link length

link parameters, since the adjustable link parameters effects the overall workspace volume more than another parameters. The link length of this parallel mechanism can be adjusted passively or actively. A mechanism with actively adjusted link length is a kind of redundantly actuated mechanism and it would not be feasible due to its high cost of many actuators. Hence, the proposed idea is to adjust the link length without actuators. There are two methods of adjusting link length; manually and automatically. We have attempted to adjust the link length automatically because manual adjustment requires the mechanism to be off.

Therefore, we need to control passive joints to adjust the link length automatically. The control of serial manipulator with passive joint using the dynamics has been studied [2][3]. We have applied this method to the parallel mechanism with adjustable link parameters and control the passive linear joints to adjust the link length automatically using its dynamics.

ALGORITHM OF LINK LENGTH CONTROL

2-DOF Planer Rotary Actuated Parallel Mechanism with Adjustable Link Length

We will discuss the control algorithm using a 2-DOF planer rotary actuated parallel mechanism with adjustable link length as shown in Figure 2. This mechanism has passive linear joints on each links with a lock. When the lock is put ON or OFF, the passive linear joints can be fixed or released respectively.

The joint r11 and r21 are active rotary joints. θ_{11}, $\dot{\theta}_{11}$, $\ddot{\theta}_{11}$, τ_{11}, θ_{21}, $\dot{\theta}_{21}$, $\ddot{\theta}_{21}$ and τ_{21} show the displacement, velocity, acceleration and torque of each active rotary joint. The joint l11 and l21 are the passive linear joints and L_{12}, $\dot{L}_{12}$, $\ddot{L}_{12}$, f_{12}, L_{22}, $\dot{L}_{22}$, $\ddot{L}_{22}$ and f_{22} show the displacement, velocity, acceleration and force of each passive linear joint. o_0 shows the origin of the base frame and the position of the joint r11. L_0 shows the position of the joint r21. L_{11} shows the length between the joint r11 and r12. L_{21} shows the length between the joint r21 and r22. $x, \dot{x}$ and $\ddot{x}$ show the displacement, velocity and acceleration vector of the end-effector.

Formulization of the Algorithm of the Passive Linear Joints Control to Adjust Link Length

The equations of motion of the 2-DOF planer rotary actuated parallel mechanism with adjustable link length is shown as follows.

$$\begin{bmatrix} \tau_a \\ f_p \end{bmatrix} = \begin{bmatrix} M_{11} & M_{12} \\ M_{21} & M_{22} \end{bmatrix} \begin{bmatrix} \ddot{q}_a \\ \ddot{q}_p \end{bmatrix} + \begin{bmatrix} B_a \\ B_p \end{bmatrix} \qquad (1)$$

M_{11}, M_{12}, M_{21} and M_{22} shows the element of the acceleration related matrix. B_a and B_p shows the element of the matrix of the Coriolis force, centrifugal force and friction force. $\tau_a = [\tau_{11} \quad \tau_{21}]^T$ shows a torque vector of the active rotary joint r11 and r21. $f_p = [f_{11} \quad f_{21}]^T$ shows a force vector of the passive linear joint l12 and l22. The subscript T shows the transposed matrix. Moreover, the acceleration of the active rotary joint shows $\ddot{q}_a = [\ddot{\theta}_{11} \quad \ddot{\theta}_{21}]^T$, the acceleration of the passive linear joints shows $\ddot{q}_p = [\ddot{L}_{12} \quad \ddot{L}_{22}]^T$.

When the locks on the passive linear joints are put OFF, the forces $f_p = [f_{11} \quad f_{21}]^T$ become zero. We also define that $\ddot{q}_{p,d}$ shows the desired acceleration of the passive linear joints. Eqn. is solved about the torques of the active rotary joints.

$$\ddot{q}_a = -M_{21}^{-1} M_{22} \ddot{q}_{p,d} - M_{21}^{-1} B_p \tag{2}$$

$$\tau_a = \left(-M_{11} M_{21}^{-1} M_{22} + M_{12}\right) \ddot{q}_{p,d} + B_a - M_{11} M_{21}^{-1} B_p \tag{3}$$

We can input the desired acceleration of the passive linear joints to the Eqn. 2, 3, to obtain the torques and accelerations of the active rotary joints.

SIMULATION

Simulations were used to investigate the algorithm of link length adjustment. First, we set up the desired trajectories of the displacements, velocities and accelerations of the passive linear joints. We estimate the torques and acceleration of the active rotary joints to obtain the desired transformation of the passive linear joints. Their torques and acceleration are given to the forward dynamics equation to estimate the acceleration of the passive linear joints. We also confirm whether the estimated acceleration of the passive linear joints is the same as the desired acceleration. If the estimated acceleration is same as the desired acceleration, we can calculate the desired link length by using their acceleration.

Kinematic parameter definition

We will discuss simulations using the planer rotary actuated parallel mechanism shown in Figure 2. The parameter values are as follows: $L_0 = L_{11} = L_{21} = 0.15$[m] and the initial link length L_{12} and L_{22} are 0.24[m] respectively. All of the link mass is 0.50[kg]. The center of mass for each link is the center of itself, and we assume that all of link frame axis as its principal axis of inertia are in the same direction. Gravity is ignored because this mechanism is fixed parallel to the ground.

Desired Trajectory of the Transformation of the Passive linear joints

We set up the desired displacement of the passive linear joints $L_{12,d}$ and $L_{22,d}$ are 0.265[m]. Here, we also set the control time is 0.15[s], because the trajectories of the transformation of the passive linear joints are given arbitrarily, as shown in Figure 3, Figure 4 and Figure 5. Figure 3 shows the desired displacements [cm] of the passive linear joints, Figure 4 shows the desired velocities [cm/s] of the passive linear joints, Figure 5 shows the desired accelerations [cm/s^2] of the passive linear joints in each vertical axis. In these figures, the horizontal axis is time [s].

Simulation Results

We simulated the desired acceleration of the passive linear joints with Eqn. 2, 3. The estimated the torques and accelerations of the active rotary joints to realize the desired transformation of the passive linear joints. Next, we give these torques and acceleration to the forward dynamics to estimate the acceleration of the passive linear joints. Figure 6 shows its results. The horizontal axis indicates time [s] and the vertical axis indicates the acceleration [cm/s^2] of the passive linear joint l12 and l22. Next, we give these torques and acceleration to forward dynamics to estimate the acceleration of the passive linear joints. Figure 6 shows its results. The horizontal axis indicates time [s] and vertical axis indicates the acceleration [cm/s^2] of the passive linear joint l12 and l22.

Figure 6 shows the same as Figure 5. If this mechanism is given the estimated torque and the acceleration of the active rotary joints, it is possible to calculate the desired passive linear joint displacements, velocities, and accelerations. Therefore, we can ultimately calculate the desired link length.

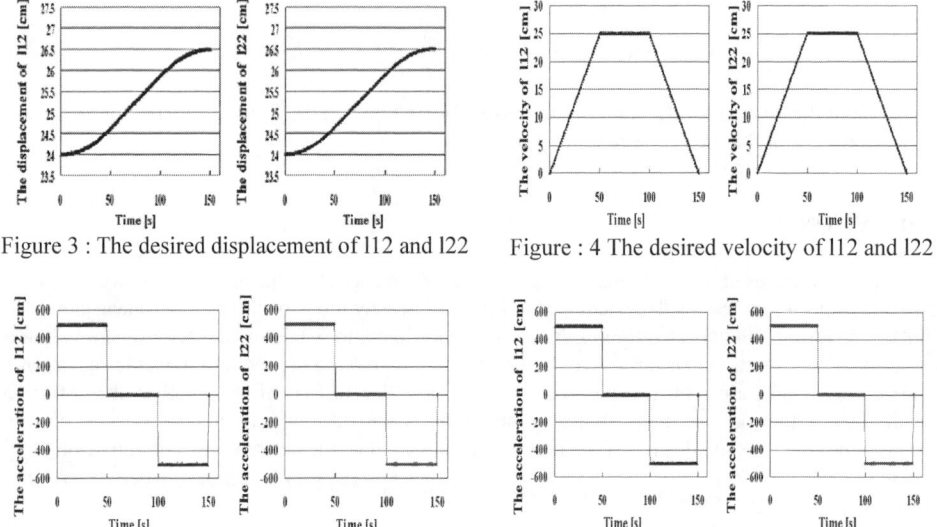

Figure 3 : The desired displacement of l12 and l22

Figure : 4 The desired velocity of l12 and l22

Figure 5 : The desired acceleration of l12 and l22

Figure 6 : The estimated acceleration of l12 and l22

CONCLUSIONS

In this paper, we have discussed a method for link length adjustment using dynamics for the 2-DOF planer rotary actuated parallel mechanism with passive linear joints on each link.

The procedure of this method is as follows: first, estimate the trajectories between the present and desired displacement, velocity and acceleration of the passive linear joints. Then, estimate the torques and accelerations of the active rotary joints in order to solve the accelerations, velocities and displacements of passive linear joints. By setting the torques and accelerations to the active rotary joints, the desired acceleration is generated on the passive linear joints. By achieving the desired trajectories of the passive linear joints, we obtain the desired link length.

References

[1] T. Arai, et al.(2000). Parallel Mechanisms with Adjustable link length. *Proc. IEEE/RSJ 2000 International Conference on Intelligent Robotics and Systems*, 671-676.
[2] H. Arai, et al.(1991). Position Control System of a Two Degree of Freedom Manipulator with Passive Joint. *IEEE Trans. Industrial Electronics*, **38:1**, 15-20.
[3] H. Arai, et al.(1991). Position Control System of a Manipulator with Passive Joints Using Dynamic Coupling. *IEEE Trans. Robotics and Automation*, **7:4**, 528-534.

A PROPOSAL OF THE MULTIMEDIA ARCHIVE SYSTEM WITH WATERMARK INDEX FOR PREVENTION OF DISASTERS

Fumio Maehara[*1] *and Yoshikazu Tanno*[*2*3]

*1. Matsushita Electric Industrial Co., Ltd.   2-15 Matsubacho Kadoma-shi, Osaka, Japan
*2   Osaka University, Machikaneyama-cho Toyonaka-shi Osaka, Japan.
*3 Yamagata Digital Contents Center for Research and Promotion, Matsuei, Yamagata, Japan
*1) E-mail: maehara.fumio@jp.panasonic.com
URL: http://www.archive.gr.jp

ABSTRUCT: According to the rapid expansion of the digital broadcasting and networks, the large scale contents creation and multimedia archives are eagerly discussed. Multimedia database or archives are very useful not only the entertainment but also the social security or prevention of disasters such as the records of earthquake or typhoon etc. In such database the retrieval method and preservation of the explanation to the situation related to the recorded video are highly important. Here we introduce the large scale archive system, the retrieval interface and index format with watermark and Mpeg-7 metadata to apply the prevention of the disaster database. And also the information security on the system, based on ISO17799 recommendation, will be discussed briefly.

KEYWORDS: Archive, ISO15004, ISP17799, Metadata, Multimedia database

1. INTRODUCTION

With the rapid progress in broadband network and terrestrial broadcasting, the environment for facilitating digital contents, database, archiving and their distribution channels become increasingly indispensable.

On the other hand multimedia database for the social security or prevention of disasters are coming to the important stage. The important issues to built the multimedia database are to make user friendly retrieval interface and ensure the physical and informational security.

To make user friendly retrieval system, generalized efficient indexing method, which has the robustness in the disaster environment is required.

We have developed the large scale archive which capacity is 7200 cassettes and 3000 contents have been shoot and stored in it The system is locate in Yamagata. System development and researches have been carried out under the management of Telecommunication Advancement Organization (TAO, National Institute of Information and Communication Technology: NIST, now) which is on of the independent legal organization of Ministry of Public Management, Home Affairs, Post and Telecommunications.

The system has the tape feed archive robot, DVD server and HD (Hard Disc) server and 62 sub-divided motion picture thumb nail retrieval interface. The major issue of the system is to give the universal indexing method in the aspects of format and data preservation, Mpeg-7 metadata format has been proposed and the watermark technology to embed the information on the video data has newly been proposed.

Here we report the overview of the system, watermark technology and indexing method of Mpeg-7 metadata and the application of the metadata to the prevention of disaster database. Information security will also be discussed briefly as the future issues.

This paper covers the contents as followings.

First, authors mentioned the research background in this section and then the overview of the multimedia archive system in Yamagata Video Archive Research Center is introduced, watermark technology to embed the information data on the video will be mentioned in section 3, The index format based on Mpeg-7 and metadata format will be discussed in section 4 and the database of prevention of disaster in section 5 and the archive security based on ISO17799 and 15408 are briefly mentioned in section 6, then conclusion and future issues will mention in the last section.

2. ARCHIVE SYSTEM

The core of the archive system is a large scale tape feed robot that can store 7200 professional use video cassettes of DVC-PRO format. This system uses a DVD server of medium speed and medium capacity and an HDD (Hard Disc Drive) server for high speed and small capacity storage. Fig.1 shows the total archive and delivery system and Fig.2 shows the outlook of the tape feed robot.

This archive system uses video sequences in DB format (SMPTE 306m, 25Mbps and 50Mbps) with high picture quality suitable for broadcasting use to store pictures on both the large scale tape feed robot and the high sped access type HDD server. Then this video contents are distributed through the network, data stream of this sequence is formatted with MPEG-1 or MPEG-2, 4. Thus DV formatted data are used for video image storage and MPEG-1 formatted data with low bit volume, are used for distribution of image sequences for browsing of digest videos both in-house and on the outside network. The data stream from the archive system in converted from SD (Standard Definition Video Stream) to MPEG-2 on MPEG –TS (Transfer Stream) using the FTP (File Transfer Protocol) and this stream is run through the distribution network for contents distribution to the other security network [1-4].

The system has the visual retrieval interface, shown in Fig. 3, which shows 32 or 64 thumbnail video simultaneously hence 64 contents can check at the same time.

3. WATER MARK

The digital watermark technology is one of the important issues to embed additional information such as the metadata to the video .
In the large amount of video database, separate database manager such as oracle system or SQL server are used.
But in such system, the data mismatch caused by human dispersing or operation error. One of the effective method to preserve the additional information relate to the video signal is digital watermarking. In this method additional data such as metadata, time-code, and copy-right information are added on the video data themselves on the lower bit area of the DCT coefficient of the video data.
Fig. 4 shows one of our original water marking method in which low space frequency area on DCT is aimed at the data embedding as well as high frequency area [9, 10].
The lower and higher part of the spatial frequency area are not very sensitive for the human eyesight hence the embedding noise is less visible and the S/N ratio can be improved on human eyesight. Fig 4 shows the experimental result of embedded disaster information on the recorded image. Fig. 3 shows an example of the S/N ratio of the video on which watermark is embedded[5-8].

4. MPEG-7 OVER CONTENTS ID

In order to solve above issues, we focused on the application of MPEG-7 description and the cid compliant to cIDf, and we decided to adopt MPEG-7 descriptor to describe contents feature on our archive system. The main purposes are to realize efficient retrieval for anybody who want the contents. In the conventional method, keywords that were easy to understand for that familiar to regional Yamagata area, where YRC is located, were attached to the contents. However, in general, assuming that those from other outside area would try to retrieve the contents, easy retrieval will not always be done from these localized keywords.
To realize the system allowing reuse of the content, we envisage the interoperability such as moving whole contents in our system to other databases. MPEG-7 is the content description system standardized in ISO/IEC JTC-1 SC29/WG11 (MPEG), and is the important technology for the era of broadband digital network. Furthermore, it is the only tool that can offer standardized representation scheme of content related metadata.[9-11].

5. FORMAT OF THE INDEX
   OF PREVENTION FOR DISASTER

In order to apply Mpeg-7 metadata to the disaster database, we considered both the fact that the video contents we currently have is the center of our system and the practicability, and decided to adopt Method b.

that makes description using (Instance Identifier) in Media Instance DS. By adopting this method, we intended to enhance possibility to realize smoother content distribution from retrieval to acquisition of copyright information of content. For this purpose, necessary first step is to retrieve contents with retrieval metadata attached to them using MPEG-7 description. The next is to display multiple screens of moving pictures that show scene change points of each picture sequence to check the retrieved contents visually to further narrow down the target content, and obtain its cid. Final step is to obtain information on copyright and distribution of the content from IPR database compliant to cIDf by using cid as a key. Here, we adopt feature description of only high level so user can recognize the description of content. By adopting such content description, MPEG-7 can further provide various original information and functions to secure interoperability of contents data.

Fig.4 shows an example of the disaster information which is embedded on the video data by watermark. The earthquake information's are shown in the figure. Fig 5 shows the example of the index format of the time code which is linked to the contents ID. This information shows the exact time of the disaster and time history can exactly traced afterward. Hence much analytic research for the disaster can be done. As is shown in Fig.5 the EDL (Editing Data Line) information by the watermark on the disaster video record. Hence the situation of the disaster can simultaneously be observed. Fig.8 shows the overall multimedia database tree of the disaster. Hence several informations can be retrieved visually by this information and several scientific analyses can be done by this.

6. CONCLUSION

Here we report the large scale archive system and the application to the security database for prevention of disaster. Mpeg-7 metadata format and the watermark method have been proposed for the efficient indexing method for the prevention fo9r disaster. In the future, we plan to utilize the verified MPEG-7 over Contents ID and make structure to consolidate he retrieval profile unique to YRC and to adapt the control and management method of content and database using the method of content and database using the method to requirement of wider content distribution world. We plant to aspire to integrated content management that can adapt to professional need from retrieval of archived material content, distribution and program production, and establish the system model that verifies content distribution from the viewpoint of users. Thus, we are committed to contribute to the development of content base industry.

Authors thanks to the members of TAO (NIEST), Ministry of Public Management, Home Affairs, Posts and Telecommunications, Yamagata city and prefecture government, Sony corporation, Mitsubishi Electric Corporation, NTT, Osaka University, Tokyo University and Tohoku University of art and design of their kind corporation. And also Midorikawa Lab. at Tokyo Institute of technology for presenting disaster image data.

REFERENCES

[1] MPEG-7: http://www.itscj.ipsj.or.jp/mpeg7/
[2] cIDf: http://www.cidf.org/
[3] Ministry of Public Management, Home Affairs, Posts and Telecommunications: 2001 White Paper Information and Communications in Japan, p138-141, 2001.
[4] Department of Cultural Information Related industry, Ministry of Public Management, Home Affairs, Posts and Telecommunications: "Establishment of investigative committee on promotion of content distribution" July 2001.
[5] Maehara, et al.: "Peta-byte Class Video Archive System by Large-scale Automatic Robot", Proc. IEICE Fall Conf., Sept.1.
[6] Fumio Hasegawa, Fumio Maehara, Suguru Kawabata, Haruo Hiki, Yoshikazu Tanno, Manabu Ito: "Program production system utilizing large-scale archive", Technical Report of IEICE, IIE99-83, Nov.1999.
[7] Haruo Hiki, Fumio Maehara, Yoshikazu Tanno, Manabu Ito, Fumio Hasegawa: "Remote Retrieval and Editing System for Large-scale archive", Technical Report of IEICE, EID99-137, IE99-13, Feb.2000.
[8] Masaru Kawabata and Fumio Hasegawa," A study for picture-quality distortion by digital watermarking", D-11-86, IEICE 2001
[9] Masaru Kawabata and Fumio Hasegawa," A study for Picture-quality Distortion by Digital Water-marking(2)", 11-2, ITE Annual Convention 2002
[10] Manabu Ito, Koichi Ito, Yoichi Ishibashi, Takuyo Kogure, Fumio Hasegawa: "Retrieval Profile Using MPEG-7/MDS for the Large Scale Video Archive", Proc. IEICE Conf., SD-3-6, pp385, March 2002.

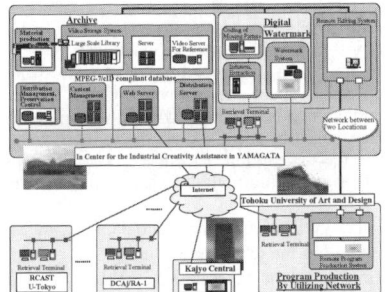

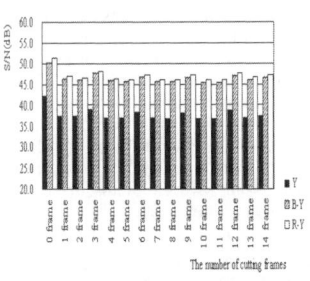

Figure 3: S/N Ratio of Watermark with the number of cutting frames

Figure 1: Schematic diagram of the archive system

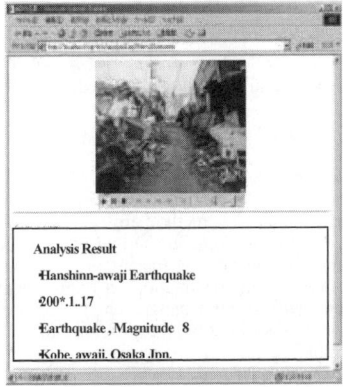

Figure 2: The outlook of the Archive System

Figure 4: An example of the watermark data

ジャンル (cid)	タイトル (オフセット)	代表静止画	状況	開始タイムコード	長さ	計算で得られる 終了タイムコード
地震	阪神淡路 大震災 (01:19:00:29)		ビルの倒壊1 三宮駅前	PODTOH0M20S 2002N30000F (00:00:20:02)	929 (00:00:30:29)	(00:00:51:01)
			建物の倒壊1	PODTOH2M55S 16016N30000F (00:02:55:16)	2102 (00:01:10:02)	(00:04:05:22)
			火災	PODTOH6M57S 14014N30000F (00:06:57:14)	1131 (00:00:37:21)	(00:07:35:07)
			阪神高速	PODTOH9M01S 19019N30000F (00:09:01:19)	2408 (00:01:20:08)	(00:10:21:27)

( )内は:MM:SS:FF

Figure 5: An proposal of EDL list Watermark Data (Midorikawa Lab. Tokyo Inst. of Tech.)

KEYWORD INDEX

3D Design	27
3D-CG Simulation	293
5-Axis Controlled Machining Center	365
6DOF	145
Accelerated Reliability Test	395
Acceleration	13
Acceleration Compensation	347
Accelerometer	145,149
Action Acquisition	385
Active Control of Sound	343
Active Suspension	333
Activity Based Costing	213
Actor-Critic	385
Adaptive Digital Filter	333
Adjustable Mechanical Parameters	431
Advanced Factory Governance	281
Agent System	133
Aging	13
Aloha Protocols	313
Analytical Model	409
Angular Accelerometer	145
Antennas	317
Architecture	265
Archive	435
Artifacts	253
Assembly Planning	129
Assembly Sequences	129
Assist System	43
Assistance Dog	53
Atomi	427
Attendant Propelling Wheelchair	43
Attendant's Model	43
Attenuation by Human Body	309
Automated Assembly	125
Automation	301,313,317,323
Barcode Mark	19
Behavioral Design	173
Bio Cell Manipulation	241
Biped-Walking	189
Blood Fluidity Evaluation	163
Blood-Compatibilities	163
Bluetooth	305,309
Building Construction Process Management	271
Camera	415
Capital Assets Management	281
Cell Phone	181
Chamfering	169
Chemical Resistively	381
Circular Polarisation	389
Clean Up Curve	109

Closed-Loop	297
Collision Avoidance	97
Communication System	31,323,317
Compensation	309
Computer Aided Design (CAD)	103,129
Computer Aided Manufacturing (CAM)	129
Computer Aided Process Planning (CAPP)	113,129
Computer Assisted Surgery	57
Computer Mannequin	63
Computer Simulation	353
Connection	395
Construction Management	293
Consumer Electronics Rental Service	253
Continua	81
Control	297
Control Policy	1
Control Valve	377
Cooperative Transportation	177
Coordinate Measuring Machine	371
Coordination	195
Corner	77
Cost	213
CPN-TOOLS	253
Current Limiter	329
Curved Hole	157
Cutting Error	153
Cyber Manufacturing	293
Damper	353
Data Transfer	297
Decision-Objects Hierarchy	281
Deformable Object	77
Deformation	81
Design Information	103
Design Intention	103
Design Methodology	63,103
Design Structure Matrix	287
Disassembly System	133
Distributed Simulation	261
Distributed Virtual Factory	213
Driving Performance	181
Driving Simulator	181,401
Drunk Driving	181
Dual-Band	389
Dynamic Characteristics of Servo Motor	329
Dynamic Interaction of Alternatives	257
Dynamic Modelling	401
Dynamic Programming	201
Dynamic Seal	377
Dynamics	431

Easy	265	Home-Entertainment Robot	189
Edge of Clothes	77	Human Behavior	9,37,181,185
Edge Pixel Selection	219	Human Behavior Analysis	31,63
Edge Tracing	77	Human Body Database	67
Elderly People	9	Human Body Equivalent Phantom	409
Electric Cart	9	Human Body Modeling	67
Electric Wheelchair	53	Human Cooperative	103
Electrical Discharge Machining	157	Human Directed Manufacturing System	73
Electrically Conductive Fibre	395	Human Friendly	343
Electromagnetic Band-Gap	389	Human Friendly Machine	53
Embedded Controller	339	Human Interactive Manipulation	177
Embedded Object	265	Human Mimetics	185
Embedded Systems	265,427	Human Motion Simulation	67
End Mill	153	Human Robot Collaboration	31
Environment Recognition	177	Human-Centered Manufacturing	281
Environmental Burden	119	Human-Machine Interface	31
EPC Tree Algorithm	313	Human-Machine System	37
Ergonomics	67	Human-Robot Interaction	177
Error Compensation	169,371	Hybrid Controller	339
Error Space	371	Hydraulic Boom	339
Fast	265	Hydraulic Servo	297
Feedback Control	81	Image Feature Generation	235
Filtering	359	Image Feature Selection	235
Fixture	125	Image Processing	19,77,169
Flatness	371	Impeller	169
Flexible System	207	Indoor Positioning	309
Four Parallel Links	9	Industrial Robot	169
FTA	1	Information Equipment	323
Fuzzy ART	385	Information Storage	281,323
Gait Analysis	13	Information Systems Design Methodology	281
Gauge Factor	139	Infrared LED	415
Genetic Algorithm	173,207,219,287	Inheritance of State-Value	385
Genetic Programming	173	Instantaneous Cutting Force	153
Geometric Modeling	103	Integrated Process Management System	271
Gland Packing	377	Intelligent Manufacturing System (IMS)	293
Grease	377	Intelligent Servo Actuator (ISA)	189
Guidance Control	53	Intensional and Extensional Wholes	257
Gyroscope	23	Intensive Data Management System	271
Handicapped People	9	Interactive Design	173
Haptic Force Generator	23	Interactive Teaching	229
Haptic Illusion	23	Intra-Body Communication	409
HCPN	253	Inverted-F Antenna	305
Hearing Protection	305	ISO15004	435
Hearing Protector	343	ISP17799	435
Helical Compression Spring	157	ITER Vacuum Vessel	421
Hemiplegic Gait	27	Job Shop Type Production	125
Heuristic Rule	201	Kalman Filter	347
Hidden Markov Model	1	Kalman Filtering	359
HLA	261	Kalman Smoothing	359
Holonic Manufacturing	201	K-Nearest-Neighbor Rule	129
Holonic Manufacturing System	195	Knotting	87
Home Service Robot	77	Lane Queuing	401

Laparoscopic Surgery	57	Musculoskeletal Model	67
Latency	297	NC Program Diagnosis	119
LCD	405	Non Contact Rotation Mechanism	241
Life Cycle Assessment	119	Nonlinear Time-Series Analysis	13
Lift	47	Non-Rotational Tool	163
Linear Motor	347	Object-Oriented	207,265
Linear Objects	87	Off-Line Programming	277
Lower Leg Swelling	73	Offset	109
Machine Tool Operation	119	Offset Loops	109
Machining	153,421	ON/OFF Valve	333
Machining Sequencing	113	On-Machine Measurement	365
Magnetorheological Fluids	353	Open Lab.	5
Manipulation	81,87	Operation Slip	1
Manipulator	385	Optimum Control	53
Manual Operation Process	63	ORiN	261
Manufacturing Feature	113	Orthosis	27
Manufacturing System	261,371	Outdoor Mobile Robot	223
Manufacturing Systems Design Decomposition	281	Pair of Curvatures	219
		Parallel Mechanism	431
Master-Slave System	57	Parallel Robot	421
Measurement	27,323,359,371	Part-Manufacturing Process Management	271
Measuring Device	73	Parylene	149
Mechanism	47	Passive Joint Control	431
Mechatronics	353	Passive Tags	313
Medical Application	13	Path Planning	97
Medical Robot	57	Peg-In-Hole Task	125
Metadata	435	Personal Area Network	409
Metallic Microchannel Array	163	Phase Measurement	409
Method of Measurement	145	Piezoelectric Actuator	241
Micro Needle	381	Piezoresistor	139
Micro Robot	241	Pivot Motion	57
Microgroove	163	PLA (Poly Lactic Acid)	381
Micromachining	139	Planning	87,185
Micro-Molding	381	Planning-Based Method	93
Microscopic Image Recognition	241	Pocket	109
Microstructure	139	Portable Instrument	13
Minimally Invasive Surgery	57	Positioning	81
Möbius Surface	257	Positioning System	329
Mobile Robot	31,185,235,427,229	Potentials and Events	257
Mobile Tracking and Control	301	Power Assist	37,47
Modular Mechatronics	427	Power Assist Device	1
Mold Machining	365	Prediction	153
Motion Control	189	Pressure	377
Motion Planning	93	Preview Control	333
Motion Restitution	401	Process Engineering	293
Multi-Axis Control Ultraprecision Machining Center	163	Process Optimization	287
		Process Planning	201
Multi-Fingered Hand	125	Product Design	293
Multi-Link Mobile Robot	385	Product Recovery	133
Multimedia Database	435	Production Cost	119
Multi-Objective Optimization	195	Production Preparation	287
Multiple Object Identification	313	Production System Design	293

Project Scheduling	287	Setting a Meal	19
Propagation Loss	309	Shutter	405
Property Service	427	Simulator	97
Prototype Casting	277	Size Reduction	157
PWM Control	333	Smart Clothing	395
RCM Mechanism	57	Smoothing	359
Reactive Method	93	Socio-Technical Systems	281
Real-Time Scheduling	195	Spasticity	27
Recovery	207	Spiral Beam	149
RECS Concept	19	Stability	343
Rectifying Antenna	389	Standing Work	73
Redundant Robot	97	State-Space Construction	385
Rehabilitation	27	Stereo Projection	405
Reinforcement Learning	385	Stewart Platform	145
Relation Diagram	287	Stiffness	149
Rescue	415	Straightness	371
Resonant Frequency	149	Strain Gauge	73
RFID	133,313,317	Studio	5
RFID Systems	301	Support System	47
RoboCity CoRE	5	Supporting Robot	31
RoboCup	5	Surgical Robot	57
RoboCupJunior	5	Sweeping Task	93
RoboCupRescue	5	Tactile Sensor	139
RoboCupSoccer	5	Take an Elevator	229
Robot	19	Talking Driving	181
Robot Cleaner	93	Tardiness of Job	207
Robot Milling	277	Task Models	229
Robotics	173	Task-Oriented	235,235
Rotational Tool	163	Tensile Stress	149
Rough Map	223	Tool Path	109
Rubble	415	Tool Swept Volume	153
Safe and Low Load Design	43	Ubiquitous	133
Safe Driving	181	Ultraprecision Cutting	163
Safety	305,343	Uncut Region	109
Safety Design	189	Unknotting	87
Safety Verification Field	5	Velocity Control	347
Scheduling	201,207,213	Verhulst Equation	257
SDR-4XII	189	Vibration	353
Search	415	Vibration Suppression	347
Search and Pose Recognition	219	Virtual Cane	23
Security	301	Virtual Factory	261
Selective Attention	235	Virtual Machining	119
Sensing Systems	377	Virtual Reality	401,405
Sensor	145,415,317	Vision-Based Navigation	223
Sequential Two Points Method	371	Visual Navigation	229
Service CAD	247	Visually Impaired	23
Service Design	247	Voting Process	219
Service Design Tool	247	Walking Stability	13
Service Engineering	247,253	Washout	401
Service Modeling	247	Water Channel	157
Service Robot	1,31	Water Hydraulic	421
Servomotor	157	Wearable	305,343

Wearing Comfort	395
Welding	421
Welfare Tools	47
Wet Chemical Etching	381
Wheelchair	47
Wheelchair Propulsion	37
Wire Feeding	157
Wireless	297
Wireless Communication	301,305,415
Wireless Power Transmission	389
Work Piece Measurement	365
Workstation System Design	293
Wrist	9

AUTHOR INDEX

Aiyama, Y.	177	Hibino, H.	261
Amouri, A.	401	Higuchi, M.	125
An, M.	173	Hikizu, M.	47,77,329
Ando, H.	309	Hirai, S.	81,87
Andou, F.	163	Hirao, M.	169
Aoki, E.	57	Hirota, T.	309,377,409
Aoki, T.	381	Hirvonen, M.	347
Aoyagi, S.	19,97,139,149,381	Hölttä, V.	359
Aoyama, H.	241	Horio, K.	153
Arai, E.	63,87,103,113,257,271	Hosaka, H.	309,377,409
Arai, T.	247,253	Ihara, Y.	365
Arai, T.	415,431	Ikeda, T.	23
Arif, M.	13	Inamori, Y.	287
Arioui, H.	401	Inooka, H.	13
Arisawa, H.	67	Inoue, K.	415,431
Asada, M.	5,235	Inoue, Y.	9
Asakawa, N.	169	Inukai, T.	261
Azuma, Y.	181	Iribe, M.	189
Goossenaerts, J.B.M	281	Irie, K.	145
Chaya, Y.	47	Ishida, T.	157,163
Chen, L.	119	Ishiwatari, Y.	177
Cortes-Ramírez, J.A.	353	Ito, K.	409
Eguchi, T.	129	Iwamura, K.	195,207
Eijnatten, F.M.V.	281	Iwasaki, T.	309
Engels, D.	323	Iwase, K.	229
Enomoto, M.	309	Izumi, H.	381
Espié, S.	401	Izumi, K.	377
Ezawa, M.	93	Izutani, J.	139
Fichiwaki, O.	241	Jarvinen, T.	395
Fujii, K.	409	Kabe, A.	31
Fujii, S.	213	Kaihara, T.	213
Fujimoto, H.	119,185	Kamiya, Y.	47,77,329
Fujimoto, H.	27	Kaneko, J.	153
Fujita, K.	13	Karhu, O.I.	297
Fukuda, Y.	261	Kato, T.	87
Fukuda, Y.	63	Kawai, T.	163
Fukumoto, Y.	93	Kawamoto, J.	97
Fukushima, T.	189	Kawano, H.	19
Goto, K.	219	Kawano, T.	73,181
Hachisuka, K.	409	Keskilammi, M.	317
Han, G.	113	Kheddar, A.	401
Handroos, H.	347,421	Kim, H.	109
Hännikäinen, J.	395	Kishi, Y.	409
Hara, T.	247,253	Kita, M.	157
Harada, K.	207	Kivikoski, M.	297,301,305,313,317,323,343,389
Hasebe, T.	119	Kiyoshige, T.	169
Hashizume, M.	57	Kobayashi, E.	57
Haverinen, J.	265	Kobayashi, Y.	47
He, W.M.	371	Koike, M.	113
Heikkinen, J.	389	Kondoh, S.	133

Konishi, K.	57	Onosato, M.	109
Koseki, Y.	431	Palovuori, K.	405
Kosola, H.	405	Penttilä, K.	313
Kurahashi, Y.	47	Pessi, P.	421
Kurata, J.	9,37,43,53,145	Raittinen, H.	343
Kuroki, Y.	189	Ritamäki, M.	305
Lammila, M.	339	Röning, J.	265,427
Lehto, E.	339	Sagara, M.	371
Mae, Y.	415,431	Saito, Y.	93
Maeda, S.	385	Saitoh, F.	219
Maeda, Y.	139	Sakaguchi, T.	207
Maehara, F.	435	Sakamoto, K.	23
Maekawa, M.	47	Sakao, T.	247
Martínez-Martínez, M.	353	Sakuma, I.	57
Matsuda, H.	23	Salleh, K.	77
Matsumoto, S.	257	Sallinen, M.	277
Matsumoto, Y.	63	Sasaki, K.	309,409
Matsuzaki, H.	371	Sashio, K.	213
Minato, T.	235	Sato, H.	371
Mine, Y.	27	Sato, S.	287
Misaki, D.	241	Sato, T.	67
Mitsuyuki, K.	63	Seki, H.	47,77,329
Miura, J.	223,229	Seki, T.	177
Miura, T.	253	Seki, Y.	195
Miyajima, N.	287	Seo, M.	109
Mizuno, T.	73	Serikitkankul, P.	329
Mochiyama, H.	185	Shimizu, Y.	23
Mohellebi, H.	401	Shimomura, Y.	247,253
Moridaira, T.	189	Shiota, Y.	23
Moriwaki, T.	181	Shirai, Y.	223,229
Murayama, T.	129	Shirase, K.	113
Muto, T.	333	Shrestha, R.	201
Myllymäki, P.	305	Sirviö, M.	277
Nagai, K.	125	Soini, M.	301,323
Nagatomi, R.	13	Sone, T.	371
Naito, M.	377	Sugimura, N.	195,201,207
Nakajima, T.	157	Suzuki, A.	13
Nakamura, S.	365	Suzuki, T.	57
Nakano, M.	287	Suzuki, T.	43
Naoto, C.	241	Suzuki, T.	287
Narita, H.	119	Sydänheimo, L.	313,301,317,323
Norihisa, T.	119	Tai, Y.C.	149
Nunobiki, M.	385	Takano, M.	19,97
Oba, F.	129	Takashima, T.	27
Ogata, H.	93	Takata, M.	271
Ohmori, H.	163	Takemoto, T.	201
Ohta, Y.	37	Takeuchi, K.	103
Ohtaki, Y.	13	Takeuchi, Y.	153,157,163
Oinonen, M.	305,343	Takubo, T.	415,431
Okuda, K.	385	Tamaki, K.	293
Okutsu, R.	377	Tanaka, H.	169
Ono, T.	63	Tanaka, S.	309

Tanaka, W.	431
Tani, Y.	371
Tanimizu, Y.	195,207
Tanno, Y.	435
Tashiro, K.	97
Tateno, T.	133
Taura, T.	173
Teramoto, K.	153,157
Terauchi, Y.	409
Tian, G.	253
Tikanmäki, A.	427
Tomiyama, T.	247
Torige, A.	93
Tsuji, T.	57
Tsumaya, A.	87,103,113
Uchiyama, H.	9,37,43,53,145
Uemoto, T.	53
Ukkonen, L.	323
Umeda, K.	371
Usuda, T.	241
Vähäkuopus, K.	395
Vallius, T.	265,427
Vanhala, J.	395
Villarreal-González, L.S.	353
Virvalo, T.	297,339
Vladimirov, B.	185
Vuorela, T.	395
Wada, M.	377
Wakamatsu, H.	87,103,113
Wu, H.	421
Yagi, J.	257,271
Yamada, H.	333
Yamada, Y.	1
Yamaguchi, T.	19
Yamaguchi, T.	125
Yamamoto, A.	163
Yamamoto, M.	415
Yoshikawa, D.	149
Yun, J.	223